THE MATHEMATICS OF FINITE ELEMENTS AND APPLICATIONS IV

Based on the proceedings of a conference held at Brunel University from 28 April to 1 May, 1981

THE MATHEMATICS OF FINITE ELEMENTS AND APPLICATIONS IV

MAFELAP 1981

Edited by

J. R. WHITEMAN

Institute of Computational Mathematics,
Brunel University, Uxbridge,
Middlesex, England

1982

ACADEMIC PRESS

A Subsidiary of Harcourt Brace Jovanovich, Publishers

London New York

Paris San Diego San Francisco

São Paulo Sydney Tokyo Toronto

ACADEMIC PRESS INC. (LONDON) LTD.
24/28 Oval Road
London NW1

United States Edition published by
ACADEMIC PRESS INC.
111 Fifth Avenue
New York, New York 10003

British Library Cataloguing in Publication Data
The mathematics of finite elements and applications
IV—MAFELAP 1981.
1. Finite element method—Congresses
I. Whiteman, J.R.
515.3'53 TA347.F5

ISBN 0-12-747254-1

LCCCN 72-7713

Printed in Great Britain

CONTRIBUTORS

Akin, J.E., *Department of Engineering Science & Mechanics, University of Tennessee, Knoxville, Tennessee 37916, U.S.A.*
Ashworth, J., *Department of Ceramics, Glass & Polymers, University of Sheffield, Elmfield, Northumberland Road, Sheffield, S10 2TZ, U.K.*
Babuska, I., *Institute for Physical Science & Technology, Space Sciences Building, University of Maryland, College Park, Maryland 20742, U.S.A.*
Barrett, J.W., *Department of Mathematics, Imperial College of Science & Technology, Huxley Building, Queens Gate, London, SW7 2BZ, U.K.*
Bercovier, M., *Division of Applied Mathematics, The Hebrew University of Jerusalem, Givat Ram, Jerusalem, Israel.*
Berger, M.H., *Department of Engineering Science & Mechanics, University of Tennessee, Knoxville, Tennessee 37916, U.S.A.*
Bernadou, M., *INRIA, Domaine de Voluceau-Rocquecourt, B.P. 105, 78150 Le Chesnay, France.*
Blum, H., *Institut für Angewandte Mathematik, Universität Bonn, Wegelerstrasse 6, D-5300 Bonn 1, West Germany.*
Bossavit, A., *Electricité de France, 1 avenue du Général-de-Gaulle, BP 408, 92141 Clamart Cedex, France.*
Brezzi, F., *Istituto di Matematica Applicata, Universita di Pavia, Piazza L. de Vinci, 27100 Pavia, Italy.*
Bruch, J.C., Jr., *Department of Mechanical & Environmental Engineering, University of California, Santa Barbara, California 93106, U.S.A.*
Burley, D.M., *Department of Applied & Computational Mathematics, University of Sheffield, Elmfield, Northumberland Road, Sheffield, S10 2TZ, U.K.*
Campbell, J.S., *Department of Civil Engineering, University College, Cork, Eire.*
Carling, J.C., *Department of Ceramics, Glass & Polymers, University of Sheffield, Elmfield, Northumberland Road, Sheffield, S10 2TZ, U.K.*

Caussignac, Ph., *Ecole Polytechnic Fédérale de Lausanne, Lausanne, Switzerland.*
Charchafchi, T.A., *Department of Civil & Structural Engineering, University College Cardiff, Newport Road, Cardiff, CF2 1TA, Wales, U.K.*
Collatz, L., *Institut für Angewandte Mathematik, Universität Hamburg, Bundesstrasse 55, 2 Hamburg 13, West Germany.*
Delves, L.M., *Department of Computational & Statistical Science, University of Liverpool, P.O. Box 147, Liverpool, L69 3BX,U.K.*
Depeursinge, Y., *Laboratoire Suisse de Recherches Horlogères, Rue A.-L. Breguet 2, Case postale 42, 2000 Neuchatel, Switzerland.*
Dinkler, D., *Institut für Statik, TU Braunschweig, Beethovenstrasse 51, D-3000 Braunschweig, West Germany.*
Engelman, M.S., *CIRES, University of Colorado, Campus Box 449, Bolder, Colorado 80303, U.S.A.*
Fix, G.J., *Department of Mathematics, Carnegie-Mellon University, Schenley Park, Pittsburgh, Pennsylvania 15213, U.S.A.*
Fried, I., *Department of Mathematics, College of Liberal Arts, Boston University, Boston, Massachusetts 02213, U.S.A.*
Fujii, H., *Istituto di Matematica Applicata, Universita di Pavia, Piazza L. de Vinci, 27100 Pavia, Italy.*
Gago, J., *Department of Civil Engineering, University of Wales Swansea, Singleton Park, Swansea, SA2 8PP, Wales, U.K.*
Gare, C.A., *Department of Ceramics, Glass & Polymers, University of Sheffield, Elmfield, Northumberland Road, Sheffield S10 2TZ, U.K.*
Gregory, J.A., *Department of Mathematics, Brunel University, Uxbridge, Middlesex, UB8 3PH, U.K.*
Griffiths, D.F., *Department of Mathematics, University of Dundee, Dundee, DD1 4HN, Scotland, U.K.*
Gunzburger, M.D., *Department of Mathematics, University of Tennessee, Knoxville, Tennessee 37916, U.S.A.*
Hackbusch, W., *Mathematisches Institut, Ruhr-Universität, Postfach 10 21 48, D-4630 Bochum 1, West Germany.*
Hasbani, Y., *Division of Applied Mathematics, The Hebrew University of Jerusalem, Givat Ram, Jerusalem, Israel.*
Hendry, J.A., *Computer Centre, University of Birmingham, Birmingham, B15 2TT, U.K.*
Hennart, J.P., *IIMAS-UNAM, Apartado Postal 20-726, Mexico 20, D.F., Mexico.*
Hertling, J., *Institut für Numerische Mathematik, Technische Universität Wien, Gusshausstrasse 27-29, A-1040 Wien,Austria.*
Hinton, E., *Department of Civil Engineering, University of Wales Swansea, Singleton Park, Swansea, SA2 8PP, Wales, U.K.*
Hughes, T.J.R., *Department of Mechanical Engineering, Stanford University, Stanford, California 94305, U.S.A.*
Ise, G., *Lehrstuhl für Schiffbau, Konstruktion und Statik, RWTH Aachen, Templergraben 55, D 51 Aachen, West Germany.*
Janovský, V., *Katedra Numericke Mathematiky, Charles University, Malostranské nam. 25, 118 00 Praha 1, Czechoslovakia.*

Jeandrevin, S., *Ecole Polytechnique Fédérale de Lausanne, Lausanne, Switzerland.*
Kant, T., *Department of Civil Engineering, University of Wales, Swansea, Singleton Park, Swansea, SA2 8PP, Wales, U.K.*
Kelly, D.W., *Department of Civil Engineering, University of Wales Swansea, Singleton Park, Swansea, SA2 8PP, Wales, U.K.*
Kohn, R., *Division of Engineering & Applied Science, California Institute of Technology, Pasadena, California 91125, U.S.A.*
Kröplin, B.-H., *Institut für Statik, TU Braunschweig, Beethovenstrasse 51, D-3300 Braunschweig, West Germany.*
Livne, E., *Division of Applied Mathematics, The Hebrew University of Jerusalem, Givat Ram, Jerusalem, Israel.*
Lo, C.S., *Department of Civil Engineering, University of Wales Swansea, Singleton Park, Swansea, SA2 8PP, Wales, U.K.*
Lyle, S., *Deptment of Electronic Engineering and Computer Science, Kingston Polytechnic, Penrhyn Road, Kingston-upon-Thames, KT1 2EE, U.K.*
Manoranjan, V.S., *Department of Mathematics, University of Dundee, Dundee, DD1 4HN, Scotland, U.K.*
Mitchell, A.R., *Department of Mathematics, University of Dundee, Dundee, DD1 4HN, Scotland, U.K.*
Mohamed, J., *Department of Computational & Statistical Science, University of Liverpool, P.O. Box 147, Liverpool L69 3BX,U.K.*
Morton, K,W., *Department of Mathematics, University of Reading, Whiteknights, Reading, RG6 2AX, U.K.*
Nicolaides, R.A., *Department of Mathematics, Carnegie-Mellon University, Schenley Park, Pittsburgh, Pennsylvania 15213, U.S.A.*
Nichols, N.K., *Department of Mathematics, University of Reading, Whiteknights, Reading, RG6 2AX, U.K.*
Oden, J.T., *The Texas Institute for Computational Mechanics, University of Texas at Austin, Austin, Texas 78712, U.S.A.*
Oliveira, E.R. de Arantes e, *Instituto Superior Técnico, Centro de Mecânica e Engenharia Estruturais, Universidade Técnica de Lisboa, Av. Rovisco Pais, 1096 Lisboa, Portugal.*
Pironneau, O., *INRIA, Domaine de Voluceau-Rocquencourt, B.P.105, 78150 Le Chesnay, France.*
Rappaz, J., *Département de Mathématiques, Ecole Polytechnique Fédérale de Lausanne, 61 Av. de Cour, CH-1007 Lausanne, Switzerland.*
Raugel, G., *Département de Mathemâtiques, Ecole Polytechnique Fédérale de Lausanne, 61 Av. de Cour, CH-1007 Lausanne, Switzerland.*
Rawson, H., *Department of Ceramics, Glass and Polymers, University of Sheffield, Elmfield, Northumberland Road, Sheffield, S10 2TZ, U.K.*
Remar, J., *Department of Mechanical & Environmental Engineering, University of California, Santa Barbara, California 93106, U.S.A.*
Rohrbach, R., *Lehrgebiet für Baumechanik, Universität Hannover, Callinstrasse 32, D-3000 Hannover 1, West Germany.*

Sabir, A.B., *Department of Civil & Structural Engineering, University College, Newport Road, Cardiff, CF2 1TA., Wales,U.K.*

Schomburg, U., *Institut für Technische Mechanik, Rheinisch-Westf. Technischen Hochschule Aachen, Templergraben 64, D 51 Aachen, West Germany.*

Sloss, J.M., *Department of Mechanical & Environmental Engineering, University of California, Santa Barbara, California 93106, U.S.A.*

Stein, E., *Lehrgebiet für Baumechanik, Universitat Hannover, Callinstrasse 32, D-3000 Hannover, West Germany.*

Stokes, A., *Department of Mathematics, University of Reading, Whiteknights, Reading, RG6 2AX, U.K.*

Strang, G., *Department of Mathematics, Massachusetts Institute of Technology, Cambridge, Massachusetts 02139, U.S.A.*

Taylor, R.L., *Department of Mechanical Engineering, Stanford University, Stanford, California 94305, U.S.A.*

Tessler, A., *Northrop Corporation, Structural Mechanics Research Department 3852/82, 3901 West Broadway, Hawthorne, California 90250, U.S.A.*

Vickers, G.T., *Department of Applied & Computational Mathematics, University of Sheffield, Elmfield, Northumberland Road, Sheffield, S10 2TZ, U.K.*

Voirol, W., *Ecole Polytechnic Fédérale de Lausanne, Lausanne, Switzerland.*

Wendland, W.L., *Fachbereich Mathematik, Technische Hochschule Darmstadt, Schlossgartenstrasse 7, 6100 Darmstadt, West Germany.*

Whiteman, J.R., *Institute of Computational Mathematics, Brunel University, Uxbridge, Middlesex, UB8 3PH, U.K.*

Wilhelmy, V., *Det Noske Veritas, Box 300, 1322 Høvik, Oslo, Norway.*

Wrixton, G.T., *Solid State Laboratory, University College, Cork, Eire.*

Zienkiewicz, O.C., *Department of Civil Engineering, University of Wales, Singleton Park, Swansea, SA2 8PP, Wales, U.K.*

PREFACE

Over a decade ago I reached the conclusion that there existed a very definite and detrimental lack of communication between those mathematicians, who were at that time working in increasing numbers on the mathematical theory of finite elements, and engineers who were routinely using finite element methods to solve practical problems. In order to try to improve this situation the idea was put forward that a conference involving both theory and applications of finite element methods should be held at Brunel University. Thus the idea of MAFELAP was born and the first Conference was held in 1972. We were of course in 1972 totally unaware that a series of triennial conferences was beginning. With the passing of time, although communication between mathematicians and engineers in the finite element field has without doubt improved, there remains a continuing need for forums to be set up which enable people from these different disciplines to come together to discuss finite element techniques, their theory and their practice. In this context MAFELAP 1981, the fourth conference on The Mathematics of Finite Elements and Applications, was held at Brunel University during the period 28 April to 1 May 1981.

MAFELAP 1981 was organised by BICOM (the Brunel University Institute of Computational Mathematics) with the willing assistance of a committee consisting of M.J.M. Bernal, J. Crank, J.A. Gregory, A.R. Mitchell, K.W. Morton, E.H. Twizell, J.R. Whiteman, A.L. Yettram and O.C. Zienkiewicz. The main task of this committee was in fact to select, from the large number of submitted papers, those which were to be presented at the conference, either as contributed papers or as papers in poster sessions. The aims of the conference were once more to bring together mathematicians and engineers whose main interests were finite elements, and at the same time to give exposure to those topics which were considered to have been the significant advances in the finite element field of the previous three years. Accordingly F. Brezzi, G. Fix, I. Fried, T.J.R. Hughes, A.R. Mitchell, J.T. Oden, G. Strang, W. Wendland, J.R. Whiteman and O.C. Zienkiewicz presented invited lectures.

This book contains the complete texts of the invited and contributed lectures together with the abstracts of the poster session papers. All authors were asked to prepare their manuscripts in camera-ready form. This was a departure from the practice with previous MAFELAP proceedings and the result has been that my task as editor has been considerably reduced. However, in some cases quite heavy editing and retyping has been undertaken, in the interest of clarity and in order to produce a fairly uniform presentation.

In connection with the conference my thanks go to the members of the Conference Committee, to the Chairmen of the Sessions, and to all the speakers and poster session authors for their friendly co-operation. The continuing success of the MAFELAP conferences is due to the unstinting efforts of the members of BICOM, and my thanks go also to D. Harrison and T.J.W. Ward and to the Institute secretary, Ms M.E. Demmar, for their contributions which were much beyond the call of duty. Finally my thanks go to Ms Demmar and Mrs. M. Reece for their retyping work on the manuscripts, and also to my wife for compiling the subject index of this book.

The conference was financed in part from a generous grant from the United States Army European Research Office, London, which is acknowledged with great pleasure.

J.R. Whiteman
Institute of Computational Mathematics
Brunel University
April 1982

CONTENTS

ANALYSIS OF A CLASS OF PROBLEMS WITH FRICTION BY FINITE ELEMENT METHODS

J. T. ODEN

The University of Texas

1. INTRODUCTION

This year marks the 200-th anniversary of the publication of the memoir of the French engineer-scientist A. C. Coulomb, "Thêorie des machines simples" in which he presented his law of friction of different bodies slipping on one another. This work, which was based on results of Coulomb's own personal experiments with simple bodies, earned him the prize of the French Academy of Science in 1781 and cast his name indelibly in the pages of the history of mechanics. The classical Coulomb law of dry friction, of course, asserts that a relative sliding of two bodies in contact will occur whenever the net tangential forces on the contact surface reach a critical value proportional to the force normal to the contact surface. The constant of proportionality is known as the coefficient of friction. There have since been proposed several modifications of this law to take into account a distinction between static and dynamic processes, lubrication of the contact surface, etc.

Coulomb himself must have perceived his law as applicable to static situations involving bodies which, for all practical purposes, are rigid. Indeed, the foundations of continuum mechanics and its theories of deformable bodies were laid down many decades after Coulomb proposed his law. In its origin form, the law was regarded as describing, for instance, the tendency of a rigid block of one substance to slide down an inclined plane constructed of another substance as the angle of inclination is increased. It should not be totally surprising, therefore, that when Coulomb's law was extended so as to apply to static frictional phenomena in elasticity, some two centuries after it was first proposed, it led to classes of boundary-value problems for which no existence theory is yet available, for which uniqueness of solutions (when they exist) cannot be guaranteed, and which may represent poor mathematical models of dry static friction between deformable bodies.

One such class of problems, considered by Duvaut and Lions [3], describes the equilibrium of a linearly elastic body in contact with a rigid foundation on which Coulomb's law holds. A variational principle for problems of this type is given as follows:

Find $\underset{\sim}{u} \in K$ such that

$$a(\underset{\sim}{u},\underset{\sim}{v}-\underset{\sim}{u}) + c(\underset{\sim}{u},\underset{\sim}{v}) - c(\underset{\sim}{u},\underset{\sim}{u}) \geq f(\underset{\sim}{v}-\underset{\sim}{u}) \qquad \forall\ \underset{\sim}{v} \in K \tag{1.1}$$

Here $\underset{\sim}{u}$ is the displacement field in a subset K of the space V of admissible displacements,

$$\left.\begin{aligned} V &= \{\underset{\sim}{v} = (v_1,v_2,\ldots,v_n) \in (H^1(\Omega))^n \mid \\ &\qquad \underset{\sim}{v} = \underset{\sim}{0} \quad \text{a.e. on } \Gamma_D \subset \partial\Omega\} \\ K &= \{\underset{\sim}{v} \in V \mid \underset{\sim}{v}\cdot\underset{\sim}{n} \leq 0 \quad \text{a.e. on } \Gamma_C\} \end{aligned}\right\} \tag{1.2}$$

$a(\cdot,\cdot) : V\times V \to \mathbb{R}$ is the bilinear form

$$a(\underset{\sim}{u},\underset{\sim}{v}) = \int_\Omega E_{ijk\ell}u_{k,\ell}v_{i,j}\,dx \tag{1.3}$$

representing the virtual work done by the stress

$$\sigma_{ij}(\underset{\sim}{u}) \overset{\text{def}}{=\!=} E_{ijk\ell}u_{k,\ell} \tag{1.4}$$

on the virtual strains $\varepsilon_{ij}(\underset{\sim}{v}) = \frac{1}{2}(v_{i,j} + v_{j,i})$, $1 \leq i,j,k,\ell \leq N$; and $c : V\times V \to \mathbb{R}$ is the virtual work term due to the Coulomb frictional forces:

$$c(\underset{\sim}{u},\underset{\sim}{v}) = \int_{\Gamma_C} \nu|\sigma_n(\underset{\sim}{u})|\ |\underset{\sim}{v}_T|\,ds \tag{1.5}$$

Finally, $f \in V'$ represents the work of the external forces, assumed here to be given by

$$\left.\begin{aligned} f(\underset{\sim}{v}) &= \int_\Omega \underset{\sim}{f}\cdot\underset{\sim}{v}\,dx + \int_{\Gamma_F} \underset{\sim}{t}\cdot\underset{\sim}{v}\,ds \\ f &\in (L^2(\Omega))^N \ ; \quad \underset{\sim}{t} \in (L^\infty(\Gamma_F))^N \end{aligned}\right\} \tag{1.6}$$

In (1.2), Ω is a smooth open bounded domain in $\mathbb{R}^N$ with boundary Γ consisting of three disjoint parts: Γ_D on which displacements are prescribed, Γ_C which contains the contact surface, and Γ_F on which the prescribed tractions $\underset{\sim}{t}$ are applied. The values of $v \in V$ on Γ are, of course, interpreted in the sense of the trace of V onto $(H^{1/2}(\Gamma))^N$ (we assume $\overline{\Gamma}_D \cap \overline{\Gamma}_C = \phi$) and $\underset{\sim}{n}$ is the unit vector exterior and normal to Γ. In (1.3) and thereafter, we employ standard index notations and the summation convention, $u_{k,\ell} \overset{\text{def}}{=\!=} \partial u_k/\partial x_\ell$ ($\underset{\sim}{x} = (x_1, x_2, \ldots, x_n)$ being a point in $\overline{\Omega}$ with Cartesian coordinates x_i) and the elasticities $E_{ijk\ell}$ are constants with the usual symmetries and positive-definiteness. Thus, the bilinear form $a(\cdot,\cdot)$ can be assumed to be symmetric, continuous, and V-elliptic; i.e., positive constants m and M exist such that

$$a(\underset{\sim}{u},\underset{\sim}{v}) \leq M \; \|\underset{\sim}{u}\|_1 \; \|\underset{\sim}{v}\|_1 \quad , \; a(\underset{\sim}{v},\underset{\sim}{v}) \geq M \; \|\underset{\sim}{v}\|_1^2 \tag{1.7}$$

for all $\underset{\sim}{u},\underset{\sim}{v}$, V, where

$$\|\underset{\sim}{v}\|_1^2 = \int_\Omega v_{i,j} v_{i,j}\, dx \tag{1.8}$$

In (1.5), ν is the coefficient of friction, $\nu \in L^\infty(\Gamma_C)$, $\nu \geq \nu_0 > 0$ a.e. on Γ_C, $\sigma_n(\underset{\sim}{u})$ is the normal stress on the contact surface; $\sigma_n(\underset{\sim}{u}) \overset{\text{def}}{=\!=} \sigma_{ij}(\underset{\sim}{u}) n_i n_j$, and $\underset{\sim}{v}_T$ is the tangential component of the trace of the displacement vector $\underset{\sim}{v}$ on Γ_C.

It is not known whether solutions exist to problem (1.1) in general. The existence of solutions to very special cases of such problems (for Ω unbounded and N=2) has been established by Nečas et al [8] and also in Demkowicz and Oden [2], but it is believed that there may not, in fact, exist solutions for some more general situations. Complications arise from the fact that the functional c is non-convex and non-differentiable and from the fact that if a solution $\underset{\sim}{u}$ were to exist, it would necessarily have components in $H^1(\Omega)$; then $\sigma_n(\underset{\sim}{u}) \in H^{-1/2}(\Gamma_C)$ is defined by duality and $|\sigma_n(\underset{\sim}{u})|$ may have no meaning. We note that if smooth solutions $\underset{\sim}{u}$ to (1.1) were to exist, they would be solutions of the classical elastostatics problem,

$$\text{I.} \quad \left.\begin{aligned} &\sigma_{ij}(\underset{\sim}{u})_{,j} + f_i = 0 \\ &\sigma_{ij}(\underset{\sim}{u}) = E_{ijk\ell} u_{k,\ell} \end{aligned}\right\} \quad \text{in } \Omega$$

$$\underset{\sim}{u} = \underset{\sim}{0} \qquad \text{on } \Gamma_D$$

$$\sigma_{ij}(\underset{\sim}{u}) n_i = t_i \qquad \text{on } \Gamma_F$$

$$\left.\begin{array}{l}
\text{II. } u_n = \underset{\sim}{u}\cdot\underset{\sim}{n} \le 0 \;;\; \sigma_n(\underset{\sim}{u}) = \sigma_{ij}(\underset{\sim}{u})n_i n_j \le 0 \\
\qquad\qquad \sigma_n(\underset{\sim}{u})u_n = 0
\end{array}\right\} \text{ on } \Gamma_C$$

$$\left.\begin{array}{l}
\text{III. } |\underset{\sim}{\sigma}_T(\underset{\sim}{u})| < \nu|\sigma_n(\underset{\sim}{u})| \Longrightarrow \underset{\sim}{u}_T = \underset{\sim}{0} \\
\qquad |\underset{\sim}{\sigma}_T(\underset{\sim}{u})| = \nu|\sigma_n(\underset{\sim}{u})| \Longrightarrow \exists\lambda \in \mathbb{R},\ \lambda > 0, \text{s.t.} \\
\qquad\qquad \underset{\sim}{u}_T = -\lambda\underset{\sim}{\sigma}_T(\underset{\sim}{u})
\end{array}\right\} \text{ on } \Gamma_C$$

$$\left.\begin{array}{l}
(\underset{\sim}{\sigma}_T(\underset{\sim}{u}))_i \equiv \sigma_{ij}(\underset{\sim}{u})n_j - \sigma_n(\underset{\sim}{u})n_i \\
\underset{\sim}{u}_T \equiv \underset{\sim}{u} - u_n\underset{\sim}{n}
\end{array}\right\} \text{ on } \Gamma_C \qquad (1.9)$$

In the present paper, several special cases of (1.1) [or (1.9)] are considered. These include the Signorini problem without friction, the Coulomb friction problem with prescribed normal pressures, a regularized version of this latter problem, and the general problem itself. In addition, the possibility of using a non-local generalization of Coulomb's law is explored. Finite element approximations of these problems are described, and several new algorithms for the numerical solution of certain cases are given. Finally, the results of preliminary numerical experiments are presented.

2. VARIOUS CONTACT PROBLEMS IN ELASTOSTATICS

There are certain special cases and modifications of problem (1.1) which are tractable and which can lead to reasonable models of certain friction phenomena. We shall now discuss some of these which have been studied recently using finite-element methods.

I. Unilateral Contact without Friction. When the bodies in contact are sufficiently lubricated, frictional forces can be neglected. Then (1.1) reduces to the classical Signorini problem of unilateral contact of an elastic body with a rigid frictionless foundation. This problem is characterized by the variational problem of finding $\underset{\sim}{u} \in K$ such that

$$a(\underset{\sim}{u},\underset{\sim}{v}-\underset{\sim}{u}) \ge f(\underset{\sim}{v}-\underset{\sim}{u}) \qquad \forall\ \underset{\sim}{v} \in K \qquad (2.1)$$

It is well known that under the conditions assumed earlier (particularly (1.7)), there exists a unique solution $\underset{\sim}{u}$ to (2.1) for any choice of $f \in V'$ (see Fichera [5] or Kikuchi and Oden [7]). Clearly, (2.1) is derived from (1.1) by setting $c \equiv 0$. Moreover, if the solution $\underset{\sim}{u}$ to (2.1) is sufficiently smooth, it is also the solution of the classical problem defined by subsets I and II of the system of equations (1.9).

II. Unilateral Contact with Prescribed Normal Pressures. This special class of contact problems of some importance involves problems in which the normal "contact" pressure is prescribed and only the tangential friction forces are unknown. In this case, we set

$$j(\underset{\sim}{v}) = \int_{\Gamma_C} \tau|\underset{\sim}{v}_T|ds \tag{2.2}$$

$$\left.\begin{aligned} &\tau \in L^2(\Gamma_C),\ \tau \geq 0 \\ &\bar{f}(\underset{\sim}{v}) = f(\underset{\sim}{v}) + \int_{\Gamma_C} F_n v_n ds \end{aligned}\right\} \tag{2.3}$$

where τ is given data, F_n is also a prescribed normal force on Γ_C, and $v_n = \underset{\sim}{v}\cdot\underset{\sim}{n}$. We then have a special variational inequality of the form,

$$\underset{\sim}{u} \in K :$$

$$a(\underset{\sim}{u},\underset{\sim}{v}-\underset{\sim}{u}) + j(\underset{\sim}{v}) - j(\underset{\sim}{u}) \geq f(\underset{\sim}{v}-\underset{\sim}{u}) \quad \forall\ \underset{\sim}{v} \in K \tag{2.4}$$

The functional j in (2.2) is convex and lower semicontinuous, but it is not differentiable (in the Gâteaux sense). As a result, it is not difficult to show that there exists a unique solution $\underset{\sim}{u}$ to (2.3) for each τ (and $\bar{f}$). Moreover, the solution of $\underset{\sim}{u}$ is also the unique minimizer of the non-differentiable energy functional,

$$J(\underset{\sim}{v}) = \frac{1}{2}a(\underset{\sim}{v},\underset{\sim}{v}) - f(\underset{\sim}{v}) + j(\underset{\sim}{v}) \ , \quad \underset{\sim}{v} \in K \tag{2.5}$$

III. A Regularization of Problem (2.4). For $\underset{\sim}{v} \in K$, let

$$\phi_\varepsilon(\underset{\sim}{v}) = \left\{\begin{array}{ll} \frac{\varepsilon}{2} - |\underset{\sim}{v}_T| & \text{if } |\underset{\sim}{v}_T| \geq \varepsilon \\ \frac{\varepsilon}{2}\ \underset{\sim}{v}_T\cdot\underset{\sim}{v}_T & \text{if } |\underset{\sim}{v}_T| \leq \varepsilon \end{array}\right\} \tag{2.6}$$

and

$$j_\varepsilon(\underset{\sim}{v}) = \int_{\Gamma_C} \tau\phi_\varepsilon(\underset{\sim}{v})ds \ ; \ \tau \in L^2(\Gamma_C),\ \tau \geq 0 \tag{2.7}$$

where ε is a positive number. Then it can be shown (see Campos, Oden, and Kikuchi [1]) that the functional j_ε represents a regularization of the functional j defined in (2.3) which is Gâteaux-differentiable, with

$$< Dj_{\varepsilon}(\underset{\sim}{u}),\underset{\sim}{v} > = \int_{\Gamma_C} \tau \left. \frac{\partial \phi_{\varepsilon}(\underset{\sim}{u} + t\underset{\sim}{v})}{\partial t} \right|_{t=0} ds \tag{2.8}$$

and for which $j_{\varepsilon} \to j$ as $\varepsilon \to 0$. Here $<\cdot,\cdot>$ denotes duality pairing on $H^{-1/2}(\Gamma_C)\times H^{1/2}(\Gamma_C)$. We can then consider the regularized problem,

$$\left.\begin{array}{l} \text{Find } u_{\varepsilon} \in K \text{ such that} \\ a(\underset{\sim}{u}_{\varepsilon},\underset{\sim}{v}-\underset{\sim}{u}_{\varepsilon}) + < Dj_{\varepsilon}(\underset{\sim}{u}_{\varepsilon}),\underset{\sim}{v}-\underset{\sim}{u}_{\varepsilon} > \geq f(\underset{\sim}{v}-\underset{\sim}{u}_{\varepsilon}) \quad \forall \ \underset{\sim}{v} \in K \end{array}\right\} \tag{2.9}$$

This problem has a unique solution $\underset{\sim}{u}_{\varepsilon}$ for each $\varepsilon > 0$. Moreover, if $\underset{\sim}{u}$ is the solution of (2.4), then

$$\| \underset{\sim}{u}-\underset{\sim}{u}_{\varepsilon} \|_1 \leq C\sqrt{\varepsilon} \tag{2.10}$$

The regularized problem (2.8) provides a more useful basis for the construction of approximate methods than does a direct use of (2.4) owing to the non-differentiability of j.

IV. Non-local Friction Laws. As noted earlier, the principal source of difficulties with the general equibilibrium problem with Coulomb friction (1.1) is that $\underset{\sim}{u},\underset{\sim}{v} \in K \Longrightarrow \sigma_{ij}(\underset{\sim}{u}) \in L^2(\Omega)$. Then $\sigma_n(\underset{\sim}{u})$ is defined by duality as a member of $H^{-1/2}(\Gamma_C)$ and and $|\sigma_n(\underset{\sim}{u})|$ has no meaning. Duvaut [3] has observed that this particular mathematical difficulty is overcome if one replaces $|\sigma_n(\underset{\sim}{u})|$ in the definition of $c(\cdot,\cdot)$ by a suitable regularization which is a non-negative function in $L^2(\Gamma_C)$. Similar ideas have been explored in some detail by Demkowicz and Oden [2]. In certain instances, such regularizations might be interpreted physically as non-local friction laws.

The essential ideas are outlined as follows. Let S denote a linear continuous function from $H^{-1/2}(\Gamma_C)$ into V which preserves non-negativeness:

$$S: H^{-1/2}(\Gamma_C) \to V \ ; \ \tau \leq 0 \Longrightarrow S(\tau) \leq 0 \tag{2.11}$$

Then we consider the non-local friction problem,

$$
\left.\begin{aligned}
&\underset{\sim}{u} \in K: \\
&a(\underset{\sim}{u},\underset{\sim}{v}-\underset{\sim}{u}) + \int_{\Gamma_C} \nu\, S(\sigma_n(\underset{\sim}{u}))(|\underset{\sim}{v}_T| - |\underset{\sim}{u}_T|)ds \\
&\qquad\qquad \geq f(\underset{\sim}{v}-\underset{\sim}{u}) \quad \forall\ \underset{\sim}{v} \in K
\end{aligned}\right\} \tag{2.12}
$$

The resolution of this problem depends intimately on problem (2.4). For each $\tau \in L^2(\Gamma_C)$, we know from (2.4) that there exists a unique $\underset{\sim}{u}_\tau \in V$ such that

$$
\left.\begin{aligned}
&\underset{\sim}{u}_\tau \in K : a(\underset{\sim}{u}_\tau,\underset{\sim}{v}-\underset{\sim}{u}_\tau) + \int_{\Gamma_C} \tau(|\underset{\sim}{u}_T| - |\underset{\sim}{v}_{\tau_T}|)ds \\
&\qquad\qquad > f(\underset{\sim}{v}-\underset{\sim}{u}_\tau) \quad \forall\ \underset{\sim}{v} \in K
\end{aligned}\right\} \tag{2.13}
$$

This establishes a correspondence $B : L^2(\Gamma_C) \to V$ defining

$$
\underset{\sim}{u}_\tau = B(\tau) \tag{2.14}
$$

which is strongly continuous.

We will now show that the problem can be reduced to one of finding a fixed point of the map,

$$
T : L^2(\Gamma_C) \to L^2(\Gamma_C) \ ; \ T = \nu\ So\text{-}\sigma_n oB \tag{2.15}
$$

Indeed, if ψ^* is a fixed point of T and if we set

$$
\underset{\sim}{u}^* = B(\psi^*) \ , \text{ then } \ \nu\ S(-\sigma_n(\underset{\sim}{u}^*)) = \psi^* \tag{2.16}
$$

which means that $\underset{\sim}{u}^*$ is a solution of (2.13) with $\tau = \psi^*$. But since $\psi^* = \nu\ S(-\sigma_n(\underset{\sim}{u}^*))$, we must conclude that $\underset{\sim}{u}^*$ is also a solution of the general problem (2.12).

Duvaut [3] has shown that the map T, in fact, has a unique fixed point for coefficients of friction ν sufficiently small. Demkowicz and Oden [2] have proved that T always has at least one fixed point for any ν and that this fixed point is unique for sufficently small ν.

3. FINITE ELEMENT APPROXIMATIONS

Let $\{V_h\}_{0<h\leq 1}$ be a family of finite-dimensional subspaces of of the space V constructed using conforming (C^0-) piecewise polynomial basis functions defined in the usual way over a finite-element mesh Ω_h approximating Ω. Here h is the mesh parameter: $h = \max_e h_e$, $h_e = \text{dia}\ \Omega_e$, Ω_e being a finite element, $\Omega_e \subset \overline{\Omega}_h$, $1 \leq e \leq E$. We suppose that the members of the family $\{V_h\}$ are

constructed using uniform or quasi-uniform refinements of regular meshes so that interpolation properties of V_h are of the standard form

$$\left.\begin{aligned} &\text{Given } \underset{\sim}{v} \in (H^m(\Omega))^N \cap V\ ,\ m > 0,\ \text{there} \\ &\text{exists } \tilde{\underset{\sim}{v}}_h \in V_h \text{ such that} \\ &\qquad \|\underset{\sim}{v} - \tilde{\underset{\sim}{v}}_h\|_s \le Ch^{\min(k+1-s,m-s)} \|\underset{\sim}{v}\|_m \\ &\qquad\qquad s = 0,1 \end{aligned}\right\} \tag{3.1}$$

where k is the degree of the largest complete polynomial contained in the finite-element shape functions and $\|\underset{\sim}{v}\|_m^2 = \sum_{j=1}^N \|v_i\|_{m,\Omega}^2$.

More specifically, we have in mind finite element approximations of two-dimensional elastostatics problems (N=2) using 9-node biquadratic elements (k=2) or 4-node bilinear elements (k=1).

For each choice of V_h, we must also construct an approximation of the constraint set K_h. Let $\sum_h$ denote a finite set of points on the approximate contact surface $\Gamma_C^h \approx \Gamma_C$. Typically, $\sum_h$ will merely denote the set of boundary nodal points on Γ_C^h or the set of Gaussian quadrature points used in evaluating $j(\underset{\sim}{v}^h,\underset{\sim}{v}^h)$ or $j(\underset{\sim}{v}^h)$, $\underset{\sim}{v}^h \in V_h$. Then we may define

$$\begin{aligned} K_h = \{\underset{\sim}{v}^h \in V_h \mid \underset{\sim}{v}^h\cdot\underset{\sim}{n}(\underset{\sim}{\xi}) \le 0 \\ \forall\ \underset{\sim}{\xi} \in \textstyle\sum_h\} \end{aligned} \tag{3.2}$$

In other words, the unilateral contact condition is to be applied only at discrete points in our finite-element approximations. Clearly, $K_h \not\subset K$, in general.

A direct approximation of the general problem (11) in K_h is embodied in the discrete problem,

$$\left.\begin{aligned} &\text{Find } \underset{\sim}{u}^h \in K_h \text{ such that} \\ &a(\underset{\sim}{u}^h,\underset{\sim}{v}^h-\underset{\sim}{u}^h) + c(\underset{\sim}{u}^h,\underset{\sim}{v}^h) - c(\underset{\sim}{u}^h,\underset{\sim}{u}^h) \\ &\qquad \ge f(\underset{\sim}{v}^h-\underset{\sim}{u}^h) \quad \forall \quad \underset{\sim}{v}^h \in K_h \end{aligned}\right\} \tag{3.3}$$

However, it is seldom advisable to attempt to solve the general

Coulomb friction problem by a direct assault on (3.3). Frequently, the construction of a numerical solution of one or more of the special problems described earlier as an intermediate step toward the analysis of (3.3) is effective. We comment further on one such scheme in the next section.

The non-local friction problem (2.12) deserves some comment. Consider a nonlocal friction law of the form,

$$\left.\begin{array}{l} |\sigma_T(u(x))| < \nu\, S(\sigma_n(u))(x) \Longrightarrow u_t(x) = 0 \\ |\sigma_T(u(x))| = \nu\, S(\sigma_n(u))(x) \Longrightarrow \exists\, \lambda \in \mathbb{R}\,, \\ \qquad \lambda > 0 \text{ s.t. } u_T(x) = -\lambda\sigma_T(u(x)) \end{array}\right\} \tag{3.4}$$

where, for example, if $\sigma_n(u)\quad L^2(\Gamma_C)$,

$$S(\sigma_n(u))(x) = \int_{-\infty}^{\infty} \omega_\rho(|x-y|)E_{ijk\ell}u_k(y),_\ell n_i n_m dy \tag{3.5}$$

where ω_ρ is a C_0^∞- mollifier kernel of the type

$$\omega_\rho(r) = \begin{cases} 0 & , r \geq \rho \\ c \exp\,[\rho^2/(\rho^2-r^2)] & , r \leq \rho \end{cases} \tag{3.6}$$

The radius ρ of the non-local friction law must be regarded as a new mechanical "roughness" property of the surfaces in contact which must be determined experimentally along with ν.

Suppose that $\{\phi_\alpha\}_{\alpha=1}^L$ denotes the global finite-element basis for V_h and that the absolute values of the normal contact pressures are approximated as combinations of functions $\{\psi_\Delta\}_{\Delta=1}^M$.

$$v_i^h = \sum_{\alpha=1}^{L} v_i^\alpha \phi_\alpha \ ; \ P^h = \sum_{\Delta=1}^{M} P^\Delta \psi_\Delta$$

Set

$$\Gamma(v^h,w^h) = (|v_T^h| - |w_T^h|)(x) \ ; \ x \in \Gamma_C^h \tag{3.7}$$

To obtain a finite element approximation of the non-local problem (2.12), we first solve

$$\underset{\sim}{w}^h \in K_h:$$

$$a(\underset{\sim}{w}^h, \underset{\sim}{v}^h - \underset{\sim}{w}^h) + \int_{\Gamma_C} P^{h(1)} \Gamma(\underset{\sim}{v}^h, \underset{\sim}{w}^h) ds \geq \bar{f}(\underset{\sim}{v}^h - \underset{\sim}{w}^h) \quad \forall \ \underset{\sim}{v}^h \in K_h \tag{3.8}$$

where $P^{h(1)}$ is an initial estimate of P^h, and then we correct P^h by choosing a $P^{h(2)}$ to satisfy

$$\int_{\Gamma_C} P^{h(2)} \Gamma(\underset{\sim}{v}^h, \underset{\sim}{w}^h) ds = \int_{\Gamma_C} \nu\, S(\sigma_n(\underset{\sim}{w}^h))\, \Gamma(\underset{\sim}{v}^h, \underset{\sim}{w}^h) ds \quad \forall \ \underset{\sim}{v}^h \in K_h \tag{3.9}$$

The corrected $P^{h(2)}$ is introduced into (3.8) and the process is repeated. At this writing, this algorithm is untested, but coding is underway.

4. ALGORITHMS FOR FRICTION PROBLEMS

We shall now describe several algorithms for the numerical solution of various types of friction problems described earlier. Some of these are standard, others are new, but each is suggested by the structure of the variational problem being analyzed.

I. The Signorini Problem without Friction. One of the simplest algorithms for solving problems of the form (2.1) is the standard iteration (relaxation) with projection. We have successfully used algorithms of this type for a wide range of contact problems without friction. The basic algorithm is:

1. Select a starting vector $\underset{\sim}{u}_\mu^{h(1)} = \underset{\sim}{0}$

2. Compute

$$\alpha_t(\underset{\sim}{u}_\mu^{h(t1)}) = \underset{\sim}{u}_\mu^{h(t)} \cdot \underset{\sim}{n} \quad (\text{on } \Gamma_C^h)$$

3. Set

$$\beta_{t+1}(u_\mu^{h(t)}) = \begin{cases} \alpha_t(u_\mu^{h(t)}) & \text{if } \alpha_t < 0 \\ 0 & \text{if } \alpha_t \geq 0 \end{cases}$$

4. Solve the linear system,

$$a(u_{\mu}^{h(t)}, v^{h}) + \mu^{-1} \int_{\Gamma_C^h} \beta_t(u_{\mu}^{h(t)}) v_n^h ds$$

$$= f(v^h), \quad v^h \in V_h$$

5. Return to step 1 or terminate this process when the relative error

$$e^{t+1} = \| u_{\mu}^{h(t+1)} - u_{\mu}^{h(t)} \|_1 / \| u_{\mu}^{h,t+1} \|_1$$

is less than a preassigned value.

Here μ is an exterior penalty parameter, and $u^{h,t}$ should approximate the solution of (2.1) as $h, \mu \to 0$. This algorithm is used (with several others) in Kikuchi and Oden [7]; see also Oden, Kikuchi, and Song [9].

II. Problems with Prescribed Normal Pressures. To analyze problem (2.4) numerically, we construct a finite element approximation of the regularized problem (2.9) for small $\varepsilon > 0$. This leads to a system of nonlinear algebraic equations of the form

$$Ku_{\varepsilon} + J_{\varepsilon}(u_{\varepsilon}) = f$$

where K is the usual stiffness matrix of the linearized problem, f is the load vector, and $J_{\varepsilon}(u_{\varepsilon})$ is a vector of nonlinear functions of the vector of nodal values of the approximation u_{ε}^h. Standard Newton-Raphson iteration has proved to be adequate for solving this system in many numerical experiments. For a variety of other algorithms for problems of this type, see Glowinski, Lions, and Tremolieres [6]

III. Contact Problems with Coulomb Friction. Campos, Oden and Kikuchi [1] have presented numerical solutions of certain cases of the general problem (1.1) obtained using the following algorithm:

1. First consider the case in which no friction is present. We then use algorithm I to obtain a first iterate $u(1)$ of the vector of nodal values of u^h.

2. Using the approximation u^1 computed in step 1, compute normal contact pressures $\sigma_n(u^{h(1)})$.

3. Treat the computed normal pressures as data in a contact problem with friction in which the normal pressures are prescribed. Use an algorithm of the type II above to compute tangential friction forces. By comparing $|\sigma_T(u^h)| - \nu |\sigma_n(u^h)|$ with ε, an estimate of the portions of the contact surface on which full

adhesion or sliding can occur.

4. Having calculated tangential frictional forces in step 3, we return to the non-frictional case treated in step 1 and treat these as prescribed forces. This new Signorini problem is solved and a second iterate is obtained.

There is no guarantee that this process is convergent, but it has proved to be convergent and to yield reasonable results in cases in which a large portion of the contact area is in full adhesion (no slipping).

IV. Non-Local Problems. An algorithm for treating finite-element approximations of the non-local friction problem (2.12) is suggested by the proof of the existence theorem and the development in Section 3:

1. Choose an arbitrary starting vector $\underset{\sim}{p}$ defining a non-negative contact pressure P^h on Γ_C^h. This becomes the data $\tau = P^h$ in a finite-element approximation of problem (2.4) (specifcally, see (3.8)).

2. Use algorithm II above to solve for a corresponding displacement field $\underset{\sim}{u}^{h(1)}$.

3. Compute $\sigma_n(\underset{\sim}{u}^{h(1)})$ and $S(\sigma_n(\underset{\sim}{u}^{h(1)})(\underset{\sim}{x})$, $\underset{\sim}{x} \in \Gamma_C^h$.

4. Use a recurrence formula such as (3.9) to compute a correction $P^{h(2)}$ of P^h.

5. Return to step 2. and compute a corrected $\underset{\sim}{u}^{h(2)}$. Repeat this process until the relative error is less than some preassigned tolerance.

5. NUMERICAL EXPERIMENTS

At this writing, extensive numerical experiments have been performed using algorithms I, II, and III but coding of algorithm IV is yet to be completed. We shall cite one example of III, discussed in greater detail in [1].

Consider the problem of indentation of a rigid cylindrical punch into an elastic half space. This problem, treated as a problem of plane strain, is modeled by a finite-element mesh of 9-node biquadratic elements shown in Fig. 1a. The material is taken to be isotropic and linearly elastic with modulus E= and Poisson's ratio $\mu = 0.3$. A coefficient of friction of $\nu = 0.6$ was used in our calculations.

The computed deformed shape for a prescribed centerline indentation of $\delta = 0.8$ is indicated in Fig. 2a and the computed tangential stresses on the contact surface is plotted in Fig. 2b. A fine mesh was needed in the vicinity of the boundary between regions of full adhesion and sliding in order to obtain proper resolution of the frictional stresses. The computed results are physically reasonable: the peak frictional stresses are 0.6 times

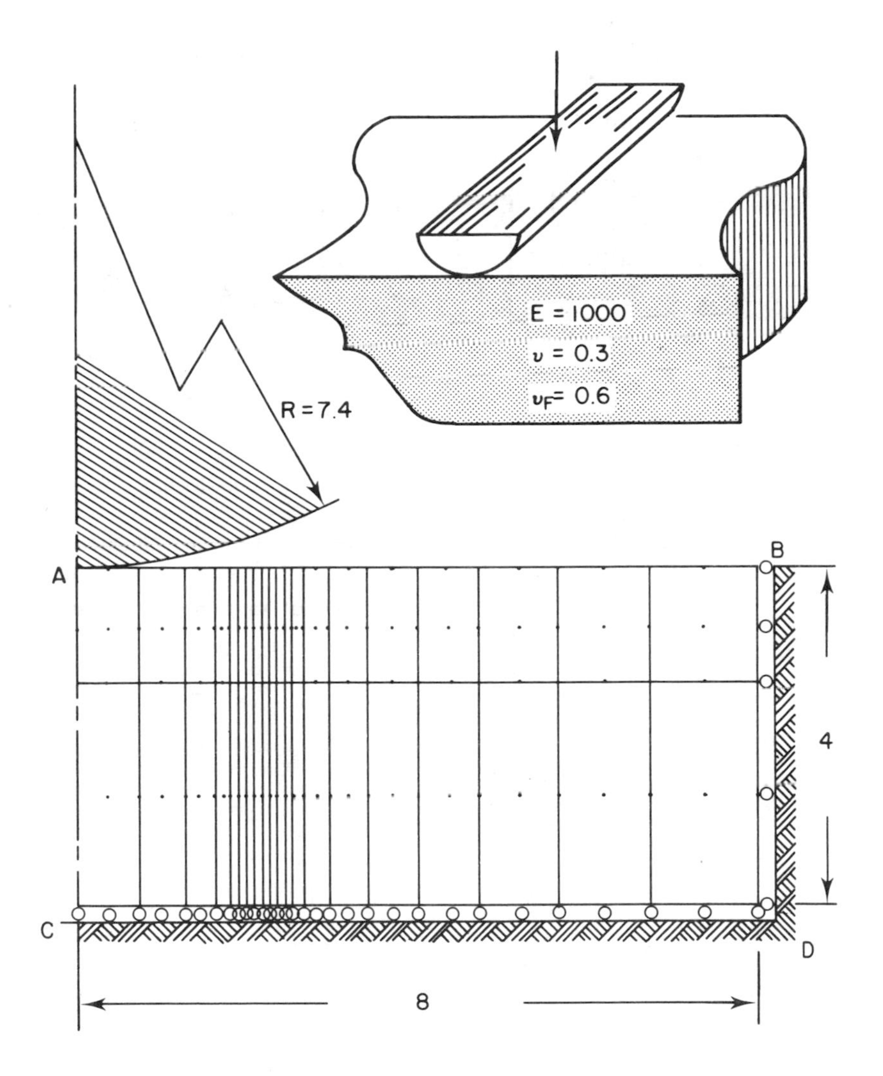

Fig. 1. A finite-element model of an elastic foundation indented by a rigid cylindrical punch. The mesh, which models half the body thanks to symmetry, consists of 40 9-node biquadratic elements.

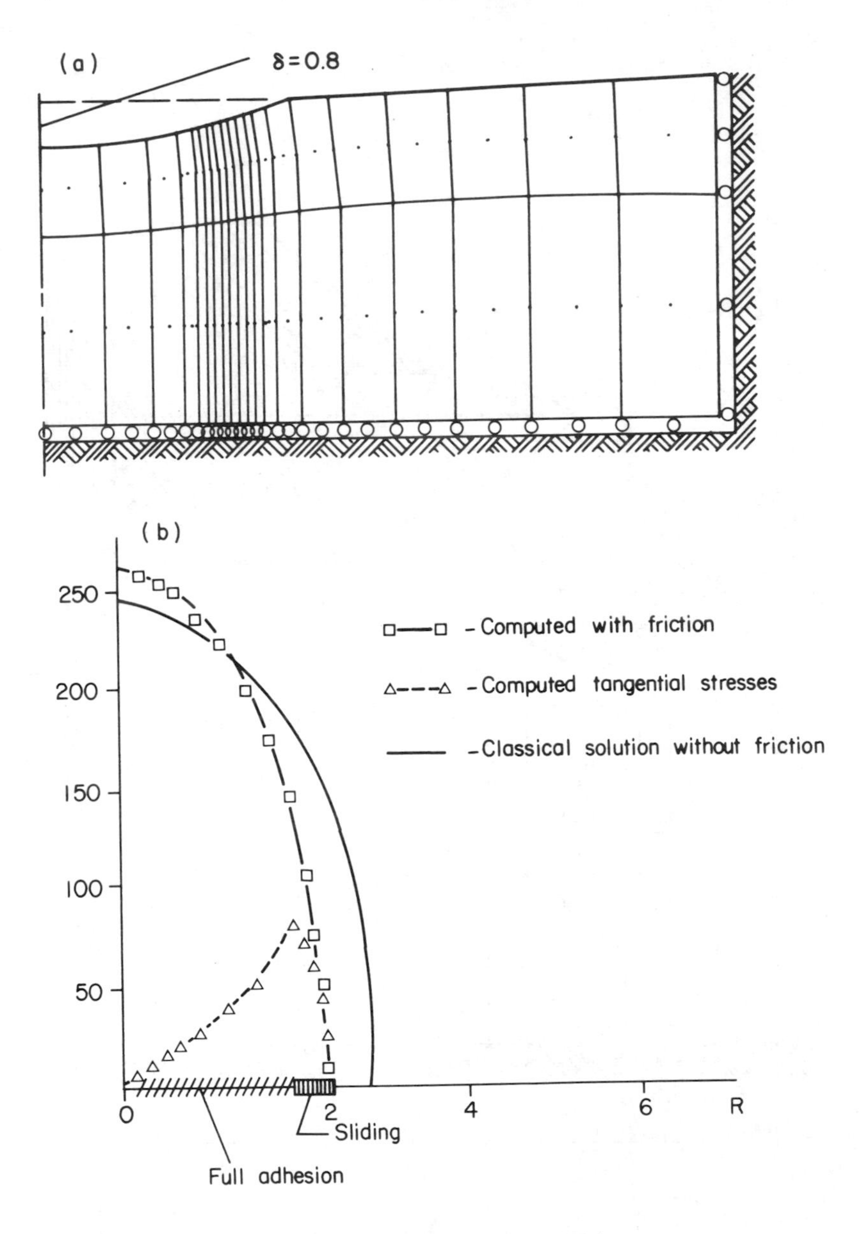

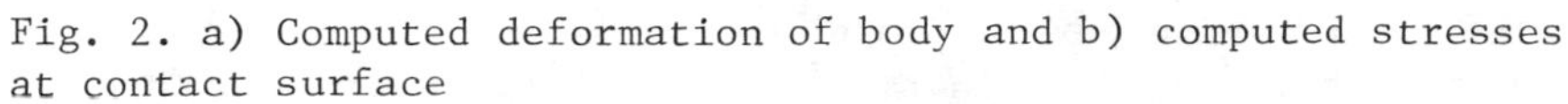
Fig. 2. a) Computed deformation of body and b) computed stresses at contact surface

the normal stress at the stick-slide interface. Comparisons of computed normal stresses in the body are in good agreement with analytical solutions available for the non-friction case.

Acknowledgement. The support of the work described here by the U.S. Air Force Office of Scientific Research under Contract F49620-78-0083 is gratefully acknowledged.

REFERENCES

1. CAMPOS, L., ODEN, J.T. and KIKUCHI, N., A numerical analysis of a class of contact problems with friction in elastostatics. *Comp. Math. Appl. Mech. Eng.* (to appear).

2. DEMOKOWICZ, L. and ODEN, J.T., On some existence and uniqueness results in contact problems with Coulomb friction. (in preparation)

3. DUVAUT, G., Équilibre d'un solide élastique avec contact unilatêral et frottement de Coulomb. *Compte-Rendus*, Acad. Sc. Paris t.290, 263-265 (1980).

4. DUVAUT, G. and LIONS, J.L., *Inequalities in Mechanics and Physics*, Springer-Verlag, N.Y. (1976).

5. FICHERA, L., Boundary-value problems in elasticity with unilateral constraints. *Encyclopedia of Physics, Vol.IV a/2-Mechanics of Solids II*, edited by C. Truesdell. Springer-Verlag, Berlin, Heidelberg, N.Y. (1972).

6. GLOWINSKI, R., LIONS, J.L. and TRÉMOLIERES, T., *Analyse Numérique des Inéquations Variationnelles, Vols. I and II.* Dunod, Paris (1976).

7. KIKUCHI, N. and ODEN, J.T., *Contact Problems in Elasticity.* SIAM Publications, Philadelphia, Penn. (1981). (to appear)

8. NEČAS,J., JARUŠEK, J. and HASLINGER, J., On the solution of the variational inequality to the Signorini problem with small friction. *BOLLETINO U.M.I.* (5)17-B, 796-811 (1980).

9. ODEN, J.T., KIKUCHI, N. and SONG, Y.J., Reduced integration and exterior penalty methods for finite element approximations of contact problems in incompressible elasticity. *TICOM Report* 80-2, Austin, Texas (1980).

FINITE ELEMENT STUDIES OF
REACTION-DIFFUSION

A. R. Mitchell and V. S. Manoranjan

University of Dundee

1. INTRODUCTION

In areas such as biology, chemistry and physiology, systems involving several reacting species with diffusion as a transport mechanism are common phenomena. The mathematical model of a process involving reaction and diffusion usually consists of a system of second order non-linear partial differential equations of the form

$$\frac{\partial u}{\partial t} = D\nabla^2 u + \sum_{j=1}^{m} M_j(x,u)\frac{\partial u}{\partial x_j} + f(u) \tag{1.1}$$

where $u(x,t)$ is an $\mathbb{R}^n$-valued function of $x \in \mathbb{R}^m$, and $t \in \mathbb{R}^+$. The diffusion matrix D has non-negative constant entries and is usually diagonal, the coefficients $M_j(j = 1,2,--,m)$ are continuous matrix valued functions, the non-linear function f describes the reaction of the system, and ∇^2 is the Laplacian operator. If (1.1) holds in a region $\Omega \times [t \geq 0]$ where Ω is in m-space with boundary $\partial\Omega$, along with (1.1) we have the initial condition

$$u(x,0) = u_o(x) \,, \quad x \in \Omega \tag{1.2}$$

and for initial-boundary value problems, the boundary condition

$$P\frac{\partial u}{\partial n} + Qu = a \,, \ (x,t) \in \partial\Omega \times [t \geq 0] \,, \tag{1.3}$$

where P,Q are matrix valued functions of x and t, a depends on x and t, and $\frac{\partial}{\partial n}$ denotes normal derivative. The nonlinear term $f(u)$ describes the kinetics of the medium and $D\nabla^2 u$ the diffusivity. The basic problem is to understand how the dynamical and diffusive terms interact to produce *persistent non-equilibrium solutions*. The convective term

$\sum_{j=1}^{m} M_j(x,u)\frac{\partial u}{\partial x_j}$ although important in some problems plays little part in the main confrontation between reaction and diffusion, and will be put to zero in this numerical study.

From the practical point of view, the important questions are

(i) "How do solutions of (1.1) to (1.3) evolve with time?"

and

(ii) "What is the asymptotic behaviour of solutions as $t \to \infty$.?" [9,10].

It is known from analysis and experiment that a large class of systems of the form (1.1) to (1.3) exhibit "travelling wave" type solutions and the principal aim of this study is to compare the results of numerical experiments with some of the known analytical results concerning travelling waves. The latter include plane wave solutions of (1.1) given by

$$u(x,t) = U(x.\nu - ct)$$

where ν is a unit vector and c is a scalar velocity.

Particular cases of these plane waves which are of interest to us are

(i) Wave fronts. ($U(-\infty)$ and $U(+\infty)$ exist and are unequal)

(ii) Pulses. ($U(\pm\infty)$ exist and are equal; U not constant).

(iii) Wave trains. (U periodic),

where $U(-\infty)$ and $U(+\infty)$ in (i) and (ii) are zeros of $f(u)$. Wave fronts arise from scalar models like Fisher's equation and pulses and wave trains from the Fitzhugh-Nagumo and Hodgkin-Huxley equations. More complicated wave-like phenomena such as target patterns and spiral waves arise in models of the Belousov-Zhabotinskii reaction [26,28]. The type of travelling wave obtained in a problem depends on the particular form of $f(u)$ and the initial and boundary conditions (1.2) and (1.3), and usually on the spatial domain being unbounded.

2. FISHER'S EQUATION

This is the simplest model of reaction-diffusion and takes the form

$$\frac{\partial u}{\partial t} = \frac{\partial^2 u}{\partial x^2} + f(u) . \qquad (2.1)$$

Such equations appear in genetics, flame propagation and many other areas. We consider two types of nonlinearity viz.

(i) $f(u) = u(1 - u)$ (2.2a)

and

(ii) $f(u) = u(1 - u)(u - a)$, $0 < a \leq \frac{1}{2}$. (2.2b)

Both the pure initial and initial boundary value problems have been analysed by many authors [1,2,8,11,12,18] and the following is a selection of the results obtained.

2.1 Analytic Results

Many of these depend on replacing (2.1) by the ordinary differential equation

$$U'' + c\,U' + f(U) = 0 \tag{2.3}$$

where

$$U(\xi) = U(x - ct) = u(x,t) \ . \tag{2.4}$$

Here $c(>0)$ is the speed of a wave travelling to the right and a dash denotes differentation with respect to ξ. $(-\infty < \xi < +\infty)$. Equation (2.3) can be written as the first order system

$$\begin{aligned} V' + cV + f(U) &= 0 \\ U' - V &= 0 \end{aligned} \tag{2.3a}$$

and so the rest states are given by

$$f(U) = 0 \ .$$

For $f(u) = u(1 - u)$, no analytic solution is available for (2.3), although it has been proved that travelling waves exist for any $c \geq 2 = c^*$. As far as the partial differential equation (2.1) is concerned, the rest states $u = 0,1$ have been shown to be unstable and stable respectively and all travelling waves have a speed c^* .

For $f(u) = u(1 - u)(u - a)$, $0 < a \leq \frac{1}{2}$, (2.3) has the Huxley solution

$$U(\xi) = (1 + \exp \frac{\xi}{\sqrt{2}})^{-1} \ , \quad c = \sqrt{2}\,(\frac{1}{2} - a) . \tag{2.5}$$

If the wave travels to the left $(\xi = x + ct)$ the solution is

$$U(\xi) = (1 + \exp (- \frac{\xi}{\sqrt{2}}))^{-1} \ , \quad c = \sqrt{2}\,(\frac{1}{2} - a) . \tag{2.5a}$$

Other exact solutions of (2.3) which have been obtained are (McKean [19])

(i) $U(\xi) = 3[\{(2-a)(\frac{1}{2}-a)\}^{\frac{1}{2}} \cosh(a\xi)^{\frac{1}{2}} + (1+\frac{1}{a})^{-1}]^{-1}$

$c = 0,\ 0 < a \le \frac{1}{2},\quad U(-\infty) = U(+\infty) = 0$, and

(ii) $U(\xi) = \frac{1}{2} + k\ \mathrm{sn}(\frac{\xi}{\sqrt{2}}(\frac{1}{2}-k^2)^{\frac{1}{2}},\ k(1-k^2)^{-\frac{1}{2}}),\ 0 < k < \frac{1}{2}, a = \frac{1}{2}$

(A one parameter family of periodic solutions in terms of Jacobi Elliptic functions).

Unfortunately both of these zero speed solutions are unstable. In this case where $f(u) = u(1-u)(u-a)$ the partial differential equation (2.1) has the rest states $u = 0,1$ (stable) and $u = a$ (unstable). All travelling waves to right or left have speed $\sqrt{2}(\frac{1}{2}-a)$. Additionally any stimulus given initially or on the boundary must exceed a or $K(\int_0^K u(1-u)(u-a) = 0)$ respectively in order to attain the stable state $u = 1$ as the solution evolves with time.

2.2 Numerical Results

The numerical solution of (2.1) is based on the continuous time Galerkin method where the weak solution $u \in H^1$ satisfies

$$(\frac{\partial u}{\partial t}, v) = (\frac{\partial^2 u}{\partial x^2}, v) + (f,v) \quad \forall\ v \in H^1, \tag{2.6}$$

with (,) denoting the L_2 inner product. If we approximate u by $U \in K_N \in H^1$ where

$$U(x,t) = \sum_{i=1}^{N} U_i(t)\ \phi_i(x), \tag{2.7}$$

(2.1) becomes after integration by parts

$$(\frac{\partial U}{\partial t}, \phi_j) + (\frac{\partial U}{\partial x}, \phi_j') = (f(U),\phi_j) + [\frac{\partial U}{\partial x}\phi_j]_B, j = 1,2,--,N \tag{2.8}$$

where a dash denotes differentiation with respect to x and $[\frac{\partial U}{\partial x}\phi_j]_B$ is a boundary term which is zero unless possibly when $j = 1,N$. The term $(f(U),\phi_j)$ is evaluated by putting

$$f(U(x,t)) = \sum_{i=1}^{N} f_i(t)\ \phi_i(x) \tag{2.9}$$

a technique which cuts down the computational work considerably and is known as product approximation [4]. Evaluation of (2.8) leads to the system of ordinary differential equations

$$M\,\dot{\underline{\alpha}} + S\,\underline{\alpha} = F(\underline{\alpha}) + \underline{b} \qquad (2.10)$$

where $M = ((\phi_i,\phi_j))$, $S = ((\phi_i',\phi_j'))$, $\underline{\alpha} = (U_1,U_2,\text{--}U_N)^T$, $\underline{b} = (b_1,0,\text{--},0,b_N)^T$ and a dot denotes differentiation with respect to time. The time variable is now discretised according to

$$t = m\,k \qquad m = 0,1,2,\text{----} \qquad (2.11)$$

leading to the approximations

$$\dot{\underline{\alpha}} = \frac{1}{k}(\underline{\alpha}^{m+1} - \underline{\alpha}^{m})$$
$$\underline{\alpha} = \frac{1}{2}(\underline{\alpha}^{m+1} + \underline{\alpha}^{m}). \qquad (2.12)$$

Substitution of (2.12) into (2.10) gives the Crank Nicolson Galerkin (C.N.G.) method which is second order correct in time and requires the solution of a system of nonlinear algebraic equations at each time step. This was accomplished by a linearising technique based on a predictor corrector method. Alternative techniques based on linearisation can be found in Fairweather [6,p.14] and Cannon and Ewing [3]. The basis functions $\{\phi_i\}$ are taken as piecewise linears in all calculations.

(i) $f(u) = u(1 - u)$. Equation (2.3) is solved by writing it as the system (2.3a) and taking as initial conditions

$$U = 0.99,\ V = -0.01 \quad \text{at} \quad \xi = X(<0)\ ,$$

where X is an arbitrary point well to the left of the origin on the ξ-axis. Calculations were carried out each with a different value of c and starting at $\xi = X$ and proceeding to the right on a grid with spacing $h = 0.1$. In agreement with theory solutions were obtained for $c \geq 2$. Equation (2.1) is now solved using the C.N.G. procedure with the initial condition taken from the solution of (2.3) with $c = 4$ and boundary conditions $u = 1$ as $x \to -\infty$ and $u = 0$ as $x \to +\infty$. The grid spacings are $h = 0.5$ and $k = 0.1$. The speed of the travelling front, initially 4, converges to the asymptotic speed 2 whilst the profile converges to the solution of (2.3) when $c = 2$. Another numerical experiment shows an initial condition in the shape of a small square pulse growing uniformly to unity with the velocity of each front converging to 2 from below

(ii) $f(u) = u(1 - u)(u - a)$, $0 < a \leq \frac{1}{2}$. We now turn to (2.1) and treat a variety of initial and boundary conditions. In all calculations the C.N.G. method is used with $k = h = 0.5$ and

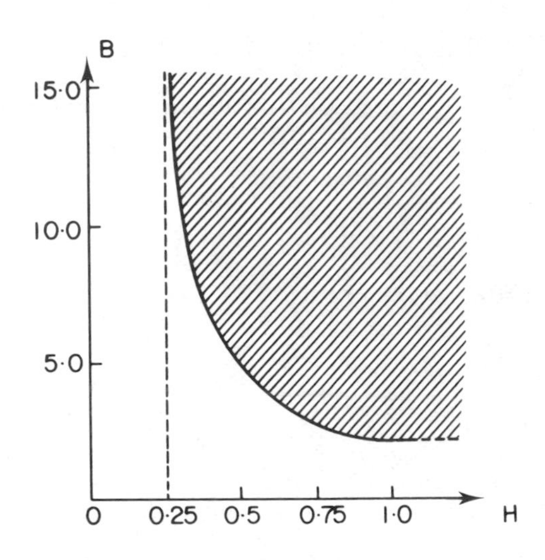

FIG. 1. Threshold curve, when the initial condition is a rectangular pulse.

$a = {}^1/4$. If we choose a rectangular pulse for initial condition, the pulse either dies away or grows to unity. In the latter case the two travelling fronts, one in each direction, obey the Huxley formula with regard to shape and velocity of propagation. The threshold curve in Fig. 1 shows the value of B(breadth of pulse) and H(height of pulse) which produce travelling fronts. The necessary condition $H > a = {}^1/4$ is substantiated by the numerical calculations. If the rectangular pulse is taken as a boundary stimulus at $x = 0$ and the problem is solved with zero initial condition in the quarter plane $x > 0$, $t > 0$, the solution again either dies away or tends to unity as time increases. The threshold curve is given in Fig. 2 and here the necessary condition

$H > K = 0.3923$ is satisfied. $\left(\int_0^K u(1 - u)(u - \frac{1}{4}) = 0 \right)$.

Before tackling numerically more difficult reaction diffusion problems it is worth assessing the accuracy of the calculations which have already taken place in Fisher's equation. The important quantities are the equilibrium state $u = 1.00$ and the asymptotic speed and shape of the travelling front. Sufficient to say that for $f(u) = u(1 - u)(u - {}^1/4)$ and the relatively course grid $h = k = \frac{1}{2}$, the equilibrium state is obtained exactly (to four decimal places), the front shape

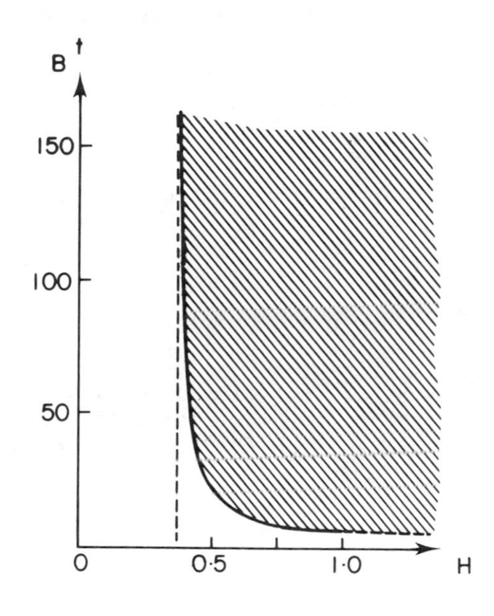

FIG. 2. Threshold curve, when the left hand boundary condition is a rectangular pulse.

likewise and the speed is 0.3544 against the theoretical value of 0.3536. These of course are limiting values of quantities which evolve with time from initial and boundary states and there is no claim that for small values of the time when the system is changing that comparable accuracy is obtained.

3. THE FITZHUGH NAGUMO (F.N.) SYSTEM

This is a more advanced model of reaction-diffusion and is given by

$$\frac{\partial u}{\partial t} = \frac{\partial^2 u}{\partial x^2} + u(1 - u)(u - a) - v , \quad 0 < a < \frac{1}{2}$$

$$\frac{\partial v}{\partial t} = b(u - dv) , \quad b \geq 0 , \ d \geq 0.$$

This system governs the conduction of electrical impulses in a nerve axon. Here $u(x,t)$ is the electrical potential across the axon and $v(x,t)$ is a type of recovery variable whose role will be explained later. The parameter a is the amount of Novocaine in the system and if $a \geq \frac{1}{2}$ the nerve goes dead. The recovery process (to the rest state) has a time scale $1/b$ and d is a measure of recovery. If $b = 0$, the system (3.1) returns to the form of Fisher's equation. At the other end of the scale (3.1) itself is a simplified model of the Hodgin-Huxley system [17].

Introduction of the variable

$$\xi = x - ct \,, \qquad c \text{ constant} \,,$$

allows (3.1) to be written as the first order system

$$\begin{aligned} W' &= U(U - a)(U - 1) + V - cW \\ V' &= -\frac{b}{c} U + \frac{bd}{c} V \\ U' &= W \end{aligned} \tag{3.2}$$

where $u(x,t) = U(\xi)$, $v(x,t) = V(\xi)$ and a dash denotes differentiation with respect to ξ . The rest states are given by $U' = V' = W' = 0$ and so from (3.2) we obtain the rest solutions

$$U = 0 \,, \quad \frac{1}{2}\,[(1 + a) \pm \{(1 - a)^2 - {}^4/d\}^{\frac{1}{2}}] \,. \tag{3.3}$$

From (3.3), if $d < (\frac{2}{1 - a})^2$, the only *real* rest state is $U = V = W = 0$. In fact from biological considerations (Hastings [15,16]) the upper limit on d is even more severe and is given by

$$d < (\max_{0 \le u \le 1} \frac{df}{du})^{-1} = 3(1 - a + a^2)^{-1} \,. \tag{3.4}$$

Many investigators have used (3.2) in an attempt to gain insight into the solutions of the F.N. system. Most solutions obtained from (3.2) are either periodic or homoclinic (i.e. tend to (0,0,0) as $\xi \to \pm\infty$). The principal weakness in the replacement of the original system by (3.2) through the assumption of wave solutions travelling at constant speeds is that the boundary and initial conditions of the original problem are not retained by the ordinary differential equation system (3.2). Nevertheless much valuable information has been obtained by studying (3.2) and numerical studies based on the original system (3.1) have leaned heavily on these results together with experimental findings.

A summary of relevant parameter limits, many of which are obtained from (3.2), now follows:

(i) $$0 \le d < \frac{4b}{(1 - a)^2} < c^2 < (1 - a)^2 \,, \quad \text{Sleeman [33]} \tag{3.5}$$

(ii) $$\max_{-\infty < \xi < +\infty} U(\xi) > \frac{1}{2}(1 + a) - \frac{1}{2}[(1 - a)^2 - \frac{4b}{c^2}]^{\frac{1}{2}},$$

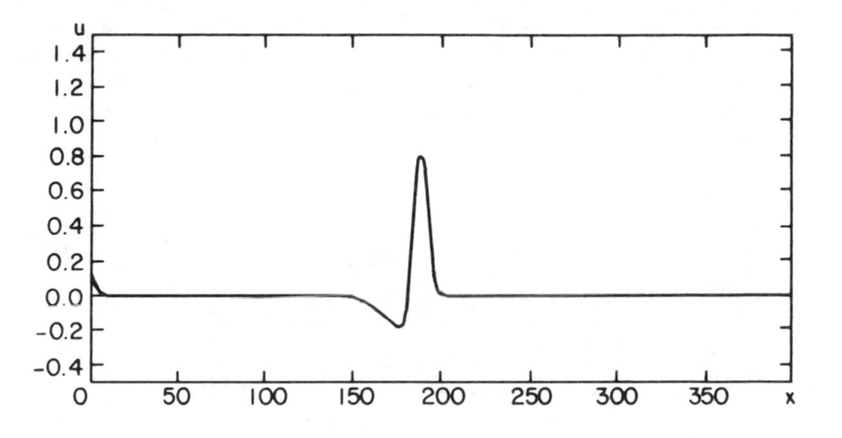

FIG. 3. Solitary pulse.

Green and Sleeman [14] (3.6)

(iii) No solitary pulse solutions for $c > 2^{-\frac{1}{2}}$. Necessary and sufficient condition for a solitary pulse is $U(\xi) \to 0$ as $\xi \to \pm \infty$.

(iv) $d < 3(1 - a + a^2)^{-1}$ (3.7)

(v) $0 < b << 1$ (slow recovery process).

3.1 Numerical Methods

We look only briefly at (3.1) with $d = 0$, a system first studied by Nagumo [27]. The recovery variable v enables the system to return to its rest state and so the travelling front solution in Fisher's equation can become a travelling pulse solution. An example of the latter is shown in Fig. 3.

A variety of alternative boundary conditions is now listed for what is effectively a semi infinite nerve axon. At $x = 0$,

(i) $u(0,t) = \begin{cases} I & 0 \le t \le T \\ 0 & t > T \end{cases}$

(ii) $u(0,t) = I \qquad t > 0$ (3.8)

(iii) $u(0,t) = \begin{cases} I & n(T_1 + T_2) \le t \le n(T_1 + T_2) + T_1 \\ 0 & n(T_1 + T_2) + T_1 < t < (n+1)(T_1 + T_2) \end{cases} \qquad n = 0,1,\text{--}$

(iv) $\frac{\partial u}{\partial x}(0,t) = -\frac{1}{2}I \qquad t > 0 \ , \ I > 0$

An alternative to (iv) is to add $I\,\delta(x)$ to the right hand side of the first equation in (3.1) and to replace the boundary condition at $x = 0$ by an homogeneous Neumann condition. In all four cases I is constant and above some threshold value. At $x = L$, where L can be as large as is required, the boundary condition is taken to be one of the following

$$\begin{aligned} &\text{(i)} \quad u = 0 \\ &\text{(ii)} \quad \frac{\partial u}{\partial x} = 0 \qquad\qquad (3.9) \\ &\text{(iii)} \quad \frac{\partial u}{\partial t} + c\frac{\partial u}{\partial x} = 0\,, \quad c \text{ a chosen constant.} \end{aligned}$$

The initial condition unless stated otherwise is

$$u(x,0) = 0 \qquad 0 \le x \le L \qquad (3.10)$$

A travelling pulse is obtained in all cases for suitable values of the parameters a,b and I. The boundary condition (iii) can even produce a train of pulses provided T_2 exceeds the duration of the refractory period following excitation of the nerve. During this period no stimulus will be able to produce a pulse. Details of finite element calculations of Nagumo's equation are available in Meiring et al. [21].

We now turn to the full F.N. system (3.1) along with boundary and initial conditions chosen from (3.8) - (3.10), and outline the numerical method used for its solution. The latter is based on the continuous in time Galerkin method

$$\begin{aligned} \left(\frac{\partial u}{\partial t}, w\right) &= \left(\frac{\partial^2 u}{\partial x^2}, w\right) + (f,w) - (v,w) \qquad \forall\, w \in H^1 \\ \frac{\partial v}{\partial t} &= b(u - dv) \end{aligned} \qquad (3.11)$$

where $(\ ,\ )$ denotes the L_2 inner product and the weak solution $u \in H^1$. Now approximate u and v by

$$U(x,t) = \sum_{i=1}^{N} U_i(t)\,\phi_i(x)$$

and (3.12)

$$V(x,t) = \sum_{i=1}^{N} V_i(t)\,\phi_i(x)$$

respectively where $\phi_i(x)$, $i = 1,2,\text{--},N$ are suitable trial functions and $U_i(t)$, $V_i(t)$, $i = 1,2,\text{--},N$ are time dependent coefficients. Substitution of (3.12) into the first equation of (3.11) followed by integration by parts leads to

$$(\frac{\partial U}{\partial t},\psi_j) + (\frac{\partial U}{\partial x},\frac{\partial \psi_j}{\partial x}) = (f,\psi_j) - (V,\psi_j) + <\frac{\partial U}{\partial x},\psi_j> \quad j = 1,2,\text{--},N, \tag{3.13}$$

where $<.,.>$ is a term involving the boundary conditions. Discretisation in time of (3.13) is now carried out according to the predictor-corrector pair

$$(\frac{1}{k}(U^{n+1*}-U^n),\psi_j)+(\frac{\partial}{\partial x}\frac{1}{2}(U^{n+1*}+U^n),\frac{\partial \psi_j}{\partial x}) = (f(U^n),\psi_j)-(V^{n+1},\psi_j)+ < \frac{\partial U^n}{\partial x},\psi_j > \tag{3.14}$$

$$(\frac{1}{k}(U^{n+1}-U^n),\psi_j)+(\frac{\partial}{\partial x}\frac{1}{2}(U^{n+1}+U^n),\frac{\partial \psi_j}{\partial x}) = (f(\frac{1}{2}(U^n+U^{n+1*})),\psi_j)- (V^{n+1},\psi_j)+ < \frac{\partial U^n}{\partial x},\psi_j > ,$$

and the second equation in (3.11) leads to

$$\frac{1}{k}(V^{n+1}- V^n) = \frac{1}{2}b \ \{(U^{n+1}+ U^n) - d(V^{n+1}+ V^n)\} \tag{3.15}$$

where $t = nk$, $n = 0,1,2,\text{---}$, with k constant. Equation (3.15) is used to eliminate V^{n+1} from (3.14). In all numerical calculations carried out, unless otherwise stated, the trial functions $\phi_i(x)$, $i = 1,2,\text{--},N$ and the test functions $\psi_j(x)$, $j = 1,2,\text{--},N$ are taken to be piecewise linear "hat" functions. It should be noted that the predictor-corrector pair (3.14) avoids the solution of a nonlinear system of equations at each time step.

3.2 Numerical Results

Numerical calculations using (3.14) with appropriate initial and boundary conditions are now carried out for ranges of the parameters a,b,d and I . We are particularly interested in determining parameter values for which trains of pulses (repetitive firing) are produced (see Fig. 4) and although we have already obtained these from Nagumo's equation ($d = 0$), using the left hand boundary condition (3.8) (iii), it is felt that this is to some extent a contrived condition and we should really consider (3.8) (iv), a boundary condition for which we were unable to obtain repetitive firing for Nagumo's equation. Consequently the following numerical results are obtained using (3.14) together with (3.8) (iv) on the left hand boundary, (3.9) (ii) on the right hand boundary, with L sufficiently large not to influence the first pulse to emerge for the

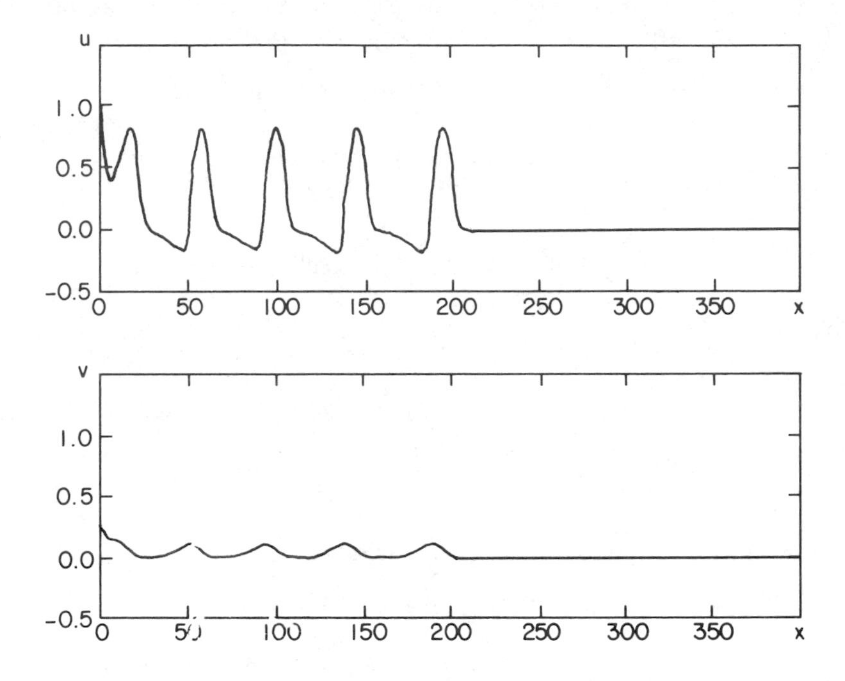

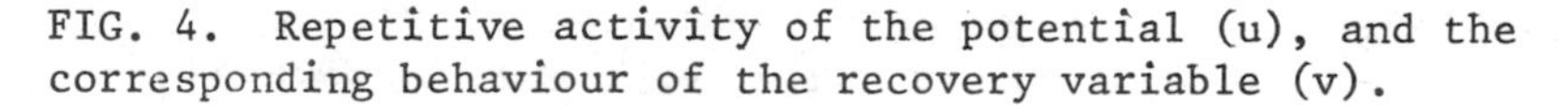
FIG. 4. Repetitive activity of the potential (u), and the corresponding behaviour of the recovery variable (v).

complete time of the calculation, and initial condition (3.10). Due to limitations imposed by computer usage, we are unable to conduct a detailed set of numerical experiments in the four dimensional space a,b,d and I even for fixed grid spacings h and k. Had this been possible we would have found the finite region in 4-space for which repetitive firing is possible. Our compromise is to take a point in 4-space for which repetitive firing has been found by other authors [30,31] and to conduct numerical experiments in the "vicinity" of this point.

The point selected in the four parameter space is

$$a = 0.139, \ b = 0.008, \ d = 2.54, \ I = 0.6.$$

These are the values chosen by Rinzel [30] to demonstrate repetitive firing and are only slightly different from those used by Fitzhugh [13]. We conduct four sets of numerical experiments in each of which $h = k = 0.5$ and the duration of each calculation is 500 time units which is equivalent to 1000 time steps. Only one of the four parameters is allowed to vary in each experiment. The number of pulses (N) in each experiment is shown in Table 1.

TABLE 1

(i)		(ii)		(iii)		(iv)	
b = 0.008 d = 2.54 I = 0.6		a = 0.139 d = 2.54 I = 0.6		a = 0.139 b = 0.008 I = 0.6		a = 0.139 b = 0.008 d = 2.54	
a	N	b	N	d	N	I	N
0.000	5	0.00	Front	1.20	0	0	0
0.050	5	0.001	1	2.00	3	.10	0
0.139	$4\frac{1}{2}$	0.002	2	2.54	$4\frac{1}{2}$	.20	1
0.147	2	0.004	3	3.0	$4\frac{3}{4}$	.30	1
0.165	1	0.006	4	3.54	5	.3375	$3\frac{3}{4}$
0.190	0	0.008	$4\frac{1}{2}$	4.00	$4\frac{1}{2}$	.35	4
	Fig. 5	0.0088	1	5.0	4	.40	4
		0.009	0	5.6	3	.60	$4\frac{1}{2}$
				5.8	Fig. 7	.65	$3\frac{3}{4}$
						.70	$2\frac{3}{4}$
						.75	3
						.90	3
						1.05	3
						1.10	$2\frac{3}{4}$
						1.15	2
						1.2	1
						2.0	1
							Pulse

A few comments are now made concerning the results in Table 1.

In (i), as a the amount of Novacaine in the axon increases, the number of pulses tends to zero as expected. (Fig. 5).

In (ii), as b tends to zero, the Fitzhugh Nagumo system tends to Fisher's equation and so the last surviving single pulse tends to a travelling front. (see Fig. 6). The increase in pulse breadth for values of b close to zero is significant. At the other end of the range as b increases above 0.008, the number of pulses falls steeply to zero.

In (iii), as d decreases below 3.54, the number of pulses tends to zero, and as d increases above 3.54, the train of pulses tends to a single pulse which degenerates into the shape

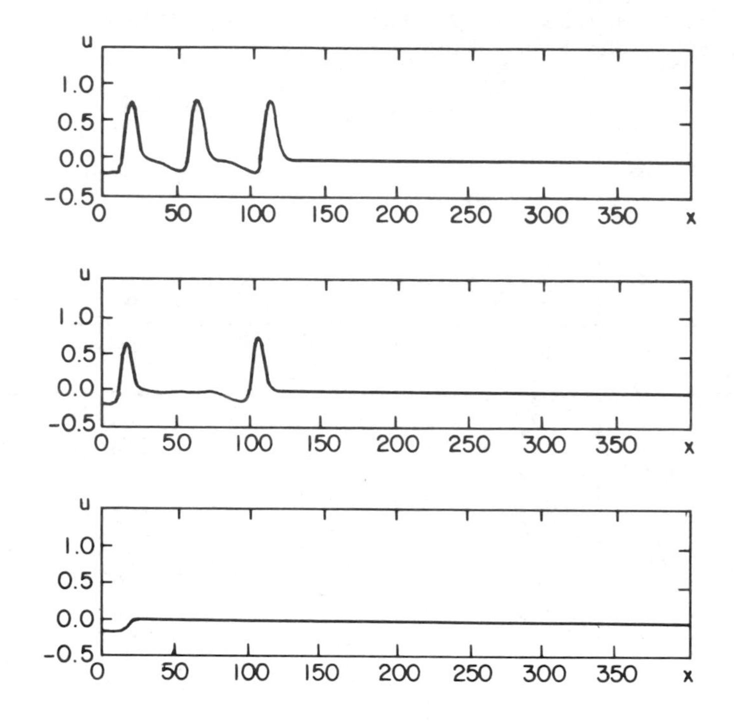

FIG. 5. Extinction of the train of pulses when a increases from 0.139.

shown in Fig. 7. The explanation for this lies in equation (3.3) where we see that for $d > \left(\frac{2}{1-a}\right)^2 = 5.4$, the system has three real rest states, the extreme ones being stable and the intermediate state being unstable. We see in Fig. 7 that the solution drops to the first stable rest state before eventually dropping further to the second stable rest state, the zero state. The fact that the calculations show this happening somewhere between the values 5.6 and 5.8 for d whereas theoretically it takes place at $d = 5.4$ can be accounted for by the numerical error. The present numerical experiments were conducted with $h = k = 0.5$ and reduction of these grid sizes brings the numerical result in line with the theoretical value.

In (iv), as I tends to zero, the number of pulses tends to zero, and as I increases the train of pulses (repetitive firing) tends to a <u>solitary pulse</u>. This limiting state is achieved at $I = 1.20$.

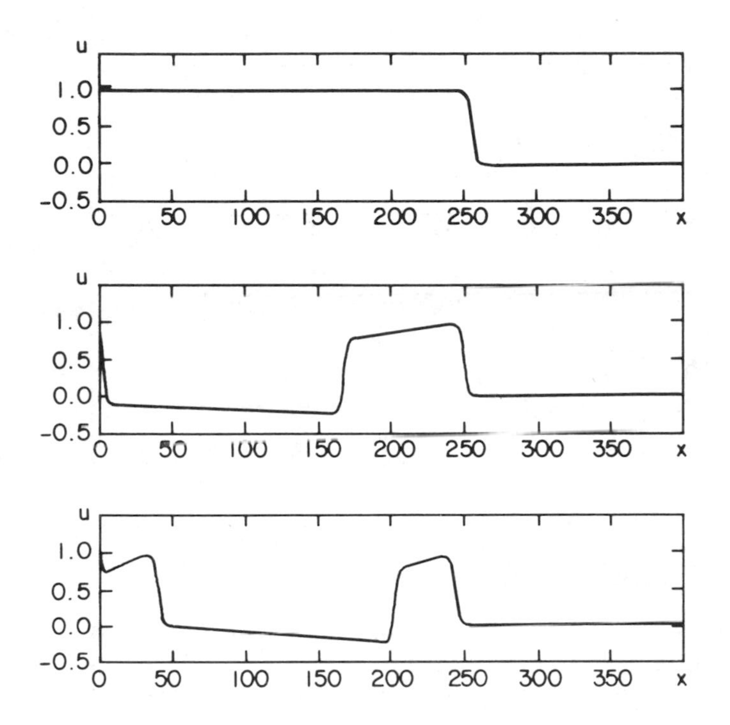

FIG. 6. Formation of repetitive activity as b increases from zero.

3.2.1 Appearance and disappearance of pulses. A brief word is called for concerning the appearance and disappearance of pulses in Table 1 as the parameters are altered. Every pulse is formed in the vicinity of the left hand boundary $x = 0$ and keeps its position in the line. The disappearance of pulses presents a more complicated picture and depends on the reduction taking place. For example if four pulses are reducing to three, one of the two inside pulses fades away while the other inside pulse readjusts to a central position between the two outside pulses.

3.2.2 Velocity of pulses. Our aim here is to find out how the pulse speed varies with changes in the parameters. The theoretical bounds on the velocity so far obtained are given in (3.5). For case (iv) where only the current I is altering, the theoretical limits are approximately $0.21 < c < 0.86$. We find for the complete range of values of I tested that the velocity maintains a <u>constant</u> value of 0.39. In case (iii), for a restricted range of values of d we find the following

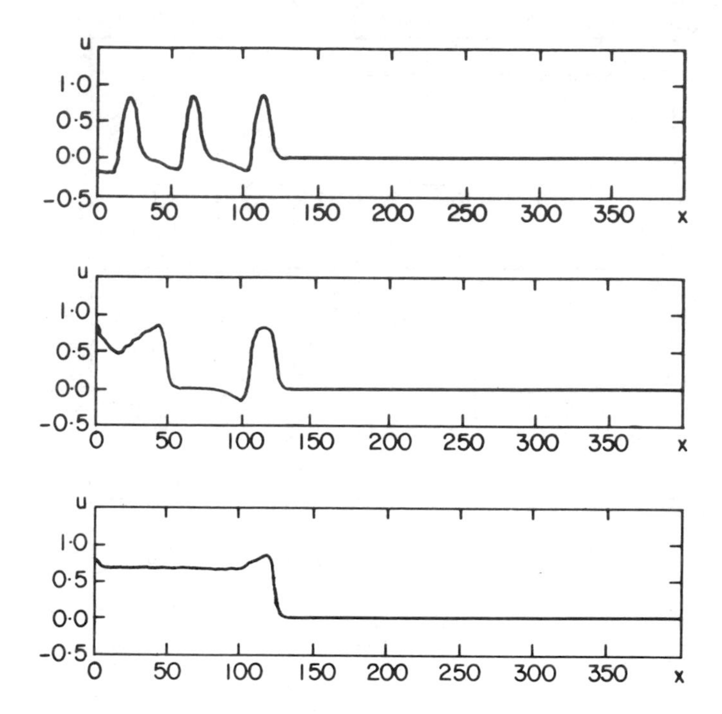

FIG. 7. Change of shape in the wave as d increases to 5.8.

results

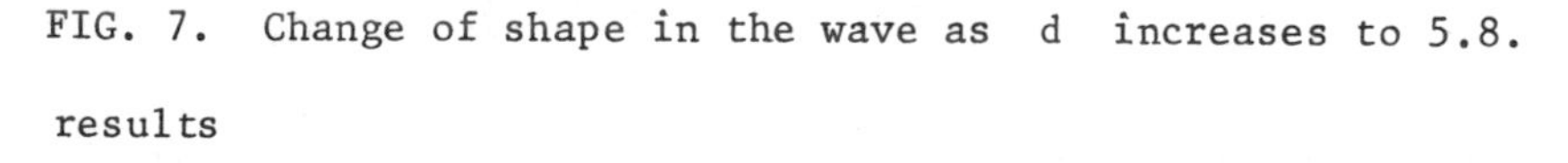

d	2.54	3.0	4.0
Speed	.3900	.3950	.4025

It therefore appears from limited tests that the pulse speed depends on a,b, and d but not on the current I .

3.2.3 Frequency curve. An interesting feature of travelling wave solutions is the dispersion relation $\omega = \omega(k)$ which relates the frequency ω to the wave number k in the system. We have already seen in (iv) where only the current I is varying in the reaction-diffusion system that the dispersion relation is linear and is given by

$$\omega = ck \qquad c \text{ constant velocity} \tag{3.16}$$

In Fig. 8 we show the graph of ω against I and note the sudden drop in frequency each time there is a disappearance of

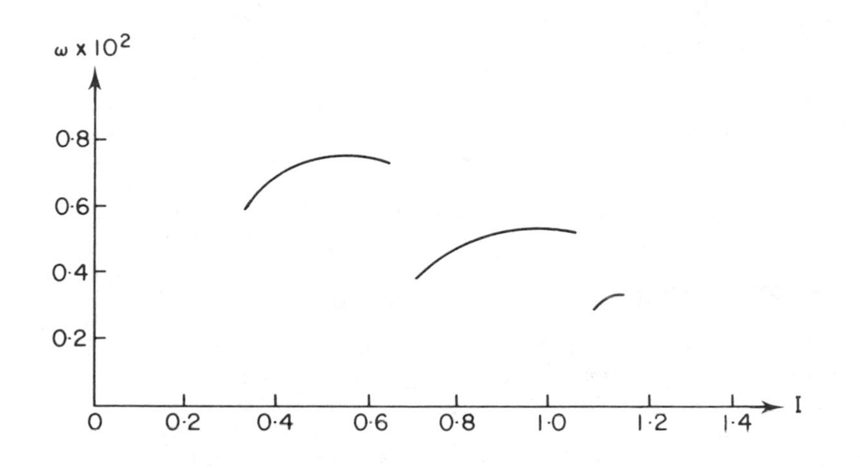

FIG. 8. Frequency curve.

one pulse. The frequency is calculated from (3.16) with $c = 0.39$ and $k = {}^1/\delta$, where δ is the distance between two adjacent pulses. In the other cases (i), (ii), and (iii), the pulse velocity is not constant and so the dispersion relation $\omega = \omega(k)$ is no longer linear. The present study does not involve sufficient numerical experiments to enable us to give reliable estimates of the influence of the parameters a,b, and d on the level of dispersion of the pulses.

3.2.4 Improved accuracy. So far all numerical experiments carried out have used the grid sizes $h = k = {}^1/2$. In order to guage the accuracy of these calculations we have repeated numerical experiment (iv) with the following alternative changes:

(i) The ratio ${}^k/h^2$ is held fixed at a value of 2 whilst h and k are both reduced, and

(ii) The linear *test* function in the Galerkin method is replaced by a cubic spline.

In both cases the effect is to increase marginally the range of values of I for which repetitive firing is achieved and to move slightly to the right the jumps in the frequency curve in Fig. 8.

3.2.5 Comparison with finite difference results. Comprehensive numerical results have been obtained by Rinzel both for Nagumo's equation [29] and for the F.N. system [30,31]. In the former paper Rinzel and Keller have replaced the cubic function $u(1 - u)(u - a)$ by a piecewise linear caricature which enables the ordinary differential equation in the variable $x - ct$ to be solved by a combination of analytical and numerical means. In the latter paper, Rinzel has used the Crank Nicolson finite difference method to solve the F.N. equation numerically. Although our numerical results using finite element Galerkin methods agree broadly with those of Rinzel, there are several important points of difference, and it is left to the interested reader to judge these for himself.

ACKNOWLEDGEMENTS

The authors acknowledge considerable assistance given by Professor B. D. Sleeman, Dr D. F. Griffiths and Miss A. F. Meiring during the preparation of this paper. Financial support of this research by the University of Dundee is also acknowledged by the second author. Finally grateful thanks are due to Miss M. Strachan for her expert typing of the manuscript.

REFERENCES

1. ARONSON, D.G. and WEINBERGER, H.G., *Nonlinear Diffusion in Population Genetics*. Lecture Notes in Mathematics. 446, Springer Verlag, Berlin (1975).

2. ARONSON, D.G. and WEINBERGER, H.F., Multidimensional Nonlinear Diffusions arising in Population Genetics. *Advances in Maths*. 30, 33-76 (1978).

3. CANNON, J.R. and EWING, R.E., Galerkin Procedures for Systems of Parabolic Partial Differential Equation related to the transmission of Nerve Impulses. *Nonlinear Diffusion* (eds. W.E. Fitzgibbon and H.F. Walker) Research Notes in Mathematics 14. Pitman (1977).

4. CHRISTIE, I., GRIFFITHS, D.F., MITCHELL, A.R., and SANZ SERNA, J., Product Approximation for Nonlinear Problems in the Finite Element Method. Report NA/42 University of Dundee, Department of Mathematics (1980).

5. COOLEY, J.W. and DODGE, F.A., Digital Computer Solutions for Excitation and Propagation of the Nerve Impulse. *Biophys. J.* 6, 583-599 (1966).

6. FAIRWEATHER, G., *Finite Element Galerkin Methods for Differential Equations*. Marcel Dekker, Basel (1978).

7. FIFE, P.C. and PELETIER, L.A., Nonlinear Diffusion in Population Genetics. *Arch. R. Mech. Anal.* 64 (2), 93-109, (1977).

8. FIFE, P.C. and McLEOD, J.B., The Approach of Solutions of Nonlinear Diffusion Equations to Travelling Front Solutions. *Arch. Rat. Mech. Anal.* 65, 335-361, (1977).

9. FIFE, P.C., Asymptotic Analysis of Reaction-Diffusion Wave Fronts. *Rocky Mountain Journal of Maths.* 7, 389-415, (1977).

10. FIFE, P.C., Asymptotic States for Equations of Reaction and Diffusion. *Bull. Amer. Math. Soc.* 84, 693-726, (1978).

11. FIFE, P.C. and McLEOD, J.B., A Phase Plane Discussion of Convergence to Travelling Fronts for Nonlinear Diffusion. M.R.C. Technical Summary Report 1986, (1979).

12. FISHER, R.A., The Advance of Advantageous Genes. *Ann. of Eugenics* 7, 355-369, (1937).

13. FITZHUGH, R., Impulse Propagation in a Nerve Fiber. *J. Applied Physiol.* 25, 628-630, (1968).

14. GREEN, M.W. and SLEEMAN, B.D., On Fitzhugh's Nerve Axon Equation. *J. Math. Biol.* 1, 153-163, (1974).

15. HASTINGS, S.P., Some Mathematical Problems from Neurobiology. *Amer. Math. Monthly* 82, 881-895, (1975).

16. HASTINGS, S.P., Single and Multiple Pulse Waves for the Fitzhugh-Nagumo Equations. (Preprint) (1981).

17. HODGKIN, A.L. and HUXLEY, A.F., A Qualitative Description of Membrane Current and its Application to Conduction and Excitation in Nerves. *J. Physiol.* 117, 500-544, (1952).

18. KOLMOGOROV, A.N., PETROVSKII, I.G. and PISKUNOV, N.S., A Study of the Equation of Diffusion with Increase in the Quantity of Matter and its Application to a Biological Problem. *Bjul Moskouskogo Gos Univ.* 1, 1-26, (1937).

19. McKEAN, H.P., Nagumo's Equation. *Advances in Maths.* 4, 209-223, (1970).

20. MEIRING, A.F., Numerical Studies of Reaction-Diffusion in Physiological Processes. Ph.D. Thesis. University of Pretoria, (1980).

21. MEIRING, A., MITCHELL, A.R., and SLEEMAN, B.D., Numerical Studies of Reaction-Diffusion. Report NA/39, University of Dundee, Department of Mathematics, (1980).

22. MITCHELL, A.R. and WAIT, R., *The Finite Element Method in Partial Differential Equations*. John Wiley and Sons, New York,(1977).

23. MITCHELL, A.R. and GRIFFITHS, D.F., *The Finite Difference Method in Partial Differential Equations*. John Wiley and Sons, New York,(1980).

24. MITCHELL, A.R., GRIFFITHS, D.F., and MEIRING, A., Finite Element Galerkin Methods for Convection-Diffusion and Reaction-Diffusion. Report NA/41, University of Dundee, Department of Mathematics, (1980).

25. MUIRA, R.M., Accurate Computation of Travelling Wave Solutions I. The Fitzhugh-Nagumo Equations - Stable Solitary Wave (Preprint) (1979).

26. MURRAY, J.D., On Travelling Wave Solutions in a Model for the Belousov-Zhabotinskii Reaction. *J. Theor. Biol.* 56, 329-353, (1976).

27. NAGUMO, J., ASIMOTO, S., and YOSHIZAWA, S., An Active Pulse Transmission-line Simulating Nerve Axon. *Proc. Inst. Radio Engrs.* 50, 2061-2070, (1962).

28. QUINNEY, D.A., On Computing Travelling Wave Solutions in a Model for the Belousov-Zhabotinskii Reaction. *J. Inst Maths. Applics.* 23, 193-201, (1979).

29. RINZEL, J. and KELLER, J.B., Travelling Wave Solutions of a Nerve Conduction Equation. *Biophys. J.* 13, 1313-1337, (1973).

30. RINZEL, J., Repetitive Nerve Impulse Propagation: Numerical Results and Methods. *Nonlinear Diffusion* (eds. W.E. Fitzgibbon and H.F. Walker) Research Notes in Mathematics. Pitman, London (1977).

31. RINZEL, J., Repetitive Activity and Hopf Bifurcation Under Point Stimulation for a Simple Fitzhugh-Nagumo Nerve Conduction Model. *J. Math. Biol.* 5, 363-382, (1978).

32. SCHONBECK, M.E., Boundary Value Problems for the Fitzhugh-Nagumo Equations. *J. Diff. Equns.* 30, 119-147, (1978).

33. SLEEMAN, B.D., Fitzhugh's Nerve Axon Equations. *J. Math. Biol.* 2, 341-349 (1975).

FINITE ELEMENTS FOR SINGULARITIES

IN TWO- AND THREE-DIMENSIONS

J.R. Whiteman*

Institute of Computational Mathematics,
Brunel University

1. INTRODUCTION

This paper is concerned with the finite element treatment of singularities in linear elliptic boundary value problems, and extends the discussion given by Whiteman and Akin in [39]. In the intervening period since [39] progress has been made on both the theoretical analysis of finite element methods for singularities and on the production of effective practical methods for their treatment. Particular advances are those which have produced L_∞-error bounds for the approximations to the solutions of certain two dimensional problems involving singularities, as well as bounds on the approximations to *stress intensity factors* in these cases, and those which have provided methods for the treatment of singularities for certain problems in three space dimensions. It is the purpose here to review some of these advances, bringing together theoretical results and practical methods, and at the same time to propose a new practical mesh refinement strategy for singularity treatment, the approach of which is motivated by previously derived theoretical results. We start in Section 2 by defining the two- and three-dimensional problems involved and for these give some results concerning the forms of the singular solutions. The determination of the singular forms is the subject of continuing research, so that it has seemed advisable to indicate cases where the singular forms are not yet available or where the actual form is the subject of some disagreement. This inconclusive situation is of importance from the point of view of constructing special finite techniques for the effective treatment of the singularities. Some special methods are given in Section 3, together with a discussion of relevant error analysis. A two-dimensional mesh refinement

*This work was supported in part by the United States Army under grant No.DAERO-78-G-069 and by N.A.T.O. under Research Grant 1374.

strategy, motivated by the error bounds of Section 3, is presented in Section 4. This strategy is then extended to treat three-dimensional problems. Some remarks about alternative methods and about the shortcomings of presently available methods are given in Section 5.

2. BOUNDARY VALUE PROBLEMS, SINGULARITIES

2.1 *Problem Formulations*

The boundary value problems which are considered here fall into two classes; one consisting of problems from potential theory and one of problems from elastostatics. Both classes are discussed for two- and three-space dimensions.

The *potential* problems in two- and three-space dimensions have respectively the general forms

$$\begin{aligned} -\Delta[u(x,y)] &= f_1(x,y) , \quad (x,y) \in \Omega , \\ u(x,y) &= g_1(x,y) , \quad (x,y) \in \partial\Omega_1, \\ \frac{\partial u(x,y)}{\partial \nu} &= g_2(x,y) , \quad (x,y) \in \partial\Omega_2, \end{aligned} \tag{2.1}$$

and

$$\begin{aligned} -\Delta[u(x,y,z)] &= f_2(x,y,z) , \quad (x,y,z) \in \Omega , \\ u(x,y,z) &= g_3(x,y,z) , \quad (x,y,z) \in \partial\Omega_1 , \\ \frac{\partial u(x,y,z)}{\partial \nu} &= g_4(x,y,z) , \quad (x,y,z) \in \partial\Omega_2 , \end{aligned} \tag{2.2}$$

where $\Omega \subset \mathbb{R}^2$ in (2.1) is a simply connected open bounded polygonal domain, $\Omega \subset \mathbb{R}^3$ in (2.2) is an open bounded polyhedral domain and each has boundary $\partial\Omega$. In (2.1) the polygonal boundary $\partial\Omega$ consists of disjoint parts $\partial\Omega_1$ and $\partial\Omega_2$ so that $\partial\Omega \equiv \partial\Omega_1 \cup \partial\Omega_2$, whilst a similar property holds for the polyhedral boundary $\partial\Omega$ in (2.2) which consists of plane surfaces. In each case $\partial/\partial\nu$ is the derivative in the direction of the outward normal to the boundary.

Although (2.1) and (2.2) have been defined for general g_i, $i = 1,2,3,4$, which must of course satisfy suitable smoothness conditions, the most studied forms of these problems are those in which there are homogeneous Dirichlet boundary conditions so that in each case $\partial\Omega_2 = \emptyset$, with the result that $\partial\Omega = \partial\Omega_1$, and $g_1 = g_3 = 0$. The homogeneous Dirichlet forms of (2.1) and (2.2) can both be written as

$$\begin{aligned} -\Delta u &= f_i , \quad \text{in } \Omega , \quad i = 1,2 , \\ u &= 0 \quad \text{on } \partial\Omega . \end{aligned} \tag{2.3}$$

In the usual Sobolev space setting the weak solution $u \in H_0^1(\Omega)$ of (2.3) satisfies

$$a(u,v) \equiv \int_\Omega \nabla u \nabla v \, d\Omega = \int_\Omega f_i \, v \, d\Omega \equiv F_i(v) \quad \forall \quad v \in H_0^1(\Omega) \ . \tag{2.4}$$

Many two dimensional problems of *linear elasticity* can be formulated in terms of the biharmonic operator so that

$$\begin{aligned} \Delta^2[u(x,y)] &= f_3(x,y) \ , \quad (x,y) \in \Omega \ , \\ u(x,y) &= g_5(x,y) \ , \quad (x,y) \in \partial\Omega \ , \\ \frac{\partial u(x,y)}{\partial \nu} &= g_6(x,y) \ , \quad (x,y) \in \partial\Omega \ , \end{aligned} \tag{2.5}$$

where Ω, $\partial\Omega$ and $\partial/\partial\nu$ are as for (2.1).

Examples of two dimensional linear elastic problems are those of the bending of a thin plate, for which $u(x,y)$ is the transverse deflection from the equilibrium position under the action of a load, and of plane strain in which $u(x,y)$ is the Airy stress function. For the case of a clamped plate $g_5 = g_6 = 0$ so that problem (2.5) becomes

$$\begin{aligned} \Delta^2 u &= f_3 \ \text{ in } \Omega \ , \\ u &= \frac{\partial u}{\partial \nu} = 0 \quad \text{on } \partial\Omega \ . \end{aligned} \tag{2.6}$$

The weak solution $u \in H_0^2(\Omega)$ of (2.6) satisfies

$$a(u,v) = \int_\Omega f_3 \, v \, d\Omega = F_3(v) \qquad \forall \quad v \in H_0^2(\Omega) \ , \tag{2.7}$$

where

$$a(u,v) \equiv \int_\Omega \Delta u \Delta v - (1-\mu)\left(\frac{\partial^2 u}{\partial x^2}\frac{\partial^2 v}{\partial y^2} - 2\frac{\partial^2 u}{\partial x \partial y}\frac{\partial^2 v}{\partial x \partial y} + \frac{\partial^2 u}{\partial y^2}\frac{\partial^2 v}{\partial x^2}\right) dxdy$$

$$u,v \in H^2(\Omega) \ , \tag{2.8}$$

in which μ is the Poisson's ratio of the plate material.

Problems of linear elasticity, in two- and three-dimensions can also be considered via the *potential energy* functional defined in terms of displacements. For three dimensions this can be written, where the x-,y-,z-displacements are respectively U,V,W, as

$$I[U,V,W] = \int_\Omega D(U,V,W) \, d\Omega - \int_{\partial\Omega^*} (U\overline{T}_x + V\overline{T}_y + W\overline{T}_z) \, ds \ , \tag{2.9}$$

over the space $H_U^1(\Omega) \times H_V^1(\Omega) \times H_W^1(\Omega)$, where $D(.,.,.)$, the strain energy density, is a quadratic function of the first derivatives of the displacements and $\overline{T}_x$, $\overline{T}_y$ and $\overline{T}_z$ are given surface tractions applied over the part $\partial\Omega^*$ of $\partial\Omega$.

2.2 *Singularities in Two- and Three-Dimensions*

Our main interest in all the above problems lies in the singularities present in the solutions when the boundary of Ω contains re-entrant corners, particularly slits. Typical two- and three-dimensional situations are shown respectively with the regions of Figs. 1 and 2. It will be noticed that whereas the polygon of Fig. 1 can contain re-entrant corners, the polyhedron of Fig. 2 can contain both re-entrant corners and re-entrant edges.

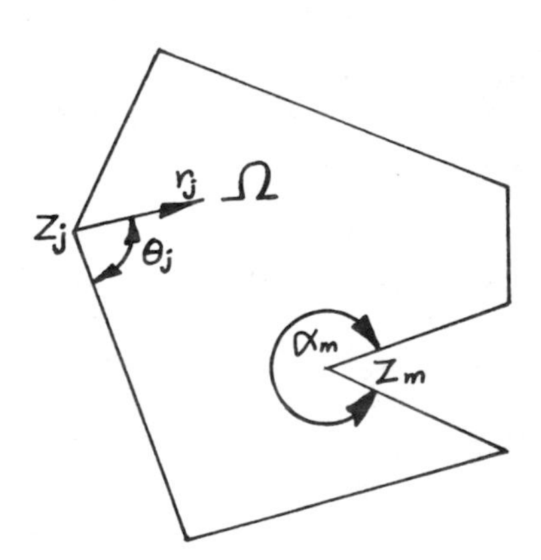

FIG. 1

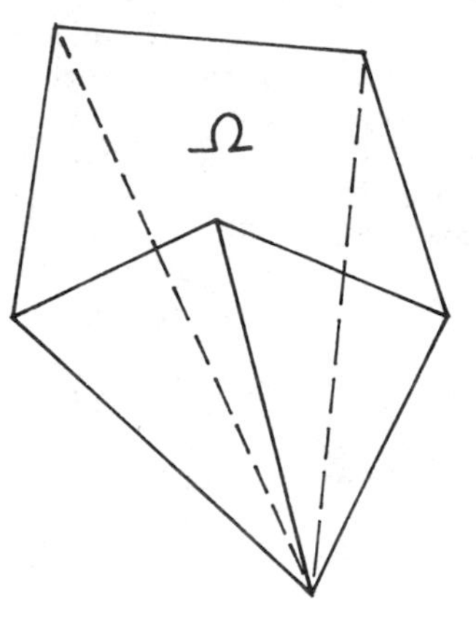

FIG. 2

We consider first *two-dimensional potential problems* of the type (2.3) and refer to Fig. 1. Suppose that the boundary $\partial\Omega$ has vertices z_j, $j = 1,2,\ldots,M$ with associated interior angles α_i, where $0 < \alpha_1 \leq \alpha_2 \leq \ldots \leq \alpha_M \leqq 2\pi$, and that at the jth vertex Ω_j denotes the intersection of Ω with a disc centred on z_j and containing no other corner. Let $\Omega_0 \equiv \Omega \setminus (\bigcup_{j=1}^{M} \Omega_j)$. In terms of local polar co-ordinates (r_j, θ_j) centred on the jth corner, the solution u of (2.4) has locally in the neighbourhood of z_j, except in the exceptional cases where $\alpha_j = \pi/2, \pi/3, \pi/4, \ldots$, see Kondrat'ev [20] and Lehman [21], the form

$$u(r_j,\theta_j) = \tilde{a}_j\, r_j^{\pi/\alpha_j} \sin\frac{\pi\theta_j}{\alpha_j} + \tilde{w}(r_j,\theta_j)\ , \qquad (2.10)$$

where $\tilde{a}_j$ is an unknown constant and $\tilde{w}$ is a smooth function. The term $r_j^{\pi/\alpha_j}\sin\pi\theta_j/\alpha_j$ has a *singularitiy* in the form of an unbounded first derivative at $r_j = 0$. The w does of course contain terms which have weaker singularities than the $r^{\pi/\alpha_j}\sin\pi\theta_j/\alpha_j$, so that the right hand side of (2.10) can be rearranged to become a sum of these terms plus another smooth function. It follows from (2.10) that *locally* $u(r_j,\theta_j) \in H^{1+\pi/\alpha_j-\varepsilon}(\Omega_j)$ for any $\varepsilon > 0$. (For the special cases $\alpha_j = \pi/2, \pi/3, \ldots$, which are ignored here, the first term in (2.10) contains an extra logarithmic factor.)

It has also been shown by Grisvard [15] that globally over Ω the solution of (2.4) can be written as

$$u = \sum_{j=1}^{M} a_j \chi_j(r_j,\theta_j) r^{\pi/\alpha_j} \sin\frac{\pi\theta_j}{\alpha_j} + w\ , \qquad (2.11)$$

where now the summation is over the M corners of the region Ω, the χ_j are smooth *cut-off* functions and $w \in H^2(\Omega)$. It is clear that $u \in H^{1+\pi/\alpha_M-\varepsilon}(\Omega)$.

It can be seen from (2.11) that the solution u contains a singularity at any corner for which $\beta_j \equiv \pi/\theta_j < 1$. For the case of a slit, where $\alpha_j = 2\pi$ so that $\beta_j = \frac{1}{2}$, it follows that $\partial u/\partial r$ has an $r^{-\frac{1}{2}}$ form so that u has an "$r^{-\frac{1}{2}}$" type singularity. More generally the singularities are of r^{β_j-1} form.

Knowledge of the actual $r^{\beta-1}$ form for any singularity is extremely useful as it enables special finite element trial functions to be constructed which are suitable for approximating the particular singular form. It is also important to be able to approximate a_j's accurately as these determine the *strength* of the singularity at any corner.

Three-dimensional potential problems of the type (2.3) have been considered by Stephan [30] who has produced expressions similar to (2.11) for the solution of the three-dimensional form of (2.4), except that they contain two summation terms, one for the vertices and one for the edges. The forms of solutions near re-entrant edges in these problems have been considered by Stephan and Whiteman [31], who show that for certain cases near an edge formed by two surface planes intersecting so that the internal angle in a plane orthogonal to the edge is α_j, see Fig. 3, in terms of cylindrical polar co-ordinates (ρ,ϕ,z) with the z-direction along the edge OF, the dominant singular part of the solution has the form

$$\rho^{\pi/\alpha_j} \sin \frac{\pi\phi}{\alpha_j} z \ . \tag{2.12}$$

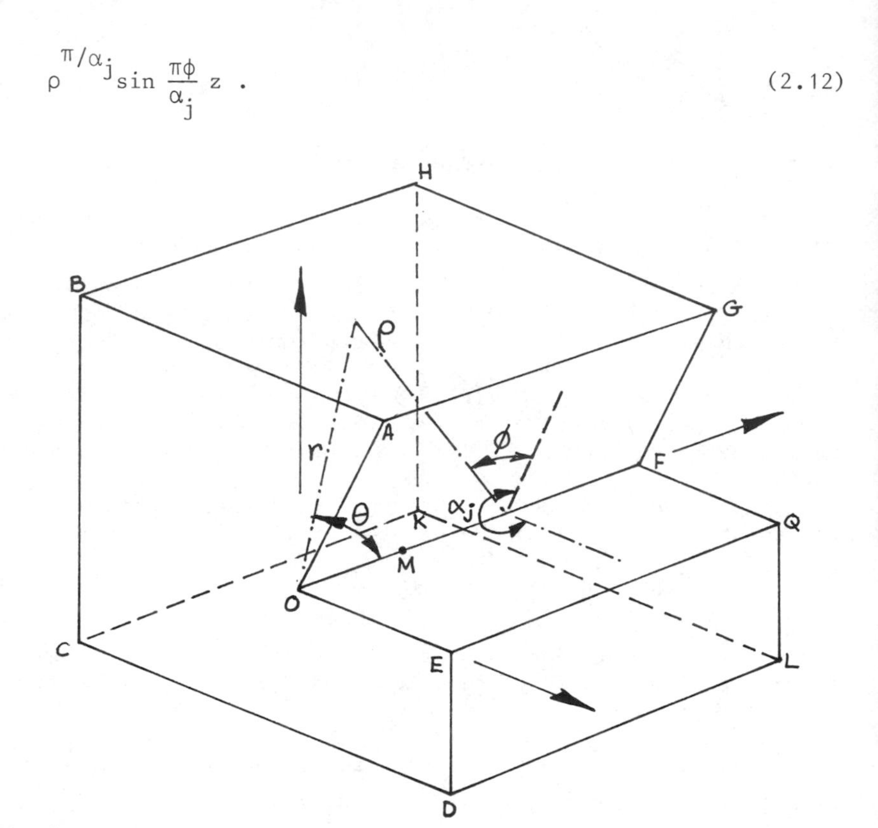

FIG. 3

For fixed z, that is for planes orthogonal to the edge, the expression (2.12) has of course the same form as the singular terms of (2.10) and (2.11), indicating that in these cases the two dimensional results carry over to three-dimensions.

Finally for potential problems we note that the methods of [31] cannot be applied to problems containing re-entrant vertices formed by the intersection of three planes. Such problems have, however, been studied by Fichera [10], who, although not being able to produce an exact β for the r^{β} form of the solution, has obtained upper and lower bounds for the β.

Boundary value problems of the form (2.6) involving the *biharmonic operator* in domains with corners can also be treated by the theory of Kondrat'ev [20], see also Blum and Rannacher [9]. The solution of (2.7) can near a corner be expressed in a form similar to that of (2.10), except that now $\tilde{w} \in H^4(\Omega)$ and the exponents of r are zeros of transcendental functions.

2.3 *Plane Strain Fracture Problem*

The case of a plane strain Mode I linear elastic fracture problem, in which the rectangular region with a crack of Fig. 4 is loaded in plane in outward normal directions to the sides BC and EF, can be formulated as in (2.5) with $f_3 = 0$ and the stress free boundary conditions $u = \partial u/\partial \nu = 0$ on the crack arms OA and OG. Williams [40] has shown that the biharmonic function which satisfies these stress free boundary conditions, in terms of local polar co-ordinates (r,θ) centred on 0 so that $\theta = 0$ for OD, contains a term involving $r^{3/2}$. This function is of course the Airy stress function and the coefficient of the $r^{3/2}$-term is related to the opening mode stress intensity factor K_I. The stresses are obtained by differentiating this function twice, so

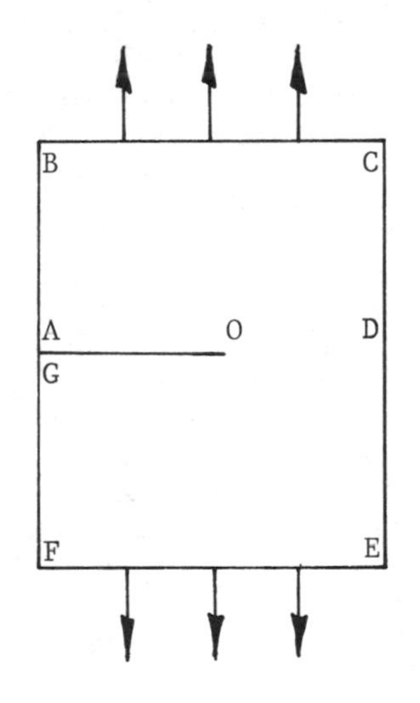

FIG. 4

that in particular near the crack tip the stress has the form $r^{-\frac{1}{2}}$; i.e. there is an $r^{-\frac{1}{2}}$ stress singularity. This same stress form has been obtained by Steinberg [29], using two harmonic functions $P(r,\theta)$ and $Q(r,\theta)$ and seeking biharmonic functions of the form $u = r^2P + Q$.

This plane strain problem can of course be considered using the potential energy functional (2.9) in terms of the displacements U and V, see [27]. For this type of fracture it has been shown, see Irwin [17] and Rice [23], that near the crack tip the singular stress field has the form

$$\begin{Bmatrix} \sigma_{xx} \\ \tau_{xy} \\ \sigma_{yy} \end{Bmatrix} = \frac{K_I}{(2\pi r)^{\frac{1}{2}}} \cos\theta/2 \begin{Bmatrix} 1 - \sin\theta/2 \sin 3\theta/2 \\ \sin\theta/2 \cos 3\theta/2 \\ 1 + \sin\theta/2 \sin 3\theta/2 \end{Bmatrix}, \tag{2.13}$$

where again K_I is the opening mode stress intensity factor. The stress intensity factor is used as a criterion for the onset of rapid fracture, and its approximation using numerical techniques is therefore of importance.

2.4 *Three-Dimensional Fracture*

For three-dimensional linear elastic fracture problems, such as those of a thick plate with a through internal crack or edge notch, there are *line* singularities along the crack fronts. If in Fig. 3 the interior angle $\alpha_j = 2\pi$, then the resulting solid has an edge notch and OF is the (straight) line of singularity. A Mode I situation arises when the faces ABHG and CDKL are loaded in the outward normal directions. It is well known that, for surface cracks with straight fronts, at an internal point (such as M of Fig. 3) on the front the singularity has the appropriate $(r^{-\frac{1}{2}})$ form in the plane through the point normal to the front. It is also known that the stress intensity factor varies with z so that $K_I = K_I(z)$. Examples of variation of $K_I(z)$ with z for problems of this type are given in [32]. However, the form of the singularity is much less certain at the end points of the crack front (end-points); for example at O and F of Fig.3 where the crack meets the stress free surfaces OABCDE and FGHKLQ. This uncertainty regarding the form of singularity arises on account of the complexity of analysing the stresses at points of this type.

The singular form has been analysed by Bentham [7] who, for the end-point of a quarter-plane crack in a hemisphere, shows that the stresses have the form $r^{-\beta}$, $0 \leqq \beta \leqq \frac{1}{2}$. In [12] and [13] Folias has, for a Mode I through crack in a thick plate, derived expressions for the end-point stresses of the form $r^{-(\frac{1}{2}+2\mu)}F_{ij}(\theta,\phi)$ where μ is the Poisson's ratio of the material. These forms as they stand indicate that for certain values of Poisson's ratio the *displacement* field is infinite at the end-point, which is unacceptable within the theory of linear elasticity. However, Folias is unable to derive the expressions $F_{ij}(\theta,\phi)$ explicitly and in [13] suggests that they may in fact vanish at the end-point so that the singularity should be regarded as being *at most* of the $r^{-(\frac{1}{2}+2\mu)}$ form. Kawai and his co-workers [18], [19], using a tensor stress function, see Sneddon and Berry [28], have derived the form $r^{(\beta(\mu)-2)}$, where $\beta(\mu) \in (1,2)$ and can be obtained by solving a pair of nonlinear equations whose coefficients involve associated Legendre functions. Kawai et al. conclude that there are in fact three possible values for β, that the singular stress is a combination of r^{δ_i-2}, $i = 1,2,3$, and that the smallest δ_i is less than 1.5 so that the singularity is stronger than $r^{-\frac{1}{2}}$. This term is dominant near $r = 0$. The other two values appear to correspond to those of Folias and Bentham.

It is clear from the above that the end-point singular forms are not as yet completely understood.

3. FINITE ELEMENT METHODS

In this Section we turn to finite element methods and their use for approximating the solutions of the problems containing singularities, as described in Section 2. Our interest is in effective techniques and in theoretical error analysis appropriate to the singular situations. As this latter has been developed only for two-dimensional problems, we restrict ourselves first to the two-dimensional case.

When conforming finite element methods are applied to the two-dimensional form of (2.4) and to (2.7), the solutions $u \in H_0^m(\Omega)$, $m = 1,2$, are approximated by $u_h \in S^h$, where $S^h \subset H_0^m(\Omega)$ is a finite dimensional space and u_h satisfies

$$a(u_h,v_h) = F_i(v_h) \ , \quad i = 1,3, \quad \forall \ v_h \in S^h \ . \tag{3.1}$$

If $a(u,v)$ is continuous over $H_0^m(\Omega)$ and H_0^m-elliptic, then it is well-known that

$$\| u - u_h \|_{H_0^m(\Omega)} \leqq C \| u - v_h \|_{H_0^m(\Omega)} \qquad \forall \ v_h \in S^h \tag{3.2}$$

and further that, if S^h consists of piecewise polynomial conforming trial functions of degree p on a quasi-uniform triangular partition of Ω with mesh size h then the right-hand side of (3.2) can be bounded so that

$$\| u - u_h \|_{H_0^m(\Omega)} \leqq K h^{\gamma} |u|_k \ , \tag{3.3}$$

where γ depends on both k and p. The major determining factor for γ is the regularity of the solution u, and this has motivated the discussions of Section 2.2 for the cases involving singularities.

If we restrict ourselves further to *two-dimensional second order* problems of the form (2.4), for the bound (3.3) to be O(h) the solution u must be in $H^2(\Omega)$. When re-entrant corners are present it is clear from Section 2.2 that this condition is not satisfied so that in (3.3) a lower order of convergence $O(h^{\gamma})$ results, where $0 < \gamma < 1$. Finite element error analysis is increasingly being undertaken using L_∞-norms, see e.g. Nitsche [22] and the references contained therein, and here again the presence of the singularities reduces the rate of convergence of

$\| u-u_h \|_{L_\infty(\Omega)}$. Schatz and Wahlbin [26] have for the two-dimensional problems of the type (2.4) shown that, for domains Ω with corners $\alpha_1, \alpha_2, \ldots, \alpha_M$ and the definitions given previously,

$$\| u - u_h \|_{L_\infty(\Omega_j)} \leqq Ch^{\min(\pi/\alpha_j, p+1, 2\pi/\alpha_M) - \varepsilon}, \quad j = 1,2,\ldots,M,$$

$$\| u - u_h \|_{L_\infty(\Omega_0)} \leqq Ch^{\min(p+1, 2\pi/\alpha_M) - \varepsilon}. \tag{3.4}$$

The bounds (3.4) indicate that the singularity causes a reduction in the rate of convergence both in the neighbourhood of the singularity and also away from it. Taking the example of an L-shape region with corners $\alpha_1 = \alpha_2 = \ldots = \alpha_5 = \pi/2, \alpha_M \equiv \alpha_6 = 3\pi/2$, the respective rates of convergence, in the case of S^h consisting of piecewise linear functions, are $O(h^{2/3})$ and $O(h^{4/3})$. These should be compared with the $O(h^2)$ which one expects when no singularities are present.

The analysis for the two-dimensional Poisson problem shows that some special adaptation of the finite element method is necessary in the neighbourhood of a singularity. A comprehensive survey of such adaptations is given in [39], for both potential problems and problems of linear elastic fracture in two- and three-space dimensions. Although much work has been done recently on methods involving the so-called *quarter-point* elements, see [5], [6], [16], [32], [33], and [34-36], we do not discuss these here but rather consider two adaptations for which mathematical results exist that can guide the manner of implementation. These are (a) the method of augmenting the approximating space, and (b) the method of local mesh refinement.

3.1 *Augmentation of Approximating Space*

In problems where the form of the singularity is known use can be made of this by augmenting the approximating space S^h with functions having the form of the singularity. Thus for the problems of Section 2 the solution u is in this case approximated by $\hat{u}_h \in \mathrm{Aug}S^h$. The technique, proposed by Fix [11] and used by Barnhill and Whiteman [2], [3] and Stephan and Whiteman [31], enables improved error bounds to be obtained and also increases the accuracy of the approximation. It also has the great advantage that it produces automatically approximations $\hat{a}_i$ to the coefficients a_j of the singular functions, as for example in (2.10). The importance of these coefficients has already been discussed.

If we consider again the two-dimensional form of problem (2.4), in an L-shaped domain so that $\alpha_M = 3\pi/2$ then it has been shown by Schatz [25] that, on a uniform partition into triangles and augmenting S^h with singular functions as in (2.10), with one singular function and S^h consisting either of piecewise linear or piecewise quadratic functions

$$\left.\begin{aligned} &\| u - \hat{u}_h \|_{L_\infty(\Omega)} = O(h^{4/3}) , \\ &|a_1 - a_1^h| = O(h^{2/3}) , \end{aligned}\right\} \tag{3.5}$$

with two singular functions and S^h consisting of piecewise linear functions

$$\left.\begin{aligned} &\| u - \hat{u}_h \|_{L_\infty(\Omega)} = O(h^{4/3}) , \\ &|a_1 - a_1^h| = O(h^{2/3}) , \quad |a_2 - a_2^h| = O(1) , \end{aligned}\right\} \tag{3.6}$$

with two singular functions and S^h consisting of piecewise quadratic functions

$$\left.\begin{aligned} &\| u - u_h \|_{L_\infty(\Omega)} = O(h^2) , \\ &|a_1 - a_1^h| = O(h^{4/3}) , \quad |a_2 - a_2^h| = O(h^{2/3}) . \end{aligned}\right\} \tag{3.7}$$

Expressions (3.5) - (3.7) indicate that in order to increase the rates of convergence one has to augment not only with a suitable number of singular functions, but also to use a space of piecewise polynomial functions of suitable degree.

The augmentation method has been largely ignored by engineers on account of the detrimental effect that it has on the structure of the matrix in the final equation system. This clearly causes an increase in the amount of computation required to set up and solve this system. The above results are therefore important in that for the Poisson problem they indicate the manner of increase in the rate of convergence, given that there will be an increase in the amount of calculation required to compute the numerical solution.

3.2 *Local Mesh Refinement*

The second paper of [26] is concerned with local mesh refinement in the neighbourhood of a singularity and gives theoretical error results for approximations to the solution of the two-dimensional problem (2.4) derived using systematically refined meshes. In particular, using the method of Rice [24], Schatz and Wahlbin indicate the type of mesh grading necessary to produce a particular rate of convergence in the neighbourhood of a corner. They set up concentric circular annuli, centred on the corner, and estimate the size that the elements in each annulus must be in order to produce the specific rate of convergence. These results thus indicate what the mesh size should be at a particular distance from the singularity. They are of significance if one seeks to retrieve a global convergence rate equal to that achieved when no singularity is present.

As a consequence of the work of Schatz and Wahlbin [26] it seems desirable that one should be able to produce easily a refined mesh which can conform to their results. The annuli are polar, centred on the singular point, and the mesh points along any radial mesh line from the singularity are graded according to a specified pattern. Thus one wants a mesh which has locally this form. A mesh refinement scheme such as that of [27] is therefore clearly not suitable. At the same time, as the distance from the singular point increases the local mesh will have either to be compatible with a *global* mesh or to adapt to the geometry of the region of the problem. A strategy for producing locally refined meshes of this type is given in the next Section.

4. A SCHEME FOR LOCAL MESH REFINEMENT

The theoretical results of [26] are of course limited to the two-dimensional problem (2.4). However, their existence demands that meshes of the required type should be produced, which can immediately be used for this type of problem. It is also reasonable to expect that similar theoretical results for more complicated problems, in both two- and three-dimensions, will in time become available, and this in turn indicates a need for suitable flexible mesh refinement schemes. Towards this end, without specifying the actual boundary value problem that we are considering but keeping the results of [26] in mind, we set out here with the aim of producing first a two-dimensional local mesh refinement scheme for which the mesh near a singular point has *polar form*, whilst away from the singularity it is compatible with a global mesh or with a specific (cartesian) global geometry. The way that this can be extended to the case of line singularities in three-dimensions will then be discussed.

By way of illustration let us consider the L-shaped region of Fig. 5. It is assumed that for the types of problem under consideration it is possible to split the domain Ω, or some subdomain of Ω which includes the corner point, into a number of non-overlapping subdomains which completely fill the interior of the corner and which can be treated separately but similarly. We emphasise that Fig. 5 can be thought of either as the domain of a problem or as a subdomain local to a re-entrant corner. The geometry of Fig. 5 suggests that OABCO, OCDEO and OEFGO are suitable subdomains, each involving O. These subdomains of course here all have the same shape, being squares. It is not in general necessary that the angles of the subdomains at O should be the

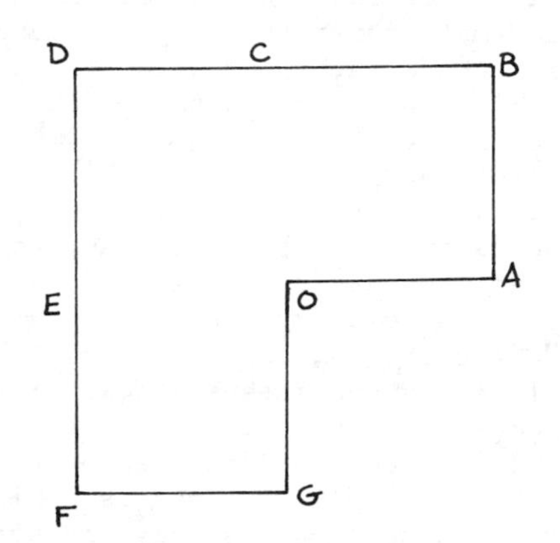

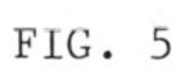
FIG. 5

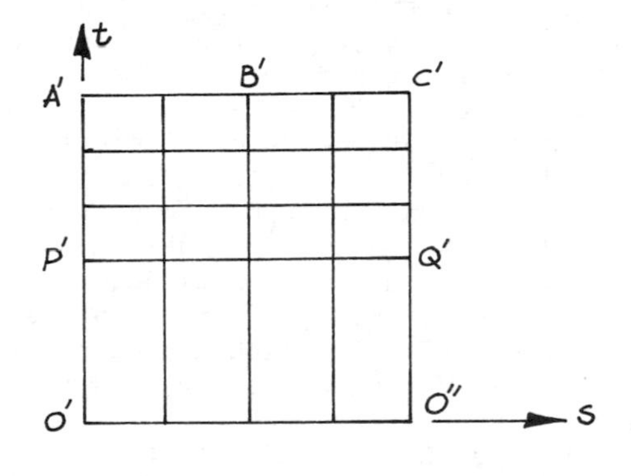

FIG. 6

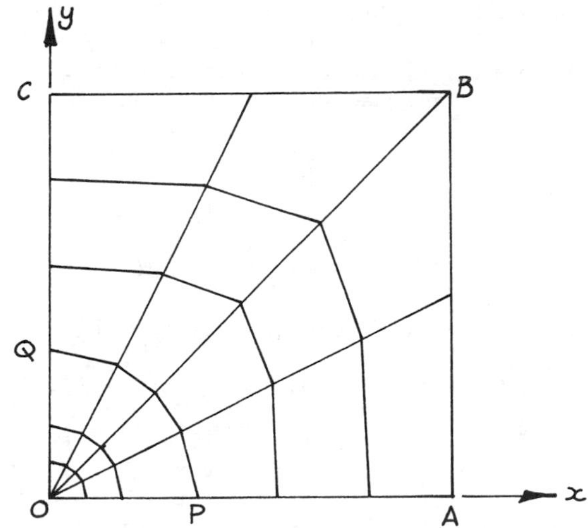

FIG. 7

same or be right angles, or that the subdomains should have similar geometries. The method of producing the mesh is now described for a single subdomain, after which it is assumed that the remaining subdomains and subdomains of other regions with different shapes can be treated in a similar manner.

Let us consider the subdomain OABCO of Fig. 5, for which we seek a mesh of the type described earlier having *polar form* near O. The requirement is that it should be easy to generate and that it must be capable of having any specified radial grading near O. The approach is to start with a *standard* unit square, such as O'A'B'C'O"O' in the (s,t)-plane, Fig. 6, in which there are mesh points defined as the intersections of the lines s = constant, t = constant. Such mesh points are easily defined. The mesh in OABCO is produced by a mapping of the standard square and its mesh onto the (x,y)-plane. This is achieved in two stages. The first of these concerns only the perimeter of the standard square. For this the mapping $\underline{F} = \underline{F}(s,t) \equiv (x,y)^T$, where x = x(s,t), y = y(s,t) must be defined so that the perimeter O'A'B'C'O"O' is mapped onto the perimeter OABCO in such a way that O' and O" coalesce to become O, whilst the line A'B'C' becomes the two lines AB and BC. This perimeter mapping is critical to the eventual production of suitable mesh. Initial attempts at mesh generation such as reported in Whiteman [37], proceeding as above, lead to unsuitable meshes. It has been found necessary additionally to include on the boundary sides O'A' and O"C' of the standard square points such as $P' \equiv (0,\frac{1}{2})$ and $Q' \equiv (1,\frac{1}{2})$, see Fig. 6, which are mapped respectively onto P and Q as in Fig. 7. The reasons for this becomes clear in the second stage of the process. In this a transfinite blending function $\underline{U}(s,t)$, see Gordon and Hall [14], which interpolates to the perimeter mapping $\underline{F}(s,t)$ is used to map the interior of the standard square onto the interior of OPABCQO. Due to the presence of the points P' and Q' we can take $\underline{U}(s,t)$ to be a function which is linear in s/quadratic in t having the form

$$\begin{aligned}\underline{U}(s,t) = {} & (1-s)\underline{F}(0,t) + s\underline{F}(1,t) + (1-t)(1-2t)\underline{F}(s,0) \\ & + 4t(1-t)\underline{F}(s,\tfrac{1}{2}) + t(2t-1)\underline{F}(s,1) \\ & - (1-s)(1-t)(1-2t)\underline{F}(0,0) - s(1-t)(1-2t)\underline{F}(1,0) \\ & - 4(1-s)t(1-t)\underline{F}(0,\tfrac{1}{2}) - 4st(1-t)\underline{F}(1,\tfrac{1}{2}) \\ & - (1-s)t(2t-1)\underline{F}(0,1) - st(2t-1)\underline{F}(1,1) \; . \qquad (4.1)\end{aligned}$$

The mapping (4.1) also enables the line P'Q' ($t = \frac{1}{2}$ in Fig. 6) to be mapped onto the quarter circle through P and Q of Fig. 7. In addition the intersections of lines t = constant ($t>\frac{1}{2}$) with lines s = constant map onto the points of intersection of image radial lines from O with peripheral curves joining points of PA to points of QC. In Fig. 7 these points have been joined with straight lines to form quadrilateral elements. The mesh in OABC outside the sector OPQ has thus been formed. Inside the sector, which is the zone where the major refinement is required, the mesh points are generated directly as the intersections of the (already defined) lines radiating from O with quarter circles centred at O. These points are again joined with straight lines to form the elements. As many quarter circles as required and with any desired radial grading, such as for example that for the Poisson problem based on the results of [26], can be used in the sector OPQ.

This two-dimensional strategy of mesh refinement has been defined with ease of implementation in mind. It can be very simply extended to three-dimensions to be used in the case of line singularities. The three-dimensional results of Section 2 indicate that the two-dimensional refinement can be used in planes orthogonal to the line of singularity and that the resulting mesh points can be joined by lines parallel to the line of singularity. Such a mesh is shown in Fig. 8 for the case of a straight line singularity. Clearly the mesh can be adapted to contain elements with curved edges for use with curved line singularities.

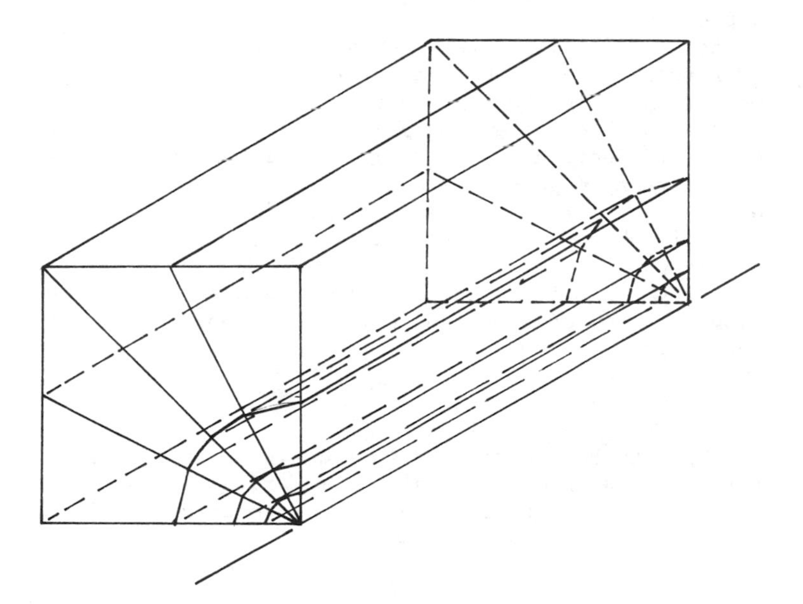

FIG. 8

5. REMARKS

Although the theoretical results of Section 3 have been used as motivation for the general mesh refinement techniques of Section 4, it must be emphasised that the theory is at present restricted to two-dimensional Poisson problems. It has been recognised for some time that *weighted* Sobolev spaces are useful for the theoretical finite element treatment of singular problems. Some very recent error estimates for finite element approximations to the solutions and to the coefficients of the singular functions obtained by Blum and Dobrowolski [8] make use of weighted spaces and involve the use of a new variant of the augmentation method of Section 3.1, again for the two-dimensional Poisson problem. It is thus clear that for problems with boundary singularities, even in the linear cases considered here, theory lags far behind the usual practice.

The results described in Section 2 on the forms of the singularities in three-dimensions indicate areas where more research is needed. In particular for the three-dimensional fracture problems the lack of a completely rigorous derivation of the singular form, at the points of intersection of the lines of singularity with stress free faces, is clearly a glaring deficiency which casts doubt on the meaning and the usefulness of the numerical results near such points. One must also question the whole idea of producing finite element approximations which reflect an r^{β}-form for the case when the β itself may only be an approximation.

As has been indicated the importance of singularities has over recent years caused a large number of special finite element adaptations to be proposed for their treatment. The two adaptations, space augmentation and local mesh refinement described in Section 3, were deliberately chosen as they belong to two different classes of method; one in which the coefficients of the singular functions are approximated automatically and the other in which these approximations must be obtained through a process of retrieval by the post-processing of already calculated results. This is another interesting area, for which the total amount of computation needed for the opposing techniques needs to be compared.

ACKNOWLEDGEMENT

The author is most grateful to R.E. Barnhill and A. Schatz for numerous informative and stimulating discussions and to J. Galliara for writing the program to derive the mesh points of Fig. 7.

REFERENCES

1. AKIN, J.E., Elements for the analysis of line singularities. pp.65-75 of J.R. Whiteman (ed.), *The Mathematics of Finite Elements and Applications III, MAFELAP 1978*. Academic Press, London (1979).
2. BARNHILL, R.E. and WHITEMAN, J.R., Error analysis of finite element methods with triangles for elliptic boundary value problems. pp.83-112 of J.R. Whiteman (ed.), *The Mathematics of Finite Elements and Applications*. Academic Press, London (1973).
3. BARNHILL, R.E. and WHITEMAN, J.R., Error analysis of Galerkin methods for Dirichlet problems containing boundary singularities. *J. Inst. Maths. Applics.* 15, 121-125 (1975).
4. BAZANT, Z.P., Three dimensional harmonic functions near termination or intersection of gradient singularity lines: a general numerical method. *Int. J. Eng. Sci.* 12, 221-243 (1974).
5. BARSOUM, R.S., On the use of isoparametric finite elements in linear fracture mechanics. *Int. J. Numer. Meth. Eng.* 10, 25-37 (1976).
6. BARSOUM, R.S., Triangular quarter point elements as elastic and perfectly plastic crack tip elements. *Int. J. Numer. Meth. Eng.* 12, 85-98 (1977).
7. BENTHAM, J.P., State of stress at the vertex of a quarter-infinite crack in a half-space. *Int. J. Solids and Structures*, 479-492 (1977).
8. BLUM, H. and DOBROWOLSKI, M., On finite element methods for elliptic equations with corners. *Preprint 446*, Sonderforschungsbereich 72, Approximation und Optimierung, Universität Bonn (1981).
9. BLUM, H. and RANNACHER, R., On the boundary value problem of the biharmonic operator on domains with angular corners. *Preprint 353*, Sonderforschungsbereich 72, Approximation und Optimierung, Universität Bonn (1980).
10. FICHERA, G., Asymptotic behaviour of the electric field and density of the electric charge in the neighbourhood of singular points on a conducting surface. *Uspekhi Mat. Nauk.* 30, 105-124 (1975).
11. FIX, G., Higher-order Rayleigh-Ritz approximations. *J. Math. Mech.* 18, 645-657 (1969).
12. FOLIAS, E.S., On the three-dimensional theory of cracked plates. *J. Applied Mechanics* 42, 663-673 (1975).
13. FOLIAS, E.S., Method of solution of a class of three dimensional problems under mode I loading. *Int. J. Fracture* 16, 335-348 (1980).
14. GORDON, W.J. and HALL, C.A., Construction of curvilinear coordinate systems and their applications to mesh generation. *Int. J. Numer. Meth. Eng.* 7, 461-477 (1973).

15. GRISVARD, P., Behaviour of the solutions of an elliptic boundary value problem in a polygonal or polyhedral domain. pp.207-274 of B. Hubbard (ed.) *Numerical Solution of Partial Differential Equations III, SYNSPADE 1975*. Academic Press, New York (1976).
16. HENSHELL, R.D. and SHAW, K.G., Crack tip elements are unnecessary. *Int. J. Numer. Meth. Eng.* 9, 495-509 (1975).
17. IRWIN, G.R., Analysis of stresses and strains near the end of a crack traversing a plate. *Trans. ASME J. Appl. Mech.* 24, 361-364 (1957).
18. KAWAI, T. and FUJITANI, Y., On the singular solution of three dimensional crack problems. 2nd Int. Conf. of *Mechanical Behaviour of Materials*, Boston (1976).
19. KAWAI, T., FUJITANI, Y. and KUMAGAI, K., Analysis of singularity at the root of the surface crack problem. pp.1157-1163 of G.C. Sih and C.L. Chow (eds.) *Proc. Int. Conf. on Fracture Mechanics and Technology*. Noordhoff (1977).
20. KONDRAT'EV, V.A., Boundary problems for elliptic equations in domains with conical or angular points. *Trans. Moscow Math. Soc.* 16, 227-313 (1967).
21. LEHMAN, R.S., Development at an analytic corner of solutions of elliptic partial differential equations. *J. Math. Mech.* 8, 727-760 (1959).
22. NITSCHE, J.A., L_∞-error analysis for finite elements. pp.173-186 of J.R. Whiteman (ed.), *The Mathematics of Finite Elements and Applications III, MAFELAP 1978*. Academic Press, London (1979).
23. RICE, J.R., Mathematical analysis in the mechanics of fracture. pp.191-311 of H. Liebowitz (ed.), *Fracture, Vol.2*. Academic Press, New York (1968).
24. RICE, J.R., On the degree of convergence of nonlinear spline approximation, pp.349-365 of I.J. Schoenberg (ed.), *Approximation with Special Emphasis on Spline Functions*. Academic Press, New York (1969).
25. SCHATZ, A., *Private communication*. Brunel University (July 1980).
26. SCHATZ, A. and WAHLBIN, L., Maximum norm estimates in the finite element method on plane polygonal domains. Parts I and II. *Math. Comp.* 32, 73-109 (1978) and *Math. Comp.* 33, 465-492 (1979).
27. SCHIFF, B., WHITEMAN, J.R. and FISHELOV, D., Determination of a stress intensity factor using local mesh refinement. pp.55-64 of J.R. Whiteman (ed.), *The Mathematics of Finite Elements and Applications III, MAFELAP 1978*. Academic Press, London (1979).
28. SNEDDON, I.N. and BERRY, D.S., The classical theory of Elasticity. pp.1-126 of S. Flügge (ed.), *Handbuch der Physik VI, Elastizität und Plastizität*. Springer Verlag, Berlin (1958).

29. STEINBERG, J., *Private Communication*, Brunel University (August 1979).
30. STEPHAN, E., A Fix method for the Laplacian in a polyhedral domain. (to appear)
31. STEPHAN, E. and WHITEMAN, J.R., Singularities of the Laplacian at corners and edges of three-dimensional domains and their treatment with finite element methods. *Technical Report* BICOM 81/1, Institute of Computational Mathematics, Brunel University (1981).
32. THOMPSON, G.M. and WHITEMAN, J.R., Solution of problems with linear elastic fracture using special finite elements. pp. 363-374 of *Proc. Int. Congress on Finite Element Methods, Baden-Baden*. IKOSS, STUTTGART (1981).
33. THOMPSON, G.M. and WHITEMAN, J.R., An analysis of strain representation by both singular and non-singular finite elements.(to appear)
34. WAIT, R., Singular isoparametric finite elements. *J. Inst. Maths. Applics.* 20, 133-141 (1977).
35. WAIT, R., A note on quarter point triangular elements. *Int. J. Numer. Meth. Eng.* 12, 1333-1337 (1978).
36. WAIT, R., Finite element methods for elliptic problems with singularities. *Comp. Meth. Appl. Mech. Eng.* 13, 141-150 (1978).
37. WHITEMAN, J.R., Numerical techniques for elliptic problems containing boundary singularities. pp.67-79 of H.Kardestuncer (ed.), *Finite Elements - Finite Differences and Calculus of Variations*. University of Connecticut (1980).
38. WHITEMAN, J.R., Finite element methods for elliptic problems containing boundary singularities. pp.199-206 of J. Albrecht and L. Collatz (eds.), *Numerische Behandlung von Differentialgleichungen, Band 3, I.S.N.M. 56*. Birkhauser Verlag, Basel (1981).
39. WHITEMAN, J.R. and AKIN, J.E., Finite element, singularities and fracture. pp.35-54 of J.R. Whiteman (ed.), *The Mathematics of Finite Elements and Applications III, MAFELAP 1978*. Academic Press, London (1979).
40. WILLIAMS, M.L., Stress singularities resulting from various boundary conditions in angular corners of plates in extension. *J. Appl. Mech.* 24, 526-528 (1952).

A SIMPLE AND ACCURATE METHOD FOR THE DETERMINATION OF STRESS INTENSITY FACTORS AND SOLUTIONS FOR PROBLEMS ON DOMAINS WITH CORNERS

H. Blum

University of Bonn

1. INTRODUCTION

On a bounded polygonal domain $\Omega \subset \mathbb{R}^2$, we consider the semilinear boundary value problem

$$\Delta^2 u - \Sigma_{k=1}^2 \partial_k F_k(.,u,\nabla u,\nabla^2 u) + F_o(.,u,\nabla u,\nabla^2 u) = f \text{ in } \Omega, \qquad (1.1)$$

$$u = u_n = 0 \text{ on } \partial\Omega. \qquad (1.2)$$

The nonlinear functions are sufficiently smooth with respect to their arguments and satisfy the condition $F_k(x,0,0,0)\equiv 0$. Moreover, the following growth conditions

$$|F_o(.,\xi,\zeta,\eta)| \leq c(1 + g(\xi) + |\xi|^m + |\eta|^2) , \qquad (1.3)$$

$$|F_k(.,\xi,\zeta,\eta)| \leq c(1 + h(\xi) + |\zeta|^r + |\eta|^{1+\varepsilon}), \ k=1,2, \qquad (1.4)$$

hold uniformly for $x \in \Omega$, $\xi \in \mathbb{R}$, $\zeta \in \mathbb{R}^2$, and $\eta \in \mathbb{R}^{2\times 2}$, where $m,r \in \mathbb{N}$ and $g,h \in C(\mathbb{R})$. ε can be chosen arbitrarily close to 0. Analogous restrictions are required for the partial derivatives of the F_k, $k=0,1,2$.

Two well known problems in continuum mechanics are covered by (1.1),(1.2). The von Kármán equations in plate bending theory which describe "large" deflections of a plate, are of type (1.1) with $F_k = 0$, $k=1,2$. On the other hand there is the two-dimensional steady state Navier-Stokes equation which takes the above form in the stream function formulation with $|F_k| = |u_j \Delta u|$, $k \neq j$, $k = 1,2$.

We make use of the standard notation $L^p = L^p(\Omega)$ for the Lebesgue spaces with norm $\| \ \|_p$, and $H^{m,p}$ for the Sobolev spaces of functions having generalized derivatives up to order m with the natural norm $\| \ \|_{m,p}$. Furthermore, $H_o^{2,2}$ denotes the subspace of $H^{2,2}$ of functions that satisfy the boundary conditions (1.2) in a generalized sense.

By (1.3) and (1.4), the nonlinear form

$$b(u;v) = (F_o(.,u,\nabla u,\nabla^2 u),v) + \Sigma_{k=1}^2 (F_k(.,u,\nabla u,\nabla^2 u),\partial_k v) \quad (1.5)$$

is well defined on $H_o^{2,2} \times H_o^{2,2}$. The problem (1.1),(1.2) has the standard weak formulation

$$A(u;\phi) = a(u,\phi) + b(u;\phi) - (f,\phi) = 0 \ , \ \forall \phi \in H_o^{2,2} \ , \quad (1.6)$$

where $a(.,.)$ is the bilinear form associated with Δ^2. In the following, we assume that (1.6) has an isolated solution $u \in H_o^{2,2}$ which implies that the linearized form $A'(u;.,.)$ is regular, i.e.

$$k\| v \|_{2,2} \leq \sup\{|A'(u;v,\phi)|, \ \phi \in H_o^{2,2}, \ \| \phi \|_{2,2} = 1\}, \quad (1.7)$$

where $v \in H_o^{2,2}$.

It is well known that the usual shift theorems for linear problems which are valid for smooth boundaries, carry over to the semilinear case by a bootstrapping argument. The presence of angular corners, however, causes local singularities in the weak solution of (1.6). By the low regularity, the effectiveness of the usual Finite Element Method is considerably reduced. In this paper an improved method is presented which is based on the singular expansion of the solution and the use of the singular functions of the problem adjoint to (1.1),(1.2).

2. SINGULAR EXPANSIONS

The structure of solutions of linear problems near angular corners of the domain has been extensively studied in the literature. The most general results are contained in Kondrat'ev [6] and Maz'ja-Plamenevskij [7] which provide the basic tools for the singular analysis in this chapter.

For the sake of simplicity we assume that there exists only one corner O with inner angle ω which destroys the validity of the shift theorems and that u is smooth outside a neighbourhood of O. $\Omega \cap B(O,R)$ coincides with an infinite cone C where $B(O,R)$ is a ball with radius R. By $\tilde{s}_i$ we denote the singular functions on C, i.e. the solutions of the homogeneous linear problem (1.1),(1.2) on C of the form

$$\tilde{s}_i = r^{1+\delta_i} \Psi_i(\phi) \ , \quad (2.1)$$

where $\delta_i \in \mathbb{C}$ and $\Psi_i \in C^\infty(\mathbb{R})$ depend on ω. (We suppress possible singular functions containing $\ln r$.) The singular functions of the adjoint linear problem have the form

$$\tilde{s}_{-i} = r^{1-\delta_i} \Psi_i(\phi) \ , \quad (2.2)$$

as is shown in [7]. Using the notation $s_{\pm i} = \tau \tilde{s}_{\pm i}$ with a sufficiently smooth cut-off function $\tau = \tau(r)$ the following results hold for the biharmonic problem

Theorem 1. If $F_k \equiv 0$, $k = 0,1,2$ and $f \in L^2$ then the weak solution u of (1.6) admits the expansion

$$u = \Sigma_{i=1}^{n} k_i s_i + w \tag{2.3}$$

with $w \in H^{4,2} \cap H_o^{2,2}$, provided that no singular function s_j exists with $\operatorname{Re}\delta_j = 2$. The sum contains all i with $0<\operatorname{Re}\delta_i<2$. Furthermore, the coefficients k_i have the integral representation

$$k_i = ((f,s_{-i}) - (u,\Delta^2 s_{-i}))/c_i \tag{2.4}$$

with

$$c_i\delta_{ij} = ((\Delta^2 s_j, s_{-i}) - (s_j, \Delta^2 s_{-i})) . \tag{2.5}$$

Note that all integrals in (2.4) and (2.5) exist since $\Delta^2 s_{\pm i}$ vanishes in a neighbourhood of 0 . An elementary proof for (2.4) can be found in [1]. The representation formula for the stress intensity factors is not meaningless, as will be seen in §4, although it involves the solution itself.

The distribution of the δ_i is analyzed e.g. in [2] and tables of the values depending on ω are contained in [8]. It is shown that in regular the sum in (2.3) does not vanish if the angle $\omega > 126,...^o$, and $\delta_1 \to 0.5$ for $\omega \to 2\pi$. (We assume that the s_i are ordered by increasing $a_i = \operatorname{Re}\delta_i$.)

Singular expansions for the semilinear case $F_k \not\equiv 0$ have been analyzed in [2]. They are obtained by a bootstrapping argument as long as the linear form $A(s_1;.)$ has the same regularity as the right hand side f . Since this is always true for the considered problem we obtain for the general case (1.1), (1.2)

Theorem 2. The statements of Theorem 1 carry over to the semilinear case with the same singular functions s_i . The representation formula (2.4) takes the analogous form

$$k_i = [(f,s_{-i}) - b(u;s_{-i}) - (u,\Delta^2 s_{-i})]/c_i \tag{2.6}$$

3. APPROXIMATION WITH FINITE ELEMENTS

In order to approximate (1.6) by the usual finite element displacement method let Γ_h be a quasiuniform family of subdivisions of Ω into trinangles Λ_h . By $S^m \subset H^{2,2}$, $m \geq 4$, we denote the usual conforming finite element spaces of piecewise polynomial functions of degree $\leq m-1$ with respect to Γ_h , and $S_o^m = S^m \cap H_o^{2,2}$, with the well known modification for $m = 4$.

Because of the low regularity of the solution u we have the following reduced L^2-convergence results

Theorem 3. If $0 < a_1 = \operatorname{Re}\delta_1 < 2$, then, for h sufficiently small, there exists a isolated solution $P_h u \in S_o^m$ of the problem

$$A(P_h u;\phi_h) = 0 \ , \ \forall \phi_h \in S_o^m \ , \tag{3.1}$$

satifying the error estimates

$$\|u-P_h u\|_2 + h^{a_1}\|u-P_h u\|_{2,2} \leq ch^{2a_1-\gamma}\|\nabla^k u\|_p \ , \tag{3.2}$$

$\gamma > 0$, for suitably chosen $p < 2$, $k \in \{3,4\}$.

Proof The result can be obtained by a deformation argument introduced by Dobrowolski and Rannacher [4]. For $t \in [0,1]$, the problems

$$tA(u_h^t,\phi_h) + (1-t)A'(u;u_h^t-u,\phi_h) = 0 \ , \ \forall \phi_h \in S_o^m \tag{3.3}$$

are considered. It has to be shown that, for some K_o, the set

$$\theta_h(K_o) = \{t \in [0,1]: \text{(3.3) has an isolated solution } u_h^t \in S_o^m \text{ satisfying } \|e\|_2 + h^{a_1}\|e\|_{2,2} < 2K_o h^{2a_1-\gamma}\|\nabla^k u\|_p \} \tag{3.4}$$

coincides with $[0,1]$, where $e = u-u_h^t$. For $t=0$ the linear problem (3.3) has a unique solution. Denoting the interpolate of u by $I_h u$ and using a discrete analogue of (1.7), we obtain

$$\|e\|_{2,2} \leq c\|u-I_h u\|_{2,2} + \sup\{|A'(u;e,\phi)|, \phi \in S_o^m, \|\phi\|=1\} \tag{3.5}$$

By (1.3) and (1.4) and the well known inverse properties of finite elements, the second term can be estimated by $ch^{\alpha}\|u-u_h^o\|_{2,2}$ for some $\alpha > 0$. For small h this leads to the energy estimate. The L^2-estimate follows by a standard duality argument. Let now $t_i \in \theta_h(K_o)$, $t_i \to t$, then the limit u_h^t is a solution of (3.3) and satisfies the inequality in (3.4) with "$\leq$". Again using (3.5) for u_h^t even the sharp inequality can be shown, for h sufficiently small, i.e. $\theta_h(K_o)$ is closed. Since it is also open to the arguments used in [4] the theorem is proved.

4. AN IMPROVED METHOD

In this paragraph we want to present a method based on the singular expansion for the solution which has improved convergence properties compared with (3.2) and leads to good approximations to the stress intensity factors. It is obtained by the following natural iterative procedure

Determine approximate solutions u_h^j and stress intensity factors k_i^j , $j \geq 1$, by

$$u_h^j = P_h u + \Sigma_{i=1}^n k_i^{j-1}(s_i - P_h s_i) , \tag{4.1}$$

$$k_i^j = [(f,s_{-i}) - b(u_h^j;s_{-i}) - (u_h^j,\Delta^2 s_{-i})]/c_i , \tag{4.2}$$

where $k_i^o = 0$ and $P_h s_i$ is the usual Ritz projection of s_i . If $P_h w$ denotes the approximation to the regular remainder w in (2.3), we easily obtain the following representations for the errors of (4.1),(4.2)

$$u - u_h^j = (w - P_h w) + \Sigma_{i=1}^n (k_i - k_i^{j-1})(s_i - P_h s_i) , \tag{4.3}$$

$$k_i - k_i^j = -[(b(u;s_{-i}) - b(u_h^j;s_{-i})) + (u - u_h^j,\Delta^2 s_{-i})] . \tag{4.4}$$

In order to estimate the first difference in (4.4) we apply the mean value theorem and the growth conditions for the derivatives of the F_k, $k=0,1,2$. By induction it follows that we have

$$|k_i - k_i^j| \leq c\| u - u_h^j\|_{2,2} , \tag{4.5}$$

with a constant c independent of j . From this, we easily obtain error bounds for the iterative procedure in various norms.

Since an estimate of type (4.5) is also true for the difference of two iterates we have the following

Theorem 4. There exist solutions $u^h \in S_o^{m,n}$, k_i^h , $1 \leq i \leq n$, of the system

$$u^h = P_h u + \Sigma_{i=1}^n k_i^h (s_i - P_h s_i) , \tag{4.6}$$

$$k_i^h = [(f,s_{-i}) - b(u^h;s_{-i}) - (u^h,\Delta^2 s_{-i})]/c_i , \tag{4.7}$$

which satisfy the error estimates

$$\|u-u^h\|_{2,2} + \Sigma_{i=1}^n |k_i-k_i^h| \leq c\|w-P_h w\|_{2,2} , \tag{4.8}$$

where $S_o^{m,n}$ denotes the spline space S_o^m augmented by the first n singular functions. Moreover, the solutions are the limits of the iterative procedure (4.1),(4.2) for $j \to \infty$.

The method defined by (4.6),(4.7) has been proposed and analized in Blum-Dobrowolski [1] for linear problems. Since it is characterized by the use of the adjoint singular functions it has been called the dual singular function method (DSFM).

Proof The difference of the stress intensity factors for two iterates can be estimated by

$$\begin{aligned}\Sigma_{i=1}^n |k_i^{j+1} - k_i^j| &\leq c\|u_h^{j+1} - u_h^j\|_{2,2} \\ &\leq c \max\|s_i - P_h s_i\|_{2,2} \Sigma_{i=1}^n |k_i^j - k_i^{j-1}|\end{aligned} \tag{4.9}$$

By this, we see that (4.1),(4.2) converges to a fixed point u^h, k_i^h, which satisfies the characteristic equations (4.6),(4.7). Using (4.3),(4.4) for $u_h^j = u_h^{j-1} = u^h$, the desired estimate follows by (4.5).

A more refined analysis of the error $k_i - k_i^h$ leads to sharper estimates than (4.5). For instance it is bounded by the L^2 norm of $u-u^h$, as can be seen using an appropriate duality argument. Thus, we see that the DSFM essentially behaves like the approximation of smooth functions on polygonal domains.

As an example, we get by the above remarks

Corollary For $0 < a_1 = \mathrm{Re}\,\delta_1 < 2$, $f \in L^2$, and $w \in H^{4,2}$ the following L^2 results hold for the DSFM

$$\|u-u^h\|_2 + h^{a_1}\|u-u^h\|_{2,2} + \Sigma_{i=1}^n |k_i - k_i^h| = O(h^{2+a_1}) \tag{4.10}$$

Optimal error orders can be obtained if w additionally satisfies certain growth conditions near the corner, as is shown in [3] by use of weighted norm techniques. Thus, perhaps adding some more singular functions, even $O(h^4)$ can be reached for the stress intensity factors and the L^2 error.

The analysis for the DSFM is very much simplified for the linear case $F_k = 0$, $k=0,1,2$. Here, the values k_i^h can easily be calculated by solving the linear system

$$\Sigma_{j=1}^n k_j^h a_{ij} = [(f,s_{-i}) - (P_h u, \Delta^2 s_{-i})]/c_i \ , \text{ with} \tag{4.11}$$

$$a_{ij} = [(\Delta^2 s_j, s_{-i}) - (P_h s_j, \Delta^2 s_{-i})]/c_i \quad . \tag{4.12}$$

By (2.5) the matrix a_{ij} is positive definite for small h. Once having calculated the values a_{ij}, for any right hand side f only one Ritz projection must be determined to get k_i^h. An additional one for the right hand side $f - \Sigma k_i^h \Delta^2 s_i$ yields u^h.

From (4.4) we see that in the linear case the error $k_i - k_i^h$ can be estimated by negative norms of $u-u^h$. Using the better approximation properties of the finite element method in negatinorms, even $O(h^{2m-4})$ for the stress intensity factors can be obtained if sufficiently many singular functions are used and the right hand side is smooth enough, see [3].

5. NUMERICAL RESULTS

In this paragraph we want to present a comparison of numerical results for linear problems for the DSFM and the well known singular function method (SFM) which consists in augmenting the spline space by several singular functions, see e.g. [5]. We use two singular functions in both methods so that the numerical amount of work is about the same. For a theoretical analysis of the SFM we refer to [3] where it is shown that the error for the stress intensity factor k_i essentially depends on the difference $\operatorname{Re}\delta_i - \operatorname{Re}\delta_{i+1}$, not on the regularity of s_i or w. If s_i is "too regular" compared with the order of the spline space, no convergence can be expected for k_i.

First we consider the biharmonic problem on an L-shaped region with $f = 1$ $(y \geq 0)$ and $f = 0$ $(y < 0)$. The Ritz-projections are determined by a mixed method using piecewise linear elements. We compare the relative errors for k_1 and k_2 for the SFM and the first iterate in (4.1),(4.2).

TABLE 1

Errors for the biharmonic problem

h^{-1}	$k_1 - k_1^1$	$k_1 - k_1^{SFM}$	$k_2 - k_2^1$	k_2^{SFM}
10	.6292(-2)	1.2969	.7845(-1)	-.4891(-4)
20	.1469(-1)	.7730	.4030(-1)	.4394(-3)
30	.1498(-1)	.3843	.2597(-1)	.1332(-2)
40	.1068(-1)	.4245	.1993(-1)	.1313(-2)
50	.8194(-2)	.3340	.1605(-1)	.1570(-2)
60	.6655(-2)	.2243	.1335(-1)	.1855(-2)
70	.5662(-2)	.1989	.1134(-1)	.1917(-2)

In column 2 and 4 we see nearly exactly the theoretical convergence rate $O(h^{1.1})$ $(a_1 \approx .55)$. The rate for the SFM is not quite clear and the absolute values are very bad. Since the value for k_2 is about .546(-2), we have no convergence for the second stress intensity factor in the SFM.

As a second example we consider the model problem $-\Delta u = 1$ on a slit domain with mixed boundary conditions $u_n = 0$ $(\phi = 0)$ and $u = 0$ $(\phi = 2\pi)$. For the calculation of the Ritz projection piecewise linear elements are used.

TABLE 2

Errors for the model problem

h^{-1}	$k_1 - k_1^{DSFM}$	$k_1 - k_1^{SFM}$	$k_2 - k_2^{DSFM}$	$k_2 - k_2^{SFM}$
10	.1131	.5800	.3358(-1)	.9987
20	.4352(-1)	.2293	.1185(-1)	.7832
30	.2000(-1)	.1392	.5345(-2)	.7360
40	.1145(-1)	.9618(-1)	.3031(-2)	.7108
60	.5358(-2)	.5452(-1)	.1352(-2)	.6731
80	.3014(-2)	.3741(-1)	.0744(-2)	.6468

These results have been obtained by Dobrowolski [3] who has also analyzed the DSFM for interface problems. For the DSFM the convergence rate $0(h^2)$ can be seen from columns 2 and 4, in perfect accordance with the theory. The absolute values for the SFM again are much worse. For the second stress intensity factor one does not even have the theoretical rate $0(h^{.5})$.

REFERENCES

1. BLUM, H. and DOBROWOLSKI, M., On Finite Element Methods for Elliptic Equations on Domains with Corners. *Preprint*, University of Bonn (1981).
2. BLUM, H. and RANNACHER, R., On the boundary value problems of the biharmonic operator on domains with angular corners. *Math. Meth. Appl. Sci.* 2, 556-581 (1980).
3. DOBROWOLSKI, M., Numerical approximation of elliptic interface and corner problems. *Habilitationsschrift*, University of Bonn (1981).
4. DOBROWOLSKI, M. and RANNACHER, R., Finite element methods for nonlinear elliptic systems of second order. *Math. Nachr.* 94, 155-172 (1980).
5. FIX, G., GULATI, S. and WAKOFF, G.I., On the use of singular functions with finite element approximations. *J. Comp. Phys.* 13, 209-238 (1973).
6. KONDRAT'EV, V.A., Boundary value problems for elliptic equations in domains with conical or angular points. *Trans. Mosc. Mat. Soc.* 16, 227-313 (1967).
7. MAZ'JA, V.G. and PLAMENEVSKIJ, B.A., Coefficients in the asymptotics of the solutions of elliptic boundary-value problems in a cone. *J. Sov. Math.* 9, 750-764 (1978).
8. MELZER, H. and RANNACHER, R., Spannungskonzentrationen in Eckpunkten der Kirchhoff'schen Platte. *Der Bauingenieur* (1980).

SOME NUMERICAL TEST PROBLEMS WITH SINGULARITIES

L. Collatz

University of Hamburg

1. INTRODUCTION

In order to test numerical methods for treating boundary value problems involving partial differential equations and to compare methods, it can be useful to know intervals in which the unknown solutions of the problems are included. One can of course construct example problems for which the solutions are known explicitly, but these tend to be over simple and somewhat artificial. It is therefore useful to have example problems of a more general type and in the following we describe how one can find such examples, using monotonicity methods which provide inclusion intervals for the solutions. The technique is illustrated on some examples one of which, a problem involving slits, was posed by Babuska and Whiteman in 1980.

2. A THREE DIMENSIONAL EXAMPLE WITH A HARMLESS SINGULARITY

This example is described in detail because it is perhaps typical and provides a good test for comparing discretisation and parametric methods. Boundary singularities in three-dimensional problems of this type are discussed in detail by Stephan and Whiteman in [11].

2.1 *Distribution of Temperature* $w(x,y,z)$

The problem is defined in the region B in (x,y,z)-space where

$$B \equiv \{(x,y,z)/|x|<2, |y|<1, |z|<1\} , \tag{2.1}$$

as in Fig. 1. We consider the boundary value problem

$$\Delta w \equiv \frac{\partial^2 w}{\partial x^2} + \frac{\partial^2 w}{\partial y^2} + \frac{\partial^2 w}{\partial z^2} = -1 \quad \text{in } B ,$$

$$w = 0 \quad \text{on the boundary } \partial B .$$

Using the transformation $u = w + \frac{1}{2}x^2$, we find that

$$\Delta u = 0 \quad \text{in } B \ , \quad u = \tfrac{1}{2}x^2 \text{ on } \partial B \ . \tag{2.2}$$

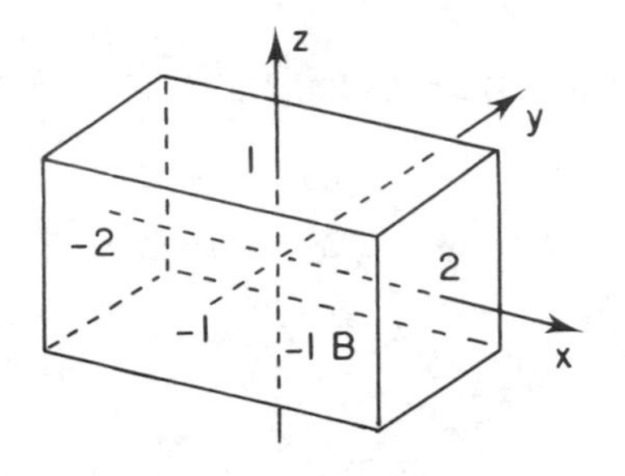

FIG. 1

2.2 *Inclusion Interval for the Solution*

We look for an approximate solution v(x,y,z) of the form

$$u \approx v = \sum_{\nu=0}^{p} a_\nu v_\nu(x,y,z) \tag{2.3}$$

with unknown constants a_ν and harmonic polynomials v_ν with $\Delta v_\nu = 0$. Using the abbreviation that $\sigma_k = y^k + z^k$, we find that

$$\left.\begin{aligned}
v_0 &= 1 \\
v_1 &= 2x^2 - y^2 - z^2 = 2x^2 - \sigma_2 \\
v_2 &= x^4 - 3x^2\sigma_2 + 3y^2z^2 \\
v_3 &= \sigma_4 - 6y^2z^2 = \mathrm{Re}(y+iz)^4 \\
v_4 &= 2x^6 - 15x^4\sigma_2 + 15x^2\sigma_4 - \sigma_6 \\
v_5 &= 4x^6 - 30x^4\sigma_2 + 180x^2y^2z^2 - 15y^2z^2\sigma_2 + \sigma_6 \\
v_6 &= \sigma_8 - 28y^2z^2\sigma_4 + 70y^4z^4 = \mathrm{Re}(y+iz)^8 \\
v_7 &= 2x^8 - 28x^6\sigma_2 + 70x^4\sigma_4 - 28x^2\sigma_6 + \sigma_8 \\
v_8 &= 3x^8 - 42x^6\sigma_2 + 35x^4(\sigma_4 + 12y^2z^2) - 210x^2y^2z^2\sigma_2 \\
&\quad + 35y^4z^4 \\
&\ldots
\end{aligned}\right\} \tag{2.4}$$

The parameters a_ν are determined in such a way that the maximum δ of the modulus of the difference between v and u on the boundary ∂B becomes as small as possible. (This is Tschebyscheff approximation; see Meinardus [8], Bredendiek [1] and Collatz [3]-[5].) The classical boundary maximum principle then gives us the inclusion theorem for the error $\varepsilon \equiv v - u$:

From $|v-u| \leq \delta$ on B it follows that $|\varepsilon| = |v-u| \leq \delta$ in the whole domain B. (2.5)

The Tschebyscheff approximation is written as semi-infinite optimization:

$$-\delta \leq v(x,y,z) - u(x,y,z) \leq \delta \text{ on } \partial B, \ \delta = \text{Min} \qquad (2.6)$$

and is solved by well known algorithms for optimization. The following table gives error bounds on δ of different values of p.

p	$\lvert u-v \rvert < \delta$; δ =
0	1
1	2/9 = 0.2222..
2	0.1361
3	0.05594
5	0.01473
8	0.007343
11	0.003 494
15	0.002 010

The table means that

$$\left| u - \sum_{\nu=0}^{15} a_\nu v_\nu(x,y,z) \right| \leq 0.00201 \ ,$$

so that for example $|u(0,0,0) - 0.28671| \leq 0.00201$ or

$$0.28469 \leq u(0,0,0) = w(0,0,0) \leq 0.28872 \ . \qquad (2.7)$$

If we select additionally the point x = 1, y = z = 0, we find that

$$|u(1,0,0) - 0.25889| \leq 0.00201 \ .$$

2.3 *Finite Elements on a Three Dimensional Grid*

We use a bilinear interpolant v to the values a,b,c,d,e,f of the function $\phi(x,y,z)$ respectively at the points (0,0,0),(h,0,0), (0,k,0), (0,0,ℓ), (h,0,ℓ), (0,k,ℓ) which are the vertices of the prism $\hat{B}$ of Fig, 2, where

$$\hat{B} \equiv \{(x,y,z) \ , \quad x \geqq 0 \ , \quad y \geqq 0 \ , \quad \frac{x}{h} + \frac{y}{k} \leqq 1 \ , \quad 0 \leqq z \leqq \ell\}.$$

Thus

$$v(x,y,z) = a + \frac{x}{h}(b-a) + \frac{y}{k}(c-a) + \frac{z}{\ell}(d-a) + \\ + \frac{xz}{h\ell}(e+a-d-b) + \frac{yz}{k\ell}(f+a-d-c)$$

and the Dirichlet-Integral is calculated as

$$\frac{12}{hk\ell}\int_{\hat{B}} (\text{grad } v)^2 \, dxdydz = 5(a^2+d^2) + 3(b^2+c^2+e^2+f^2)$$

$$+ 2ad + (b-e)(c-f) - 3(a+d)(b+c+e+f) \; .$$

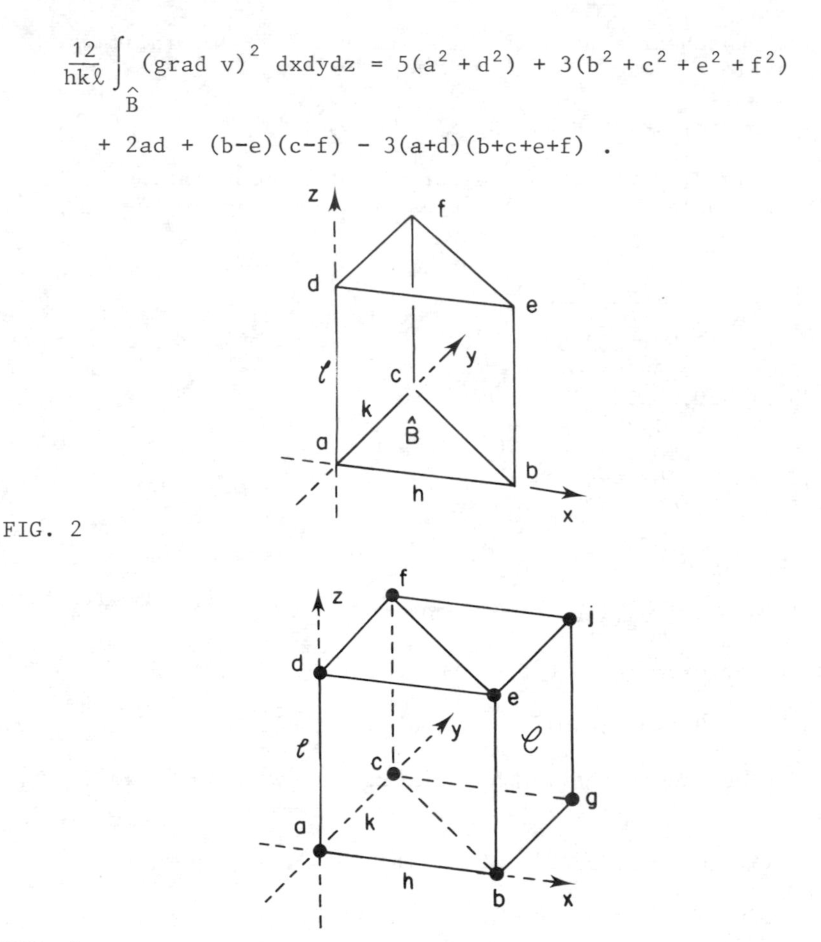

FIG. 2

FIG. 3

If we now add the points (h,k,0) and h,k,ℓ) with values g and j to produce the cuboid C of Fig. 3, where

$$C \equiv \{(x,y,z)/0 \leqq x \leqq h,\ 0 \leqq y \leqq k,\ 0 \leqq z \leqq \ell\}$$

and now interpolate ϕ on C by the corresponding bilinear interpolant v, we obtain the Dirichlet integral

$$\frac{12}{hk\ell}\int_C (\text{grad } v)^2 \, dxdydz = 5(a^2+d^2+g^2+j^2) + 6(b^2+c^2+e^2+f^2)$$

$$+ 2ad + 2gj + 2(b-e)(c-f) - 3(a+d+g+j)(b+c+e+f) \; .$$

For details of the finite element method see Whiteman [13], [14], Mitchell and Wait [9], Strang and Fix [12], Zienkiewicz [15], Babuska and Aziz [1] and Schwarz [10].

The domain B of (2.1) can be divided with prisms in different ways. We select two partitions, Methods (I) and (II), as indicated in Fig. 4. If Method (I) is used with the above approximation, then the values obtained are too large, whilst those of Method (II) are too small. The arithmetic mean of these two values is better, as can be seen from Table 1.

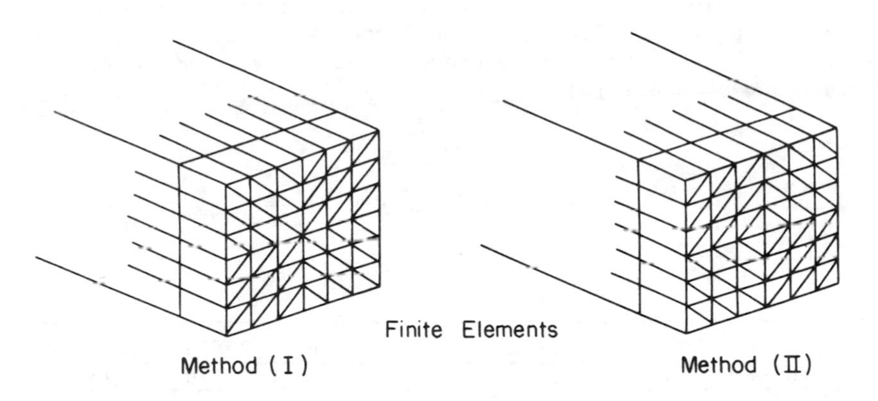

FIG. 4

2.4 *Difference Method and Mehrstellen-method*

These methods are so well known that we state only the relevant Mehrstellen formulae for the Laplace equation $\Delta u = 0$ using the notation of Fig. 5.

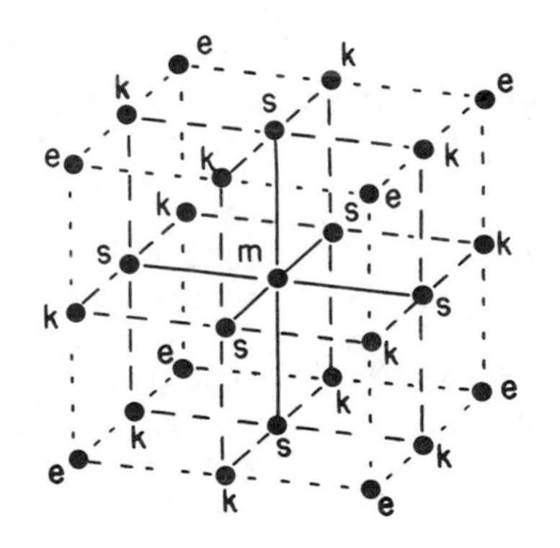

FIG. 5

These are: Improved Difference Method: "Mehrstellen-Method"

Method (I): $24u_m - 2\Sigma u_s - \Sigma u_k = 0$,

Method (II): $128u_m - 14\Sigma u_s - 3\Sigma u_k - \Sigma u_e = 0$.

2.5 *Comparison*

We take $h = k = \ell$ for all the discretisation methods. Table 1 gives the approximate values at the origin $x = y = z = 0$ for the mesh sizes $h = 1/q$, $q = 1,2,3,4,5$. The classical difference method and Methods (I) and (II) with bilinear finite elements give values outside the interval (2.7). However, it is easy to get better values by using finer grids or improved finite elements or improved finite difference methods, as can be seen from Table 1 with the Mehrstellen-method; compare with Collatz, Günther and Sprekels [6].

TABLE 1

Approximate Values for $u(0,0,0)$

		$h = 1$	$h = \frac{1}{2}$	$h = \frac{1}{3}$	$h = \frac{1}{4}$	$h = \frac{1}{5}$
Finite	FEM (I)	0.3333	0.3077	0.2984	0.29429	0.29210
Elements	FEM (II)	0.1633	0.2471	0.2669	0.27479	0.27872
$\frac{1}{2}$[FEM(I)+FEM(II)]		0.2483	0.2774	0.2827	0.28454	0.28541
Classical Diff.Meth.		0.2353	0.2714	0.2798	0.28286	0.28432
Mehrstellen-	(I)	0.2958	0.2875	0.2871	0.28700	0.28698
Method	(II)	0.2926	0.2873	0.2870	0.28698	0.28697

3. THE SLIT-PROBLEM

The following problem was posed by I. Babuska and J.R. Whiteman as a test problem for numerical procedures at a podium discussion at the Symposium on Finite Elements for Nonlinear and Singular Problems in Durham in 1980. It is a simplified model for the displacement $u(x,y)$ of an elastic beam under forces in the (x,y)-plane. The beam cross-section may have the form of a rectangle $B \equiv \{(x,y)/|x| < 2,\ |y| < 3\}$ and there are slits S for $1 \leqq |x| \leqq 2$, $y = 0$, see Fig. 6. In the interior of B we take as differential equation

$$\Delta u = \frac{\partial^2 u}{\partial x^2} + \frac{\partial^2 u}{\partial y^2} = 0 \ ,$$

and the boundary conditions are

$$u = 1 \text{ on } \Gamma_1, (y = 3) \ ; \quad u = -1 \text{ on } \Gamma_2, (y = -3) \ ;$$

$\frac{\partial u}{\partial n} = 0$ on Γ_3, the remainder of the boundary including the slits,

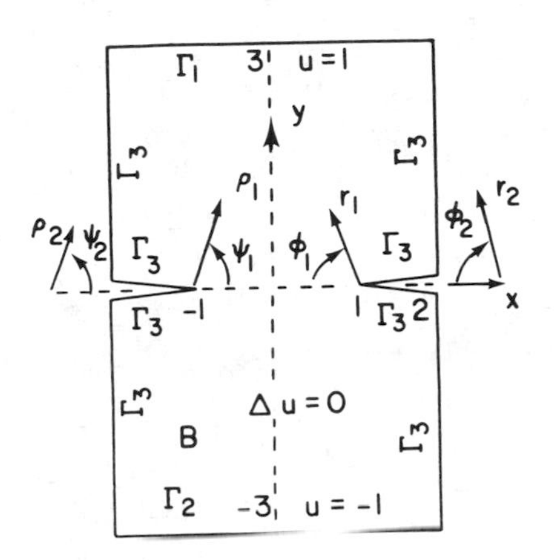

FIG. 6

where n is the outward normal. Other interpretations for u may be the distribution of temperature in a room with temperatures prescribed at the walls $y = \pm 3$ or potential. The potential problem has been solved using conformal mapping techniques by Papamichael.

On account of the symmetry of the problem, it is necessary to consider only a quarter of the region B. Fig. 7 shows some lines u = constant for the solution in this region with a table of specific values. We look for an approximation of the form $u \approx w = w_1 + w_2 + w_3$ with

$$w_1 = \sum_{\nu=1}^{p} a_\nu P_\nu(x,y) ,$$

$$w_2 = \sum_{\nu=1}^{q} b_\nu S_\nu(x,y) ,$$

$$w_3 = \sum_{\nu=1}^{s} d_\nu \hat{S}_\nu(x,y) ,$$

where

$$P_\nu = \mathrm{Re}(y - ix)^{2\nu-1},\ P_1 = y,\ P_2 = y^3 - 3x^2y, \ldots ,$$

$$S_\nu = (r_1^{\frac{1}{2}c_\nu})\sin\left(\frac{c_\nu \phi_1}{2}\right) + (\rho_1^{\frac{1}{2}c_\nu})\sin\left(\frac{c_\nu \psi_1}{2}\right) ,\quad c_\nu = 2\nu-1 ,$$

$$\hat{S}_\nu = r_\nu^{1/2}\sin\left(\frac{\phi_\nu}{2}\right) + \rho_\nu^{1/2}\sin\left(\frac{\psi_\nu}{2}\right) .$$

and ρ_i, ψ_i, θ_i and r_i are as in Fig. 6.

w (x,y) [with |w-u| 0.00261]

y \ x	0	1	2
1	0.38285	0.41697	0.45887
0.5	0.20091	0.26316	0.36424
0	0	0	0.32619

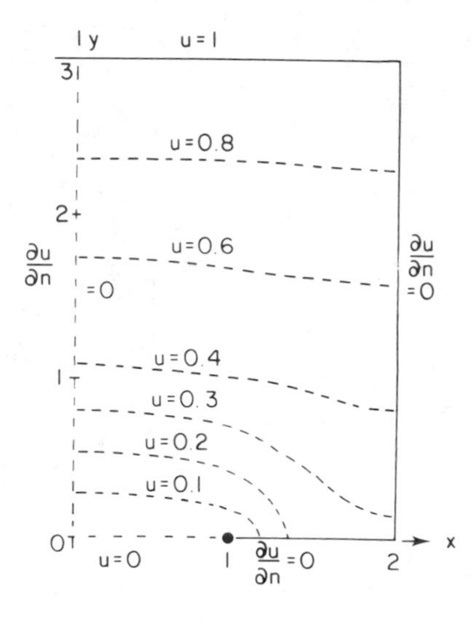

FIG. 7

In a first attempt at an approximation the terms with higher singularities in w_2 were omitted and we took q = 1. The resulting numerical results were unsatisfactory. The term w_3 was then omitted, higher singularities were included in w_2 and successful results were obtained. For different numbers of terms in w_1 and w_2, Table 2 gives the maximum δ of the modulus of the difference between upper and lower bounds for the solution. The arithmetic average of these bounds has then the error bound $\frac{1}{2}\delta$.

TABLE 2

Values for δ

p= \ q=	1	2	3	4	5
0	∞	1.95	0.58	0.160	0.0495
1	0.491	0.135	0.060	0.0384	0.012 0
2	0.433	0.051	0.033 5	0.016 0	0.011 5
3	0.407	0.049 2	0.012 9	0.012 3	0.004 26
4	0.371	0.045 5	0.010 3	0.006 8	0.003 92
5	0.358	0.040 8	0.010 1	0.002 61	0.002 57

The comparison with discretization methods is similar to those for the example of Section 2. Therefore only a few results are given.

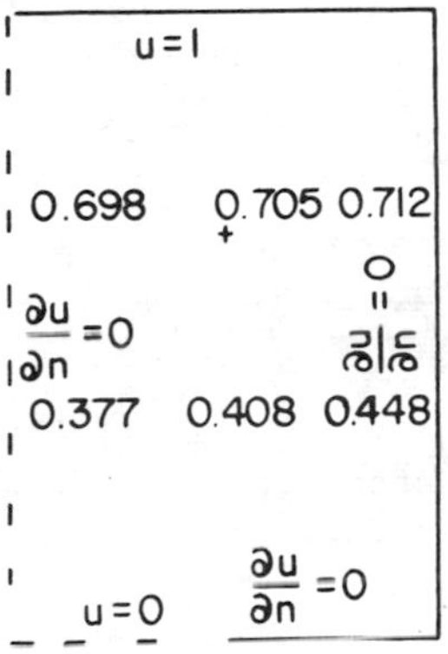

u=1
0.698 0.705 0.712
$\frac{\partial u}{\partial n}=0$
$\frac{\partial u}{\partial n}=0$
0.377 0.408 0.448
u=0
$\frac{\partial u}{\partial n}=0$

FIG. 8

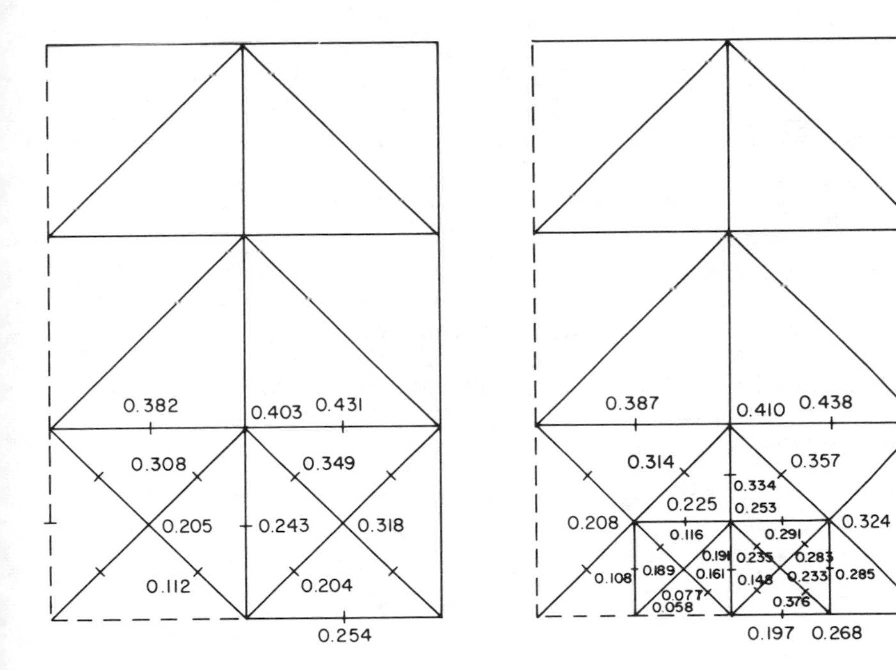

FIG. 9

Those obtained with classical finite differences with mesh sizes $h = 1/n$, $n = 1, 0.5$ and using Richardson extrapolation are given in Fig. 8, whilst those in Fig. 9 are obtained with finite elements using a quadratic polynomial in each triangular element. As with the example of Section 2 one cannot judge a numerical method on the basis of so few results.

4. OTHER TEST EXAMPLES

Many other examples containing singularities have been computed. Only two more are described here.

4.1 *Fluid Flow*

Streamlines u = constant for the ideal flow of a river with a breakwater are given in Fig. 10.

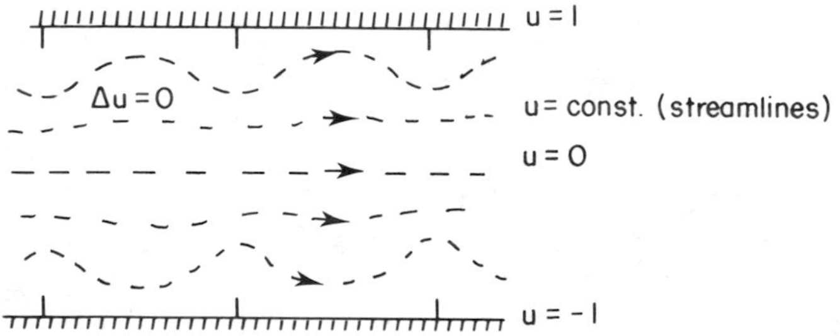

$u = 1$

$\Delta u = 0$

$u =$ const. (streamlines)

$u = 0$

$u = -1$

FIG. 10

4.2 *Condensor Problem*

This was also stated at the Durham Symposium in 1980 and is described in Fig. 11. The computational methods of Grothkopf, [7], have been used for the calculations of this paper. I thank Mr. U. Grothkopf and Mr. C. Maas for their numerical calculations.

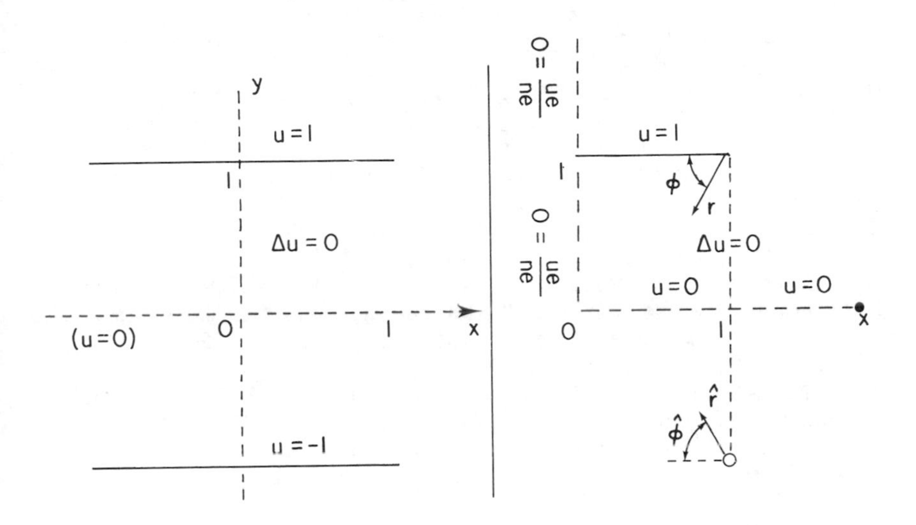

FIG. 11

REFERENCES

1. BABUSKA, I. and AZIZ, A.K., Survey lectures on the mathematics of the finite element method. pp.359 of A.K. Aziz (ed.) *The Mathematical Foundations of the Finite Element Method with Applications to Partial Differential Equations.* Academic Press, New York (1972).
2. BREDENDIEK, E. and COLLATZ, L., Simultan Approximation bei Randwertaufgaben. *I.S.N.M.* 30, 147-174. Birkhauser Verlag, Basel (1976).
3. COLLATZ, L., Aufgaben monotoner Art. *Arch. Math.* 3, 366-376 (1952).
4. COLLATZ, L., *Funktional Analysis und Numerische Mathematik.* Springer Verlag, Berlin (1968).
5. COLLATZ, L., Monotonicity and free boundary value problems. pp.33-45 of G.A. Watson (ed.) *Lecture Notes in Mathematics 773.* Springer Verlag, Berlin (180).
6. COLLATZ, L., GÜNTHER, H. and SPREKELS, J., Vergleich zwischen Diskretisierungsverfahren und parametrischen Methoden an einfachen Testbeispielen. *Z. Angew Math. Mech.* 56, 1-11 (1976).
7. GROTHKOPF, U., *Simplex-Verfahren und gleichmässige Approximation.* Diplomarbeit Hamburg (1979).
8. MEINARDUS, G., *Approximation of Functions, Theory and Numerical Methods.* Springer Verlag, Berlin (1967).
9. MITCHELL, A.R. and WAIT, R., *The Finite Element Method in Partial Differential Equations.* Wiley, London (1977).

10. SCHWARZ, H.R., *Methode der Finiten Elemente*. Teubner, Stuttgart (1980).
11. STEPHAN, E. and WHITEMAN, J.R., Singularities of the Laplacian at corners and edges of three dimensional domains and their treatment with finite element methods. *Technical Report* BICOM 81/1, Institute of Computational Mathematics, Brunel University (1981).
12. STRANG, G. and FIX, G.J., *An Analysis of the Finite Element Method*. Prentice Hall, New Jersey (1973).
13. WHITEMAN, J.R., Some aspects of the mathematics of finite elements. pp.25-42 of J.R. Whiteman (ed.) *The Mathematics of Finite Elements and Applications II, MAFELAP 1975*. Academic Press, London (1976).
14. WHITEMAN, J.R., Finite element methods for elliptic problems containing boundary singularities. pp.199-206 of J. Albrecht and L. Collatz. (eds.) *Numerische Behandlung von Differentialgleichungen Band 3. I.S.N.M.* 56. Birkhauser Verlag, Basel (1981).
15. ZIENKIEWICZ, O.C., *The Finite Element Method in Engineering Science*. McGraw Hill, London (1977).

ITERATIVE SOLUTION OF THE GEM EQUATIONS

J.A. Hendry
University of Birmingham

L.M. Delves and J. Mohamed
University of Liverpool

1. INTRODUCTION

The Global Element Method (GEM, see [2]) for the solution of elliptic partial differential equations can be viewed as a variant of the Finite Element family using large, variable order, nonconforming elements with special provision for curved boundaries, inter-element continuity and a variety of boundary conditions and singularities. The basic philosophy of the method is to use a fixed number M of elements and to improve the solution accuracy by increasing the degree (N-1) of the approximation in each element. In practice very rapid convergence can be obtained as N is increased (see [6], [4]). In [3], special techniques were described for setting up and solving the defining equations in a time $O(MN^4)$. We here present results obtained for a corrected version of the $O(MN^4)$ scheme and demonstrate an extension which yields an $O(MN^3)$ solution scheme.

2. GENERAL DESCRIPTION OF THE ITERATIVE SOLUTION SCHEME

The "block sparse" system to be solved in the GEM has the general form

$$L\underline{u} = \underline{b}. \tag{2.1}$$

We seek a splitting of the matrix L,

$$L = L_0 + \delta L$$

which will yield a rapidly convergent iterative scheme of the standard form,

$$\underline{u}^{(n+1)} = \underline{u}^{(n)} + \underline{\delta u}^{(n+1)} \qquad n = 0, 1, \ldots \quad . \tag{2.2}$$

$$L_0\underline{\delta u}^{(n+1)} = \underline{b}-L\underline{u}^{(n)} = \underline{r}^{(n)} \qquad \underline{u}^{(0)} \text{ given}$$

and here discuss the choice of L_0 and the solution of (2.2). With M elements in two dimensions, the matrix L of (2.1) is $MN^2 \times MN^2$ and can be partitioned naturally into a MxM matrix of $N^2 \times N^2$ blocks, giving for the partitioned solution vectors $\underline{u}_i$ in each element i, i = 1, ... , M

$$L_{ii}\, \underline{u}_i = \underline{b}_i - \sum_{j \in C(i)} L_{ij}\, \underline{u}_j \ , \tag{2.3}$$

where C(i) is the set of elements connected to element i. Within each element, a product form of trial functions in mapped coordinates (s,t), $-1 \leq s, t \leq 1$ is used,

$$h_{pq}(s,t) = P_p(s)\, Q_q(t) \qquad p,q = -2, -1, \ldots , N-3. \tag{2.4}$$

Using a y-ordering (see [3]) each L_{ij} is then $N^2 x N^2$ (and can be considered naturally as a block NxN matrix, each block itself of size NxN) with elements,

$$(L_{ij})_{q'p',\, qp} \qquad p',q',\ p,q = -2,-1, \ldots , N-3$$

for an appropriate mapping $(qp) \to 1, \ldots, N^2$.
The diagonal blocks have both "volume" and "surface" contributions,

$$L_{ii} = V_{ii} + S_{ii}$$

while the off-diagonal blocks have only "surface" contributions,

$$L_{ij} = \begin{cases} S_{ij} & j \in C(i) \\ 0 & \text{otherwise} \end{cases} \ .$$

For each element i, and for suitable constants V and t_p, $t_q > 0$ (and large hopefully!),

$$|(V_{ii})_{q'p',qp}| < V|p'-p|^{-t_p}\, |q'-q|^{-t_q} .$$

The matrices S_{ij} are of low block rank structure with two parts in general:

$$S_{ij} = S_{ij}^{(1)} + S_{ij}^{(2)}$$

where (omitting the suffices i, j)

$$S^{(k)} = \sum_{\ell=1}^{L_k} \alpha_\ell^{(k)} \gamma_\ell^{(k)} \beta_\ell^{(k)T} \qquad k=1,\ 2 \tag{2.5}$$

$\alpha_\ell^{(k)}$, $\beta_\ell^{(k)}$ are block N vectors (with each component an NxN matrix) and $\gamma_\ell^{(k)}$ is an NxN matrix. L_1 and L_2 depend on i,j but *not* on N,M. More details can be found in [3]. For the choice of trial functions in [3], a detailed examination of the structure of $S_{ij}^{(1)}$ shows that it can only have non-zero entries in the first two block rows and block columns (and one term in (2.5) is restricted to the top 2 x 2 blocks). Similarly $S_{ij}^{(2)}$ has only non-zero entries in the first two rows and columns of each block of the NxN block matrix.

We define

$$V_{ii} = V_{i0} + \delta V_i$$

where for some integer μ,

$$(V_{i0})_{q'p',\,qp} = \begin{cases} (V_{ii})_{q'p',\,qp}, & |p'-p| \leq \mu,\ |q'-q| \leq \mu, \\ 0 \quad \text{otherwise} & . \end{cases}$$

V_{i0} is thus block banded of block semibandwidth μ (and overall semibandwidth μN) while each block is banded of semibandwidth μ. This splitting effectively defines the desired iteration scheme,

$$[V_{i0} + S_{ii}^{(1)} + S_{ii}^{(2)}]\,\underline{\delta u}_i^{(n+1)} = \underline{r}_i^{(n)} - \sum_{j\varepsilon C(i)} (S_{ij}^{(1)} + S_{ij}^{(2)})\,\underline{\delta u}_j^{(n+1)}. \tag{2.6}$$

In [3], it was suggested that a simple iteration of (2.6) be carried out treating the S terms by a low rank modification scheme. This approach requires V_{i0} to be invertible. Unfortunately V_{i0} can be singular (and indeed is for the example presented later). The singularity can be removed by adding to V_{i0} special terms which do not affect the overall structure of V_{i0} far from the diagonal blocks and at the same time retain the overall form of (2.5). This is achieved by including a term which contributes only to the top 2x2 blocks and whose effect is limited to the (1,1) element of each of these 4 blocks (i.e. the γ matrix of (2.5) has only the (1,1) element non-zero). With this modification, the (effective) V_{i0} is now non-singular and the overall form of the equations to be solved is as in (2.6). We assume subsequently that, if necessary, such a correction has been made. From (2.5), dropping the superfix k, for any N vector $\underline{\delta u}_j$,

$$S_{ij}\,\underline{\delta u}_j = \sum_{\ell=1}^{L} \alpha_\ell \gamma_\ell \beta_\ell^T\,\underline{\delta u}_j = \sum_{\ell=1}^{L} \alpha_\ell\,\underline{z}_{\ell,j} \tag{2.7}$$

where $\underline{z}_{\ell,j} = \gamma_\ell \beta_\ell^T\,\underline{\delta u}_j$. It thus follows that (2.6) can be solved for each i as

$$\underline{\delta u}_i = \underline{\tilde{r}}_i - \sum_{m=1}^{P} \sum_{j\varepsilon C(i)} \tilde{\alpha}_m\,\underline{z}_{m,j} \tag{2.8}$$

where $\underline{\tilde{r}}_i = V_{i0}^{-1}\,\underline{r}_i$, $\tilde{\alpha}_m = V_{i0}^{-1}\,\alpha_m$ and $P=0(2L)$, $L=\max\,(L_1,L_2)$. Note that P is independent of N. Equation (2.8) is uncoupled in i and provides the solution $\underline{\delta u}_i$ if the $\underline{z}_{m,j}$ are known. These can be found from the solution of an auxiliary matrix equation (of size 0(PMN)) constructed by substituting (2.8) into (2.7) to obtain

$$\underline{z}_{\ell,i} = \gamma_\ell \beta_\ell\,\underline{\tilde{r}}_i - \sum_{m=1}^{P} \sum_{j\varepsilon C(i)} (\gamma_\ell \beta_\ell \tilde{\alpha}_m)\,\underline{z}_{m,j}. \tag{2.9}$$

3. COST OF THE ITERATIVE SCHEME

The iterative scheme described above can be split into a preliminary stage and an iterative loop. Considering the V_{i0} to be banded (of overall semibandwidth μN) gives the following operation counts for the preliminary stage:-

t1: Triangulate V_{i0} at a cost $0(\mu^2 N^2) \times 0(N^2) \times M = \quad 0(M\mu^2 N^4)$

t2: Back substitute for the $\tilde{\alpha}_m$ (2.8) at cost
$0(\mu N) \times 0(N^2) \times N \times M = \quad 0(M\mu N^4)$

t3: a) Assemble auxiliary 0(PMN) matrix (2.9) at cost
$0(N^2) \times 0(N^2) \times P^2 \times M^2 = \quad 0(P^2 M^2 N^4)$

b) Triangulate this matrix at cost $= \quad 0(P^3 M^3 N^3)$

For each occurrence of the iterative loop, the costs are:

t4: a) compute required residuals at cost $= 0(MN^4)$

b) back substitute for $\tilde{\underline{r}}_i$ at cost $0(\mu N)x0(N^2)xM= 0(M\mu N^3)$

c) back substitute for $\underline{z}_{m,i}$ at cost $= 0(P^2M^2N^2)$

d) assemble solution (2.8) at cost $0(N^3)xPxM = 0(PM N^3)$

Since μ,P are independent of N (and in practice M<< N), the overall cost is $0(MN^4)$ per iteration. Provided that convergence is sufficiently rapid the overall cost is $0(MN^4)$. The above scheme has been applied to a simple example (with 3 elements) constructed from the Motz problem [5] by removing the semicircular element of radius a and applying suitable boundary conditions on the boundary so created. Figure 1 shows the region. Dirichlet boundary conditions apply on AA', BC while Neumann conditions apply on A'D'C'B', B'B,CD,DA.

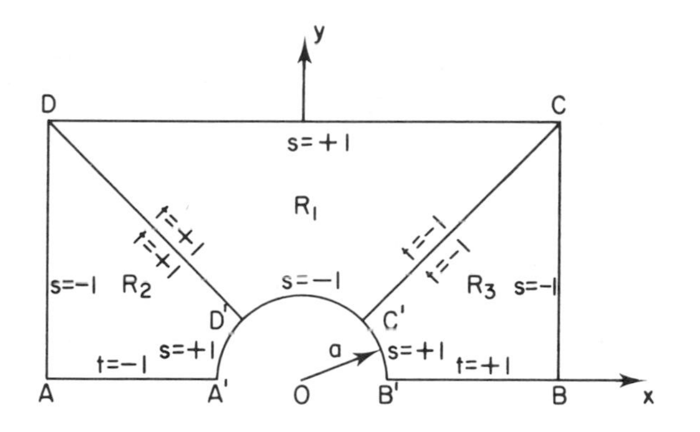

FIG. 1

The results obtained are shown in Table I. The semibandwidth μ=8 for these results, the iteration being terminated when

$$||\underline{\delta u}_i^{(n+1)}|| \leqslant 10^{-9} \; ||\underline{u}_i^{(n)}|| \qquad i = 2,3,4.$$

The timings (t1, t2 etc.) correspond to the various stages of the algorithm as laid out above. Additionally the total solution time and a direct (Gauss-Elimination) solution time (of cost $0(N^6)$) are given (where these could be obtained). Note that for $\mu=8$, only 1 iteration is needed for $N \leq 9$. Nevertheless, despite the apparent complexity of the above scheme, it is in fact faster than the direct solution due to the efficient treatment of the off-diagonal blocks. For $N \geq 9$, the timings obey (approx.) the predicted operation counts and show that the "preliminary" stages completely dominate. For $N>9$ the number of iterations is approximately independent of N (as predicted in [3]).

TABLE 1

N	Iterations for Convergence	Times (seconds)					
		t1	t2	t3	t4	Total	Direct
8	1	0.726	0.902	0.803	0.087	2.518	3.75
9	1	1.461	1.598	1.165	0.129	4.353	7.33
10	4	2.580	2.622	1.642	0.181	7.569	-
11	4	4.154	4.036	2.252	0.244	11.420	-
12	4	6.378	5.960	3.008	0.325	16.645	-
13	5	9.427	8.538	3.928	0.418	23.983	-

4. EXTENSION TO AN $O(MN^3)$ SCHEME

The basic scheme of Section 3 can be improved in (at least) three ways:-

i) V_{i0} is block banded, each block itself being banded. In a conventional elimination, the banded blocks fill in. This fill in can be ignored, the justification being that V_{i0} is block W.A.D and is thus expected to have a block W.A.D inverse (see [3]). In meaningful terms, elements being ignored in the inverse should be small if the elements already set to zero in V_{i0} were small in the original matrix.

ii) The γ matrices of (2.5) are obtained from an integration along a boundary/element interface and are thus expected to be asymptotically diagonal. The full matrices can therefore be replaced by banded matrices of semibandwidth μ.

iii) By using FFT techniques it is possible to compute the required residuals (without explicitly forming the diagonal blocks) in a time $O(MN^2 \ell n\ N)$.

With these improvements, the new scheme then has costs:-

t1:		$O(\mu^4) \times O(N^2) \times M$	$= O(M\mu^4 N^2)$
t2:		$O(\mu^2) \times O(N^2) \times NxM$	$= O(M\mu^2 N^3)$
t3:	a)	$O(N^2) \times O(\mu N) \times P^2 xM^2$	$= O(P^2 M^2 \mu N^3)$
	b)		$= O(P^3 M^3 N^3)$
t4:	a)		$= O(MN^2 \ell n\ N)$
	b)	$O(\mu^2) \times O(N^2) \times M$	$= O(M\mu^2 N^2)$
	c)		$= O(P^2 M^2 N^2)$
	d)		$= O(PMN^3)$

The corresponding results are shown in Table II. For $N > 9$, the individual timings are smaller than their earlier counterparts, though the overall gain is diminished by the increased number of iterations. Unfortunately, the overall operation counts do not have the hoped for $O(N^3)$ behaviour, but behave rather like $O(N^4)$ due to the large value of μ required. (The code does have the prediction costs for small μ - but no convergence!).

5. CONCLUSIONS

We here demonstrated two iterative techniques for the solution of the GEM equations. At first sight the $O(N^3)$ scheme is disappointing. However, since it does not require the full assembly of the diagonal matrix, the consequent reduction in storage (and assembly costs!) could prove useful in practice.

TABLE 2

N	Iterations for Convergence	Times (seconds)				
		t1	t2	t3	t4	Total
8	1	0.729	0.902	0.804	0.087	2.522
9	1	1.465	1.599	1.166	0.128	4.358
10	5	2.491	2.574	1.631	0.177	7.582
11	6	3.793	3.856	2.206	0.234	11.259
12	7	5.459	5.507	2.896	0.303	15.986
13	8	7.540	7.620	3.713	0.380	21.923

REFERENCES

1. DELVES, L.M., Unpublished work.
2. DELVES, L.M. and HALL, C.A., An implicit matching procedure for global element calculations. *J. Inst. Maths. Applics.* 23 223-234 (1979).
3. DELVES, L.M. and PHILLIPS, C., A fast implementation of the Global Element Method. *J. Inst. Maths. Applics.* 25, 177-197 (1980).
4. HENDRY, J.A., Singular problems and the Global Element Method. *Comp. Methods in Appl. Mech. and Engng.* 21, 1-15 (1980).
5. HENDRY, J.A. and DELVES, L.M., The Global Element Method applied to a mixed harmonic boundary value problem. *J. Comp. Phys.* 33, 33-44 (1979).
6. HENDRY, J.A., DELVES, L.M. and PHILLIPS, C., Numerical experience with the Global Element Method. pp.341-348 of J.R. Whiteman (ed.), *The Mathematics of Finite Elements and Applications III, MAFELAP 1978*. Academic Press, London (1978).

ON FINITE ELEMENTS
FOR THE ELECTRICITY EQUATION

A. Bossavit

Electricité de France, Clamart, France

1. INTRODUCTION

We are concerned here with a class of problems occurring in many electrotechnical applications, where the wave-length greatly exceeds the characteristic dimensions of the system, and for this reason, the so-called "displacement currents" in Maxwell equations can be neglected [5]. Depending on symmetries and boundary conditions, various forms of boundary value problems then appear, all parabolic, without apparent common formulation [7]. Such a formulation does exist, as was shown in [1] and [2], and may be referred to as "the electricity equation". It takes the field j of current densities as configuration variable. It is recalled in Section 2. But, though quite appropriate for the finite element treatment (so that a general code could be built on this basis), it leads to the solution of full-symmetric linear systems of large size. Here, an alternative approach is proposed (Section 3), using the magnetic field h as unknown quantity. It has this "local" character which leads naturally to sparse matrices. A discrete analog of this formulation is then derived (Section 4). The problem with this h-formulation as opposed to the j one is that the magnetic field exists in the whole space. Discretizing everywhere is out of the question, so a boundary integral method is used to take into account the magnetic phenomena outside the metallic parts, so that only these are divided into finite elements, just as with the j-formulation.

2 THE J-FORMULATION

2.1 Topology

Let Ω be a bounded domain, of smooth boundary Γ, and let ω be the outer domain, Ω and ω each locally on one side of Γ. In Ω, a positive function ρ, the resistivity, is given, and Ω will represent a metallic circuit. Neither Ω nor ω are necessarily connected, and their connected components can be multiply-connected (i.e. with holes). Algebraic topology considerations allow one to define a minimal set of "cutting surfaces" S_i, $i = 1, \ldots, r$, in Ω and s_i, $i = 1, \ldots, r$, in ω, so that $\Omega - \cup S_i$ and $\omega - \cup s_i$ become simply connected (Fig. 1). The boundaries ∂S_i and ∂s_i are (possibly finite sets of) closed curves drawn on Γ. (The fact that r is the same for Ω and ω is an aspect of Alexander's duality [4].)

The integer r is referred to as the number of "branches" in the circuit. Intuitively, a branch is a closed path in the circuit, crossing one of the cutting surfaces S_i.

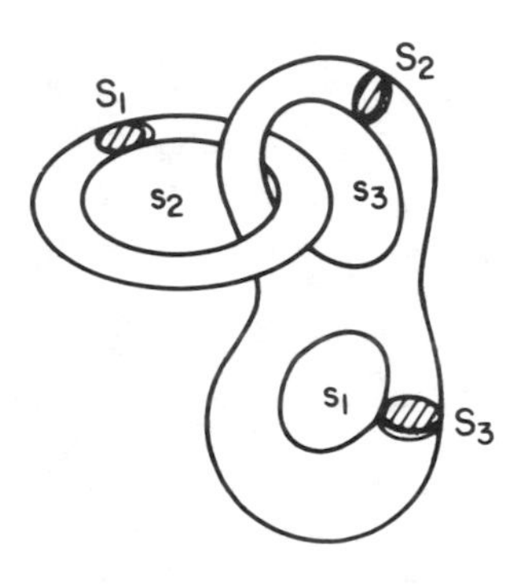

FIG. 1 A circuit with three branches (r = 3)

2.2 A Variational Formulation

Let us introduce the functional spaces

$$V = \{j \in [L^2(\Omega)]^3 | \operatorname{div} j = 0\}$$

$$V_0 = \{j \in V | j.n = 0\}$$

(where n is the outer normal) equipped with the L^2 scalar product (,). Let G be the operator so defined: First, given j, set

$$a(x) = \frac{\mu_0}{4\pi} \int \frac{j(y)}{x-y} dy \qquad (2.1)$$

(where μ_0 is the permeability of the vacuum) and let Gj be the restriction of a on Ω. Consider also the linear functionals $\Phi_i : V_0 \to R$,

$$\Phi_i(j) = \int_{S_i} j.n$$

(n is the normal on S_i, with an arbitrary but definite orientation). The j's are *current densities* and $\Phi_i(j)$ is the *total intensity in the branch i*.

The following theorem is easily established [2]:

Theorem 2.1 Let $v_i(t)$, $i = 1, \ldots, r$, be given functions in $L^2([0, T])$. The variational equation

$$d/dt\ (Gj, 1) + (\rho j, 1) = \sum_i v_i(t)\ \Phi_i(1) \quad \forall\ 1 \in V_0 \qquad (2.2)$$

with Cauchy or periodic conditions, has a unique solution in $L^2(0, T; V_0)$.

To interpret (2.2), we need some knowledge about the structure of V. The results used below can be found in [3] or [8].

2.3 Interpretation

The following decomposition into orthogonal spaces holds:

$$
\begin{aligned}
[L^2(\Omega)]^3 = & \{j \mid j = \text{grad}\ \phi,\ \phi \in H^1_0(\Omega)\} \\
& \oplus \{j \mid j = \text{grad}\ \phi,\ \phi \in H^1(\Omega),\ \Delta\phi = 0\} \\
& \oplus \{j \mid j = \text{grad}\ \phi,\ \phi = \textstyle\sum_i J_i \phi_i\} \\
& \oplus \{j \in V_0,\ \int_{S_i} j.n = 0 \quad \forall i\}
\end{aligned}
\tag{2.3}
$$

where the r functions ϕ_i are defined on $\Omega - S_i$ respectively as solutions of the following boundary value problems:

$$
\begin{aligned}
& -\Delta\phi_i = 0 \text{ in } \Omega - S_i, \quad \partial\phi_i/\partial n = 0 \text{ on } \Gamma, \\
& [\partial\phi_i/\partial n] = 0 \text{ on } S_i, \ [\phi_i] = 1 \text{ on } S_i
\end{aligned}
\tag{2.4}
$$

(where [ ] stands for the jump of a quantity through S_i).

Let us remark that, from (2.4),

$$\Phi_i(j) = \int_{\Omega - S_i} \text{grad}\ \phi_i . j \quad .$$

Using (2.3) to characterize the orthogonal of V_0 in V, we obtain, for all j satisfying (2.2),

$$\partial/\partial t\ Gj + \rho j - \textstyle\sum_i v_i \ \text{grad}\ \phi_i = \text{grad}\ \phi \tag{2.5}$$

where $\phi \in H^1(\Omega)$. Let $b = \text{curl}\ a$, $h = b/\mu_0$ and $e = \rho j$ (Ohm's law). Then (2.1) shows that curl h = j. Taking the curl of (2.5), we have $\partial b/\partial t + \text{curl}\ e = 0$. We can therefore conclude that (2.2) does solve the set of Maxwell equations where displacement currents have been discarded.

The v_i's are interpreted as electromotive forces present in the different branches of the circuit.

Using a tetrahedral paving of Ω and constant vector-valued shape functions on each tetrahedron, we can discretize (2.2) (see [2]). But the matrix associated with G is full. Moreover, magnetic non-linearities are not easily dealt with. A better formulation will now be derived.

3. THE H-FORMULATION

3.1 Primitive Form

Let us define

$$\mathcal{H} = \{h \in [\mathcal{D}(R^3)]^3 \mid \text{curl}\ h = 0,\ \text{div}\ h = 0 \text{ in } \omega\}$$

and let H be the completed space, with respect to the scalar product

$$(h, k) = \int_{R^3} h.k + \int_{\Omega} \text{curl } h.\text{curl } k$$

One can prove that the continuous image of H by curl and restriction to Ω is exactly V_0 (but curl is not an isomorphism). Consequently, the linear functionals $F_i : H \to R$,

$$F_i(h) = \int_{\partial S_i} h.dl$$

are continuous, since by Stokes theorem, $\Phi_i(\text{curl } h) = F_i(h)$.

The following result is easily proved:

Theorem 3.1 Let the v_i's be as in Th. 2.1. The variational equation, with Cauchy or periodic conditions,

$$\mu_0 \, d/dt \int_{R^3} h.k + \int_{\Omega} \rho \text{ curl } h.\text{curl } k = \textstyle\sum_i v_i(t) \, F_i(k) \qquad \forall \, k \in H \tag{3.1}$$

has a unique solution in $L^2(0, T; H)$.

The integral on R^3 will now be replaced by a more manageable (as far as dicretization is concerned) expression.

3.2 *Outer Rigidity and Revised Formulation*

The space $[L^2(\omega)]^3$ can be decomposed just as in (2.3), but the Sobolev spaces should be replaced by Beppo-Levi spaces $BL(\omega)$ and $BL_0(\omega)$. As curl h = 0 and div h = 0 in ω, h is in ω of the form $h = \text{grad } \phi + \sum_i J_i \text{grad}\phi_i$. The tangential component of h is continuous through Γ, since curl $h \in L^2$. Therefore, ϕ and $F_i(h)$ can be obtained from the restriction of h to Ω. So, for h and k in H, we have:

$$\int_{R^3} h.k = \int_{\Omega} h.k + \int_{\omega} (\text{grad } \phi + \textstyle\sum_i F_i(h) \text{grad } \phi_i)(\text{grad } \psi + \sum F_i(k) \text{grad} \psi_i)$$
$$= \int_{\Omega} h.k + \int_{\Gamma} \partial\phi/\partial n \, \psi + \textstyle\sum_{i,j} F_i(h) \, F_j(k) \displaystyle\int_{S_i} \partial\psi_j/\partial n$$

(other terms disappear, as seen from (2.4). We shall use the notations

$$L_{ij} = \int_{S_i} \partial\psi_j/\partial n$$

(notice that L is symmetric) and, using the same symbols for ϕ or ψ and their traces on Γ,

$$\int_{\Gamma} \partial\phi/\partial n \, \psi = \langle R \, \phi \, , \, \psi \rangle$$

where $<.,.>$ is the scalar product on $L^2(\Gamma)$. The symmetric operator R, or "outer rigidity", and the matrix L allow one, once computed, to obtain the quantity $\int_\omega h.k$ using information given on Ω only: for this reason, a dicretization of Ω will be all that is needed to approximate the different terms in (3.1). The revised form of (3.1) is, with obvious notations,

$$\mu_0 \, d/dt \int_\Omega h.k + <R\phi, \psi> + (L\,F(h),\, F(k)) + \int_\Omega \rho \text{curl } h.\text{curl } k = \sum_i v_i(t)\, F_i(k) \;\; \forall\, k \in H \qquad (3.2)$$

3.3 Energy

Let us assign to k in (3.1) the instantaneous value of h. We obtain

$$d/dt\,(\textit{energy of the magnetic field}) + (\textit{Joule losses in } \Omega) = \sum_i v_i(t)\; F_i(h) \qquad (3.3)$$

which confirms the interpretation of the v_i's as e.m.f.'s, since $F_i(h)$ is an intensity. The right-hand side is the power injected into the system.

It is also interesting to take $k \in [\mathcal{D}(\Omega)]^3$. This shows that

$$\mu_0\, \partial h/\partial t + \text{curl}(\rho \text{ curl } h) = 0 \qquad (3.4)$$

Multiplying (3.4) by h, integrating by parts, and comparing with (3.3), we obtain

$$\mu_0 \int_\omega dh/dt.h + \int_\Gamma (\rho \text{curl } h,\, h,\, n) = \sum_i F_i(h)$$

i.e.

$$d/dt(\textit{energy of the outer field}) + (\textit{flux of the Poynting vector}) = \textit{injected power}$$

3.4 Open Circuit Problems

Of course the previous formulation is not realistic, for it forces to take into account the main network and all the metallic bodies in space. In real problems, there is a region of Ω, Ω_g, where the current density j_0 is already known. Let $h_0 \in H$ be such that curl $h_0 = j_0$. Set

$$U = \{h \in H \mid \text{curl } h = 0 \text{ on } \Omega_g,\ \text{div } h = 0 \text{ on } \Omega_g\}$$

and look for a solution in the form $h + h_0$, with $h \in U$. Then

$$d/dt\ \mu_0 \int_{R^3} h.k + \int_{\Omega - \Omega_g} \rho \text{ curl } h.\text{ curl } k = \sum_i v_i(t)\ F_i(k) \ldots$$
$$-d/dt \int_{R^3} h_0 k \quad \forall\, k \in U \qquad (3.5)$$

This is a problem similar to (3.1), with $\Omega - \Omega_g$ as a domain, but the right-hand side is different. Some $F_i(k)$, corresponding to branches in Ω_g, are now zero, and there is an induction term which takes into account outer influences.

4. FINITE ELEMENTS

4.1 Configuration Variables

The idea is to define a finite set of scalar configuration variables, from which the integrals in (3.2) can be approximately evaluated.

Let us build a tetrahedral paving of Ω. Let E be the set of edges and E_0 the set of inner edges (those not on Γ), S the set of vertices on Ω. The configuration variables will be

$$\{h^e \mid e \in E_0\} \cup \{\phi^s \mid s \in S\} \cup \{J_1, \ldots, J_r\}$$

Let $n = \text{card}(E_0)$, $m = \text{card}(S)$. Let u, of length n+m+2, be the configuration vector.

Given u, a magnetic field can be constructed as follows. Let us first define "multivalued grid functions" $\{\phi_i^s \mid s \in S\}$ by restricting to S the ϕ_i's of (2.4), and take finite differences of the multivalued grid function $\phi + \sum_i J_i\ \phi_i^s$ between vertices. This way, an h^e is assigned to each surface edge.

Now, from the h^e's, a field h of the form $h = \alpha + \beta \times x$ (where α and β are fixed vectors of R^3 and x is the coordinate vector) is interpolated in each tetrahedron, matching the given h's on the edges. (This kind of finite element has been proposed by Nédelec for the Stokes problem; see [6].)

Let u and v be two configuration vectors. We see that all integrals in (3.2) can be easily expressed as symmetric quadratic forms in u and v. As for the term $\langle R\phi, \psi \rangle$, computing the discrete outer rigidity R is a standard problem in boundary element methods.

4.2 Approximation Scheme

The structure of the set of differential equations finally obtained is displayed below (the correspondence with terms in (3.2) is indicated). The sparsity is quite apparent. As a bonus, we have obtained a method to deal with non-linear magnetization characteristics $b = \beta(h)$: just replace $\mu_0 h$ in the d/dt term by $\beta(h)$. The differential system will be solved by a Crank-Nicolson scheme; each step requires the solution of a non-linear system of algebraic equations.

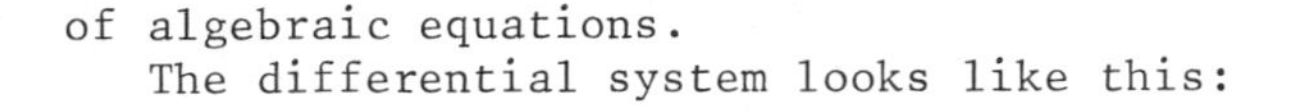

The differential system looks like this:

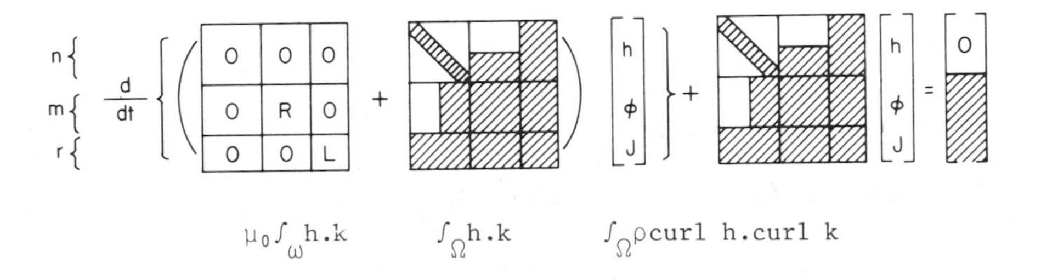

In the linear case, with time-periodic data ("steady-state" problem), one replaces d/dt by $i\omega$ and solves in the complexfield.

Due to the big size of the matrices encountered in real problems, solving such systems by efficient methods, making good use of the sparsity in particular, is still a difficult problem.

REFERENCES

1. BOSSAVIT, A., Le problème des courants de Foucault, *E.D.F.-Bulletin de la Direction des Etudes et Recherches-Série C,* 1, 5-14 (1980).

2. BOSSAVIT, A. A Variational Formulation for the Eddy-currents Problem, *Comp. Meth. Mech. & Engng.,* to appear (1981).

3. FOIAS, C., and TEMAM, R., Remarques sur les équations de Navier-Stokes stationnaires et les problèmes successifs de bifurcation, *Annali Sc. Norm. Sup. Pisa, Serie IV,* 5, 1, 29-63 (1978).

4. GREENBERG, M., *Lectures on Algebraic Topology,* Benjamin, New York (1967).

5. LANDAU, L., *Cours de Physique Théorique,* Mir, Moscow (1965).

6. NEDELEC, J.C., Mixed Finite Elements in R^3, *Numerische Matematik,* to appear (1981).

7. STOLL, R.L., *The Analysis of Eddy-Currents,* Clarendon Press, Oxford (1974).

8. WEYL, H., The Method of Orthogonal Projections in Potential Theory, *Duke Math. J.,* 7, 411-444 (1940).

FINITE ELEMENT MODELLING AND PARAMETRIC STUDY OF DC AND HIGH FREQUENCY AC BEHAVIOUR OF SCHOTTKY BARRIER DIODES

J.S.Campbell

University College, Cork

1. INTRODUCTION

The first significant paper on numerical simulation of the general transport equations for semiconductor devices(SCD's) was published by Gummel[1] only seventeen years ago.Since then numerical simulation of SCD's has increased dramatically and is now playing an increasingly important role.Several factors contribute to this increasing popularity,not the least of which is the continuing reduction in computing costs.An equally important factor, however,is the failure of simplified models as predictive tools.For real SCD's,operating under complex conditions,the behaviour is governed by a set of coupled nonlinear time dependent partial differential equations,which neccessitates a numerical simulative approach.

Gummel's paper simulated bi-polar transitors using a one-dimensional finite difference method which was subsequently applied to other devices[2].Two-dimensional finite difference algorithms soon followed for a variety of SCD applications[3-8].In these applications novel strategies were adopted whereby the equations were treated in a decoupled manner and simplifications to the general case were made based upon physical considerations for particular devices.More recently the finite element method has been utilised[9-13] and the boundary integral technique has also been proposed[14].

2. SCHOTTKY BARRIER DIODES

Schottky barrier diodes(SBD's) have been used since the early 60's in microwave mixers and the gallium arsenide(GaAs) diode is now universally used as the nonlinear mixing element in high frequency receivers[15].A SBD consists of a small Pt-Au contact deposited on epitaxial GaAs,as shown schematically in figure 1. Due to the high level of doping the conductivity varies rapidly with depth as shown in figure 2.Fabrication of diodes entails forming an ohmic contact,opening small holes in the SiO_2 passivation using electron beam lithography,and then electoplating the Pt and the Au on the top contact.

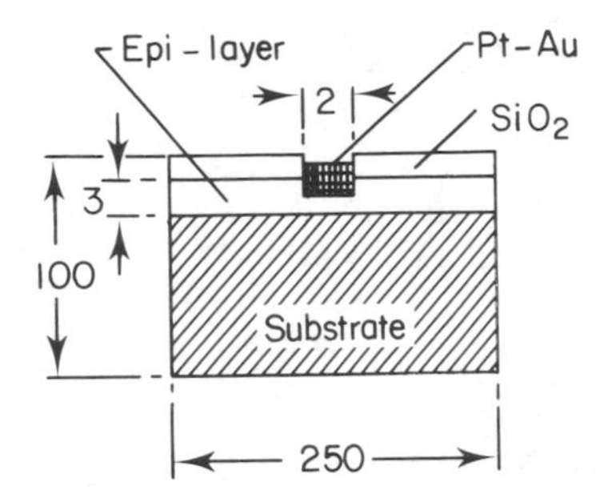

FIG.1 Schematic of SBD

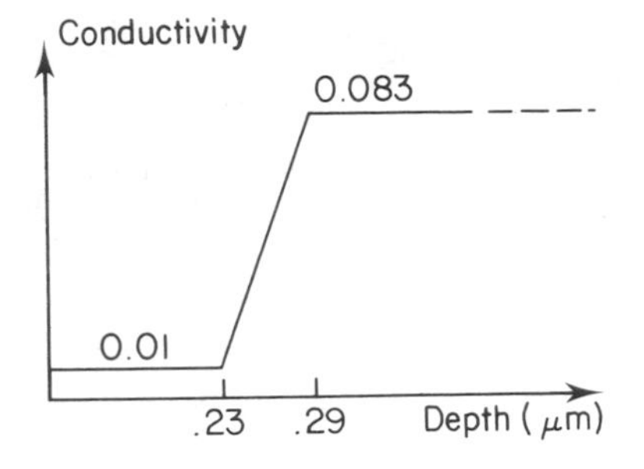

FIG.2 Variation of conductivity with depth

A DC diode may be represented by the lumped element model shown in figure 3a, in which C_B and R_B are the barrier capacitance and the nonlinear resistance respectively, and R_S is the DC series resistance. For the high frequency AC case, shown in figure 3b, the series resistance is the sum of the epi-layer resistance, the substrate resistance and the skin resistance R_{SK}.
A considerable fraction of conversion losses in diodes arises from parasitic terms associated with the series resistance. Earlier investigation[16] has indicated that it is possible to substantially reduce the resistance associated with the parasitic losses by increasing the perimeter to area ratio of the anode, whilst retaining a constant anode area (and hence maintaining a constant capacitance).

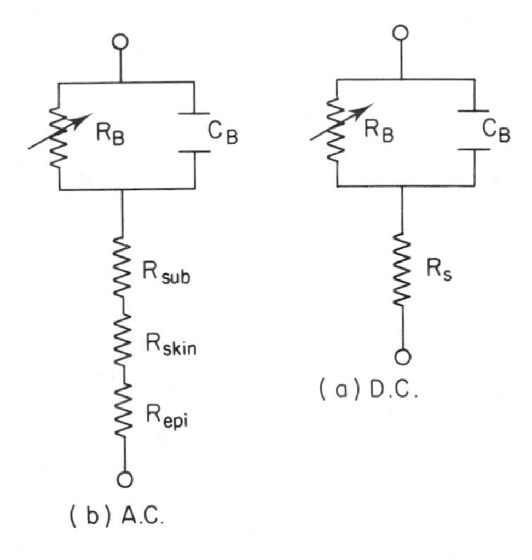

FIG.3 Lumped element models

In the operation of the diodes a scalar potential V is applied at the geometrically minute anode. In the DC case the current flows from the anode, through the bulk of the diode volume, to the ohmic contact zone where the potential is zero. For the high frequency AC case the current flows through a thin skin at the external surface of the diode, passing from the anode to the ohmic contact zone, whilst the core of the diode is effectively short-circuted. For both the DC and the AC cases, the potential gradients existing in the anode region are extremely high and, indeed, the conditions closely resemble a point singularity.

Such conditions mitigate strongly against the use of regular finite difference meshes and finite element or boundary integral approaches are a natural choice. In the work reported here finite element parametric studies of SBDs have been used to determine the effect of ohmic contact configuration and anode shape. Hence the conditions have been determined under which the DC series resistance and the AC skin effect resistance are minimal.

3. GOVERNING EQUATIONS AND FINITE ELEMENT MODELLING OF SBD'S

The general differential equations which govern the transport of mobile electrons and holes in a semiconductor material are[17]

$$\nabla . \varepsilon \nabla \phi + q(p-n+N_d) = 0$$

$$q \frac{\partial p}{\partial t} + \nabla . \underline{J}_p = q\,G$$

$$q \frac{\partial n}{\partial t} - \nabla . \underline{J}_n = q\,G$$

However, for SBD's operating under steady state conditions the above equations may be simplified, as shown by Langley[18], and these then reduce to the single Laplacian equation of electostatics

$$\nabla . \sigma \nabla \phi = 0 \tag{1}$$

in which ϕ is the scalar potential, σ is the conductivity, and Ω is the domain. It is important to note here that the domain Ω excludes the space charge region and that the associated boundary conditions are then taken as

$$\phi - V = 0 \quad \text{... on the boundary of the space charge region} \tag{2}$$

$$\phi = 0 \quad \text{... on the ohmic contact boundary} \tag{3}$$

$$\nabla \phi . \underline{n} = \frac{\partial \phi}{\partial n} = 0 \quad \text{... on all remaining parts of the boundary} \tag{4}$$

in which V is a prescribed potential.

For the finite element analyses the variational approach is employed, in which the functional $\Pi(\phi)$ is minimised with respect to the ϕ field. For the SBD's the appropriate functional is

$$\Pi(\phi) = \tfrac{1}{2} \int \sigma (\nabla \phi)^2 \, d\Omega \tag{5}$$

For the models described below, the dependent variable ϕ is discretized as

$$\phi(x) = N_i(x) a_i \tag{6}$$

in which $N_i(x)$ are prescribed polynomial shape functions of position x and a_i are discrete parameters. Minimising the functional $\Pi(\phi)$ with respect to the discrete parameters a_i, and applying the appropriate Dirichlet boundary conditions, leads to a set of linear algebraic equations.

Two finite element models for the analysis of SBD's are described here
(i) solid axisymmetric model for AC analysis
(ii) multi-plane hollow box model for high frequency AC analysis
A third model,for solid three-dimensional DC analysis was reported by Campbell,Langley & Wrixon[19] and was also fully described by Langley[18].Here the application of the multi-plane model is described briefly,as an extensive description of its application is given elsewhere[20,21].

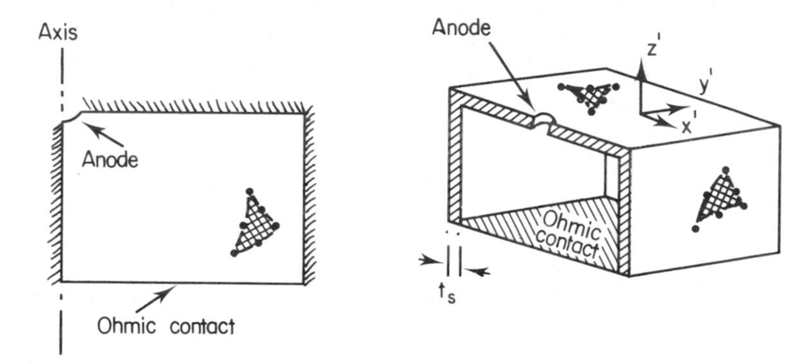

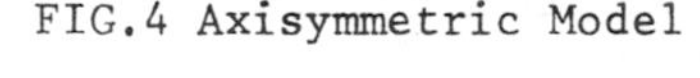

FIG.4 Axisymmetric Model FIG.5 Multi-plane hollow-box model (Half Section)

The choice of element type for these models was based upon a number of considerations.First,for the DC models,the capability of including rapid depth variation of material properties was required.Second,for both models,curved edges were neccessary for the anode region,and third,elements which facilitate a high degree of mesh grading were mandatory in order to successfully model the extremely high gradients which occur in the anode region. High order,numerically integrated isoparametric elements easily meet these requirements,with triangular elements being particularly suitable.

For the solid axisymmetric DC model the cylindrical coordinate system $x = \{r,\theta,z\}$ was used with a six noded toroidal isoparametric triangular element.Details of the polynomial shape functions,curvilinear coordinates and numerical integration of such elements are given in the text by Zienkiewicz[22] and the axisymmetric model is illustrated in figure 4.

For the high frequency AC case the current flows in the hollow-box shaped region shown in figure 5.For each plane,it may be reasonably assumed that the potential does not vary through the thickness t_S.Thus the $x'y'$ plane of figure 5 may be analysed using plane,two-dimensional finite elements of thickness t_S and six noded curved-edge isoparametric elements were used.

4. DC SERIES RESISTANCE AND AC SKIN EFFECT RESISTANCE OF SBD'S

For SBD's the DC series resistance and the AC skin effect resistance, jointly denoted here by R_*, is defined by the equation $R_* = V/I$ in which I is the total current passing from the anode to the ohmic contact. Considering a surface S, which separates the anode and ohmic contact surfaces, then the total current I is given by the surface integral of the current density flux

$$I = \int_S \sigma \nabla\phi . \underline{n}\ dS \qquad (7)$$

For the devices considered here an alternative approach is available for computing the total current I. Consider again the functional of equation (5) and apply integration by parts to obtain

$$\Pi(\phi) = \tfrac{1}{2}\int_\Gamma \phi\sigma\nabla\phi . \underline{n}\ d\Gamma - \int_\Omega \phi\ \nabla . \sigma\ \nabla\ \phi\ \ d\Omega \qquad (8)$$

The second integral is identically zero in domain Ω (eqn 1). Also, for the diodes, the boundary consists of three parts

(i) nonconducting surfaces on which $\nabla\phi . n$ is zero
(ii) ohmic contact surfaces on which ϕ is zero
(iii) anode surface on which $\phi = V =$ prescribed constant.

Hence equation (8) becomes simply

$$\Pi(\phi) = \frac{V}{2} \int \sigma\ \nabla\phi\ .\ \underline{n}\ \ d\Gamma = \tfrac{1}{2}\ V\ I \qquad (9)$$

and thus the total current may be recast in terms of the functional $\Pi(\phi)$

$$I = 2\Pi(\phi)/V = \int \sigma\,(\nabla\phi)^2\ \ d\Omega/V \qquad (10)$$

Equations (7) and (10) are surface and volumetric expressions for the total current and hence the resistance $R*$ may be written in the alternative forms

$$R* = V/\int \sigma\ \nabla\ \phi\ .\ \underline{n}\ dS$$

$$R* = V^2/\int\sigma(\nabla\phi)^2\ \ d\Omega$$

The volumetric form was used in the subsequent analyses.

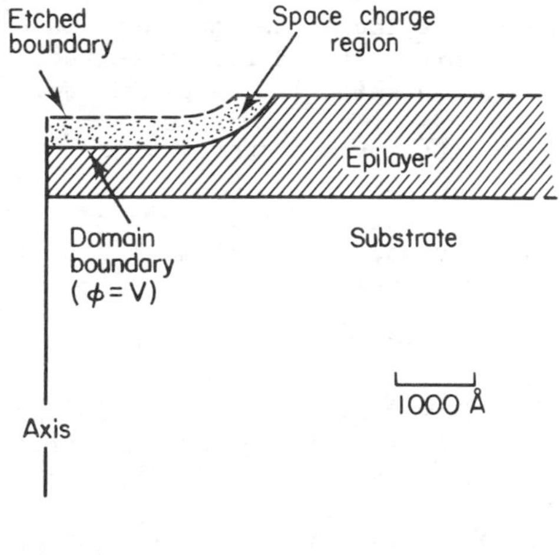

FIG.6 Anode details for DC case

5. PARAMETRIC STUDIES OF SBD'S USING FINITE ELEMENTS

Parametric studies of the DC and high frequency AC diodes were undertaken for a range of realistic diodes having anode shapes which are physically realisable using current electron beam lithographic techniques and having a range of ohmic contact configurations.

DC ANALYSES

For the DC axisymmetric model the conductivity versus depth profile of figure 2 was applicable. The shape of the etched boundary and the space charge region are illustrated in figure 6 in which the width of the space charge region was taken as 780 Å. A set of ten analyses were executed involving two etched anode diameters (1 and 2μm) and the five ohmic contact configurations shown in figure 7.

AC ANALYSES

The features of the high frequency AC analyses were: constant material properties; three ohmic contact configurations (B,BS,BST): two etched anode areas ($\pi/4$ and π $[\mu m]^2$); a range of anode shapes (circular, rectangular, cruciform, octoform) and a skin thickness of 2.37μm which corresponds to a frequency of 400 GHz.

In [21] full details are given of the precise geometric features of the anode shapes, computation of space charge region width, computation of skin thickness and full results for the DC and high frequency AC analyses.

6. DISCUSSION AND CLOSURE

The parametric studies of the SBD's support the following general conclusions:-

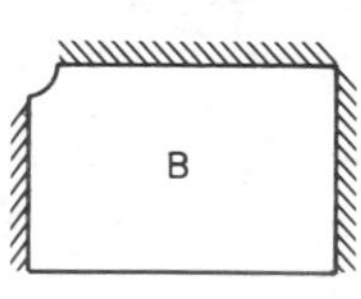

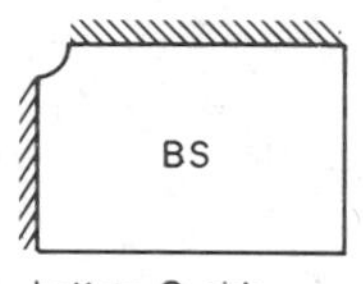

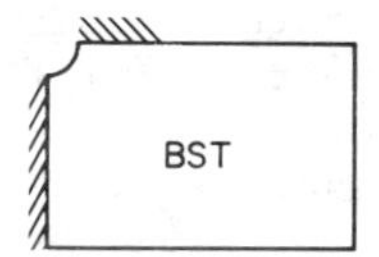

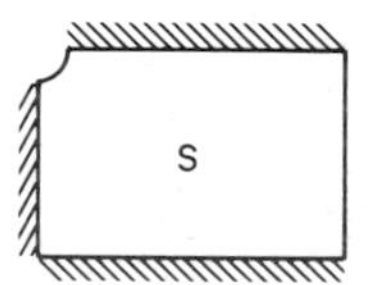

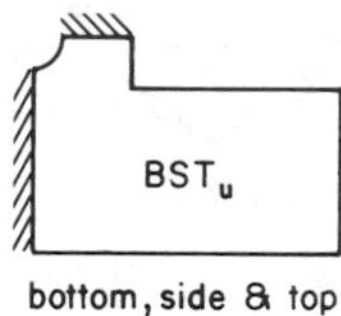

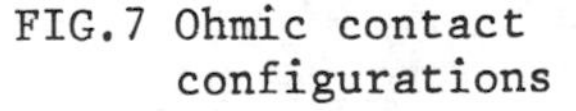

FIG.7 Ohmic contact configurations

The DC series resistance R_S is essentially independent of the ohmic contact configuration whereas the high frequency AC skin-effect resistance R_{SK} is intimately dependent upon ohmic contact conditions. R_{SK} is greatly reduced as the ohmic contact is brought closer to the anode. For both the DC and AC cases R* is reduced as the anode area is increased. R_{SK} is substantially reduced by increasing the perimeter to area ratio of the anode (the rectangle being the optimal shape).

The above applications illustrate the extreme suitability of the finite element method for modelling semiconductor devices in which the gradients are extremely large. In this respect the finite element method is vastly superior to finite difference methods with regular meshes. It is most probable that the same advantages of the f.e.m. will accrue in the solution of more general SCD's having nonlinear time dependent coupled governing equations.

7. ACKNOWLEDGEMENTS

The author gratefully acknowledges financial support from the National Board for Science and Technology, Ireland.

8. REFERENCES

1. CUMMEL, H.K., A self consistent iterative scheme for one-dimensional steady state transistor calculations. *IEEE Trans. on Electron Devices*, ED-11, 455-465 (1964).
2. SCHARFETTER, D.L. and GUMMEL, H.K., Large-signal analysis of a silicon Read oscillator. *IEEE Trans. on Electron Devices*, ED-16, 64-67 (1969).
3. KENNEDY, D.P. and O'BRIEN, R.R., *IBM J. Research and Development*. 662-674 (1969).
4. BARRON, M.B., *Stanford Electronic Laboratories Report*, SEC-69-069. Stanford, California (1969).
5. SLOTBOOM, J.W., Computer aided two-dimensional analysis of bi-polar transistors. *IEEE Trans. Electron Devices*, ED-20, 869-879 (1973).
6. VANDORPE, D. and BOREL, J., An accurate two-dimensional analysis of the MOS transistor. *Solid State Electronics*. 15, 547-557 (1972).
7. MOCK, M.S., Two-dimensional mathematical model of the insulated gate field effect transistor. *Solid State Electronics*, 16, 601-609 (1973).
8. ENGL, W.L. and MANCK, O., Two-dimensional numerical analysis of bipolar transistor transients. *Int. Electron Devices Meeting*, *Digest Tech Papers*, 381-383 (1973).
9. CALZOLARI, P., GRAFFI, S. and PIERNI, G., Finite-difference and finite-element for the semiconductor equation. Proc *Int Symp. on Finite Element Methods in Flow Problems*, UAH Press, Hunstville (1974).
10. COTTREL, P.E. and BUTURLA, E.M., Steady state analysis of field effect transistors via the finite element method. *Int. Electron Devices Meeting*, *Tech. Digest*, 51-54 (1975).
11. HACHTEL, G.D., MACK, M.H., and O'BRIEN, R.R., Semiconductor device analysis via finite elements. Conference Records, *Eight Asilomar Conf on Circuits Systems and Computers*, California, 332-338 (1974).

12. BARNES, J.J. and LOMAX, R.J., Two dimensional finite element simulation of semiconductor devices. *Electron Letters*, 10, 341-343 (1974).
13. ADACHI, T., YOAHII, A. and SADO, T., Two dimensional semiconductor analysis using the finite element method. *IEEE Trans Electron Devices*, ED-26(7), 1026-1031 (1979).
14. MEY, G. De., Integral equation techniques for the calculation of two dimensional and time dependent problems in pn junctions and solar cells. pp. 224-225 of B.T. Browne and J.J.H. Miller (Ed.), *Numerical analysis of semiconductor devices*. Boole Press, Dublin (1979).
15. KELLY, W.M. and WRIXTON, G.T., Optimization of Schottky barrier diodes for low noise, low conversion operation at near-millimeter wavelengths. in K.J. Button, *Infrared and Millimeter Wavelengths*, Vol III, Academic Press London (1980).
16. WRIXTON, G.T. and PEASE, R.F.W., Schottky barrier diodes fabricated on GaAs using electron beam lithography. *Inst. Physics Conf*, Series No 24, Chapter 2, 55-60 (1975).
17. SZE, S.M., *Physics of semiconductor devices*. Wiley, New York (1969).
18. LANGLEY II, J.B., Finite element analysis and electron beam lithographic fabrication of GaAs Schottky barrier diodes with sub-micron features. *PhD. Thesis, Dept. Elect. Eng., University College Cork* (1980).
19. CAMPBELL, J.S., LANGLEY, J.B. and WRIXTON, G.T., Finite element analysis of Schottky barrier diode structures for sub-millimeter wavelengths. *IFEE Conf on Infrared and Millimeter Wavelengths and their Applications*, Miami (1979).
20. CAMPBELL, J.S., WRIXTON, G.T., Finite element analysis of skin effect resistance in sub-millimeter Schottky barrier diodes. (To be published)(1981).
21. CAMPBELL, J.S., Finite element analysis of Schottky barrier diodes. *Internal Report, Dept. of Civil Eng, University College Cork*, (1979).
22. ZIENKIEWICZ, O.C., *The Finite Element Method*, 3rd Edition, McGraw-Hill, New York (1977).

A POSTERIORI ERROR ESTIMATES FOR FINITE ELEMENT SOLUTIONS OF PLATE BENDING PROBLEMS

R. Rohrbach

Leibniz-Rechenzentrum, München

E. Stein

University of Hannover

1. INTRODUCTION

To achieve convergence in the method of nonconforming finite elements for plate bending problems, one has to choose piecewise polynomial functions, which must satisfy not only an approximating condition but also certain continuity conditions on inter-element boundaries. Success in this *patch test* depends strongly on the type and form of the chosen discretisation, as is known from numerical experience [1], [5], [7] and theoretical error analysis [8]. Therefore an effective estimation of the accuracy of a computed finite element solution has to include *a posteriori* information. In this paper a theorem is presented which relates the existence of effective computable error indicators to the asymptotic behaviour of the discretisation error. It turns out that these error indicators, being obtainable in a post solution process, converge to certain real values if and only if the nonconforming finite element solutions themselves converge to the solution in question for $h \to 0$, where h measures the size of the elements. Various numerical investigations show that the error indicators can be used for a practical *a posteriori* assessment of the accuracy of computed finite element approximations [5].

The motivation for the outlined work came originally from observations we made with numerical investigations for several plate bending elements. In the comparison of the convergence behaviour of displacement methods it was found that some information which is easily obtainable after the solution process - such as certain parts of the finite element solution energy - can be used to give a rough survey of the error distribution. In the following we will explain how computed information can be used for the construction of error indicators.

2. MATHEMATICAL FORMULATION AND NOTATIONS

Let Ω be a bounded domain in the plane $\mathbb{R}^2$ with polygonal boundary Γ which is composed of two disjoint parts Γ_c and Γ_s

such that $\Gamma = \overline{\Gamma_c \cup \Gamma_s}$. We consider the mixed boundary value problem for the biharmonic operator :

$$\begin{aligned} \Delta^2 u &= f && \text{in } \Omega \\ u = u_n &= 0 && \text{on } \Gamma_c \\ u = \Delta u &= 0 && \text{on } \Gamma_s \end{aligned} \tag{2.1}$$

Here u_n denotes the derivative with respect to the outer normal direction n of Γ, Δ is the Laplacian. The weak solution u_o of the above problem describes the deflection of a plate which is clamped along Γ_c and simply supported along Γ_s, loaded by square integrable f. To obtain u_o, we have to introduce the space

$$U = \{ u \in H^2(\Omega) : u = 0 \text{ on } \Gamma,\ u_n = 0 \text{ on } \Gamma_c \}$$

and the bilinear form

$$a(u,v) = \int_\Omega \{\Delta u \cdot \Delta v + (1-\mu)(2u_{xy}v_{xy} - u_{xx}v_{yy} - u_{yy}v_{xx})\}\, dxdy \quad .$$

Here for integral m $H^m(\Omega)$ are the Sobolev spaces of $L_2(\Omega)$ - functions having generalized derivatives up to order m in $L_2(\Omega)$. The spaces $H^m(\Omega)$ are provided with the norm

$$||u||_m = (\sum_{k \leq m} |u|_k^2)^{1/2} \quad , \text{ where } |u|_k^2 = \sum_{|\alpha| = k} (D^\alpha u, D^\alpha u)$$

is the seminorm over $H^k(\Omega)$, $(\cdot,\cdot)$ the inner product of $L_2(\Omega)$, $|\cdot|_k$ being a norm over

$$H_o^k(\Omega) = \{u \in H^k(\Omega) : D^\alpha u_{|\Gamma} = 0 \text{ for } |\alpha| \leq k-1\} \quad ,$$

where the boundary conditions are satisfied in a generalized sense.

The bilinear form $a(\cdot,\cdot)$ is continuous on $(U, ||\cdot||)$ with $||\cdot|| := ||\cdot||_2$ and $\mu \in [0, 1/2)$, and also coercive if Ω is a regular domain [2]. It follows by the well-known LAX-MILGRAM-theorem, that for any $f \in L_2(\Omega)$ there is a unique solution $u_c \in U$ of the variational equation

$$a(u_o,\phi) = (f,\phi) \qquad \forall \phi \in U \quad . \tag{2.2}$$

3. NONCONFORMING FINITE ELEMENT METHODS

We consider continuous finite elements for the approximation of the solution u_o of the plate bending problem. Let $\{T_h\}$ be a family of regular triangulations of Ω depending on some real mesh parameter h, such that $\overline{\Omega} = \cup\{T \in T_h\}$. For any integer $k \geq 0$ let $P_k(T)$ denote the set of polynomials of degree $\leq k$ over T (cf. e.g. [3]). With $1 \geq 3$ we get a family $\{U_h\}$ of finite dimensional function spaces such that

$$U_h \subset \{\phi \in C^o(\overline{\Omega}) : \phi_{|T} \in P_1(T)\ \forall T \in T_h\} \quad \cap \quad H_o^1(\Omega) \quad . \tag{3.1}$$

The nonconforming approximation $u_h \in U_h$ of u_o is defined by the variational equation

$$a_h(u_h,\phi) = (f,\phi) \qquad \forall\phi \in U_h \tag{3.2}$$

where $f \in L_2(\Omega)$ as before and the bilinear form $a_h(\cdot,\cdot)$ is defined over U_h by

$$a_h(\phi,\psi) := \sum_{T\in T_h} \{(\Delta\phi,\Delta\psi)_T + (1-\mu)\cdot\int_T (2\phi_{xy}\psi_{xy} - \phi_{xx}\psi_{yy} - \phi_{yy}\psi_{xx})dxdy\}$$

We have the following

THEOREM 1 [5]:

Let $||\cdot||_h := (\sum_{T\in T_h} |\cdot|^2_{2,T})^{1/2}$. *Then* $||\cdot||_h$ *defines a norm over* U_h *and the bilinear form* $a_h(\cdot,\cdot)$ *is continuous and coercive on* $(U_h, ||\cdot||_h)$. *Furthermore, equation* (3.2) *has one and only one solution* u_h.

From standard error analysis for nonconforming finite element methods [3], [7] we recall the a priori error estimate

$$||u_o - u_h||_h \leq \text{const}\cdot(\inf_{\phi\in U_h} ||u_o-\phi||_h + \sup_{\psi\in U_h} \{|E_h(u_o,\psi)|/\ ||\psi||_h\}),$$

where $E_h(u_o,\psi) := (f,\psi) - a_h(u_o,\psi)$. (3.3)

If $|E_h(u_o,\psi)| = O(||\psi||_h)$, then the *patch test* is passed and convergence of the approximations u_h can be proved using the above estimate if the approximating condition

$$\inf_{\phi\in U_h} ||u_o-\phi||_h = O(1) \tag{3.4}$$

is fulfilled by the spaces U_h .

4. *A posteriori* ERROR ESTIMATES

The *a posteriori* character of the following estimates is represented by a post solution approximation of higher order derivatives of the solution u_o of the plate bending problem using the same mesh. We assume from now on that the solution is sufficiently regular, $u_o \in H^{l+1}(\Omega)$, such that the approximating condition (3.4) is satisfied and the expression E_h from (3.3) takes the form

$$E_h(u_o,\psi) = \sum_{T\in T_h} \{(-\Delta u_o,\psi_n)_{\Gamma_T} + (1-\mu)\cdot(u_{ss},\psi_n)_{\Gamma_T}\}. \tag{4.1}$$

for all $\psi \in U_h$. Here u_{ss} denotes the second order derivative of u_o with respect to the tangential direction of the boundary Γ of Ω. The subscript Γ_T indicates that the integration is to be restricted to the boundary of the element T.

Now let $\{V_h\}$ be a family of finite dimensional function spaces such that

$$U_h \subset V_h \subset \{\phi \in C^o(\bar{\Omega}) : \phi_{|T} \in P_1(T) \quad \forall T \in T_h\} \cap H^1(\Omega) .$$

We may find a function $v_h \in V_h$ with the following properties :

$$||v_h - v_o||_i = O(1) , \quad (\nabla v_h, \nabla\psi) = (f,\psi) \qquad \forall\psi \in U_h \; ; \; i=0,1$$

where $v_o := -\Delta u_o$. This can be proved by using a slight modification of the techniques described in [6]. Applying Green's formula we may write

$$E_h(u_o,\psi) = \sum_{T\in T_h} \{(v_o - v_h, \psi_n)_{\Gamma_T} + (v_h, \Delta\psi)_T + (\Delta u_h, \Delta\psi)_T \} + (1-\mu)\cdot E_h^{\mu}(u_o,\psi)$$

from which we deduce

$$|E_h(u_o,\psi)| \leq \text{const}\cdot||\psi||_h\cdot\left(\sum_{T\in T_h} ||v_h+\Delta u_h||_{o,T} + ||v_h - v_o||_1\right) + |(1-\mu)\cdot E_h^{\mu}(u_o,\psi)| \tag{4.2}$$

where the last term can be estimated by

LEMMA 1 [5]:

There exists a computable $\beta_h \in \mathbb{R}$, $\beta_h \geq 0$, *such that* $|E_h^{\mu}(u_o,\psi)| \leq c_1(\psi) + \beta_h\cdot c_2(\psi)$ *for all* $\psi \in U_h$, *and the constants have the form* $c_1(\psi) = O(||\psi||_h), c_2(\psi) = O(||\psi||_h)$. *Furthermore, if* $||u_o - u_h||_h \to 0$ *then* $\beta_h \to 0$.

Setting $\alpha_h = \sum_{T\in T_h} ||v_h+\Delta u_h||_{o,T}$ it follows that

$$\alpha_h \leq ||v_h - v_o||_o + \sum_{T\in T_h} ||v_o + \Delta u_h||_{o,T}$$

$$\leq ||v_h - v_o||_o + \text{const}\cdot||u_o - u_h||_h , \quad \text{such that } \alpha_h \to 0$$

if $||u_o - u_h||_h \to 0$. Thus, combining this inequality with Lemma 1 we obtain with $\gamma_h := \alpha_h + \beta_h$ from (4.2)

THEOREM 2 :

There exists a computable $\gamma_h \in \mathbb{R}$, $\gamma_h \geq 0$, *such that* $|E_h(u_o,\psi)| \leq O(||\psi||_h) + \gamma_h\cdot O(||\psi||_h)$ *for all* $\psi \in U_h$, *and* $\gamma_h \to 0$ *if* $||u_o - u_h||_h \to 0$.

We are now in a position to state the final result as a corollary of the above theorem, using the standard error estimate for nonconforming finite elements (3.3).

THEOREM 3 [5]*:*

There exists a computable $\gamma_h \in \mathbb{R}$, $\gamma_h \geq 0$ *such that* $\gamma_h \to 0$ *if and only if* $||u_o - u_h||_h \to 0$.

The proof uses the fact that the approximating condition (3.4) is satisfied under the assumption that the solution u_o is sufficiently regular. Note that this is no restriction to the computation of the error indicators γ_h.

5. NUMERICAL EXPERIMENTS

We consider the nonconforming triangular plate bending element of Zienkiewicz [1]. For this cubic element it is known that the *patch test* is passed if all the triangles $T \in T_h$ have their edges parallel to three given directions [8], whereas convergence cannot be proved otherwise. This theoretical result can be confirmed by various numerical investigations [1],[5].

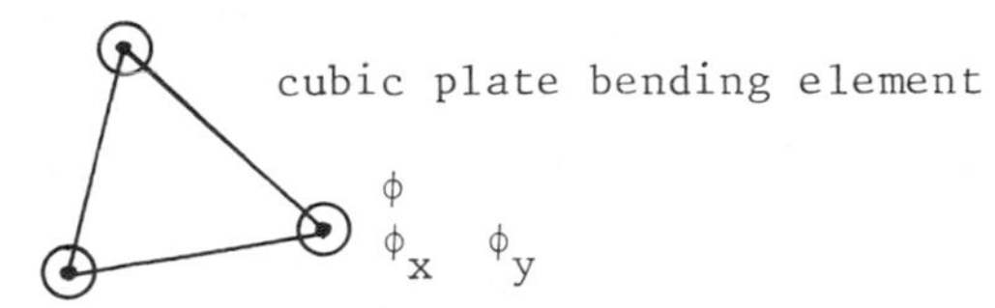

Regular mesh types (uniform) :

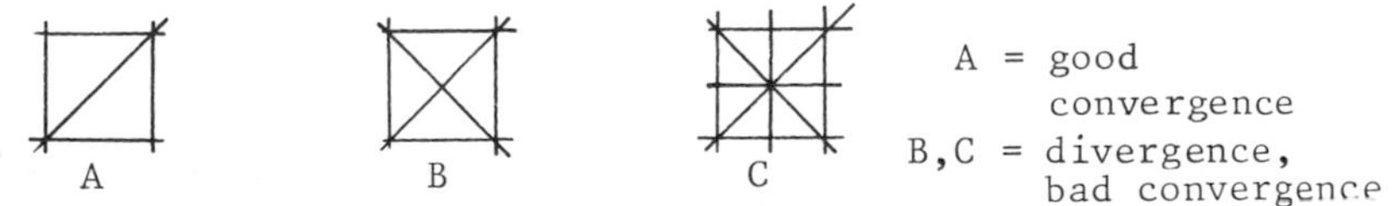

In the following we consider the behaviour of portions of the indicator α_h introduced in the last section :

$$\alpha_h^2 = \sum_{T \in T_h} \{(v_h,v_h)_T + (v_h,\Delta u_h)_T + (v_h,\Delta u_h)_T + (\Delta u_h,\Delta u_h)_T\} .$$

To test the deviation of the energy parts $(\Delta u_h,\Delta u_h)$, $a_h(u_h,u_h)$, we set ($u_h \neq 0$) :

$$\zeta_h = \sum_{T \in T_h} (-v_h,\Delta u_h)_T/(v_h,v_h) , \qquad \eta_h = \sum_{T \in T_h} (-v_h,\Delta u_h)/a_h(u_h,u_h) .$$

Note that for clamped plates the equality (see e.g. [4])

$$a(u_o,u_o) = (\Delta u_o,\Delta u_o)$$

holds, such that we can expect $\eta_h \to 1$ if u_h converges to u_o.

5.1. Clamped square plate under polynomial load

In $\overline{\Omega} = [0,1]\times[0,1]$ the function $u_o(x,y)=(x(x-1)y(y-1))^2$ is the solution of the clamped plate bending problem with $f = \Delta^2 u_o$. Table 1 shows the computed values of ζ_h and η_h , the latter being used (Fig. 1) to compute $\varepsilon_h = |1-\eta_h| \cdot 100$.

TABLE 1

h	meshtype A ζ_h	meshtype A η_h	meshtype C ζ_h	meshtype C η_h
1/4	1.0065	0.8747	1.0059	0.8914
1/6	0.9998	0.9312		
1/8	0.9989	0.9573	0.9864	0.9436

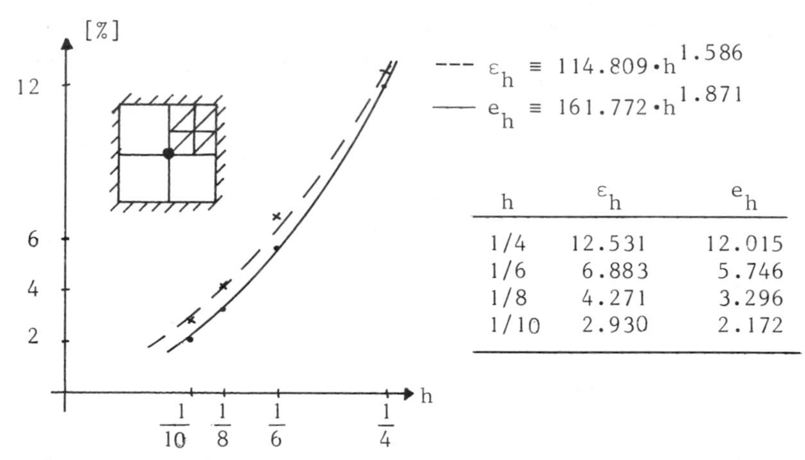

h	ε_h	e_h
1/4	12.531	12.015
1/6	6.883	5.746
1/8	4.271	3.296
1/10	2.930	2.172

Fig. 1. Relative error e_h of displacement at centre of square plate in % and behaviour of $|1-\eta_h| \cdot 100 = \varepsilon_h$, mesh A.

5.2 Clamped square plate under uniform load and simply supported square plate under uniform load

TABLE 2

(a) clamped (b) simply supported

	h	meshtype A ζ_h	meshtype A η_h	meshtype C ζ_h	meshtype C η_h
(a)	1/4	1.0069	0.8403	1.0011	0.8662
	1/6	0.9942	0.9102		
	1/8	0.9943	0.9448	0.9838	0.9374
(b)	1/4	1.0109	0.9506	0.9935	0.9249
	1/6	1.0036	0.9725		
	1/8	1.0010	0.9820	0.9852	0.9518

Tables 1 and 2 clearly show that the convergence behaviour of the indicators ζ_h and η_h reflects the theoretical predicted convergence behaviour, namely the dependence on the chosen mesh-type, of the cubic plate bending element.

5.3 *Simply supported plate under uniform load, L-shaped domain*

For the numerical treatment of a singularity problem we take the L-shaped domain $\overline{\Omega} = [0,1]\times[-1,1] \cup [-1,0]\times[0,1]$. The weak solution u_o of the corresponding boundary value problem possesses singularity parts of the form

$$r\cdot \ln r,\ r^{4/3}\ln^2 r\ ,\ r^{8/3}\cdot \ln^2 r$$

in a neighbourhood of (0,0) (cf. [9]). Computing local values

$$\zeta_h^T = (-v_h,\Delta u_h)_T/(v_h,v_h)_T$$

for each element $T \in T_h$ the indicator ζ_h can be used to "locate" the singularity. As is shown in the following figures, the mesh refinement causes the local values to change sign near the singularity point (0,0).

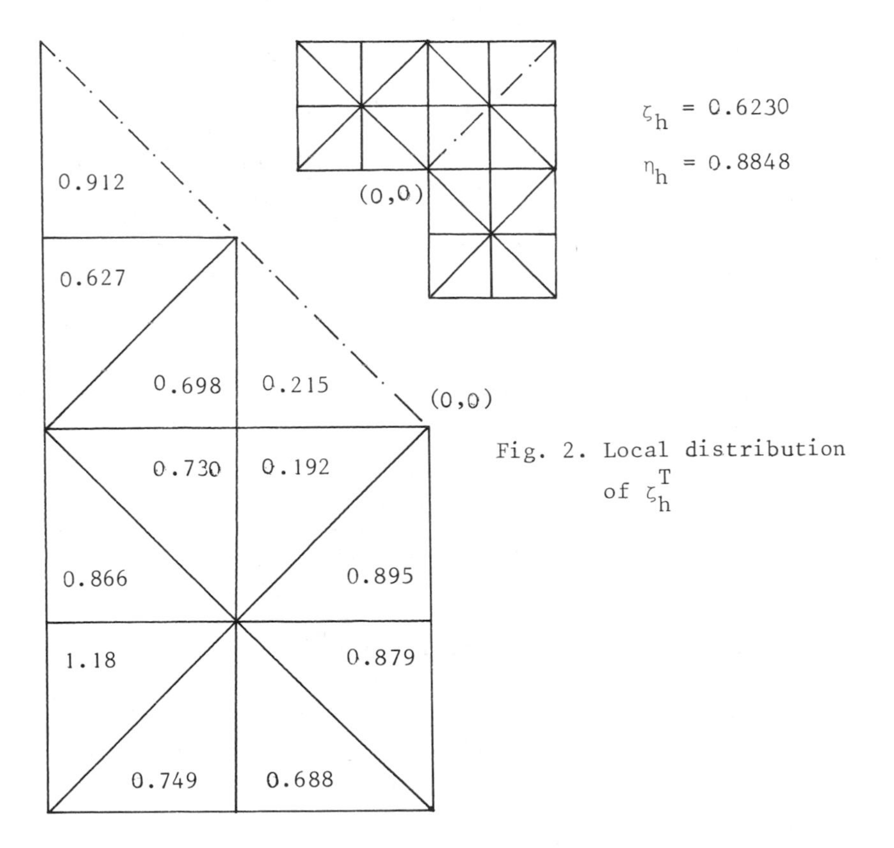

Fig. 2. Local distribution of ζ_h^T

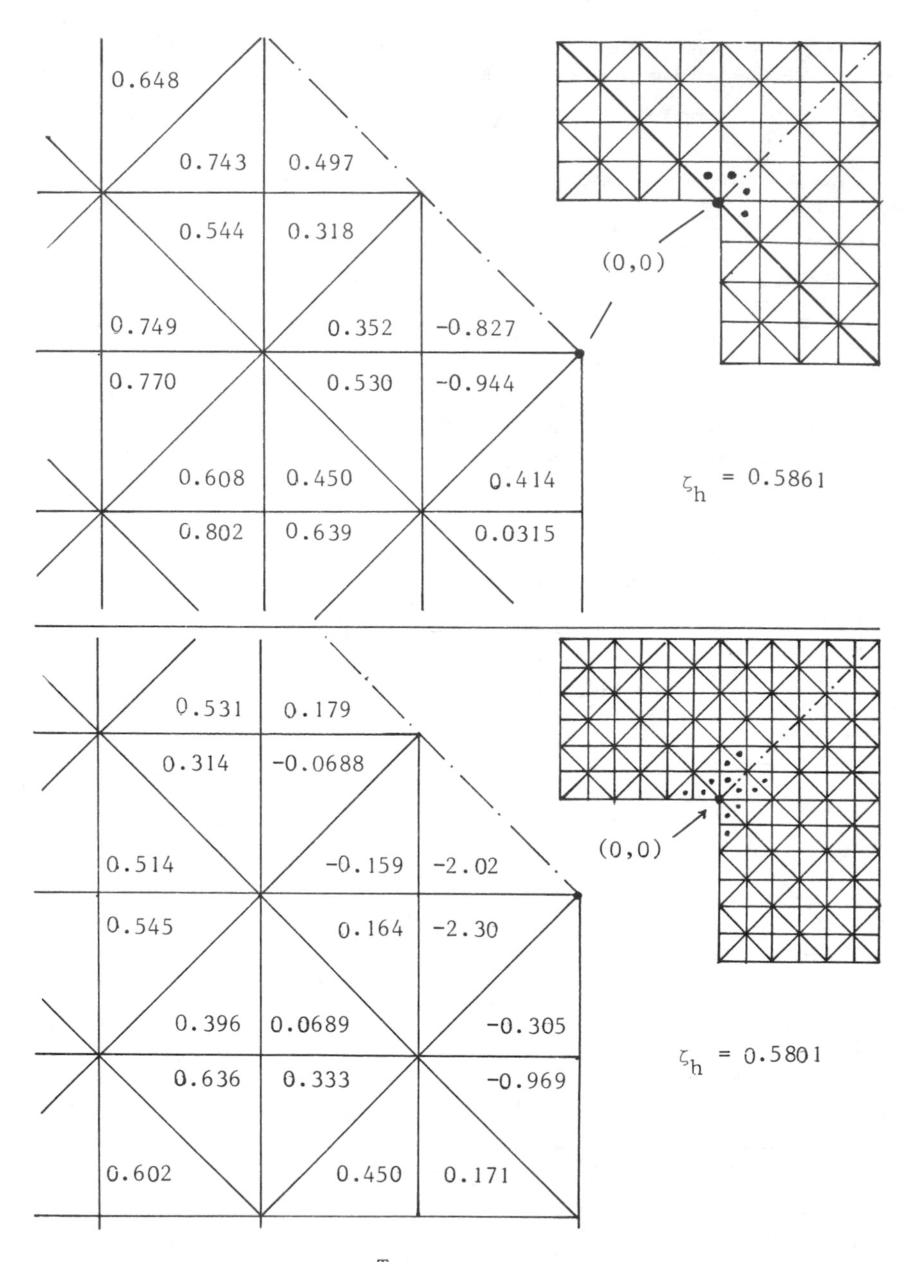

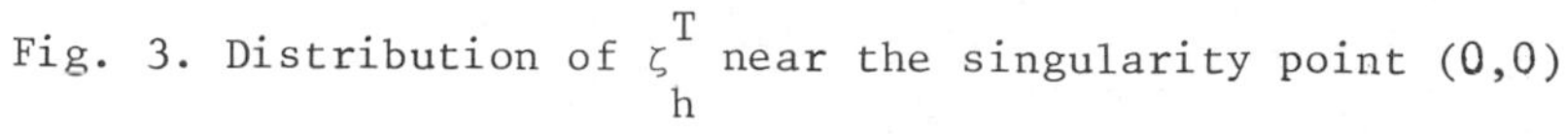

Fig. 3. Distribution of ζ_h^T near the singularity point (0,0)

REFERENCES

1. BAZELEY, G.P., CHEUNG, Y.K., IRONS, B.M. and ZIENKIEWICZ, O.C., Triangular elements in plate bending - conforming and non-conforming solutions. *Proc. 1st Conf. Matrix Methods in Structural Mechanics*, Wright-Patterson, Ohio (1965).

2. BLUM, H. and RANNACHER, R., On the boundary value problem of the biharmonic operator on domains with angular corners. *Preprint No. 353*, Universität Bonn (1980).

3. CIARLET, P.G., *The Finite Element Method for Elliptic Problems*. North Holland Publishing Co., Amsterdam (1978).

4. CIARLET, P.G., Numerical analysis of the finite element method. *Séminaire de mathématiques supérieures - été 1975*, Les presses de l'université de Montreal (1976).

5. ROHRBACH, R. and STEIN, E., Konvergenz und Fehlerabschätzung bei der Methode der finiten Elemente. *Report*, Lehrgebiet für Baumechanik, Universität Hannover (in preparation).

6. SCHOLZ, R., Approximation von Sattlepunkten mit Finiten Elementen. *Bonner Mathematische Schriften Nr. 89*, Bonn (1976).

7. STRANG, G. and FIX, G.J., *An Analysis of the Finite Element Method*. Prentice-Hall Inc., Englewood Cliffs, N.J. (1973).

8. LASCAUX, P. and LESAINT, P., Some nonconforming finite elements for the plate bending problem. *RAIRO Anal.Numer.* 9, 9-53 (1975).

9. MELZER, H. and RANNACHER, R., Spannungskonzentrationen in Eckpunkten der Kirchhoffschen Platte. *Der Bauingenieur* (1980) *and Preprint No.270, SFB 72*, University of Bonn, (1979).

LARGE DEFLECTION ANALYSIS OF IMPERFECT MINDLIN PLATES USING THE MODIFIED RIKS METHOD

E. Hinton and C.S. Lo

University College of Swansea.

1. INTRODUCTION

This paper is concerned with the post buckling and general large deflection analysis of shallow shells or plates with initial imperfections. Typically behaviour involving snap-through and snap-back must be accommodated. The equilibrium path shown in Figure 1 highlights some of the difficulties encountered in a post-buckling analysis. If loads are incremented then at point A the solution jumps suddenly to A' thereby missing a large portion of the equilibrium curve. When the displacements are incremented the solution jumps from B to B'.

There are two main aims in this paper:

(i) To develop finite element models which are suitable for use in the large deflection analysis of shallow shells and which are based on a Mindlin plate formulation with initial imperfections.

(ii) To use 'arc length' methods to obtain solutions involving snap-through and snap-back.

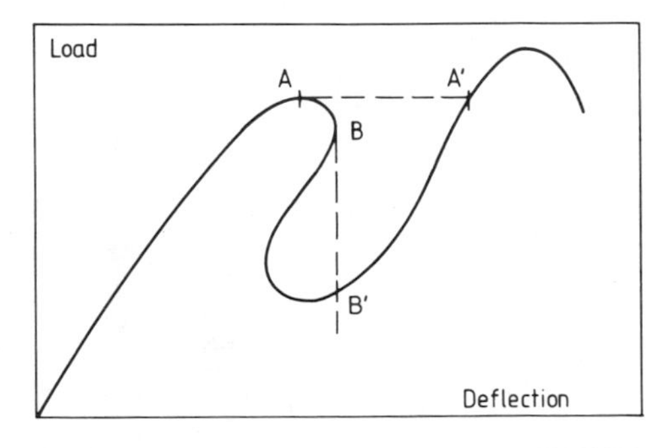

FIGURE 1 TYPICAL LARGE DEFLECTION PROBLEM

2. NONLINEAR SHALLOW SHELL FORMULATION

The nonlinear shallow shell formulation is based on Mindlin plate assumptions with a Marguerre shallow shell idealisation and the inclusion of Von Karman nonlinear strain terms (4,5). The initial imperfections are denoted by the symbol $\hat{w}$ and thus at any point within the plate the displacements $\bar{u}$, $\bar{v}$ and $\bar{w}$ in the x, y and z directions respectively may be expressed as

$$\begin{bmatrix} \bar{u}(x,y,z) \\ \bar{v}(x,y,z) \\ \bar{w}(x,y,z) \end{bmatrix} = \begin{bmatrix} u(x,y)-z\theta_x(x,y) \\ v(x,y)-z\theta_y(x,y) \\ w(x,y)+\hat{w}(x,y) \end{bmatrix} \quad (1)$$

where u, v and w are the displacements at the plate midsurface (xy-plane) in the x, y and z directions respectively. The normal rotations in the xz and yz planes are given as θ_x and θ_y respectively.

A total Lagrangian formulation is adopted and the Greens strains may be written as

$$\underline{E} = \underline{E}^L + \underline{E}^{NL} + \underline{E}^I \quad (2)$$

where the linear strains are

$$\underline{E}^L = [(u,_x-z\theta_{x,x}),(v,_y-z\theta_{y,y}),\ (u,_y+v,_x-z\theta_{x,y}-z\theta_{y,x}),\ (w,_x-\theta_x),\ (w,_y-\theta_y)]^T$$

the nonlinear strains are

$$\underline{E}^{NL} = [\tfrac{1}{2}(w,_x)^2,\ \tfrac{1}{2}(w,_y)^2,\ (w,_x w,_y),\ 0,0]^T$$

and the strains associated with the initial imperfections are

$$\underline{E}^I = [(w,_x\hat{w},_x),\ (w,_y\hat{w},_y),\ (w,_x\hat{w},_y+w,_y\hat{w},_x),0,0]^T$$

and $w,_x = \dfrac{\partial w}{\partial x}$ etc.

The elastic stress-strain relationships are given as

$$\underline{S} = \underline{D}\ \underline{E} \quad (3)$$

where $\underline{S} = [S_{xx},S_{yy},S_{xy},S_{xz},S_{yz}]^T$ is the vector of second Piola-Kirchhoff stresses and $\underline{D}$ is the matrix of elastic coefficients obtained from the three dimensional stress-strain relationships by making S_{zz}=0 and eliminating E_{zz}.

A hierarchical version of the Heterosis formulation is used to represent the main variables so that

$$[u,v,w,\theta_x,\theta_y]^T = \sum_{i=1}^{9} N_i[u_i,v_i,w_i,\theta_{xi},\theta_{yi}]^T \tag{4}$$

$$= \sum_{i=1}^{9} N_i \underline{d}_i$$

where a subscript i implies that the related variable is associated with node i. The initial imperfections are expressed in terms of the nodal values of the imperfections $\hat{w}_i$ as

$$\hat{w} = \sum_{i=1}^{8} N_i \hat{w}_i \tag{5}$$

Serendipity shape functions are used for nodes 1 to 8 and a bubble function is used at the central node 9. Thus the degrees of freedom at node 9 are hierarchical and to obtain the Heterosis element it is necessary to constrain w_9 to equal zero. Further details of the evaluation of the element stiffness matrices and applied and residual force vectors are given elsewhere (4,5).

3. SOLUTION PROCEDURE

An incremental-iterative solution procedure is adopted and at the ith. iteration of the Mth. load step the residual force vector may be expressed as

$$\underline{r}_M^{\,i} = \int_\Omega [\underline{B}_M^{\,i}]^T \, \underline{S}_M^{\,i} \, d\Omega - \underline{f}_M^{\,i} \tag{6}$$

where $\underline{B}_M^{\,i}$ is the incremental strain-displacement matrix and $\underline{f}_M^{\,i}$ is the applied load vector. Note that $\underline{f}_M^{\,i} = \lambda_M^{\,i}\underline{f}$ where $\lambda_M^{\,i}$ is the current loading parameter and $\underline{f}$ is the reference load vector. Usually a modified Newton-Raphson method is adopted in which the applied load level is held constant throughout the load step and the incremental displacements are found using the expression

$$\delta\underline{d}_r^{\,i} = \underline{K}^{-1} \, \underline{r}_M^{\,i} \tag{7}$$

where $\underline{K}$ is the tangential stiffness matrix which may be updated during the load step.

4. ARC LENGTH METHODS (MODIFIED RIKS)

In the present studies arc length or modified Riks methods are adopted in place of the modified Newton-Raphson methods. Instead of holding the applied load level constant during a load step, in an arc length method the load level is modified at each iteration so that the solution follows some specified path until convergence is attained as shown in Figure 2 (2,6,7). The path may be in a plane normal to the tangent at the beginning of the load step (normal plane) or the path may be an

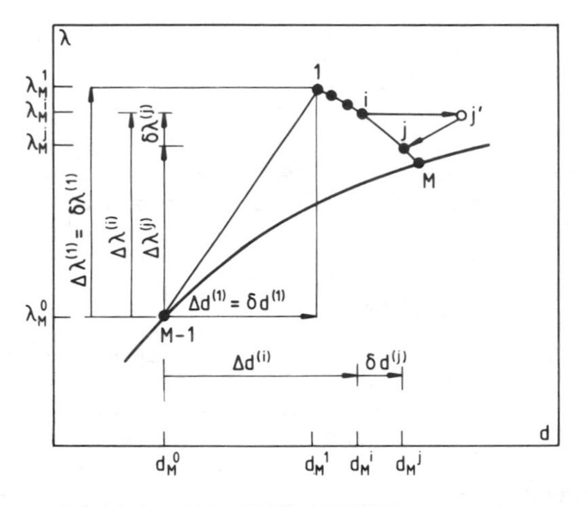

FIGURE 2 ARC LENGTH METHODS

arc of a sphere of given radius (spherical path) or have some other shape (updated normal plane). The length of the tangent in Figure 2 is called the arc length and has a magnitude equal to

$$\ell = \left(\left[\Delta \underline{d}^{(1)}\right]^T \Delta \underline{d}^{(1)} + (\Delta\lambda^{(1)})^2 \underline{f}^T \underline{f} \right)^{\frac{1}{2}} \tag{8}$$

Thus to initiate a solution it is necessary to specify either the arc length ℓ and then find $\Delta \underline{d}^{(1)}$ and $\Delta\lambda^{(1)}$ or specify $\Delta\lambda^{(1)}$ and find $\Delta d^{(1)}$ and ℓ. The increments in the loading parameters in the normal plane method may then be derived and written as

$$\delta\lambda^{(j)} = - \frac{\left[\Delta \underline{d}^{(1)}\right]^T \delta \underline{d}_r^{(j)}}{\left[\Delta \underline{d}^{(1)}\right]^T \delta \underline{d}_f^{(j)} + \Delta\lambda^{(1)}} \tag{9}$$

where $\delta \underline{d}_r^{(j)}$ and $\delta \underline{d}_f^{(j)}$ are increments in the displacement vectors due to the residual forces and the applied loads respectively. To find $\delta\lambda^{(j)}$ in the spherical path method a quadratic equation must be solved and details omitted for brevity are given in full elsewhere (2,6).

The arc length may be held constant throughout the analysis or it may be updated as suggested by Ramm (6) to equal

$$\ell_M = \ell_{M-1} (I_D / I_{M-1})^{\frac{1}{2}} \tag{10}$$

where I_D is the desired number of iterations and I_{M-1} and ℓ_{M-1}

are the actual number of iterations and the arc length respectively in the previous step.

For all load steps other than the first, the sign of the initial incremental loading parameter is chosen following an approach due to Bergan and Soreide (1) in which the sign follows that of the previous increment unless the determinant of the tangential stiffness matrix has changed sign in which case a sign reversal is applied.

5. SHALLOW ARCH EXAMPLE

The symmetrical snap-through of a shallow circular arch subjected to a concentrated point load is now considered using the arc length methods. The geometric and material properties of the arch are given in Figure 3(i). Two 9-noded, imperfect, Heterosis, Mindlin plate elements are used to model half of the arch which is treated as a narrow strip with initial imperfections.

Both the normal plane and spherical path methods give almost identical results when used in the analysis of the shallow arch example. The number of desired iterations I_D is 4 throughout the analysis in all cases. However, different values of the initial load step $\Delta\lambda^{(1)}$ and hence the basic arc length 1 are used. Figure 3(ii) compares the results obtained by Horrigmoe (3) with those obtained by the spherical path method with $\delta\lambda^{(1)} = 0.08$. Also shown for comparison is the solution obtained using the present Mindlin formulation in which the loads are incremented using Bergan's method (1). Excellent agreement is obtained but in this case, Bergan's method is more expensive.

Figure 3(iii) shows that even for a very large initial applied load of $\delta\lambda^{(1)} = 0.6$, the results are accurate. However, the analysis misses the whole part of the pre-buckled state and is thus undesirable.

Finally, it should be noted that in this example there is no difference in computer time between the two modified Riks methods.

6. CYLINDRICAL SHELL EXAMPLES

Two cylindrical shell examples are now considered in which snap-back and no-snap behaviour occurs. The basic geometry and material properties are shown in Figure 4. In both analyses the shell is treated as an imperfect Mindlin plate and by use of symmetry, one quarter of the shell is modelled by 4 Heterosis elements for the first problem and by 16 Heterosis elements in the second problem.

6.1 Snap-back behaviour

In this example, the shell has its straight edges hinged and its curved edges free. Its thickness is 6.35 mm. The shell exhibits snap-back behaviour when subjected to a concentrated

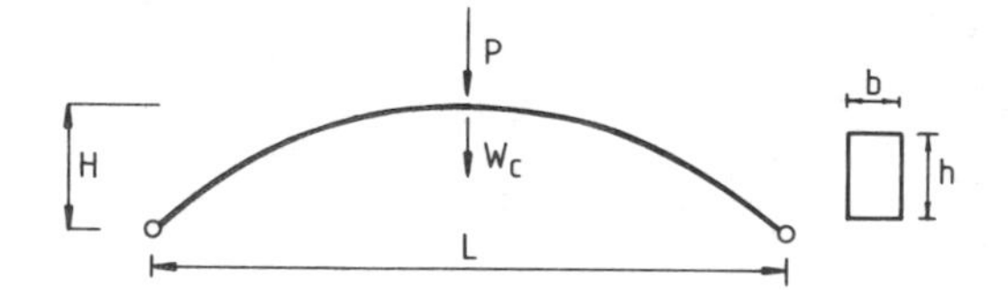

(i) Geometry and notation

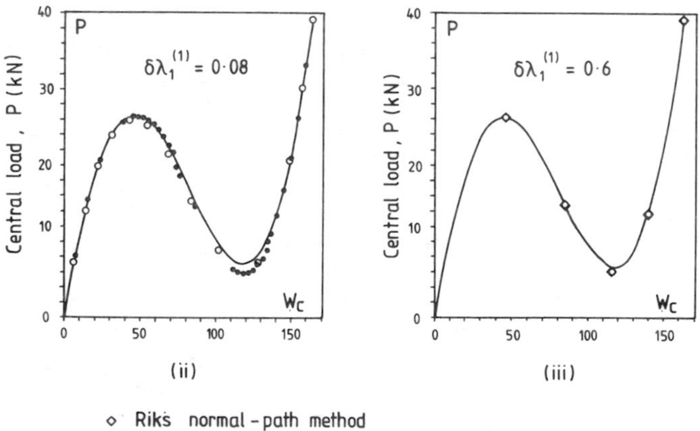

◇ Riks normal - path method
○ Riks spherical - path method
• Bergan's method
—— Horrigmoe [3]

H = 76·2 mm h = 79·2 mm L = 2540 mm E = 68·948 BN/mm²
b = 25·4 mm ν = 0 P = 102 kN

FIGURE 3 · SNAP - THROUGH BEHAVIOUR OF A HINGED CIRCULAR ARCH

point load at the centre. A comparison of results obtained using both the normal plane and the spherical path methods is shown in Figure 5. Also shown are results obtained by Ramm [6]. In this example, slightly greater accuracy is obtained using the normal plane method. However, the spherical path provides a cheaper solution.

6.2 No-snap behaviour

In this example, the shell is fully clamped on all sides and is subjected to a uniformly distributed load. The thickness of the shell is 3.175 mm. Figure 6 shows the results obtained using the spherical path method and also a modified Newton-

Raphson method in which the stiffness matrix is updated at the beginning of the load step. For comparison, results obtained by Crisfield (2) are also shown and excellent agreement is obtained. However, it should be mentioned that the modified Newton-Raphson method requires twice as many iterations as the spherical path method.

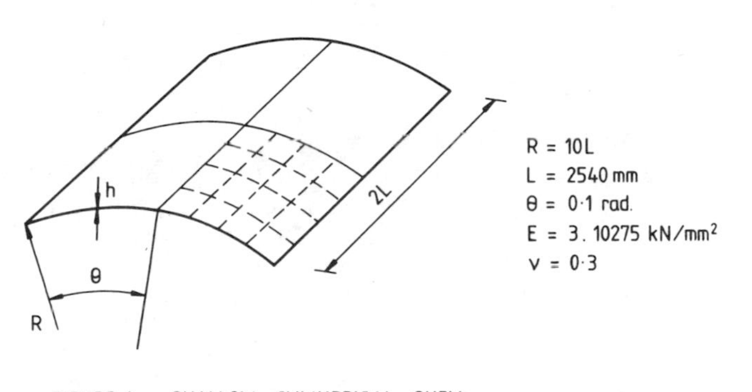

FIGURE 4 SHALLOW CYLINDRICAL SHELL

CASE (i) Central point load P, $h = 6{\cdot}35$ mm, straight edges hinged, curved edges free, 2×2 mesh

CASE (ii) Uniformly distributed load q, $h = 3{\cdot}175$ mm, all edges clamped, 4×4 mesh

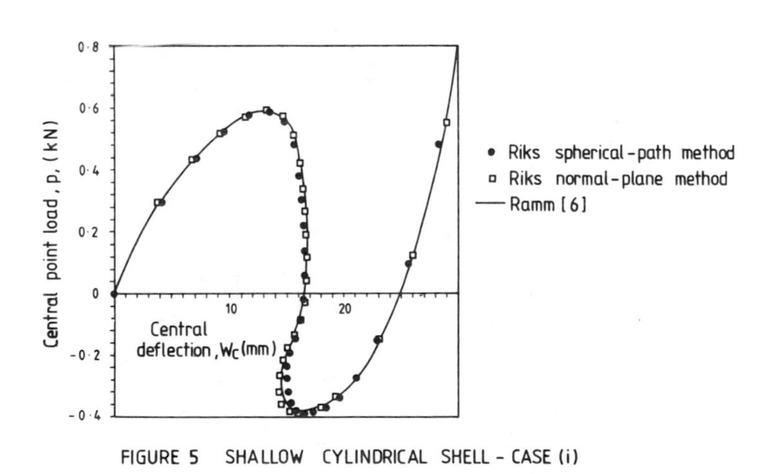

FIGURE 5 SHALLOW CYLINDRICAL SHELL - CASE (i)

REFERENCES

1. BERGAN, P.G. and SOREIDE, T.H., Solution of large displacement and instability problems using the current stiffness parameter. pp.647-669 of *Finite elements in nonlinear mechanics*. Tapir Press, Norway (1978).
2. CRISFIELD, M.A., A fast incremental/iterative solution procedure that handles "snap-through". *Computers and Structures 13*, 55-62 (1981).
3. HORRIGMOE, G., Finite element instability analysis of free-form shells. Report No. 77-2, University of Trondheim, Norway (1977).
4. PICA, A. and WOOD, R.D., Postbuckling behaviour of plates and shells using a Mindlin shallow shell formulation. *Computers and Structures 12*, 759-768 (1980).
5. PICA, A., WOOD, R.D. and HINTON, E., Finite element analysis of geometrically nonlinear behaviour using a Mindlin formulation. *Computers and Structures* ***11***, 203-215 (1980).
6. RAMM, E. Strategies for tracing nonlinear response near limit points. Europe-USA Workshop: *Nonlinear finite element analysis in structural mechanics*. Bochum, West Germany (1980), to be published.
7. RIKS, E. An incremental approach to the solution of snapping and buckling problems. *Int. J. Solids Structures*, 15, 524-551 (1979).

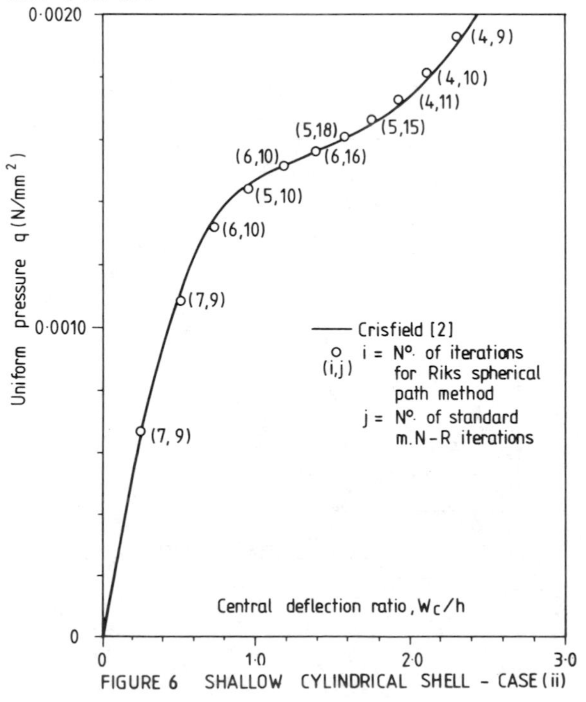

FIGURE 6 SHALLOW CYLINDRICAL SHELL - CASE (ii)

ON A CONFORMING, MINDLIN-TYPE PLATE ELEMENT

A. Tessler

University of California, Los Angeles.

1. INTRODUCTION

Extensive research efforts devoted to Mindlin-type element formulations are largely due to such attractive features as C° kinematic continuity requirement and the direct inclusion of transverse shear and rotary inertia effects. While the former provides a simple means of selecting the kinematic interpolations that ensure interelement continuities, the latter allows for a wide range of modeling situations extending from the shear-free Kirchhoff regime to the shear-weak plates of moderately thick, sandwich, and laminated construction.

Based on C° kinematic interpolations and exact integration of element energies, the generation of conforming elements proved disappointing: the bilinear four-node quadrilateral resulted in unrealistically stiff solutions; higher order serendipity elements encountered *locking* (overly stiff behavior) in the thin flexure regime; and only higher order Lagrange functions produced elements of acceptable accuracy [5]. Various nonconforming schemes were proposed to deal with these deficiencies and are discussed, for example, in [10]. Among these schemes the uniform and selective reduced integrations were addressed most extensively [1-3, 5, 6, 9].

Recently, in a formulation of Timoshenko beam elements [8], a special interpolation requirement was introduced leading to a successful hierarchy of conforming elements. This requirement, or interpolation strategy, called Interdependent Variable Interpolation (hereafter referred to as IVI), limited the choice of C° interpolations to those allowing the shear angles to obey the Kirchhoff constraints throughout the element domain.

In this paper, a Mindlin-type formulation is presented utilizing the IVI strategy for a rectangular plate element. The Mindlin equations are reviewed, and the IVI strategy akin to plate elements is discussed. The element matrices are consistently derived from Hamilton's principle where exact Gaussian quadrature is employed. The efficiency of the approach is verified by numerical tests for thin and moderately thick isotropic plates in static and free vibration analyses.

2. MINDLIN PLATE EQUATIONS

Let $\underline{x}$ denote the vector of Cartesian coordinates (x,y,z) and let the mid-plane of the plate lie in the xy-plane. The Cartesian displacements according to Mindlin theory [4] are

$$u(\underline{x},t) = z\theta_y(x,y,t), \quad v(\underline{x},t) = -z\theta_x(x,y,t), \quad w(\underline{x},t)=w(x,y,t) \tag{2.1}$$

where $w(x,y,t)$, $\theta_x(x,y,t)$ and $\theta_y(x,y,t)$ represent the weighted average transverse deflection and normal rotations, and t denotes time. The generalized in-plane (or bending) and transverse shear strain components are accordingly

$$\underline{\varepsilon}_b^T = z\left\{\frac{\partial\theta_y}{\partial x}, -\frac{\partial\theta_x}{\partial y}, \left(\frac{\partial\theta_y}{\partial y} - \frac{\partial\theta_x}{\partial x}\right)\right\} = z\{\chi_{xx},\chi_{yy},\chi_{xy}\} \tag{2.2}$$

$$\underline{\gamma}^T = \left\{\theta_y + \frac{\partial w}{\partial x}, -\theta_x + \frac{\partial w}{\partial y}\right\} = \{\gamma_{xz},\gamma_{yz}\} \tag{2.3}$$

where χ_{ij} and γ_{ij} are the curvatures and the shear angles, respectively. The components of the moment tensor and transverse shear force vector are related to their corresponding generalized strains as

$$\underline{M} = \begin{Bmatrix} M_{xx} \\ M_{yy} \\ M_{xy} \end{Bmatrix} = \begin{bmatrix} D_{11} & D_{12} & 0 \\ & D_{22} & 0 \\ \text{sym.} & & D_{66} \end{bmatrix} \begin{Bmatrix} \chi_{xx} \\ \chi_{yy} \\ \chi_{xy} \end{Bmatrix} = D_f\underline{\chi} \tag{2.4}$$

$$\underline{Q} = \begin{Bmatrix} Q_x \\ Q_y \end{Bmatrix} = \begin{bmatrix} D_{s_1} & 0 \\ 0 & D_{s_2} \end{bmatrix} \begin{Bmatrix} \gamma_{xz} \\ \gamma_{yz} \end{Bmatrix} = D_s\underline{\gamma} \tag{2.5}$$

where for an isotropic plate of thickness h, Young's modulus E, and Poisson's ratio ν, the flexural and shear rigidities take the form

$$D_{11} = D_{22} = D = \frac{Eh^3}{12(1-\nu^2)}, \quad D_{12} = \nu D_{11}, \quad D_{66} = \frac{h^3}{12}G \tag{2.6}$$

$$D_{s_1} = D_{s_2} = k^2hG, \quad G = \frac{E}{2(1+\nu)}, \quad k^2 = \frac{\pi^2}{12} \tag{2.7}$$

where k^2 is the shear correction factor. The sign convention for the kinematic variables and stress resultants is illustrated in Fig. 1.

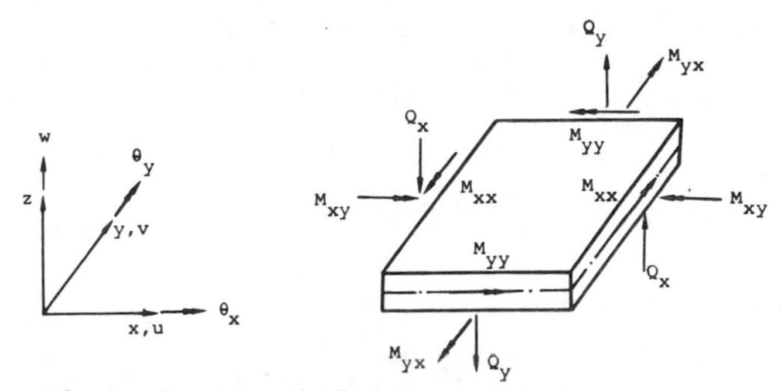

FIG. 1. Plate sign convention

Hamilton's variational equation embodying this theory can be written in the form

$$\delta \int_{t_1}^{t_2} L \, dt = \delta \int_{t_1}^{t_2} [T - U + W_E] dt = 0 \tag{2.8}$$

where the expressions for the kinetic energy T, strain energy U, and external work W_E are accordingly

$$T = \frac{1}{2} \iint_A \left[\frac{\rho h^3}{12} (\dot{\theta}_x^2 + \dot{\theta}_y^2) + \rho h \dot{w}^2 \right] dxdy \tag{2.9}$$

$$U = \frac{1}{2} \iint_A \underline{\chi}^T D_f \underline{\chi} \, dxdy + \frac{1}{2} \iint_A \underline{\gamma}^T D_s \underline{\gamma} \, dxdy \tag{2.10}$$

$$W_E = \iint_A wq dxdy \tag{2.11}$$

where dot (•) denotes differentiation with respect to time, ρ is the material density, q is the lateral load, and A is the plate's area. After integrating Eq. (2.8) by parts, the governing equations of motion and the corresponding natural boundary conditions are readily obtained.

3. IVI STRATEGY

Consider the strain energy expression (2.10) where the first and second integrals are the bending and shear energies. The shear energy can be regarded as a penalty term enforcing the Kirchhoff constraints in the thin-plate regime, i.e.,

$$\gamma_{xz} = \frac{\partial w}{\partial x} + \theta_y = 0 \text{ and } \gamma_{yz} = \frac{\partial w}{\partial y} - \theta_x = 0 \tag{3.1}$$

The IVI strategy [8] suggests that for the Kirchhoff constraints (3.1) to be satisfied everywhere in the element domain, the interpolations for the primary kinematic variables (w,θ_x,θ_y) must allow the following conditions to be met

$$\left.\begin{aligned} &\frac{\partial w}{\partial x} \text{ and } \theta_y \rightarrow \text{ same polynomial form} \\ &\frac{\partial w}{\partial y} \text{ and } \theta_x \rightarrow \text{ same polynomial form} \end{aligned}\right\} \tag{3.2}$$

These kinematic interpolations can be written in terms of the Lagrange functions as

$$w = \sum_{n=0}^{p+1} \sum_{k=0}^{p+1} a_{nk} F_{nk}, \quad \theta_x = \sum_{n=0}^{p+1} \sum_{k=0}^{p} b_{nk} F_{nk},$$

$$\theta_y = \sum_{n=0}^{p} \sum_{k=0}^{p+1} c_{nk} F_{nk}, \quad F_{nk} = x^n y^k, \quad p = 1,2,\ldots \tag{3.3}$$

where $a_{nk} = a_{nk}(w_i)$, $b_{nk} = b_{nk}(\theta_{xi})$, and $c_{nk} = c_{nk}(\theta_{yi})$ are the generalized coordinates comprised of the transverse deflection, w_i, and the normal rotation, θ_{xi} and θ_{yi}, nodal degrees of freedom (dof). With the assumptions (3.3), the shear angles (2.3) take the form

$$\gamma_{xz} = \sum_{n=0}^{p} \sum_{k=0}^{p+1} [(n+1)\, a_{(n+1)k} + c_{nk}] F_{nk} \tag{3.4}$$

$$\gamma_{yz} = \sum_{n=0}^{p+1} \sum_{k=0}^{p} [(k+1)\, a_{n(k+1)} - b_{nk}] F_{nk} \tag{3.5}$$

These shear angles may be considered to be of a *proper* kinematic form since each of their generalized coordinates is a linear combination of the deflection and normal rotation degrees of freedom. Therefore, the Kirchhoff state of zero shear is kinematically admissible throughout the element domain and is enforced as

$$(n+1)\, a_{(n+1)k} + c_{nk} = 0, \quad (n=0, \ldots p;\ k=0,\ldots, p+1) \tag{3.6}$$

$$(k+1)\, a_{n(k+1)} - b_{nk} = 0, \quad (n=0, \ldots, p+1;\ k=0,\ldots, p) \tag{3.7}$$

Hence, no *locking* should be expected in the thin-plate limit.

In this paper the focus of study is the rectangular element of order p=1. The element kinematic field is written directly in terms of its nodal dof as

$$w = \sum^{9} N_i^w w_i, \quad \theta_x = \sum^{6} N_i^{\theta_x} \theta_{xi}, \quad \theta_y = \sum^{6} N_i^{\theta_y} \theta_{yi} \tag{3.8}$$

where $N_i^w = O(\xi^2,\eta^2)$, $N_i^{\theta_x} = O(\xi^2,\eta)$, and $N_i^{\theta_y} = O(\xi,\eta^2)$ are the Lagrange interpolation functions expressed in terms of the parametric coordinates

$$\xi = (x-x_c)/a, \quad \eta = (y-y_c)/b, \quad \xi,\eta\in[-1,1] \tag{3.9}$$

By utilizing the displacement assumption (3.8) in Eqs. (2.9-11), and after employing exact integration (3x3 Gaussian quadrature) over the element area, the element stiffness matrix, mass matrix and load vector are consistently derived. The nodal configuration for this 21 dof element is depicted in Fig. 2.

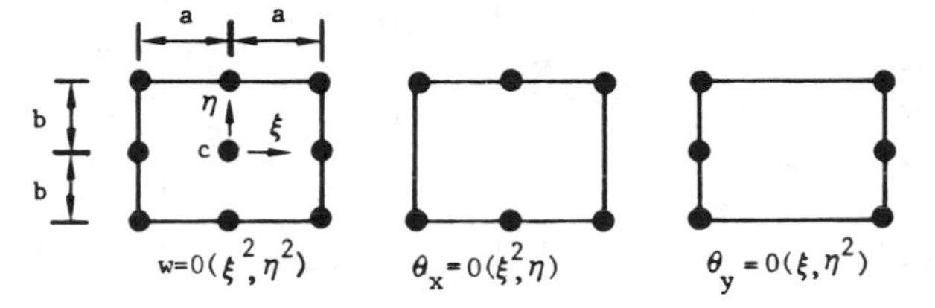

FIG. 2. IVI - 21 dof element nodal configuration

4. NUMERICAL RESULTS

Several tests were conducted for thin (plate's lateral dimension to thickness ratio $L/h=10^4$) and moderately thick ($L/h=10$) isotropic square plates in static and free vibration analyses. By taking advantage of biaxial symmetry, only one quadrant of a plate was modeled as shown in Fig. 3. The material properties were taken as

$$E = 10.92 \text{ msi}, \quad \nu = 0.3 \tag{4.1}$$

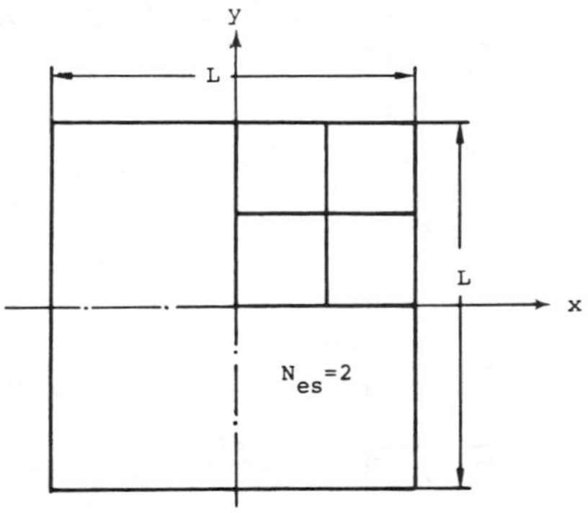

FIG. 3. Discretization of one quadrant of square plate

and the material density was chosen such that $L^2/\sqrt{\rho/Eh^2} = 1$.

The reliability of the element's kinematic behavior was verified via a spectral analysis of the stiffness matrix - three zero energy modes associated with rigid body motion were revealed.

Convergence studies were carried out for simply supported and clamped plates under central concentrated and uniformly distributed loads. The results for central deflection and central bending moment are illustrated in Fig. 4. It is seen that the element exhibits rapid convergence from below to the corresponding analytic solutions.

Free vibration analyses were conducted for simply supported plates. Rapid convergence of the fundamental frequency from above is demonstrated in Fig. 5. In Table 1, the present element frequencies for the three lowest symmetric modes are

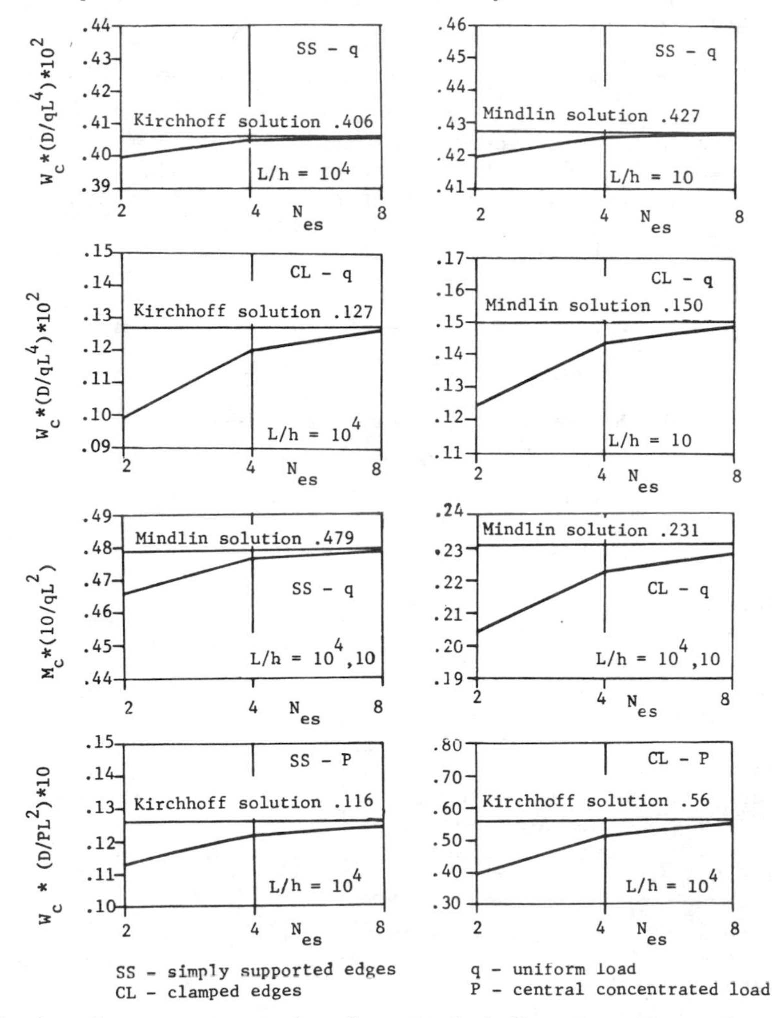

FIG. 4. Convergence study of central deflection, W_c, and central bending moment, M_c, for square plates.

compared with several analytic and nonconforming element solutions. The present element accuracy is observed to be quite competitive even with fewer degrees of freedom for this problem.

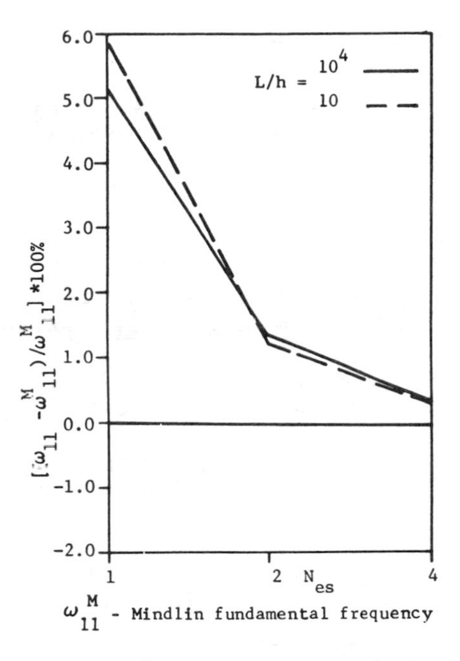

FIG. 5. Convergence of the fundamental frequency, ω_{11}, for simply supported square plates.

CONCLUSIONS

A conforming rectangular element has been derived utilizing Hamilton's principle in conjunction with exact Gaussian quadrature. The special interpolation strategy (IVI) employed herein allows the enforcement of the Kirchhoff constraints over the entire element domain. The element stiffness matrix is of a correct rank, and it suffers no deficiencies in the thin-plate limit.

The numerical examples demonstrated monotone convergence and competitive accuracy attainable with this element for thin and moderately thick plates in static and free vibration problems.

TABLE 1

Comparison of nondimensional frequencies $\lambda_{nm} = \omega_{nm}L^2\sqrt{\rho/Eh^2}$ *of a simply supported moderately thick square plate*
($\nu = 0.3$, L/h = 10)

Mode number	3-D linear elasticity solution [7]	Mindlin theory solution	Kirchhoff theory solution	MSC/NASTRAN QUAD4 $N_{es}=4$ (48 dof)	Reddy [6] $N_{es}=2$ (36 dof)	Present element $N_{es}=2$ (32 dof)
n m	λ^{3-D}_{nm}	λ^M_{nm}	λ^K_{nm}	λ_{nm}	λ_{nm}	λ_{nm}
1 1	5.780	5.769	5.973	5.854	5.920	5.842
1 3	25.867	25.734	29.867	28.760	28.133	29.840
3 3	42.724	42.383	53.868	47.130	49.723	46.990

ACKNOWLEDGEMENTS

The author wishes to express his gratitude to Professor Stanley B. Dong for his valuable recommendations during the course of this research.

REFERENCES

1. HUGHES, T.J.R., TAYLOR, R.L. and KANOKNUKULCHAI, W., A simple and efficient element for plate bending. *Int. J. Numer. Meth. Eng.* 11, 1529-1543 (1977).

2. HUGHES, T.J.R. and COHEN, M., The 'Heterosis' finite element for plate bending. *Computers and Structures* 9 445-450 (1978).

3. MACNEAL, R.H., A simple quadrilateral shell element. *Computers and Structures* 8, 175-183 (1978).

4. MINDLIN, R.D., Influence of rotatory inertia and shear on flexural motion of isotropic elastic plates. *J. Appl. Mech.* 18, 31-38 (1951).

5. PUGH, E.D.L., HINTON, E. and ZIENKIEWICZ, O.C., A study of quadrilateral plate bending elements with 'reduced' integration. *Int. J. Numer. Meth. Eng.* 12, 1059-1079 (1978).

6. REDDY, J.N., Free vibration of antisymmetric, angle-ply laminated plates including transverse shear deformation by the finite element method. *J. Sound and Vibration* 66 (4), 565-576 (1979).

7. SRINVAS, S., JOBA RAO, C.V. and RAO, A.K., An exact analysis for vibration of simply supported homogeneous and laminated thick rectangular plates. *J. Sound and Vibration* 12 (2), 187-199 (1970).

8. TESSLER, A. and DONG, S.B., On a hierarchy of conforming Timoshenko beam elements. *Computers and Structures* (to appear).

9. ZIENKIEWICZ, O.C., TAYLOR, R.L. and TOO, J.M., Reduced integration techniques in general analysis of plates and shells. *Int. J. Numer. Meth. Eng.* 3, 275-290 (1971).

10. ZIENKIEWICZ, O.C., *The Finite Element Method*. McGraw-Hill, London (1977).

THE LINEAR TRIANGULAR BENDING ELEMENT

*Thomas J.R. Hughes and +Robert L. Taylor

*Stanford University, +University of California, Berkeley

1. INTRODUCTION

The appeal of simple elements for plate and shell applications seems to be increasing [1-3,5,6,8-15,21,22]. Perhaps the main motivation for this is the current interest in nonlinear analysis, for which simpler elements often prove advantageous compared with more expensive, higher-order elements. Unfortunately, simple and effective elements for plate and shell applications have not been so easy to come by. Only in recent years has significant progress been made on this topic.

In [13], Hughes and Tezduyar systematically explored an idea first used by MacNeal [15] in the development of bending elements. It was pointed out in [13] that MacNeal's approach provides a new link between reduced/selective integration techniques and order-of-accuracy concepts. (Malkus [17] has been developing a mathematical analysis for the incompressible continuum problem based upon similar ideas.) In [13] these ideas were used to construct a new version of the bilinear quadrilateral element which behaves well and does not suffer any ostensible defect, such as rank deficiency. It was remarked upon in [13] that a similar approach could be used to generate a linear triangular element and that this might result in the simplest effective bending element yet developed. It is the purpose of this paper to explore this possibility.

In Section 2, we articulate the "Kirchhoff-mode criteria" for designing plate elements based upon Mindlin theory [18] (i.e., "thick-plate theory"). In Section 3 we describe the formulation of the linear triangular bending element and in Section 4 we consider its implementation. Numerical examples presented in Section 5 indicate that the behavior of the element is only satisfactory in the cross-diagonal mesh configuration. The improved behavior is reminiscent of the analogous situation observed for linear triangles in incompressible applications [19]. It is noted that 1-point centroidal quadrature may be used to economically calculate the element stiffness and that this does not engender any global rank deficiency in an assemblage of two or more elements. The results obtained, which are limited in scope, tentatively suggest that the linear triangular element may be competitive in certain situations.

2. KIRCHHOFF-MODE CRITERIA

The material presented in this section is abstracted from [13], which should be consulted for a more detailed presentation.

Thin plate behavior is governed by the classical Poisson-Kirchhoff theory. In this limiting situation the face rotations become equal to the slopes of the transverse displacement field. Analytically, the rotations are no longer independent kinematic variables, but become the derivatives of the transverse displacement field. To assess the ability of Mindlin-type plate elements to correctly handle limiting thin-plate behavior, we shall examine the Mindlin elements with respect to the modes of deformation emanating from the classical theory.

To be more precise, let us define a *Kirchhoff mode* by the relation

$$\theta_\alpha = w_{,\alpha} \tag{2.1}$$

where w is a given transverse displacement; θ_α is the x_α-rotation, $\alpha = 1,2$; and a comma is used to denote partial differentiation (e.g., $w_{,\alpha} = \partial w/\partial x_\alpha$).

A *Kirchhoff mode of order* m will be one in which w is taken to be a complete m^{th}-order polynomial, $P_m(x_1,x_2)$. An example of a complete polynomial is the quadratic polynomial

$$P_2(x_1,x_2) = C_1+C_2x_1+C_3x_2+C_4x_1^2+C_5x_1x_2+C_6x_2^2 \tag{2.2}$$

where the C's are arbitrary coefficients.

Criterion 1: As a measure of the effectiveness of an element, we shall ask what order Kirchhoff mode is the element able to exactly interpolate. The higher the order, the greater the ability of the element to perform accurately in the thin-plate limit.

Criterion 2: A weakened version of the above criterion, which accommodates reduced/selective integration and other procedures, asks for what order Kirchhoff mode is the strain energy calculated exactly. This is the form of the criterion originally proposed by MacNeal [15] and employed by Parisch [20].

Note that criterion 1 implies criterion 2.

Posing the criteria in terms of complete polynomials links up with order-of-accuracy concepts and may be useful in mathematical error analysis.

Criterion 1 has the advantage that it suggests element interpolation schemes which may be effective. In this regard, it is immediately apparent that, according to criterion 1, ideal interpolations may be devised by assuming w to be a polynomial one order higher than that assumed for the θ_α's. A variety of interpolatory schemes were suggested in [13] for beams and quadrilateral and triangular plates. The triangular family illustrated in Fig. 1 appears unique among two-dimensional element families in that the functions which constitute the rotational interpolations are obtained exactly from the derivatives of displacement -- no more, no less. The first element of this family is the conceptual basis of the linear triangular bonding element described in the following section.

3. THE LINEAR TRIANGULAR BENDING ELEMENT

The starting point is the straight-edged triangular element in which transverse displacement is interpolated via six-node quadratic shape functions and rotations are interpolated via 3-node linear shape functions (see Fig. 1). This element achieves

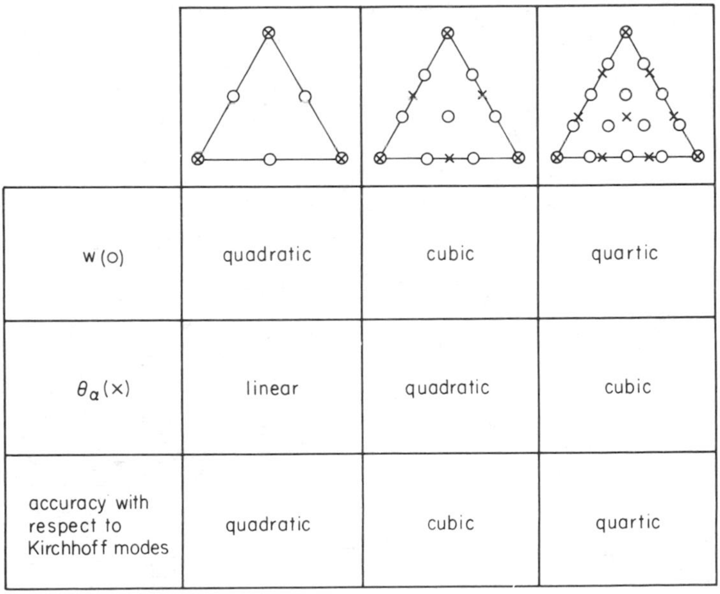

w (o)	quadratic	cubic	quartic
θ_α(x)	linear	quadratic	cubic
accuracy with respect to Kirchhoff modes	quadratic	cubic	quartic

Fig. 1 Triangular plate elements derived from the Kirchhoff-mode criterion.

quadratic accuracy with respect to Kirchhoff modes. The idea is to calculate the transverse shear strains in a special way independent of the midside displacement degrees-of-freedom. In this way, the element stiffness senses only the vertex transverse displacement degrees-of-freedom and, consequently, three-node linear shape functions may be used in place of the six-node quadratic shape functions in formulating the element arrays. Examination of the interpolations reveals that the midpoints of the sides are locations at which the transverse shear strain components parallel to the sides are independent of the aforementioned nodal values. These three scalar values will be used to define the transverse shear strain variation throughout the element. The details of the procedure follow.

3.1 *Definition of Element Transverse Shear Strains*

Geometric and kinematic data are defined in Fig. 2. Note that the direction vectors have unit length (e.g., $\|\underset{\sim}{e}_{11}\| = 1$, etc.). Let w_a and $\underset{\sim}{\theta}_a$ denote the transverse displacement and rotation vector, respectively, associated with node a. Throughout, a subscript b is defined by the relation below:

a	b
1	2
2	3
3	1

(3.1)

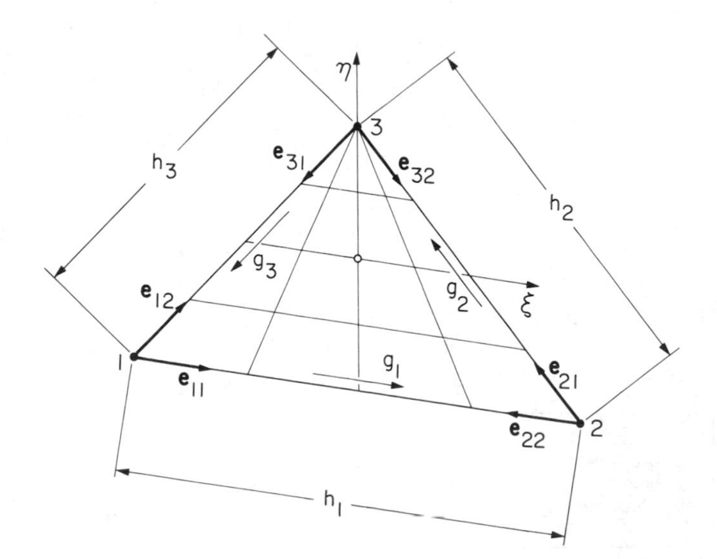

Fig. 2 Geometric and kinematic data for the three-node linear triangular element.

The definition of the element shear strains may be facilitated by the following steps:

1. For each element side define a shear strain component, located at the midpoint, in a direction parallel to the side, viz.

$$g_a = (w_b - w_a)/h_a - \underset{\sim}{e}_{a1} \cdot (\underset{\sim}{\theta}_b + \underset{\sim}{\theta}_a)/2 \quad . \tag{3.2}$$

2. For each node, define a shear strain vector (see Fig. 3 for a geometric interpretation of this process):

$$\underset{\sim}{\gamma}_b = \gamma_{b1} \underset{\sim}{e}_{b1} + \gamma_{b2} \underset{\sim}{e}_{b2} \tag{3.3}$$

$$\gamma_{b2} = (1-\alpha_b^2)^{-1} (g_{b2} - g_{b1} \alpha_b) \tag{3.4}$$

$$\gamma_{b1} = (1-\alpha_b^2)^{-1} (g_{b1} - g_{b2} \alpha_b) \tag{3.5}$$

$$\alpha_b = \underset{\sim}{e}_{b1} \cdot \underset{\sim}{e}_{b2} \tag{3.6}$$

$$g_{b1} = g_b \tag{3.7}$$

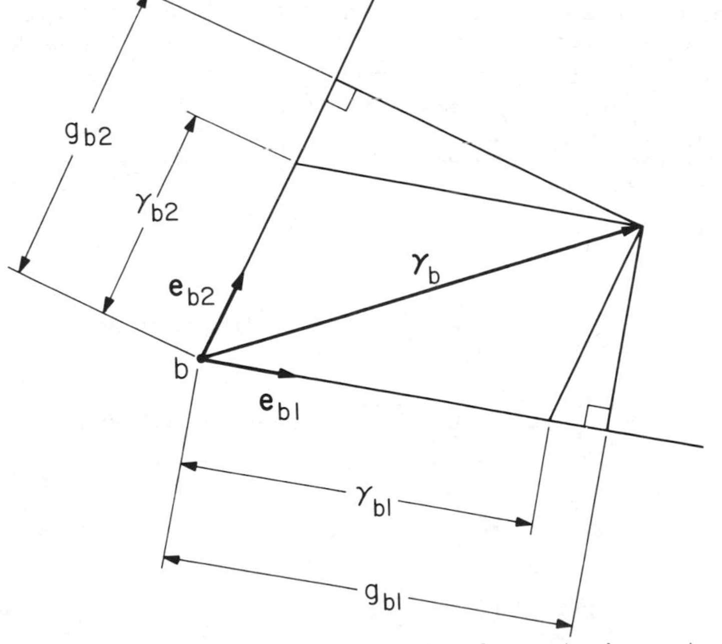

Fig. 3 Definition of nodal transverse shear strain vector.

$$g_{b2} = -g_a \quad . \tag{3.8}$$

3. Interpolate the nodal values by way of the standard linear shape functions (N_a's):

$$\chi = \sum_{a=1}^{3} N_a \chi_a \quad . \tag{3.9}$$

Remarks:

1. If the nodal transverse displacements and rotations are specified to consistently interpolate a constant transverse shear strain field, say $\bar{\gamma}$, then the preceding steps will result in $\chi = \bar{\chi}$. *That is, constant transverse shear deformation modes are exactly representable.*

2. By virtue of the way the transverse shear strains are interpolated, quadratic accuracy with respect to Kirchhoff modes is attained in the sense of criterion 2.

4. IMPLEMENTATION

In this section we consider the implementation of the linear, triangular, Mindlin plate element. It suffices for the present purposes to consider the simple case of a homogeneous, isotropic, linearly-elastic plate of constant thickness t.

Let A^e and s^e denote the area and boundary, respectively, of a typical element.

The *element stiffness matrix*, $\underset{\sim}{k}^e$, may be defined as follows:

$$\underset{\sim}{k}^e = \underset{\sim}{k}^e_b + \underset{\sim}{k}^e_s \tag{4.1}$$

$$\underset{\sim}{k}^e_b = \int_{A^e} \underset{\sim}{B}^{b^T} \underset{\sim}{D}^b \underset{\sim}{B}^b \, dA \qquad \text{bending stiffness} \tag{4.2}$$

$$\underset{\sim}{k}^e_s = \int_{A^e} \bar{\underset{\sim}{B}}^{s^T} \underset{\sim}{D}^s \bar{\underset{\sim}{B}}^s \, dA \qquad \text{shear stiffness} \tag{4.3}$$

$$\underset{\sim}{B}^b = [\underset{\sim}{B}^b_1 \; \underset{\sim}{B}^b_2 \; \underset{\sim}{B}^b_3] \tag{4.4}$$

$$\bar{\underset{\sim}{B}}^s = [\bar{\underset{\sim}{B}}^s_1 \; \bar{\underset{\sim}{B}}^s_2 \; \bar{\underset{\sim}{B}}^s_3] \tag{4.5}$$

$$\underset{\sim}{B}_a^b = \begin{bmatrix} 0 & N_{a,1} & 0 \\ 0 & 0 & N_{a,2} \\ 0 & N_{a,2} & N_{a,1} \end{bmatrix} \qquad 1 \le a \le 3 \qquad (4.6)$$

The definition of $\overline{\underset{\sim}{B}}_a^s$ is the essential ingredient in the development of an effective element. In the "normal" case, $\overline{\underset{\sim}{B}}_a^s = \underset{\sim}{B}_a^s$, which is defined by

$$\underset{\sim}{B}_a^s = \begin{bmatrix} N_{a,1} & -N_a & 0 \\ N_{a,2} & 0 & -N_a \end{bmatrix} \qquad 1 \le a \le 3 \ . \quad (4.7)$$

Without further modifications, this generally leads to an excessively stiff element.

In the present formulation the reduced/selective-integration effect is accounted for directly in the definition of $\overline{\underset{\sim}{B}}_a^s$ [7,9]. For the transverse shear strain interpolations derived in the previous section, $\overline{\underset{\sim}{B}}_a^s$ takes on the following form [recall the relation between subscripts a and b, see (3.1)]:

$$\overline{\underset{\sim}{B}}_b^s = [\overline{\underset{\sim}{B}}_{b1}^s \ \overline{\underset{\sim}{B}}_{b2}^s \ \overline{\underset{\sim}{B}}_{b3}^s] \qquad 1 \le b \le 3 \qquad (4.8)$$

$$\overline{\underset{\sim}{B}}_{b1}^s = h_a^{-1}\underset{\sim}{G}_a - h_b^{-1}\underset{\sim}{G}_b \qquad (4.9)$$

$$\overline{\underset{\sim}{B}}_{b2}^s = (e_{b2}^1\underset{\sim}{G}_a - e_{b1}^1\underset{\sim}{G}_b)/2 \qquad (4.10)$$

$$\overline{\underset{\sim}{B}}_{b3}^s = (e_{b2}^2\underset{\sim}{G}_a - e_{b1}^2\underset{\sim}{G}_b)/2 \qquad (4.11)$$

$$\underset{\sim}{G}_a = (1-\alpha_a^2)^{-1}N_a(\underset{\sim}{e}_{a1}-\alpha_a\underset{\sim}{e}_{a2})-(1-\alpha_b^2)^{-1}N_b(\underset{\sim}{e}_{b2}-\alpha_b\underset{\sim}{e}_{b1}) \qquad (4.12)$$

$$\underset{\sim}{e}_{b1} = \begin{Bmatrix} e_{b1}^1 \\ e_{b1}^2 \end{Bmatrix} \ , \ \text{etc.} \qquad (4.13)$$

The matrices $\underset{\sim}{D}^b$ and $\underset{\sim}{D}^s$, for the isotropic, linearly-elastic, constant thickness case, take on the following forms (resp.):

$$\underset{\sim}{D}^b = \frac{t^3}{12} \begin{bmatrix} 2\mu + \bar{\lambda} & \bar{\lambda} & 0 \\ & 2\mu + \bar{\lambda} & 0 \\ \text{symm.} & & \mu \end{bmatrix} \tag{4.14}$$

and

$$\underset{\sim}{D}^s = \kappa t \mu \begin{bmatrix} 1 & 0 \\ 0 & 1 \end{bmatrix} \tag{4.15}$$

where $\bar{\lambda} = 2\lambda\mu/(\lambda+2\mu)$, λ and μ are the Lamé parameters, and κ is a "shear correction factor," which is taken to be 5/6 throughout.

The *external load vector*, $\underset{\sim}{f}^e$, is given by

$$\underset{\sim}{f}^e = \{f_I^e\} \tag{4.16}$$

$$f_I^e = \begin{cases} \int_{A^e} N_a F dA + \int_{s^e \cap s_2} N_a Q ds , & I=3a-2 , \quad 1\le a\le 3 \\ \\ -\int_{A^e} N_a C_\alpha dA - \int_{s^e \cap s_2} N_a M_\alpha ds , & I=3a+\alpha-2, \quad 1\le a\le 3, \quad \alpha=1,2 \end{cases} \tag{4.17}$$

where F is the total applied transverse force per unit area, C_α is the total applied couple per unit area, Q is the applied shear force, M_α is the applied boundary moment, and s_2 is the portion of the plate boundary upon which forces and moments are prescribed.

The element stress resultants may be obtained from the following relations:

$$\begin{Bmatrix} m_{xx} \\ m_{yy} \\ m_{xy} \end{Bmatrix} = -\underset{\sim}{D}^b \underset{\sim}{B}^b \underset{\sim}{d}^e \qquad \text{bending moments} \tag{4.18}$$

$$\begin{Bmatrix} q_x \\ q_y \end{Bmatrix} = \underset{\sim}{D}^s \underset{\sim}{\bar{B}}^s \underset{\sim}{d}^e \qquad \text{shear resultants} \tag{4.19}$$

where

$$\underset{\sim}{d}^e = \{d^e_I\} \quad \text{element displacement vector} \tag{4.20}$$

$$d^e_I = \begin{cases} w_a \,, & I=3a-2 \quad, \; 1\le a\le 3 \\ \theta_{\alpha a}\,, & I=3a+\alpha-2 \;, \; 1\le a\le 3 \;, \; \alpha = 1,2 \end{cases} \tag{4.21}$$

$$\underset{\sim}{\theta}_a = \begin{Bmatrix} \theta_{1a} \\ \theta_{2a} \end{Bmatrix} \quad . \tag{4.22}$$

Remark:

Generalization of the formulation to fully-nonlinear analysis is straightforward by way of the procedures described in [7,9].

5. NUMERICAL EXAMPLES

The calculations were performed at the University of California, Berkeley, on a VAX Computer in double precision (64 bits per floating point word). Data employed in the calculations are given as follows: Poisson's ratio = .3; Young's modulus = 10.92×10^5; and the plate thickness = 0.1.

In the context of Mindlin theory, two interpretations of the classical simply-supported boundary condition are possible: SS_1, in which only the transverse displacement is set to zero; and SS_2, in which the transverse displacement and tangential rotation are set to zero. In applications to thin plates, SS_1 is generally preferable since it leads to convergent results when polygonal approximations of curved boundaries are employed. Nevertheless SS_2 corresponds to the simply-supported condition of classical thin plate theory and may be safely employed for the analysis of polygonal, and in particular rectangular, plates. See [8] for a discussion of the treatment of simply supported boundary conditions and references to pertinent literature.

We tried two different quadrature treatments of the linear triangle: 1-point centroidal quadrature; and 2×2 Gauss quadrature in the ξ, η-system (see Fig. 2) which is exact in the present circumstances. (Note that centroidal quadrature is not the same as 1-point Gauss quadrature.) A single triangle possesses correct rank (i.e., 6) when integrated exactly, but possesses one spurious mechanism when under-integrated by 1-point centroidal quadrature. However, in any assemblage of two or more elements, the spurious mechanism disappears, and thus is of no consequence in practical computing.

Calculations were performed for simply-supported, thin, square and circular plates subjected to uniform loads. The edge length of the square plate and radius of the circular plate were taken to be 10.0 and 5.0, respectively. In each case, due to symmetry, only one quadrant of the plate was discretized. Meshes are depicted in Figs. 4 and 5. Note that the cross-diagonal meshes involve approximately twice the number of unknowns as the other mesh types. The SS_1 boundary condition was used in each case. The quadrature treatment of the consistent load was the same as used for the stiffness.

Numerical results for the cases studied are presented in Tables 1 and 2. Moment results were obtained at the centroids of the triangular elements nearest the plate center. When the vertices of two triangles coincided with the plate center, moments were averaged over the two elements. It is immediately apparent from Table 1 that exact quadrature behaves very poorly for mesh types A and B.

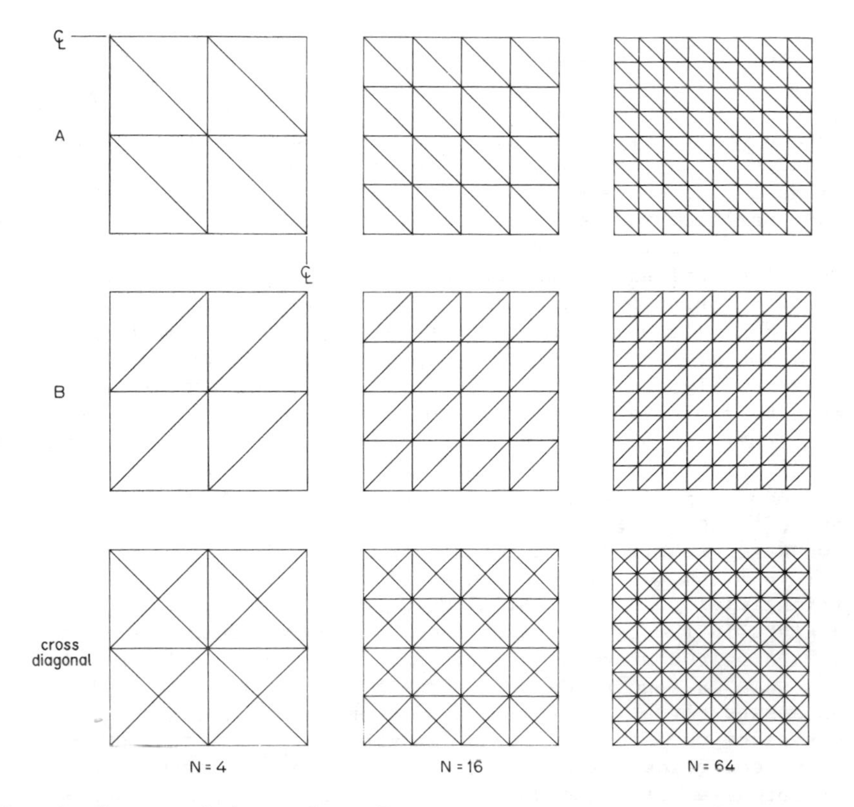

Fig. 4 Square plate meshes. Due to symmetry only one quadrant is discretized.

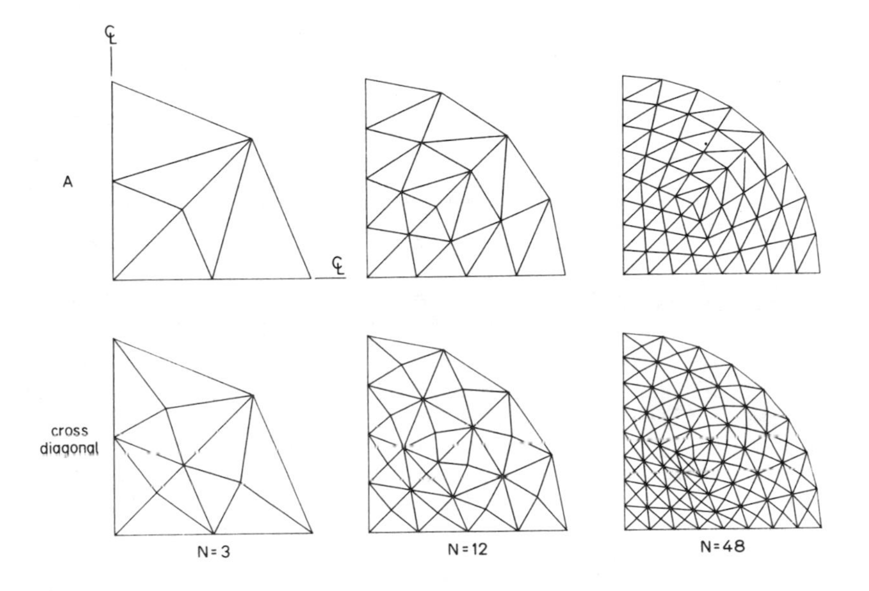

Fig. 5 Circular plate meshes. Due to symmetry only one quadrant is discretized.

TABLE 1

Center Displacement and Moment For Simply-supported, Thin, Square Plate Subjected to Uniform Load

a. Displacement

N	1-pt. centroidal quadrature A	B	cross diagonal	Exact quadrature A	B	cross diagonal
4	.681	.883	.912	.681	.002	.912
16	.784	.989	.978	.773	.036	.978
64	.947	.999	.994	.827	.363	.994

b. Moments

N	1-pt. centroidal quadrature A	B	cross diagonal	Exact quadrature A	B	cross diagonal
4	.555	.904	.919	.555	.002	.919
16	.564	1.127	.979	.539	.036	.979
64	.835	1.098	.996	.580	.386	.996

TABLE 2

Center Displacement and Moment For Simply-supported, Thin, Circular Plate Subjected to Uniform Load

Results presented are for 1-point centroidal quadrature.

	A		cross diagonal	
N	displacement	moment	displacement	moment
3	.703	.576	.927	.885
12	.912	.878	.981	.976
48	.948	.975	.976	.986

On the other hand, for the cross-diagonal meshes, exact quadrature yields quite satisfactory results. One-point centroidal quadrature represents some improvement for mesh types A and B, but still yields moments which are not satisfactory. Note that 1-point quadrature for the cross-diagonal pattern yields identical results to those obtained by exact quadrature.

It is quite clear from the results of Table 1 that only the cross-diagonal meshes yield consistently satisfactory results. Furthermore, there is no advantage to exact quadrature. The economic superiority of 1-point quadrature makes it the obvious choice for practical use.

The results for the circular plate, presented in Table 2, are satisfactory for both mesh patterns studied.

Results for the cross-diagonal meshes are compared in Figs. 6 and 7 with results obtained for a four-node bilinear quadrilateral element presented in [13]. This element is felt to be a relatively good performer, consequently the comparison serves as a benchmark for the triangular element. The SS_2 boundary condition was used with the quadrilateral element, however, this is felt to have little effect on the results. "Center moments" for the quadrilateral are computed at the Gauss point (2×2 rule) nearest the center of the plate (see [13] for further details). Overall, the quadrilateral results are somewhat superior to those for the triangles, although the latter behaves competitively.

Discussion

MacNeal [16] has pointed out to us that a linear triangular bending element, TRIA3, is currently available in the MSC version of NASTRAN. This element appears to be identical to the exactly integrated triangle described herein, except for the introduction of a "residual bending flexibility" [15].

Garnet et al. [5] have also reported upon a linear triangle in which the tendency to "lock" in transverse shear is alleviated by a so-called "c-factor" modification [4].

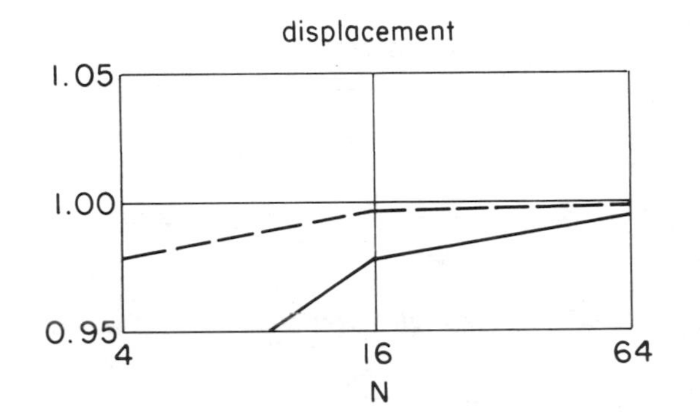

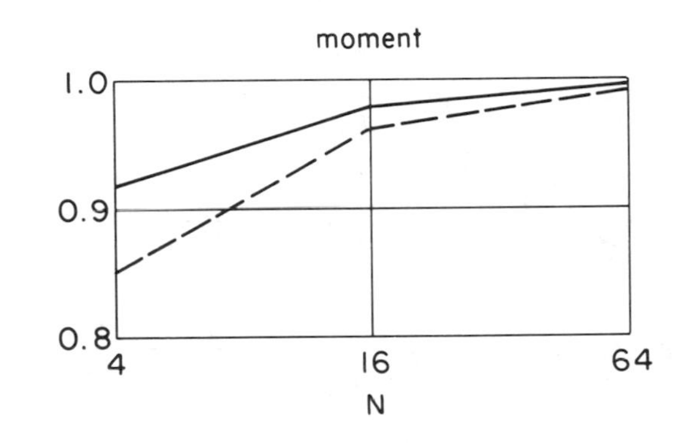

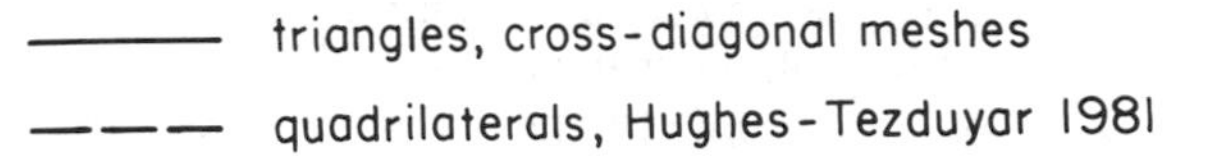

Fig. 6 Comparison of center displacement and moment for triangular and quadrilateral elements; simply-supported, thin, square plate subjected to uniform load.

Our goals in nonlinear analysis (i.e., large strains, rotations, and materially inelastic dynamic response) make "tuning" as in the above approaches, suspect. Thus we have eschewed procedures of these kinds in our work. Admittedly though, improved behavior can be attained in many circumstances by way of these devices.

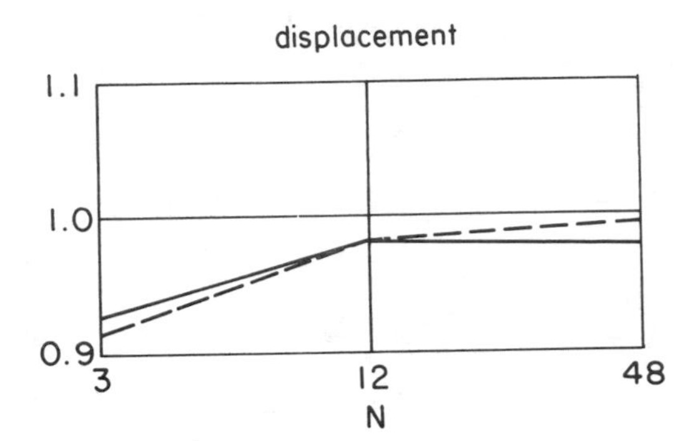

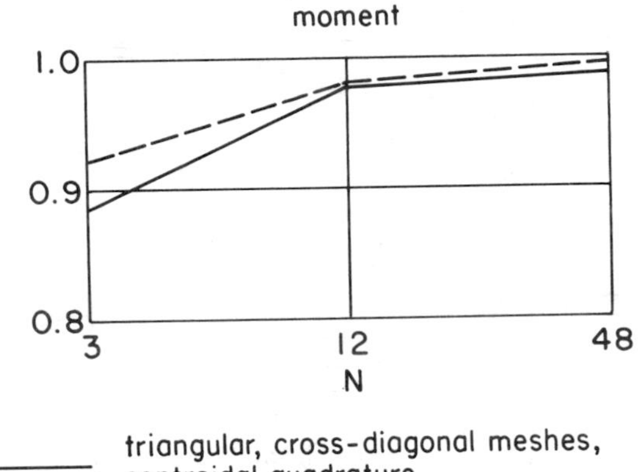

triangular, cross-diagonal meshes, centroidal quadrature.

quadrilaterals, Hughes-Tezduyar 1981

Fig. 7 Comparison of center displacement and moment for triangular and quadrilateral elements; simply-supported, thin, circular plate subjected to uniform load.

6. CONCLUSIONS

In this paper we have examined the behavior of a linear triangular bending element. Shear strains were interpolated in analogous fashion to a procedure which was used, and shown to be effective, for a bilinear quadrilateral element [13]. This procedure results in an element which attains quadratic accuracy with respect to Kirchhoff modes. However, numerical results indicated that the exactly integrated element behaved erractically for certain mesh configurations. The cross-diagonal pattern was found to be successful for both exact and reduced integration (i.e., 1-point centroidal quadrature). Although 1-point quadrature results in rank deficiency for a single element, this is not a practical detriment since the rank deficiency disappears in an assemblage of two or more elements. Due to the greater economy of 1-point quadrature, it appears the obvious choice for practical use.

Comparison of results with those for the bilinear quadrilateral presented in [13] suggests that linear triangles in the cross-diagonal pattern may be competitive. This conclusion is tentative since only limited data has been obtained so far. In the comparison, cross-diagonal triangulated meshes required approximately twice the number of unknowns as for bilinear quadrilaterals. On the other hand, the center-node degrees-of-freedom are easily eliminated via element-wise static condensation, and the 1-point evaluation over each triangle is amenable to very efficient element programming. Further numerical studies need to be performed to assess the behavior of the triangular element over a broader spectrum of test cases.

REFERENCES

1. ARGYRIS, J.H., DUNNE, P.C., MALEJANNAKIS, G.A. and SCHELKLE, E., A simple triangular facet shell element with applications to linear and nonlinear equilibrium and elastic stability problems. *Computer Methods in Applied Mechanics and Engineering*, 10, No.3, 371-403 (March 1977); 11, No.1, 97-131 (April 1977).
2. BATOZ, J.L., BATHE, K.J. and HO, L.W., A Search for the Optimum Three-Node Triangular Plate Bending Element. *Report 82448-8*, Massachusetts Institute of Technology, Cambridge, Mass. (1978).
3. BERKOVIC, M., Thin shell isoparametric elements. *Proceedings of the Second World Congress on Finite Element Methods*, Bournemouth, Dorset, England (1978).
4. FRIED, I. and YANG, S.K., Triangular, nine-degree-of-freedom, C^0 plate bending element of quadratic accuracy. *Quarterly Journal of Applied Mathematics* 31, 303-312 (1973).
5. GARNET, H., CROUZET-PASCAL, J. and PIFKO, A.B., Aspects of a simple triangular plate bending finite element. *Computers and Structures*, 12, 783-789 (1980).
6. GOUDREAU, G.L., A Computer Module for One Step Dynamic Response of an Axisymmetric Plane Linear Elastic Thin Shell. *Lawrence Livermore Laboratory Report No. UCID-17730* (1978).
7. HUGHES, T.J.R., Generalization of selective integration procedures to anisotropic and nonlinear media. *International Journal for Numerical Methods in Engineering* 15, 1413-1418 (1980).

8. HUGHES, T.J.R., COHEN, M. and HAROUN, M., Reduced and selective integration techniques in the finite element analysis of plates. *Nuclear Engineering and Design* 46, 203-222 (1978).
9. HUGHES, T.J.R. and LIU, W.K., Nonlinear finite element analysis of shells: Part I. Three-dimensional shells. *Computer Methods in Applied Mechanics and Engineering*, to appear.
10. HUGHES, T.J.R. and LIU, W.K., Nonlinear finite element analysis of shells: Part II. Two-dimensionsal shells. *Computer Methods in Applied Mechanics and Engineering*, to appear.
11. HUGHES, T.J.R., LIU, W.K. and LEVIT, I., Nonlinear dynamic finite element anlysis of shells. *Proceedings of the Europe-U.S. Workshop on Finite Element Methods in Structural Mechanics*, Bochum, West Germany, (July 28-31, 1980).
12. HUGHES, T.J.R., TAYLOR, R.L. and KANOKNUKULCHAI, W., A simple and efficient element for plate bending. *International Journal for Numerical Methods in Engineering* 11, No. 10, 1529-1543 (1977).
13. HUGHES, T.J.R. and TEZDUYAR, T.E., Finite elements based upon Mindlin plate theory with particular reference to the four-node bilinear isoparametric element, in *New Concepts in Finite Element Analysis* (ed. T.J.R. Hughes), ASME, New York (June 1981).
14. KANOKNUKULCHAI, W., A simple and efficient finite element for general shell analysis. *International Journal for Numerical Methods in Engineering* 14, 179-200 (1979).
15. MacNEAL, R.H., A simple quadrilateral shell element. *Computers and Structures* 8, 175-183 (1978).
16. MacNEAL, R.H., private communication (1981).
17. MALKUS, D.S., private communication (1980).
18. MINDLIN, R.D., Influence of rotatory inertia and shear on flexural motions of isotropic, elastic plates. *Journal of Applied Mechanics* 18, 31-38 (1951).
19. NAGTEGAAL, J.C., PARKS, D.M. and RICE, J.R., On numerically accurate finite element solutions in the fully plastic range. *Computer Methods in Applied Mechanics and Engineering* 4, 153-178 (1974).
20. PARISCH, H., A critical survey of the 9-node degenerated shell element with special emphasis on thin shell application and reduced integration. *Computer Methods in Applied Mechanics and Engineering* 20, 323-350 (1979).
21. TAYLOR, R.L., Finite elements for general shell analysis. *Preprints of the 5th International Seminar on Computational Aspects of the Finite Element Method*, Berlin (West),Germany (August 20-21, 1979).
22. ZIENKIEWICZ, O.C., BAUER, J., MORGAN, K. and ONATE, E., A simple element for axisymmetric shells with shear deformation. *International Journal for Numerical Methods in Engineering* 11, 1545-1558 (1977).

FINITE ELEMENT COMPUTATION OF
LARGE ELASTIC DEFORMATIONS

Isaac Fried

Boston University

1. INTRODUCTION

Approximate Gauss quadrature of the total potential energy[2] is showing great promise to become a universal means for the set up of nonlinear finite elements. As in the linear case also here all the numerically integrated finite elements are expressed in terms of few numerical vectors and matrices and in a form convenient for standard assembly and use in the Newton-Raphson method.

Detailed derivation and actual computation is included in this paper for the element gradient and stiffness matrices of the largely deformed beam, ring, circular plate and rubber membrane.

2. BEAM AND RING

Our starting point for the unit elastica [8], shown in Fig. 1, is its total potential energy

$$\pi(\theta) = \int_0^1 (\tfrac{1}{2}EI\theta'^2 - P\sin\theta + Q\cos\theta)ds - M\theta(1) \qquad (2.1)$$

for which the admissible θ is continuous and satisfies the fixed end condition $\theta(0)=0$.

Initial curvature in the form $\tilde{\theta} = \theta(s)$ alters the total potential energy of (2.1) into

$$\pi(\theta) = \int_0^1 [\tfrac{1}{2}EI(\theta'-\tilde{\theta}')^2 - P\sin\theta + Q\cos\theta]ds \qquad (2.2)$$

where the end moment M is assumed absent. After integration by parts

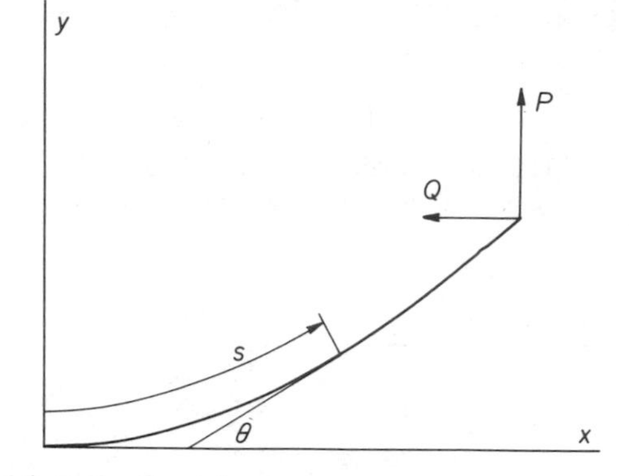

FIG. 1. Tip loaded elastica.

$$\pi(\theta) = \int_0^1 [\tfrac{1}{2}EI\theta'^2 + EI\tilde{\theta}''\,\theta - P\sin\theta + Q\cos\theta]ds - EI\tilde{\theta}(1)\theta(1) \qquad (2.3)$$

and for the circular ring $\tilde{\theta}'' = 0$ so that

$$\pi(\theta) = \int_0^1 (\tfrac{1}{2}EI\theta'^2 - P\sin\theta + Q\cos\theta)ds - EI\tilde{\theta}'(1)\theta(1) \qquad (2.4)$$

as for the straight elastica with an end moment $M=EI\tilde{\theta}'(1)$.

Through the use of $x'=\cos\theta$ and $y'=\sin\theta$ we may write the total potential energy in terms of y as

$$\pi(y) = \int_0^1 [\tfrac{1}{2}EI\frac{y''^2}{1-y'^2} + Q(1-y'^2)^{\frac{1}{2}}]ds - Py(1) \qquad (2.5)$$

where the admissible y is C^1 and satisfies the fixed end conditions $y(0) = y'(0) = 0$.

We propose a finite element discretization of $\pi(\theta)$ in (2.1) with a quadratic interpolation of θ over each element and a two point Gauss quadrature of the total potential energy. This minimal integration scheme assures the numerical stability of the finite element method and is sufficiently accurate. The resulting quadratic element is precise, efficient and easily programmable.

Interpolation of θ over the three-nodal-point element is compactly expressed by $\theta=\underline{u}_e^{\tau}\underline{\phi}$ where $\underline{u}_e^{\tau} = \{\theta_1,\theta_2,\theta_3\}$ is the element nodal values vector, and

$$\phi^{\tau} = \{\tfrac{1}{2}\xi(\xi-1), 1-\xi^2, \tfrac{1}{2}\xi(\xi+1)\} \qquad -1\le\xi\le1 \qquad (2.6)$$

the shape functions vector.

Let the typical element be of size h such that $ds = \frac{1}{2}hd\xi$ and $\theta' = 2h^{-1}\dot{\theta}$, where $\dot{(\ )} = d/d\xi$. Two Gauss point integration of $\pi(\theta)$ in (2.1) over the eth element results in the approximation

$$\pi_e = \frac{1}{2}h \sum_{j=1}^{2} 4h^{-1}\dot{\theta}_j^2 - P\sin\theta_j + Q\cos\theta_j \tag{2.7}$$

in which the index j=1,2 refers to the Gauss points $\xi_1 = -\sqrt{3}/3$ and $\xi_2 = \sqrt{3}/3$. The values of θ and $\dot{\theta}$ at the jth Gauss point are computed from $\theta_j = \underline{u}_e^\tau \underline{\phi}_j$ and $\dot{\theta}_j = \underline{u}_e^\tau \dot{\underline{\phi}}_j$, where $\underline{\phi}_j = \underline{\phi}(\xi_j)$ and $\dot{\underline{\phi}}_j = \dot{\underline{\phi}}(\xi_j)$, and from (2.6) we have that

$$\begin{aligned} \underline{\phi}_{1,2} &= \frac{1}{6}\{1 \pm \sqrt{3},\ 4,\ 1 \mp \sqrt{3}\} \\ \dot{\underline{\phi}}_{1,2} &= \frac{1}{6}\{\mp 2\sqrt{3} - 3,\ \pm 4\sqrt{3},\ \mp 2\sqrt{3} + 3\} \end{aligned} \tag{2.8}$$

where the upper sign of $\sqrt{3}$ belongs to j=1 and the lower to j=2.

From π_e in (2.7) we derive the element gradient vector

$$\underline{g}_e = \frac{\partial \pi_e}{\partial \underline{u}_e} = \sum_{j=1}^{2} 2h^{-1}\dot{\theta}_j \dot{\underline{\phi}}_j - \frac{1}{2}h\,(P\cos\theta_j + Q\sin\theta_j)\underline{\phi}_j \tag{2.9}$$

and the element stiffness matrix

$$k_e = \frac{\partial \underline{g}_e}{\partial \underline{u}_e} = \frac{\partial^2 \pi_e}{\partial \underline{u}_e^2} = \sum_{j=1}^{2} 2h^{-1}\dot{\underline{\phi}}_j \dot{\underline{\phi}}_j^\tau - \frac{1}{2}h(Q\cos\theta_j - P\sin\theta_j)\underline{\phi}_j\underline{\phi}_j^\tau \tag{2.10}$$

where

$$\dot{\underline{\phi}}_1\dot{\underline{\phi}}_1^\tau + \dot{\phi}_2\dot{\phi}_2^\tau = \frac{1}{6}\begin{bmatrix} 7 & -8 & 1 \\ -8 & 16 & -8 \\ 1 & -8 & 7 \end{bmatrix},\quad \underline{\phi}_1\underline{\phi}_1^\tau = \frac{1}{18}\begin{bmatrix} 2+\sqrt{3} & 2(1+\sqrt{3}) & -1 \\ 2(1+\sqrt{3}) & 8 & 2(1-\sqrt{3}) \\ -1 & 2(1-\sqrt{3}) & 2-\sqrt{3} \end{bmatrix},$$

$$\underline{\phi}_2\underline{\phi}_2^\tau = \frac{1}{18}\begin{bmatrix} 2-\sqrt{3} & 2(1-\sqrt{3}) & -1 \\ 2(1-\sqrt{3}) & 8 & 2(1+\sqrt{3}) \\ -1 & 2(1+\sqrt{3}) & 2+\sqrt{3} \end{bmatrix} \tag{2.11}$$

After g_e and k_e are prepared the finite element assembly and solution procedure follows for the nonlinear case precisely as for the linear: An initial guess $\underline{u}_o$ is made for the global slopes vector $\underline{u}$, all $\underline{g}_e$ and k_e are computed from it and routinely assembled into the global $\underline{g}$ and K, the essential boundary conditions and boundary work terms are included in $\underline{g}$ and K: and $\underline{u}_o$ is improved into $\underline{u}_1$ with the Newton-Raphson method $\underline{u}_1 = \underline{u}_o - K_o^{-1}\, \underline{g}_o$ until convergence.

To discretize $\pi(y)$ in (2.5) we propose a piecewise cubic, C^1, interpolation of y with two Gauss point quadrature of the element total potential energy.

$$\text{Now } y = \underline{u}_e^{\tau}\, \underline{\phi}\ , \quad \text{where } \underline{u}_e^{\tau} = \{y_1, \dot{y}_1, y_2, \dot{y}_2\}\ , \quad \text{and}$$

$$\underline{\phi}^{\tau} = \{1-3\xi^2+2\xi^3, \xi-2\xi^2+\xi^3, 3\xi^2-2\xi^3, -\xi^2+\xi^3\}\ \ 0 \le \xi \le 1 \tag{2.12}$$

and the two Gauss points are at $\xi_1 = (3-\sqrt{3})/6$ and $\xi_2 = (3+\sqrt{3})/6$ with equal weights $w_1 = w_2 = \frac{1}{2}$. Consequently the approximate element total potential energy (2.5) becomes

$$\pi_e = \frac{1}{2}h \sum_{j=1}^{2} \frac{1}{2}h^{-4}\ddot{y}_j(1-h^{-2}\dot{y}_j^2)^{-1} + Q(1 - h^{-2}\dot{y}_j^2)^{\frac{1}{2}} \tag{2.13}$$

where $\dot{y}_j = \underline{u}_e^{\tau}\dot{\underline{\phi}}_j$, $\ddot{y}_j = \underline{u}_e^{\tau}\ddot{\underline{\phi}}_j$, and

$$\begin{aligned} \dot{\underline{\phi}}_{1,2}^{\tau} &= \{-1,\ \pm\sqrt{3}/6,\ 1,\ \mp\sqrt{3}/6\} \\ \ddot{\underline{\phi}}_{1,2}^{\tau} &= \{\mp 2\sqrt{3},\ -1\mp\sqrt{3},\ \pm 2\sqrt{3},\ 1\mp\sqrt{3}\} \end{aligned} \tag{2.14}$$

From π_e in (2.13) we get

$$\underline{g}_e = \frac{\partial \pi_e}{\partial \underline{u}_e} = \sum_{j=1}^{2} a_j\ddot{\underline{\phi}}_j + b_j\dot{\underline{\phi}}_j \tag{2.15}$$

with

$$a_j = \frac{1}{2}h^{-3}\ddot{y}_j(1 - h^{-2}\dot{y}_j^2)^{-1}\ , \quad b_j = \frac{1}{2}h^{-5}\dot{y}_j\ddot{y}_j(1-h^{-2}\dot{y}_j^2)^{-2} \tag{2.16}$$

and

$$k_e = \frac{\partial \underline{g}_e}{\partial \underline{u}_e} = \frac{\partial^2 \pi_e}{\partial \underline{u}_e^2} = \sum_{j=1}^{2} a_j\ddot{\underline{\phi}}_j\ddot{\underline{\phi}}_j^{\tau} + b_j(\dot{\underline{\phi}}_j\ddot{\underline{\phi}}_j^{\tau} + \ddot{\underline{\phi}}_j\dot{\underline{\phi}}_j^{\tau}) + c_j\dot{\underline{\phi}}_j\dot{\underline{\phi}}_j^{\tau} \tag{2.17}$$

with

$$
\begin{aligned}
a_j &= \tfrac{1}{2}h^{-3}(1-h^{-2}\dot{y}_j^2)^{-1}, \quad b_j = h^{-5}\dot{y}_j\ddot{y}_j(1-h^{-2}\dot{y}_j^2)^{-2} \\
c_j &= \tfrac{1}{2}h^{-5}\ddot{y}_j(1-h^{-2}\dot{y}_j^2)^{-2} + 2h^{-7}\dot{y}_j^2\ddot{y}_j^2(1-h^{-2}\dot{y}_j^2)^{-3} \qquad (2.18) \\
&\quad -\tfrac{1}{2}Qh^{-1}[(1-h^{-2}\dot{y}_j^2)^{-\frac{1}{2}} + h^{-2}\dot{y}_j^2(1-h^{-2}\dot{y}_j^2)^{-\frac{3}{2}}]
\end{aligned}
$$

To observe the behavior of the discretization method in its computational realization we undertake the calculation of the bent elastica using (2.1) with $Q = 0$. For a tip force $P = 1.5$ the Newton-Raphson method successively computes a tip deflection $y(1)$=0.5000000, 0.4337216, 0.4124599, 0.4109994, 0.4109928, 0.4109928; having started with a zero initial sag.

When P is increased above 1.5 the Newton-Raphson method suddenly fails to converge from a zero initial guess. A better starting shape is needed then for the iterative solution, or the deflection under higher loads can be reached stepwise with the computed solution at the end of the previous step serving as an initial guess for the next iteration with the higher load. One is thus confronted with the choice of small load increments with fewer iterative corrections-an incremental method, or large load increments with more corrections-a global method. In the presence of multiple solutions to the stiffness equation $\underline{g}(\underline{u})=0$, the load history of the incremental method is what determines which one of them will be discovered, while the solution reached with the global method is determined by the initial guess $\underline{u}_o$.

To disclose the discretization accuracy of the element in (2.15)-(2.18) a varying number, Ne, of elements are employed in the computation of cantilever deflected by $P = 5$. For Ne = 1,2,3,...,9 we, respectively, compute $y(1)$ = 0.7669329, 0.7183933, 0.7143314, 0.7140374, 0.7139174, 0.7138622, 0.7138340, 0.7138185 and extrapolation to the limit provides the estimate $|\text{error in } y(1)| = 0.1\text{Ne}^{-3.75}$.

Solving the same cantilever problem with the element given in (2.9) and (2.10) we get for Ne = 3,4,5,6,7 the corresponding tip slopes $\theta(1)$ = 1.2149992, 1.2152510, 1.2153196, 1.2153444, 1.2153549; and with extrapolation to the limit we reach the estimate $|\text{error in } \theta(1)| = 0.03\text{Ne}^{-4}$.

A seven element discretization of the elastica with the element data in (2.9) and (2.10) computes the stable and unstable equilibrium configurations [4] shown in Figs. 2,3 and 4. Figure

2 shows the shapes the cantilever assumes when the tip is forced by $P = 0.5, 1.0, 1.5, \ldots, 9.0$. All eigenvalues of the global stiffness matrix are positive here and the equilibrium configurations shown in Fig. 2 are concluded to be stable. Figure 3 depicts other possible stable equilibrium states for $P = 10.5, 12.0, \ldots 24.0$. The elastica equilibrium configurations shown in Fig. 4 for a tip load $P = 14, 15, \ldots, 29$ are unstable; the lowest eigenvalue of K being negative.

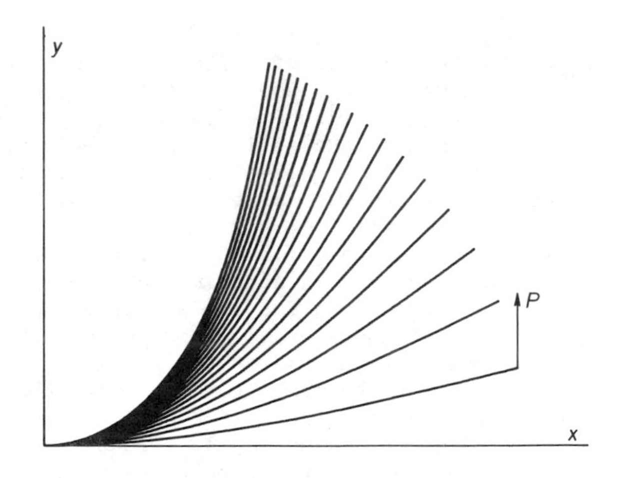

FIG. 2. Stable equilibrium states of elastica.

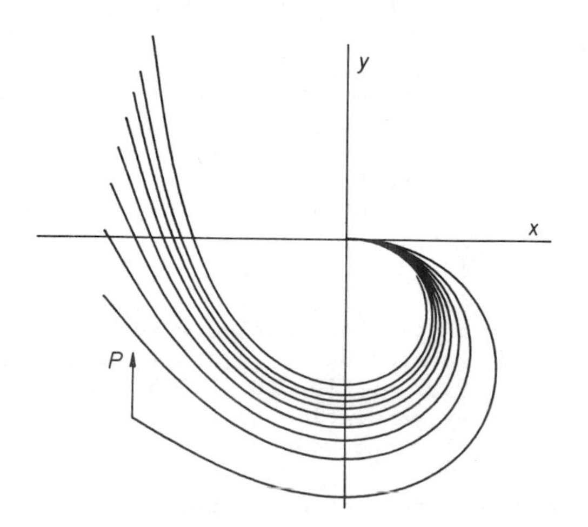

FIG. 3. Other stable equilibrium states of elastica.

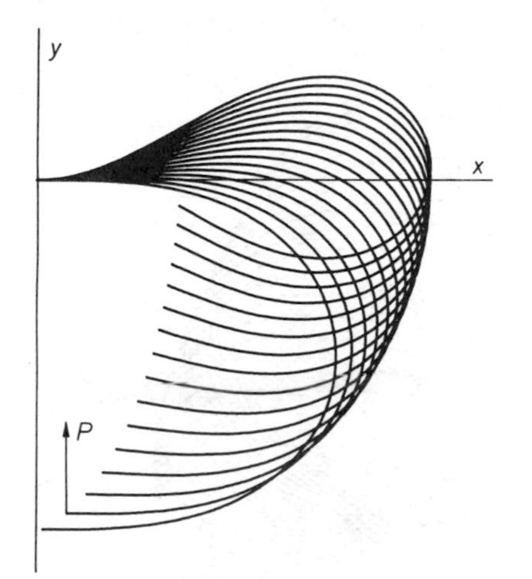

FIG. 4. Unstable equilibrium states of elastica.

Figures 5 and 6 show a similar computation for the circular ring. Figure 5 follows the opening of a C-spring [9] with a force Q = 1,2,...,10; while Fig. 6 follows the compression of a circular ring [5] squeezed by equal and opposite forces P.

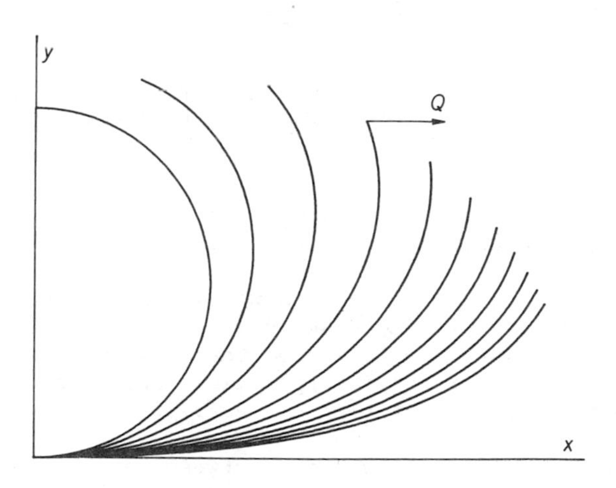

FIG. 5. Forcing of a circular C-spring.

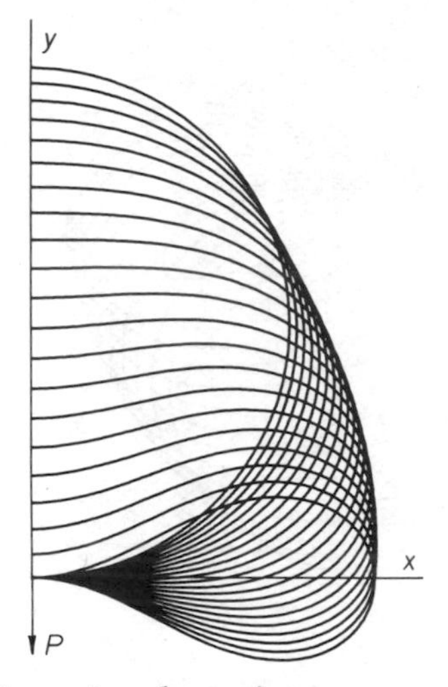

FIG. 6. Squeezing of a circular ring.

3. CIRCULAR PLATE

A unit circular plate ($\nu=0$) is highly bent [11] under the action of an axisymmetrically distributed lateral load f and a uniform edge compression p. The total potential energy of this plate reads

$$\pi(u,w) = \frac{1}{2}\int_0^1 \left[\frac{1}{12}\left(w''^2 + \frac{w'^2}{r^2}\right) + \frac{u^2}{r^2}\right] r\,dr + \frac{1}{2}\int_0^1 \left(u' + \frac{1}{2}w'^2\right)^2 r\,dr - \int_0^1 f w r\,dr + p u(1) \tag{3.1}$$

where u and w denote the inplane and lateral displacements, respectively, and where $(\ )' = d/dr$.

We propose to discretize $\pi(u,w)$ with a C^1 piecewise cubic interpolation of w, a piecewise quadratic interpolation of u, and a two Gauss point integration of the total potential energy over each element. A linear interpolation scheme for u is noticed to produce a decidedly inferior element.

Typically an element extends between $r=r_1$ and $r=r_3$, has three nodal points and is associated with the element nodal values vector $\underline{u}_e^T = \{u_1, w_1, \dot{w}_1, u_2, u_3, w_3, \dot{w}_3\}$. Interpolation of u and w inside the element is formally written as $u = \underline{u}_e^T \phi(\xi)$ and $w = \underline{u}_e^T \psi(\xi)$ with

$$\underline{\phi}^{\tau} = \{\tfrac{1}{2}\xi(\xi-1),0,0,1-\xi^2, \tfrac{1}{2}\xi(\xi+1),0,0\}$$
$$\underline{\psi}^{\tau} = \tfrac{1}{4}\{0,2-3\xi+\xi^3,1-\xi-\xi^2+\xi^3,0,0,2+3\xi-\xi^3,-1-\xi+\xi^2+\xi^3\} \tag{3.2}$$

where $-1 \le \xi \le 1$.

A two point Gauss integration produces from (3.1) the approximate

$$\pi_e = \sum_{j=1}^{2} \tfrac{1}{3}h^{-3}r_j\ddot{w}_j^2 + \tfrac{1}{12}r_j^{-1}h^{-1}\dot{w}_j^2 + \tfrac{1}{4}hr_j^{-1}u_j^2 + h^{-3}(h\dot{u}_j + \dot{w}_j^2)^2 r_j - \tfrac{1}{2}hr_jf_jw_j \tag{3.3}$$

for which the values of $u_j, w_j, \dot{u}_j, \dot{w}_j$ and $\ddot{w}_j$ are computed from $u_j = \underline{u}_e^{\tau}\underline{\phi}_j$, $w_j = \underline{u}_e^{\tau}\underline{\psi}_j$, $\dot{u}_j = \underline{u}_e^{\tau}\dot{\underline{\phi}}_j$, $\dot{w}_j = \underline{u}_e^{\tau}\dot{\underline{\psi}}_j$, and $\ddot{w}_j = \underline{u}_e^{\tau}\ddot{\underline{\psi}}_j$, with the aid of the numerical vectors

$$\begin{aligned}
\underline{\phi}_{1,2}^{\tau} &= \tfrac{1}{6}(1\pm\sqrt{3},0,0,4,1\mp\sqrt{3},0,0)\\
\dot{\underline{\phi}}_{1,2}^{\tau} &= \tfrac{1}{6}(\mp2\sqrt{3}-3,0,0,\pm4\sqrt{3},\mp2\sqrt{3}+3,0,0)\\
\psi_{1,2}^{\tau} &= \tfrac{1}{18}(0,9\pm4\sqrt{3},3\pm\sqrt{3},0,0,9\mp4\sqrt{3},-3\pm\sqrt{3})\\
\dot{\psi}_{1,2}^{\tau} &= \tfrac{1}{6}(0,-3,\pm\sqrt{3},0,0,3,\mp\sqrt{3})\\
\ddot{\psi}_{1,2}^{\tau} &= \tfrac{1}{2}(0,\mp\sqrt{3},-1\mp\sqrt{3},0,0,\pm\sqrt{3},1\mp\sqrt{3})
\end{aligned} \tag{3.4}$$

From (3.3) we produce

$$\underline{g}_e = \frac{\partial\pi_e}{\partial\underline{u}_e} = \sum_{j=1}^{2} a_j\ddot{\underline{\psi}}_j + b_j\dot{\underline{\psi}}_j + c_j\underline{\psi}_j + d_j\dot{\underline{\phi}}_j + e_j\underline{\phi}_j \tag{3.5}$$

with

$$a_j = \tfrac{2}{3}h^{-3}r_j\ddot{w}_j, \quad b_j = \tfrac{1}{6}h^{-1}r_j^{-1}\dot{w}_j + 4h^{-3}r_jw_j(h\dot{u}_j + \dot{w}_j^2)$$
$$c_j = -\tfrac{1}{2}hr_jf_j, \quad d_j = 2h^{-2}r_j(h\dot{u}_j+\dot{w}_j^2), \quad e_j = \tfrac{1}{2}hr_j^{-1}u_j \tag{3.6}$$

and

$$k_e = \frac{\partial g_e}{\partial \underline{u}_e} = \frac{\partial^2 \pi_e}{\partial^2 \underline{u}_e} = \sum_{j=1}^{2} a_j \underline{\ddot{\psi}}_j \underline{\ddot{\psi}}_j^{\tau} + b_j \underline{\dot{\psi}}_j \underline{\dot{\psi}}_j^{\tau} + c_j (\underline{\dot{\phi}}_j \underline{\dot{\psi}}_j^{\tau} + \underline{\dot{\psi}}_j \underline{\dot{\phi}}_j^{\tau}) + d_j \underline{\dot{\phi}}_j \underline{\dot{\phi}}_j^{\tau} + e_j \underline{\phi}_j \underline{\phi}_j^{\tau} \tag{3.7}$$

with

$$a_j = \tfrac{2}{3} h^{-3} r_j \ , \ b_j = \tfrac{1}{6} h^{-1} r_j^{-1} + 12 h^{-3} r_j \dot{w}_j^2 + 4 h^{-2} r_j \dot{u}_j$$
$$c_j = 4 h^{-2} r_j \dot{w}_j \ , \ d_j = 2 h^{-1} r_j \ , \ e_j = \tfrac{1}{2} h r_j^{-1} \tag{3.8}$$

To assess the performance of our element we use it to compute the deflection of the uniformly loaded (i.e. f=const.), clamped (i.e. w(1)=w'(1)=0) plate with an immovable (i.e. u(1) =0) edge. For f = 10 we compute, with Ne=2,3,...,7, a central deflection w(0)=1.138993,1.140754, 1.141714, 1.142070, 1.142220, 1.142293; meaning a relative error in w(0) equal to $0.068Ne^{-3.3}$. A Newton-Raphson solution of the nonlinear stiffness equation for f = 10 and Ne = 7 successively comes up with w(0)=1.874962, 1.394681, 1.182066, 1.143467, 1.142294, 1.142293, having started with a zero deflection.

The critical thrust for the plate is given by $p_{cr}=\alpha^2/12$ where $\alpha^2=3.39$ is the first root of Bessel's function equation $J_1(\alpha)=0$. Figure 7 traces the computed central deflection of a simply supported plate bent under the combined action [3,10] of a lateral load f and an edge thrust p that exceeds p_{cr}. When f > 0 the unstable trivial solution w = 0 for $p > p_{cr}$ is absent and a zero initial deflection can be chosen for the Newton-Raphson method. When f = 0 the Newton-Raphson method must start with a nonzero initial guess but proceeds otherwise as before to produce the typical biforcation curve in Fig. 7.

Close to a critical point at which K^{-1} is non computable, the condition of K declines and the Newton-Raphson method slows down. It is our experience, though, that by using higher precision computations and more corrective iterations one can get as close as it is only numerically meaningful to such a point.

4. RUBBER MEMBRANE

Let the generating curve of the undeformed axisymmetric membrane be given in the (r,z) plane through r=r(s) and z=z(s), s being the arc length. Under the action of applied forces and prescribed displacements the point (r,z) moves to the deformed location (x,y). An arc element ds is stretched thereby to $d\tilde{s}$

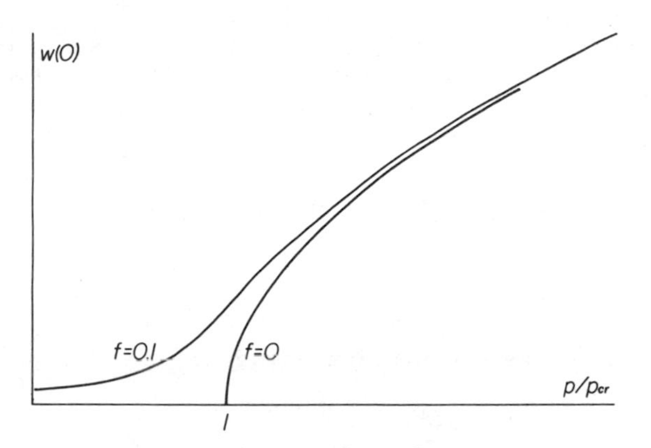

FIG. 7. Face and edge forced circular plate.

and the thickness t of the membrane is reduced to $\tilde{t}$.

The energy density function of the membrane is ultimately expressed in terms of the three principal stretch ratios

$$\lambda_1 = \frac{d\tilde{s}}{ds}, \quad \lambda_2 = \frac{2\pi x}{2\pi r}, \quad \lambda_3 = \frac{\tilde{t}}{t} \tag{4.1}$$

and if the deformation is incompressible $\lambda_1\lambda_2\lambda_3 = 1$. With $d\tilde{s} = (dx^2+dy^2)^{\frac{1}{2}}$ the stretch ratios become

$$\lambda_1 = (x'^2 + y'^2)^{\frac{1}{2}}, \quad \lambda_2 = \frac{x}{r}, \quad \lambda_3 = \frac{1}{\lambda_1\lambda_2} \tag{4.2}$$

where $(\ )' = d/ds$.

Assuming a membrane made of Mooney material its total potential energy acquires the form

$$\pi(x,y) = 2\pi\mu t \int_0^S [(I_1-3)+\alpha(I_2-3)]rds + \frac{1}{2}\frac{p}{\mu t}\int_0^S x^2 y' ds \tag{4.3}$$

where p is the pressure, μ and α material constants, and where I_1 and I_2 are the strain invariants

$$\begin{aligned} I_1 &= \lambda_1^{\,2} + \lambda_2^{\,2} + \lambda_3^{\,2} \\ I_2 &= \lambda_1^2\lambda_2^2 + \lambda_2^2\lambda_3^2 + \lambda_3^2\lambda_1^2 \end{aligned} \tag{4.4}$$

with $\lambda_3 = 1/\lambda_1\lambda_2$. From here on we replace $p/\mu t$ by p and assume that $2\pi\mu t = 1$.

We propose to discretize $\pi(x,y)$ with a quadratic-quadratic interpolation of x and y over the element and a two Gauss point quadrature of the element total potential energy. Inside each element $x = \underline{u}_e^\tau \underline{\phi}$ and $y = \underline{u}_e^\tau \underline{\psi}$, where $\underline{u}_e^\tau = \{x_1, y_1, x_2, y_2, x_3, y_3\}$ and

$$\underline{\phi}^\tau = \{\tfrac{1}{2}\xi(\xi-1), 0, 1-\xi^2, 0, \tfrac{1}{2}\xi(\xi+1), 0\} \qquad -1 \le \xi \le 1 \tag{4.5}$$
$$\underline{\psi}^\tau = \{0, \tfrac{1}{2}\xi(\xi-1), 0, 1-\xi^2, 0, \tfrac{1}{2}\xi(\xi+1)\}$$

We shall need the Gauss point values $x_j = \underline{u}_e^\tau \underline{\phi}_j$, $y_j = \underline{u}_e^\tau \underline{\psi}_j$, $\dot{x}_j = \underline{u}_e^\tau \dot{\underline{\phi}}_j$ and $\dot{y}_j = \underline{u}_e^\tau \dot{\underline{\psi}}_j$, computed from $\underline{u}_e$ and

$$\phi_{1,2}^\tau = \frac{1}{6}\{1\pm\sqrt{3}, 0, 4, 0, 1\mp\sqrt{3}, 0\}$$
$$\dot{\phi}_{1,2}^\tau = \frac{1}{6}\{\mp 2\sqrt{3}-3, 0, \pm 4\sqrt{3}, 0, \mp 2\sqrt{3}+3, 0\} \tag{4.6}$$
$$\dot{\underline{\psi}}_{1,2} = \frac{1}{6}\{0, \mp 2\sqrt{3}-3, 0, \pm 4\sqrt{3}, 0, \mp 2\sqrt{3}+3\}$$

as before.

The element total potential energy is written as the sum

$$\pi_e = J_1+J_2+J_3+\alpha(J_4+J_5+J_6)+\tfrac{1}{2}pJ_7 \tag{4.7}$$

of the seven approximate integrals

$$J_1 = \int_e \lambda_1^2 r\,ds = h^{-1}\sum_{j=1}^{2} r_j(\dot{x}_j^2 + \dot{y}_j^2)$$
$$J_2 = \int_e \lambda_2^2 r\,ds = h\sum_{j=1}^{2} r_j^{-1} x_j^2 \tag{4.8}$$
$$J_3 = \int_e \lambda_1^{-2}\lambda_2^{-2} r\,ds = h^3\sum_{j=1}^{2} r_j^3 x_j^{-2}(\dot{x}_j^2+\dot{y}_j^2)^{-1}$$
$$J_4 = \int_e \lambda_1^2\lambda_2^2 r\,ds = h^{-1}\sum_{j=1}^{2} r_j^{-1} x_j^2(\dot{x}_j^2+\dot{y}_j^2)$$

$$J_5 = \int_e \lambda_1^{-2} r\,ds = h^3 \sum_{j=1}^{2} r_j(\dot{x}_j^2+\dot{y}_j^2)^{-1}$$

$$J_6 = \int_e \lambda_2^{-2} r\,ds = h \sum_{j=1}^{2} r_j^3 x_j^{-2}$$

$$J_7 = \int_e x^2 y' ds = \sum_{j=1}^{2} x_j^2 \dot{y}_j^2$$

Repeated differentiation of π_e with respect to $\underline{u}_e$ furnishes

$$\underline{g}_e = \frac{\partial \pi_e}{\partial \underline{u}_e} = \sum_{j=1}^{2} a_j \underline{\dot{\phi}}_j + b_j \underline{\dot{\psi}}_j + c_j \underline{\phi}_j \tag{4.9}$$

with

$$a_j = 2r_j x'_j (1-\lambda_{1j}^{-4}\lambda_{2j}^{-2})(1+\alpha\lambda_{2j}^2)$$

$$b_j = 2r_j y'_j (1-\lambda_{1j}^{-4}\lambda_{2j}^{-2})(1+\alpha\lambda_{2j}^2) + \tfrac{1}{2}px_j^2 \tag{4.10}$$

$$c_j = 2h\lambda_{2j}(1-\lambda_{1j}^{-2}\lambda_{2j}^{-4})(1+\alpha\lambda_{1j}^2) + hpx_j y'_j$$

and

$$k_e = \frac{\partial \underline{g}_e}{\partial \underline{u}_e} = \frac{\partial^2 \pi_e}{\partial \underline{u}_e^2} = \sum_{j=1}^{2} a_j \underline{\dot{\phi}}_j \underline{\dot{\phi}}_j^\tau + b_j \underline{\dot{\psi}}_j \underline{\dot{\psi}}_j^\tau + c_j \underline{\phi}_j \underline{\phi}_j^\tau$$

$$+ d_j(\underline{\phi}_j \underline{\dot{\phi}}_j^\tau + \underline{\dot{\phi}}_j \underline{\phi}_j^\tau) + e_j(\underline{\phi}_j \underline{\dot{\psi}}_j^\tau + \underline{\dot{\psi}}\ \underline{\phi}_j^\tau) \tag{4.11}$$

$$+ f_j(\underline{\dot{\phi}}_j \underline{\dot{\psi}}_j^\tau + \underline{\dot{\psi}}_j \underline{\dot{\phi}}_j^\tau)$$

with

$$a_j = 2h^{-1}r_j\left(1-\frac{y_j'^2 - 3x_j'^2}{\lambda_{2j}^2\lambda_{1j}^6}\right)(1+\alpha\lambda_{2j}^2)$$

$$b_j = 2h^{-1}r_j\left(1-\frac{x_j'^2 - 3y_j'^2}{\lambda_{2j}^2\lambda_{1j}^6}\right)(1+\alpha\lambda_{2j}^2)$$

$$c_j = 2hr_j^{-1}\,(1+3\lambda_{1j}^{-2}\,\lambda_{2j}^{-4})(1+\alpha\lambda_{1j}^2) + hpy'_j \tag{4.12}$$

$$d_j = 4x_j'\lambda_{2j}(\alpha + \lambda_{1j}^{-4}\lambda_{2j}^{-4})$$

$$e_j = 4y_j'\lambda_{2j}(\alpha + \lambda_{1j}^{-4}\lambda_{2j}^{-4}) + px_j$$

$$f_j = 8h^{-1}r_jx_j'y_j'\lambda_{1j}^{-6}\lambda_{2j}^{-2}\,(1 + \alpha\lambda_{2j}^2)$$

Our first application of the rubber membrane element is to compute the inflated shape [1] of a unit disc ($\alpha = 0.1$) for a pressure $p = 5$ and an inplane stretching of the edge to $x(1) = 1.1$. The disc is substantially deformed by this high pressure, and for a uniform layout of Ne finite elements we compute, corresponding to Ne = 1,2,3,4,10, a polar rise $y(0)=1.2546$, 1.4278, 1.4299, 1.4304, 1.4304. We reach this last value of $y(0)$ with the Newton-Raphson scheme that successively computes $y(0)=2.31875$, 1.44782, 1.44302, 1.43059, 1.43040, 1.43040.

Figure 8 shows inflated shapes of the disc for a pressure $p = 1,2,\ldots,7$. Corresponding to these pressures are the polar stretch ratios λ_o = 1.144, 1.233, 1.375, 1.648, 2.430, 4.625, 6.552. The global stiffness matrix K is found to be positive definite for the shapes in Fig. 8 and we conclude that the disc is in stable equilibrium.

Figure 9 shows a torus [6] ($\alpha = 0.25$), generated by $r = 2 + \cos s$, $z = \sin s$ $0 \leq s \leq \pi$, inflated by a pressure that increases in ten equal steps from $p = 0$ to a critical $p = 2.185$. As the pressure approaches this last value of p the lowest eigenvalue of the global stiffness matrix K nears zero indicating a decline in stability.

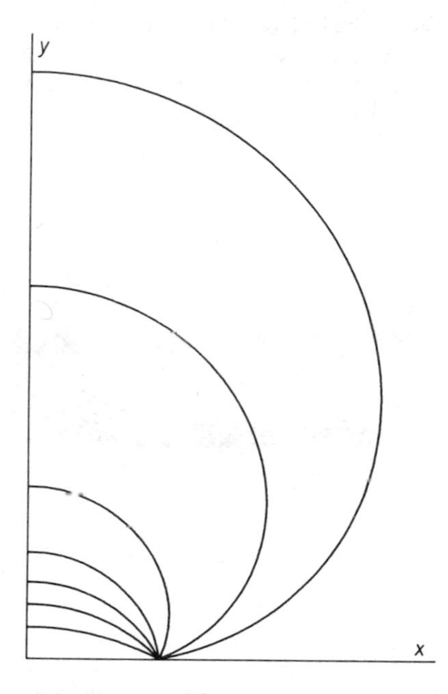

FIG. 8. Stretched and inflated disc.

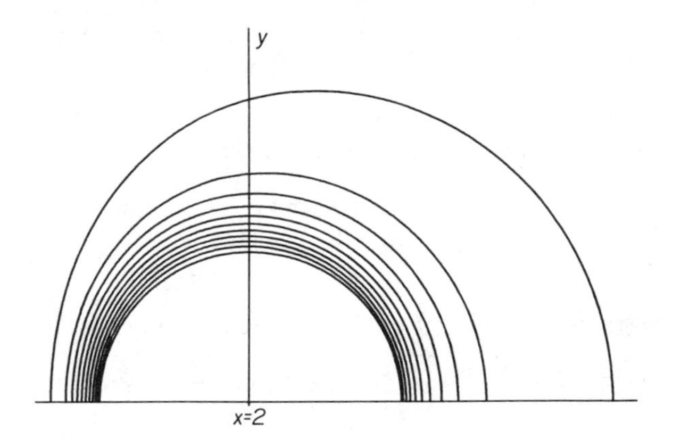

FIG. 9. Inflated torus.

Figure 10 shows bulging [6] of a tube ($\alpha = 0$), stretched both axially and circumferently, and inflated. The curves in Fig. 10 are for a pressure that increases in ten equal steps from p = 0 to the critical p = 0.95.

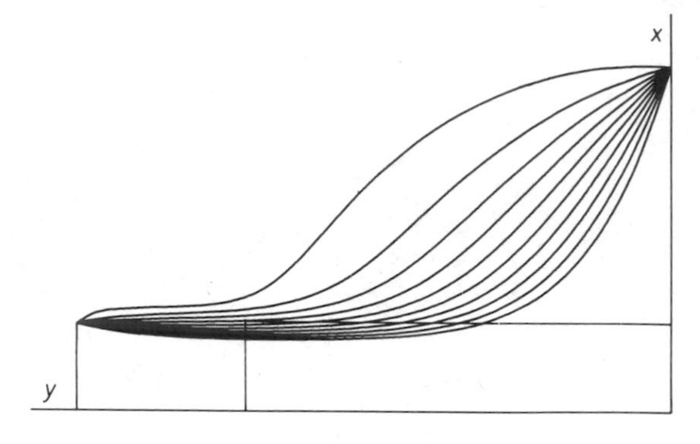

FIG. 10. Bulging of stretched and inflated tube.

ACKNOWLEDGEMENT

The results of this paper were obtained with the support of the Office of Naval Research, under contract: ONR-N00014-76C-036.

REFERENCES

1. ADKINS, J.E., and RIVLIN, R.S., Large Elastic Deformations of Isotropic Materials, IX the Deformation of Thin Shells. *Phil. Trans. R. Soc.*, Section A, 505-531 (1952).

2. FRIED, I., *The Numerical Solution of Differential Equations*. Academic Press, New York (1979).

3. FRIEDRICHS, K.O., and STOKER, J.J., Buckling of the Circular Plate Beyond the Critical Thrust. *J. Appl. Mech.* 9, A7-A14 (1942).

4. FRISCH-FAY, R., *Flexible Bars*, Butterworth, London (1962).

5. FRISCH-FAY,R. The Deformation of Elastic Circular Rings. *Aus. J. Appl. Sci.* 11, 329-340 (1960).

6. KYDONIEFS, A.D. and SPENCER, A.J.M., The Finite Inflation of an Elastic Toroidal Membrane of Circular Cross Section. *Int. J. Engng. Sci.* 15, 367-391 (1967).

7. KYDONIEFS, A.D. and SPENCER, A.J.M., Finite Axisymmetric Deformation of an Initially Cylindrical Elastic Membrane. *Quart. J. Mech. Appl. Math.* 20, 88-95 (1969).

8. LOVE, A.E., *A Treatise on the Mathematical Theory of Elasticity*. Chapters 18 and 19, Dover Publications, New York (1944).

9. SHINOHARA, A. and HARA, M. Large Deflection of a Circular C-shaped spring. *Int. J. Nonlinear Mech.* 8, 169-178 (1973).

10. TANI, J., Elastic Instability of an Annular Plate Under Uniform Compression and Lateral Pressure. *J. Appl. Mech.* 47, 591-594 (1980).

11. TIMOSHENKO, S. and WOINOWSKY-KRIEGER, S. *Theory of Plates and Shells*. McGraw-Hill Book Co., New York (1959).

OPTIMAL DESIGN OF CYLINDERS IN SHEAR

Gilbert Strang and Robert Kohn

Massachusetts Institute of Technology and New York University

This paper will study the theory rather than the finite element analysis of problems in shape optimization. The choice of topic reflects the progress of our research; we are still developing models to help us understand how optimal shapes are likely to look, and we hope that our analytic solutions may later serve as a test of the effectiveness of finite element algorithms. Our basic problem is the classical one in optimization: to minimize the cost while satisfying the constraints.

In the present case the cost is measured by the volume of material; we try to find among all admissible structures the one which has least weight for greatest strength. The strength can be decided in a number of ways: for an elastic material we use the compliance, in the plastic case there is a limit load, for a vibrating material it is the lowest frequency that is fundamental, and the resistance to twist is given by the torsional rigidity. In each of these problems, and in others, we can distinguish four steps:

(i) Analytical - to formulate the optimization problem precisely and to establish the conditions for optimality

(ii) Exploratory - to construct and solve some model problems, looking for properties that may be typical of the general case

(iii) Algorithmic - to develop an effective numerical method for the solution of more difficult problems

(iv) Practical - to put the preceding steps to use.

We have made some progress on step (i), following the lead of Prager and Mroz (and many others). It may be most useful to give here the primal and the dual forms of a particular shape optimization problem, and to describe the optimality conditions. There are important analogies with discrete problems of "network flow" -- the transportation problem and the maximal flow problem -- and we don't know what implications they may have for numerical computation. In our continuous case, the duality theory depends on the proper choice of function spaces for the unknowns, and a more complete analysis will appear in our forthcoming paper [3]. But we hope that the role of duality, and the effect of introducing inequality constraints, will be straightforward enough to make direct use of it in the applications.

The other object of the paper is step (ii). The examples completed so far come from special problems in a plane domain Ω (and they have analogies for fluid flow or traffic flow, with shear stresses replaced by velocities). In these problems the stress is a vector with two components, and its dual is a scalar displacement. There are at least two engineering situations in which this is natural, and both are associated with an infinite cylindrical rod -- whose cross-section is Ω. One is the optimization of torsional rigidity: the classical problem would compute the rigidity of a given rod [5,8], whereas our problem asks for given rigidity using least material and staying within Ω. It is the last constraint which makes the geometry not completely arbitrary and the solution not automatically a circle. We give the solution for a square cross-section in [3,4]; the underlying mathematical problem is to compute a "warping function" w whose gradients approximate the particular function $v = (y,-x)$. In elasticity it is the mean-square (L_2) distance which enters, and minimizing a quadratic gives a linear equation. In plasticity it is L_1 and L_∞, and we have minimization with inequality constraints.

The second example in which there are only the stresses $\sigma_1 = \sigma_{xz}(x,y)$ and $\sigma_2 = \sigma_{yz}(x,y)$ is the problem of antiplane shear. The rod is subject to surface tractions f acting parallel to the axis and with magnitude independent of z. This gives rise to boundary conditions in the equation for equilibrium:

$$\operatorname{div} \sigma = \frac{\partial \sigma_1}{\partial x} + \frac{\partial \sigma_2}{\partial y} = 0 \quad \text{in} \quad \Omega, \quad \sigma \cdot n = f \quad \text{on} \quad \Gamma \quad . \tag{1}$$

The yield condition on the material, which is assumed to be perfectly plastic, can be normalized to

$$|\sigma| = (\sigma_1^2 + \sigma_2^2)^{1/2} \le 1 \quad . \tag{2}$$

This is to hold at each point of the domain. The problem is, while satisfying (1) and (2), *to minimize the area in which* σ *is nonzero*. In other words, we try to achieve $\sigma = 0$ in as large a region as possible, since no material is needed where the stress vanishes.

As a specific example (to which most of this paper will be devoted) let the cross-section have the shape of a "butterfly". The vertical sides are subject to uniform shearing forces $\pm f_0$, and the sloping boundaries are unloaded. In case the external force is too large, $f_0 > w/\ell$, there will be no stress that satisfies both (1) and (2); the structure will break along the center line. This leads to a problem in *limit analysis*, to compute the largest value f_0 which the structure can withstand. The

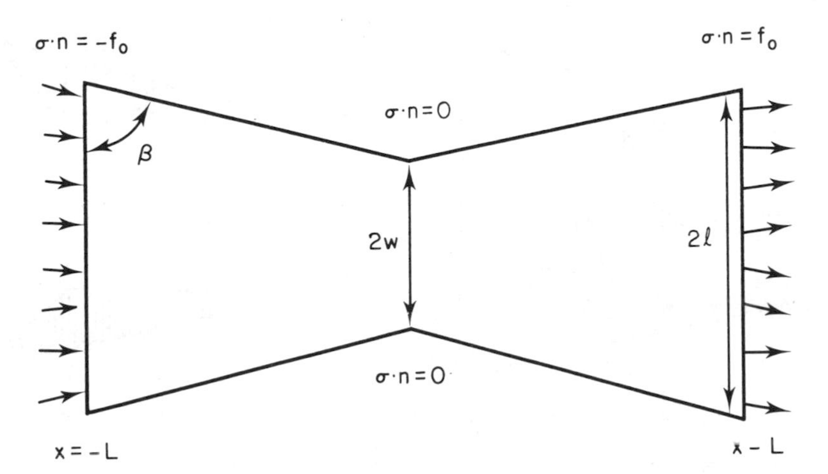

FIG.1. The butterfly.

theory is summarized below, and in this example it yields the following result: If $f_0 = w/\ell$, and if the corner angle β is not too small, then there is a stress σ which satisfies both the equilibrium equation (1) and the inequality constraint (2).

More precisely, there are infinitely many such stress distributions, and our goal is to find the one which vanishes in the largest possible set. The analogy with traffic flow suggests some possible solutions. We imagine a city of shape Ω, with traffic that enters from one side and leaves through the other; σ gives the direction and magnitude of the flow. The "continuity" equation (1) means that no cars are parked (and none are annihilated). The constraint (2) bounds the flow across a unit width of road, and limits the capacity. Our goal is to minimize the area that has to be paved; where $\sigma = 0$ we can build a park.

One possibility is to channel the flow as far as possible along the boundary, leaving a wedge-shaped hole in the center of each wing (Fig.2a). If the remaining traffic lanes have combined width $2w$, the flow is feasible; we need room near the boundary to establish a flow that goes correctly across the wing tip. (Near the center there are circular sectors of radius w, where the flow can change direction; it passes horizontally through the center line and turns until it is parallel to the boundary.) In terms of shear, the structure can still withstand the force $f_0 = w/\ell$.

A second possibility is to remove material along the boundary -- in other words, to direct the traffic straight through the center of Ω. Again it is the region at the far right that is the source of difficulty; the design in Fig.2b is not optimal, because we can introduce small holes near the wing tip without collapsing the structure or making the flow infeasible.

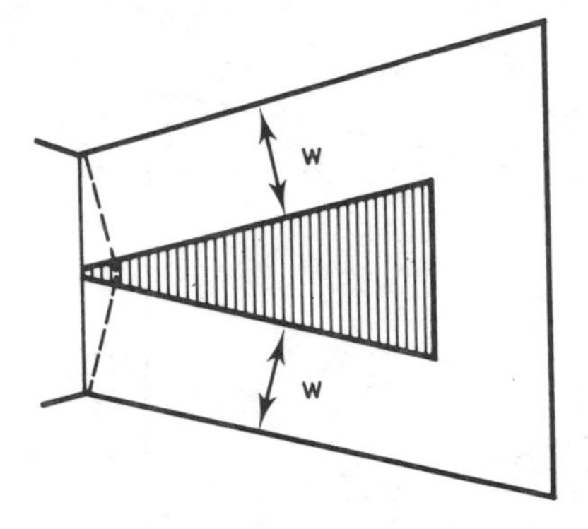

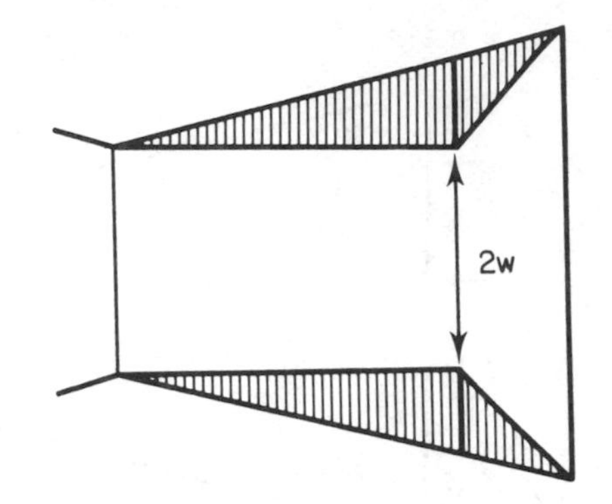

FIG.2a. Wedge-shaped hole.

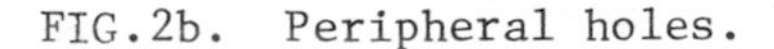
FIG.2b. Peripheral holes.

To make a better start, we alternate thin holes and thin roads (of total width w) as in Fig.3. This design is closer to optimal; the flow more nearly matches the boundary condition. In fact, *the more holes we introduce, the less space is required for adjustments at the boundary*. We seem to be designing a structure which is composed of "fibers" that are oriented so as barely to support the given load. In fact, we are approaching a totally anisotropic material, created by leaving a certain fraction, or "density" of the original one. Where the density is zero, there is a genuine hole; where it is unity, the material is solid; in between is something new.

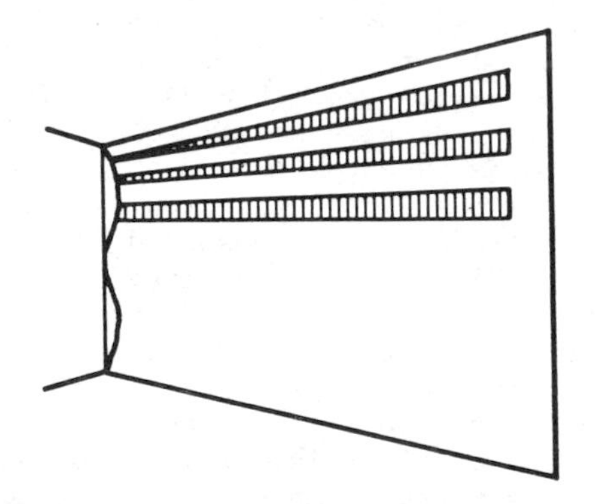

FIG.3. Thin wedges and thin roads.

We remark on the similarity to the "Michell trusses" described in 1904. Those are the solutions to a corresponding problem in plane strain, without a constraint of the type $|\sigma| \leq 1$. They are essentially impossible to manufacture (as our solutions are) but they represent an optimum design. It is even conceivable that in some biological applications, nature does come close to this optimum -- in the morphology of bone there is a qualitative rule known as *Wolff's law*, that the density changes according to exercise or disuse in such a way as to minimize the stress. In the healing of a fracture, the formation of new bone is being studied both experimentally (in the patella of a horse) and by finite element simulation [2], in order to make Wolff's law quantitative. This is a multi-axial problem that goes beyond pure shear, but the same essential features of limiting materials are appearing also in the optimal design of plates [1]; see also the work of Banichuk and Lurie. Mathematically it is a problem of "homogenization", but instead of a periodic distribution of identically shaped holes or reinforcements we encounter a foliation of fibers.

MINIMUM PRINCIPLES AND CONVEX ANALYSIS

Our problem is to find a conventional optimization problem which can yield an unconventional design. Roughly speaking, the magnitude $|\sigma(x,y)|$ gives the fraction of capacity which is actually used at each point by the given flow -- where $|\sigma| = 1$ the full capacity is reached. We may anticipate that this fraction of capacity is also close to the fraction of material required, as more and more holes are introduced (in the direction of σ). There is just enough material to withstand the stress σ, and integrating over Ω should give the total area required. Therefore our goal is to prove that the problem

(P) Minimize $\iint_\Omega |\sigma|\, dxdy$ subject to (1) and (2)

yields a lower bound on the area, and that we can come arbitrarily near this bound by a proper choice of holes in Ω. *The minimizing* σ *describes the optimal design, and it can be approached by a sequence of ordinary designs.*

We want to show how this problem (P) can be solved. The constraint $\operatorname{div}\sigma = 0$ suggests the introduction of a "stress function" ψ, with

$$\sigma = (\psi_y, -\psi_x) \quad \text{so that automatically} \quad \frac{\partial\sigma_1}{\partial x} + \frac{\partial\sigma_2}{\partial y} = 0 \quad .$$

The boundary condition on $\sigma \cdot n$ yields the tangential derivative of ψ, since

$$(\psi_y, -\psi_x) \cdot (\nu_1, \nu_2) = (\psi_x, \psi_y) \cdot (-\nu_2, \nu_1) = \nabla\psi \cdot t = \partial\psi/\partial t \quad .$$

If we fix $\psi = 0$ at some boundary point P_0, say the midpoint of the wing tip at the right, then integrating $\partial\psi/\partial t = f$ along Γ yields an ordinary Dirichlet condition on ψ:

$$\psi = g = \int_{P_0} f(s)ds \quad .$$

In our case this integration produces $g = f_0 y$ on the right wing, $g = \text{constant} = f_0 \ell = w$ on the sloping boundaries where $f = 0$ makes no contribution, and finally $g = f_0 y$ also on the other wing. The plasticity constraint is just an upper bound on the gradient of ψ:

$$|\sigma| = (\psi_y^2 + \psi_x^2)^{1/2} = |\nabla\psi| \leq 1 \quad . \tag{3}$$

Therefore the minimum principle can be restated as

(Q) Minimize $\iint_\Omega |\nabla\psi| dxdy$ subject to $|\nabla\psi| \leq 1$, $\psi = g$ on Γ .

To find the optimality conditions (Kuhn-Tucker conditions) we need to apply duality theory. But for this particular problem there is a different and rather special formula contributed by geometric measure theory. It is the *coarea formula*, which expresses $\iint |\nabla\psi|$ in terms of the lengths of the curves $\psi = \text{constant}$. The integral itself is called the total variation of ψ, and the formula (which can be directly verified for piecewise-linear functions) is

$$\iint_\Omega |\nabla\psi| dxdy = \int_{-\infty}^{\infty} (\text{length of } C_t) dt. \tag{4}$$

C_t is the level curve $\psi = t$, and both the length on the right side and the integral on the left are measured in the open set Ω. (There is a modification which applies to the closure of Ω, and to the integral of $h|\nabla\psi|$ for a bounded function h, but neither is needed here.) In case ψ is constant over a set of positive measure, as it will be on any "hole" where $\sigma = 0$, then $\nabla\psi = 0$ and the curve C_t should be interpreted as the boundary of the set where $\psi \geq t$. With this convention the formula extends to any function of bounded variation, continuous or not -- and in particular to the function χ_t which equals one where $\psi \geq t$ and zero where $\psi < t$. For this function $\iint |\nabla\chi_t|$ is exactly the length of C_t, so the coarea formula (4) can be rewritten as

$$\iint |\nabla\psi| dxdy = \int |\nabla\chi_t| dt \quad . \tag{5}$$

Now we put the formula to use, remembering the constraints $\psi = g$ on Γ and $|\nabla\psi| \leq 1$. Consider the boundary points at which $g = t$; the level curve C_t must connect these points. At the same time it must stay at least a distance $|t - t'|$ from the points where $g = t'$; this holds for all t' or $|\nabla\psi| \leq 1$ would be violated. Therefore the curve C_t has to avoid a set S_t which is a *union of open disks* -- in other words, C_t must not enter the disk of radius $|t - g(P)|$ around any point P on the boundary Γ. Our problem (Q) minimizes $\iint|\nabla\psi|$, and from the coarea formula this is achieved if we *minimize the length of each level set* C_t *while avoiding* S_t.

For the butterfly we take a typical value in the interval $0 \le t < w = g_{max}$. Then there is a point on each wing tip at which $g = t$. The two points have the same height $y = t/f_0$, since $g(y) = f_0 y$ on both vertical boundaries. The shortest curve connecting these points would be a horizontal line, but there is no guarantee that this line avoids the disks that compose S_t (or even that it stays in Ω). In fact, there is no problem with the disks around other points on the vertical boundary -- the points at height y' have $g = f_0 y' = t'$, and the circle of radius $|t - t'| = f_0|y - y'|$ doesn't reach from y' to y, since $f_0 = w/\ell$ is less than 1. The most dangerous disks are those around the points on the sloping boundary, and in particular the one around the internal corner at the center. At these points $g = w$, so the disks have radius $w - t$ and their union is a strip of that width that runs parallel to the boundary and ends in a circular sector. (The sector is to the left of the thin lines in Fig.4a.) The shortest path around this set is a straight line PQ tangent to the circle at Q, followed by a short circular arc. The other parts of the butterfly give a similar path.

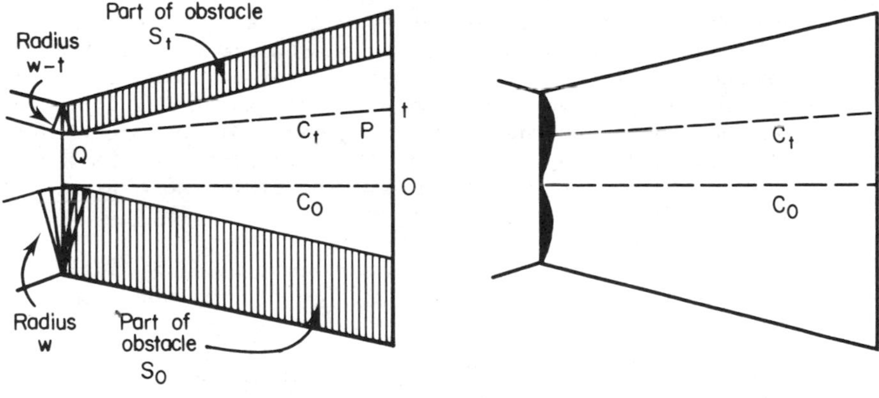

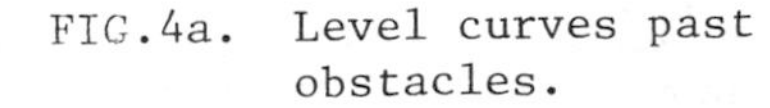
FIG.4a. Level curves past obstacles.

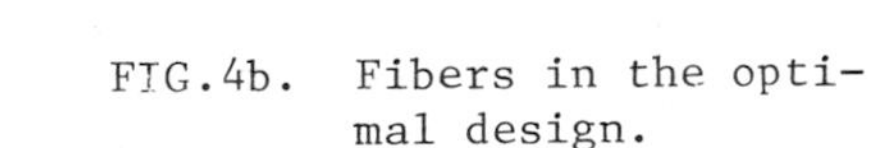
FIG.4b. Fibers in the optimal design.

We now have a complete picture of ψ, and also of the vector $\sigma = (\psi_y, -\psi_x)$ which has the same direction as the level curves of ψ and is orthogonal to $\nabla\psi = (\psi_x, \psi_y)$. In most of the wing these level curves are straight and $0 < |\sigma| < 1$; this part of the optimal design is composed of fibers. In that region the "traffic flow" is on straight roads, but not parallel to one another or perpendicular to the boundary. Near the center of the butterfly the curves become circular arcs and σ is the unit vector tangent to these circles. Since $|\sigma| = 1$ the material in that region remains solid (Fig.4b), and its exact shape can be computed.

LIMIT ANALYSIS

You may have recognized one possible obstacle to the construction of the level curves C_t. We would be in difficulty if the set S_t completely blocked the way from the two boundary points P and P' which share the same value $\psi = g = t$ and the same height $y = t/f_0$. In fact, if S_t blocks the path then there is no ψ such that $\psi = g$ on Γ and $|\nabla\psi| \le 1$. In other words, there are no feasible ψ in problem (Q) and therefore no admissible σ in problem (P). This is exactly the question of *limit analysis*, to decide whether the inequality constraint $|\sigma| \le 1$ is compatible with the equilibrium equation (1).

In our problem there are two ways in which the union of disks S_t could prevent the passage of the level curve C_t. We consider the case $t = 0$, when P and P' are the midpoints of the wing tips and the level curve is anticipated to go along the x-axis. First we must ensure that the disks of radius $f_0\ell$ around the internal corners at $x = 0, y = \pm w$, do not block the way through the center. This requires that $f_0\ell \le w$, which is exactly the condition we recognized at the beginning of this note (and barely satisfied, by choosing $f_0 = w/\ell$).

The other uncertainty, not yet faced, is whether the obstacle set S_t might reach so far down as to actually include the point P where the level curve is supposed to start. If we take $t = 0$ in Fig.4a, the question is whether P is less than a distance w from the sloping boundary (where $g = w$). Since its distance from the top corner is ℓ, and the angle there is β, the actual distance to the boundary is $\ell \sin\beta$. Therefore there is a second condition $\ell \sin\beta \ge w$ on the geometry of the butterfly if limit analysis is to permit the prescribed load $f_0 = w/\ell$.

A more standard and more coherent account of limit analysis for antiplane shear can be given in a few lines. Our primal problem is

$$(P')\ \text{Maximize } f_0 \text{ subject to } |\sigma| \le 1,\ \operatorname{div}\sigma = 0,\ \sigma\cdot n = \pm f_0 \text{ and } 0 \text{ on } \Gamma.$$

The unknowns are the scalar f_0 and the vector function σ, and in many problems it is difficult to construct σ. However, even in ignorance of the stress field, the optimal scalar f_0 can frequently be computed by duality. Proceeding as in [3,9], we impose the constraint $\operatorname{div}\sigma = 0$ by introducing a Lagrange multiplier $u(x,y)$ and reach the form known to engineers as the "upper bound" theorem:

(D′) Minimize $\iint |\nabla u|$ subject to $\int_{x=L} u\,dy - \int_{x=-L} u\,dy = 1$.

This minimum will agree with the maximum f_0 in the primal (P′). It is easier to compute, as a consequence of the coarea formula: *The extreme function* u *is a multiple of a characteristic function* [9]. Thus $u = 0$ on part of Ω, and $u = c$ on the other part, with c chosen to satisfy the constraint in (D′). One candidate is the funtion u_1 which vanishes on the left half of the butterfly and equals $1/2\ell$ on the right half. Then $\int u_1 dy = 1$, and the gradient of u_1 is zero except down the center of Ω -- where it is a "line of δ-functions" that represent the jump from 0 to $1/2\ell$. Integrating down the centerline,

$$\iint |\nabla u_1|\, dxdy = 2w/2\ell = w/\ell \quad .$$

Another candidate u_2 is zero except in the corner of the butterfly, where it equals $1/\ell$ in a right triangle whose hypotenuse is the upper half of the wing tip. This half has length ℓ, so that $\int u_2\, dy = 1$. One leg of the triangle goes down the sloping boundary, and the other (of length $\ell \sin\beta$) is perpendicular to that boundary. Along that leg, separating $u_2 = 0$ from $u_2 = 1/\ell$, the gradient is again a line of δ-functions and

$$\iint |\nabla u_2|\, dxdy = \frac{1}{\ell}\,\ell \sin\beta = \sin\beta \quad .$$

You will recognize u_1 and u_2 as representing the two ways in which S_t can block C_t, and with patience they are seen as the only serious candidates in the minimization (D′). Therefore the maximal f_0 is $\min(w/\ell, \sin\beta)$. In our example we assumed that w/ℓ was the smaller of the two, and could therefore take it as f_0. The optimal design for the opposite case $f_0 = \sin\beta < w/\ell$ will be determined by the same rule: Minimize the length of C_t while avoiding S_t.

MINIMUM MATERIAL: THE DUAL PROBLEM

We return to the original problem

(P) Minimize $\iint_\Omega |\sigma|\, dxdy$ subject to $\operatorname{div}\sigma = 0$, $\sigma\cdot n = f$, $|\sigma| \le 1$

in order to find its dual. For numerical computations this form will be of primal importance; it turns out to be an unconstrained optimization. In fact, after a period in which optimization algorithms were regarded as impossibly slow and effort was concentrated on optimality criteria, there is now a clear shift back to techniques of mathematical programming. The improvement has come by a more careful choice of the design variables and the algorithm; Schmit and Fleury [7] have shown how to combine approximations to the primal with explicit solutions of the dual in an extremely effective way. It has been estimated by Pedersen [6] that a mixture of finite elements and simplex methods can now solve a design problem in the time normally needed for ten analysis problems. For the algorithmic step (iii) mentioned in our introduction, that represents an important achievement.

To determine the dual of (P), we introduce a Lagrange multiplier $u(x,y)$ for the constraint $\operatorname{div}\sigma = 0$, and proceed formally with Green's theorem and the minimax theorem:

$$\min_{\substack{\operatorname{div}\sigma = 0 \\ \sigma\cdot n = f \\ |\sigma| \le 1}} \iint |\sigma| = \min_{\substack{\sigma\cdot n = f \\ |\sigma| \le 1}} \max_{u} \iint |\sigma| + u \operatorname{div}\sigma$$

$$= \max_{u} \min_{|\sigma| \le 1} \iint |\sigma| - \sigma\cdot\nabla u + \int_{\Gamma} uf\, ds \quad .$$

At each point we minimize $|\sigma| - \sigma\cdot\nabla u$ subject to $|\sigma| \le 1$. This makes σ parallel to ∇u, and three possibilities arise:

$$\text{if } |\nabla u| < 1 \text{ then } \sigma = 0 \quad \text{(a hole)} \tag{6a}$$

$$\text{if } |\nabla u| = 1 \text{ then } 0 \le |\sigma| \le 1 \quad \text{(fibers)} \tag{6b}$$

$$\text{if } |\nabla u| > 1 \text{ then } |\sigma| = 1 \quad \text{(solid)} \tag{6c}$$

In the first two cases the minimum is $|\sigma| - \sigma\cdot\nabla u = 0$; in the third,

$$\sigma = \frac{\nabla u}{|\nabla u|} \quad \text{and} \quad |\sigma| - \sigma\cdot\nabla u = 1 - |\nabla u| \quad .$$

Combining the three minima into the single expression $(1 - |\nabla u|)_-$, which is zero except when $1 - |\nabla u|$ is negative in case (6c), we reach the dual:

$$\text{(D)} \quad \text{Maximize} \iint_{\Omega} (1 - |\nabla u|)_- \, dxdy + \int_{\Gamma} uf\, ds \quad .$$

We establish in [3] that this maximum does equal the minimum in (P), and that it gives the greatest lower bound on the material required to withstand the surface traction f. The admissible space in (D) is BV, the class of *functions of bounded variation*. For such u the integrals $\iint |\nabla u|$ and $\int |u|$ are finite, and therefore (since $|f| \le 1$ is implied by $\sigma\cdot n = f$, $|\sigma| \le 1$) the two integrals in (D) are well defined.

Since ∇u is parallel to σ, it is perpendicular to $\nabla\psi$; the level curves $u = \text{constant}$ and $\psi = \text{constant}$ form an orthogonal family. In the butterfly, the optimal design had two different regions. One was in the center, where $\psi = \text{constant}$ on circular arcs and $u = \text{constant}$ on radial lines from the internal corners at $y = \pm w$. The other is in the wing, where ψ is constant on straight lines and the level curves of u cross these lines as they cross the wing. *One family of curves is straight exactly when the gradient of the other is a unit vector;* $|\sigma| = |\nabla\psi| = 1$ at the center, and $|\nabla u| = 1$ on the fibers. This is a general property of orthogonal curves, which must have been observed before. The curves themselves are C^1, so the traffic can flow smoothly -- and there may be a chance that minimum principles of this kind could occur in nature.

ACKNOWLEDGMENTS

This research was supported by the National Science Foundation, through contract MCS 78-12363 as well as a postdoctoral fellowship, and by the Army Research Office (DAAG29-K0033).

REFERENCES

1. CHENG, K.T., and OLHOFF, N., An investigation concerning optimal design of solid elastic plates, DCAMM Report 174, Tech. Univ. Denmark, Copenhagen (1980).
2. HAYES, W.C., Stress-morphology relationships in trabecular bone of the patella, OBL Report 80-7, Beth Israel Hospital, Boston (1980).
3. KOHN, R., and STRANG, G., Optimal design and convex analysis, In Preparation.
4. KOHN, R., and STRANG, G., Optimal design for torsional rigidity, *Proc. Int. Symp. on Hybrid and Mixed Finite Element Methods*, Atlanta (1981), to appear.
5. NADAI, A., *Theory of Flow and Fracture of Solids*. McGraw-Hill, New York (1950).
6. PEDERSEN, P., The integrated approach of FEM-SLP for solving problems of optimal design, DCAMM Report 182, Tech.Univ. Denmark, Copenhagen (1980).
7. SCHMIT, L.A., and FLEURY, C., Structural synthesis by combining approximation concepts and dual methods, *AIAA Journal* 18, 1252-1260 (1980).
8. SOKOLNIKOFF, I.S., *Mathematical Theory of Elasticity*. McGraw-Hill, New York (1956).
9. STRANG, G., A minimax problem in plasticity theory, *Springer Lecture Notes in Mathematics* 701, Heidelberg (1979).

STRESS FORMULATION WITH AUGMENTED LAGRANGE METHOD IN ELASTO-PLASTICITY

U. Schomburg

Technical University of Aachen

1. INTRODUCTION

As the stresses are in general the main design variables to predict failure or plastic flow of structures it is obvious that a direct stress formulation should be of some advantage. In particular this is realized for domains with notches, where compared with the displacement method, a finite element method with strong static connectors yields considerably more accurate results as the static boundary conditions can be taken into account exactly.

There are many different choices for strong static connectors. Stress functions present some difficulties in the formulation of boundary conditions especially in three dimensions. A single field approach with stresses can be establihed with Courant's [1,2] penalty method [1-10,12,20] based on the complementary energy [3-7] or the virtual work principle [4,18] where the equilibrium conditions in the domain and on the boundary are taken into account by weighted least square terms, i.e. penalty terms. In order to obtain a good approximation of the static constraints high weights (penalty factors) have necessarily to be chosen.

Unfortunately this sometimes leads to ill-conditioning of the final system of equations. To overcome this difficulty

Powell [8] and Hestenes [9] proposed alterations in optimization theory. In solid mechanics this method can either be understood as a special method of successive loading, resp. distortion [5,11], or as a special Lagrange multiplier method where the Lagrange functional is augmented by a penalty term, in which the constant penalty factor is chosen sufficiently small in order not to run into ill-conditioning. By an appropriate numerical saddle point analysis we can prevent some disavantages of the Lagrange method or in other words we preserve some of the advantages of the penalty method. The stress minimization process still yields a positive definite system of equations, which is not enlarged as the Lagrange multipliers are assumed to be known during the minimization process.

If one wants to improve the satisfaction of the constraints as well as the results of the stresses and displacements, a post-iteration can be started. A convenient method is a gradient method [8,9]. But for our problems the conjugate gradient method turned out to be considerably quicker, i.e. even with small penalty factors only very few iterations are needed [10]. Both methods can be arranged in such a way that the iterations effect only the right hand side of the final system of equations. So the procedure is equivalent to the iterative initial stress or strain method used in incremental plasticity analysis. This fact can be used with some advantage in programming.

2. AUGMENTED LAGRANGE METHOD

To derive the augmented Lagrange method in elasto-plasticity we start with the complementary energy principle. Minimize

$$C(\underline{s}) = \frac{1}{2}\int_V \underline{s}^T E^{-1} \underline{s} \, dV - \int_V \underline{s}^T E^{-1} \underline{s}_o \, dV - \int_{S_u} \underline{\bar{u}}^T N^T \underline{s} \, ds \qquad (1)$$

subject to the equilibrium constraints

$$L^T \underline{s} = - \overline{\underline{b}} \qquad \text{in } V \tag{2}$$

$$N^T \underline{s} = \overline{\underline{t}} \qquad \text{on } S_t \tag{3}$$

where $\underline{s}$, $\underline{s}_o$, $\underline{u}$, $\underline{u}$, $\overline{\underline{b}}$, $\overline{\underline{t}}$ are the vectors of the stresses, initial stresses, displacements, prescribed displacements on S_u, given body forces, given tractions on S_t resp., and E, N, L are the matrices of elasticity coefficients, direction cosines, differential operators [12] resp. The domain is characterized by V, the surface by S. We shall later assume that the equilibrium conditions (3) on the surface S_t can be fulfilled explicitly. Otherwise they will be treated as the volume equilibrium conditions (2).

To get rid of the static constraints (2) we can use the Lagrange multiplier method which yields the Lagrange or Hellinger-Reissner functional. It can be shown that the Lagrange multipliers are the displacements.

$$H(\underline{u},\underline{s}) = C(\underline{s}) + \int_V \underline{u}^T (L^T \underline{s} + \overline{\underline{b}}) \, dV \tag{4}$$

subject to (3).

We are looking now for the saddle point $(\underline{u}^*, \underline{s}^*)$ where $H(\underline{u}^*, \underline{s}^*)$ is minimal with respect to $\underline{s}^*$ and maximal with respect to $\underline{u}^*$. We now have a two-field principle. The discretized system of equations is indefinite and comparatively large as not only the stresses but also the Lagrange multipliers have to be discretized.

The choice of the interpolation functions for $\underline{u}$ and $\underline{s}$ has to be handled with care in order not to run into the limitation principle of Fraeijs de Veubeuke [13] or into the locking phenomina discovered by Nagtegaal [16,17].

The penalty method [1-7] is another way of releasing the stress trial functions from the static constraints (2).

$$P(\underline{s}) = C(\underline{s}) + \frac{a}{2}\int_V (L^T\underline{s} + \overline{\underline{b}})^T (L^T\underline{s} + \overline{\underline{b}})\, dV \tag{5}$$

subject to (3)

We now want to find $\underline{s}^*$, such that $P(\underline{s}^*)$ is minimal with respect to $\underline{s}$. Later on it will be shown that if the penalty factor a tends to infinity then the solution of the original problem is approached.

To show the relation between the penalty method and the Lagrange method [14,5,18,10] we consider the functional

$$H^a(\underline{u},\underline{s}) = H(\underline{u},\underline{s}) - \frac{1}{2a}\int_V \underline{u}^T\underline{u}\, dV \tag{6}$$

subject to (3)

i.e. we have added the quadratic term in the displacements to the Lagrange functional (4) in order to have a regularized functional. The first variation of (6) with respect to $\underline{u}$ has to disappear, which gives us

$$\underline{u} = a(L^T\underline{s} + \overline{\underline{b}}) \tag{7}$$

If this is inserted into the functional (6) we obtain the penalty functional (5). That means, for $a \to \infty$ the penalty functional approaches the Lagrange functional. From the numerical point of view this equivalence statement [14] is in general not true, as the discretized equations obtained from the penalty method are quite often ill-conditioned if large penalty factors are used. This is a considerable disadvantage of the penalty method.

But on the other hand it has the advantage, that no new variables are introduced and that the discretized equations are positive definite.

In order to preserve the advantages of the Lagrange and the penalty method we will mix both methods by augmenting the Lagrange functional with the penalty term. Thus

$$A(u,s) = C(s) + \int_V \underline{u}^T (L^T \underline{s} + \overline{\underline{b}})\, dV + \frac{a}{2} \int_V (L^T \underline{s} + \overline{\underline{b}})^T (L^T \underline{s} + \overline{\underline{b}})\, dV \quad (8)$$

We are looking now for the saddle point $(\underline{u}^*, \underline{s}^*)$ of $A(\underline{u},\underline{s})$, such that for all $\underline{u}$ and $\underline{s}$

$$A(\underline{u},\underline{s}^*) \leq A(\underline{u}^*,\underline{s}^*) \leq A(\underline{u}^*,\underline{s}) \quad (9)$$

This can be achieved by the following algorithm. We start with $\underline{u} = 0$. Then the functional (8) is reduced to the penalty functional. In order not to run into ill-conditioning we take a moderate penalty factor. Some indication of the choice of the penalty factor by energy-balancing is given by Fried [19]. In general the equilibrium conditions will not be sufficiently satisfied. By a post-iteration procedure we will correct this error in each step n+1. The functional (8) will be minimized with respect to $\underline{s}^{n+1}$ by asssuming the displacements to be known by

$$\underline{u}^{n+1} = \underline{u}^n + g\, (L^T \underline{s}^n + \overline{\underline{b}}) \quad (10)$$

This can be interpreted as a gradient method (with the step-length g) to calculate the maximum of the dual functional

$$D(\underline{u}) = \min_{\underline{s}} A\, (\underline{u},\underline{s}) \quad (11)$$

If we take a constant step-length a good choice will be g = a = penalty factor. Then we have linear convergence where the rate of convergence is controlled by the penalty factor. The rate of convergence can be improved considerably by using the conjugate gradient method with optimal step-length [10]. Even for small penalty factors two or three iterations are in general sufficient. The iterations can be obtained very cheaply as the stepwise known displacements influence only the right hand side of the system of equations for the stresses. If the displacements are discretized by piecewise discontinuous trial functions, they can be eliminated on the element level. Compared with the method of selective integration [17] (or collocation at a reduced number of integration points) this gives more freedom in the choice of how much the constraints should be taken into account.

3. NUMERICAL RESULTS

The method outlined has been applied to problems in elasto-plasticity using the initial stress method [12]. We took 9-node isoparametric plane stress elements with selective integration [5,15] for a double notch tension strip of 1mm thickness. Due to symmetry only a quarter of the strip has been considered (Fig. 1). In Fig. 2 the notch tip is shown in detail. In Figs. 3 and 4 the plastic data are given. Young's modulus is given by 21000N/mm^2 and Poisson's ratio by 0.33. We used 99 elements. In Fig. 5 the von Mises stress along the smallest cross-section is drawn for a load of 901.9N. The stress solution (with +) is compared with the displacement solution (solid line) where 8 node isoparametric elements have been used. The differences are due to the fact that the displacement method cannot take the static conditions at the tip boundary into account exactly.

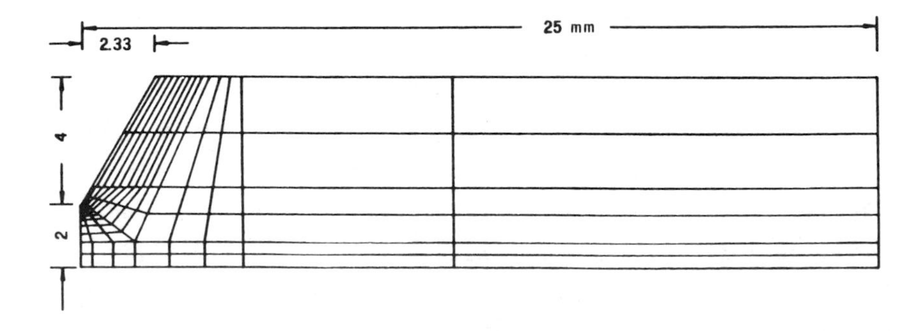

FIG. 1. Finite element net for one quarter of a double notch tension strip.

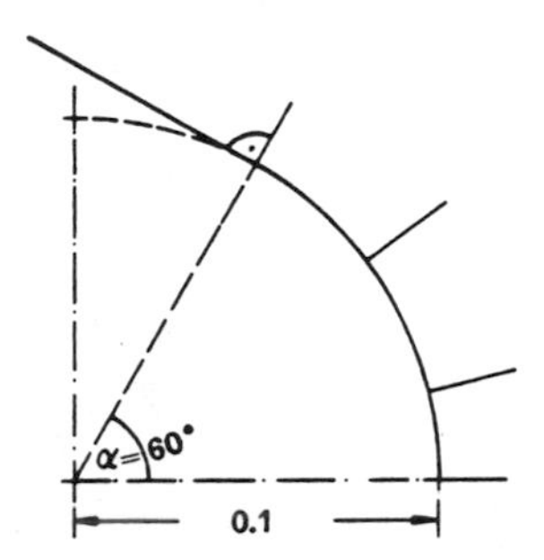

FIG. 2. Detail at the notch tip of the strip.

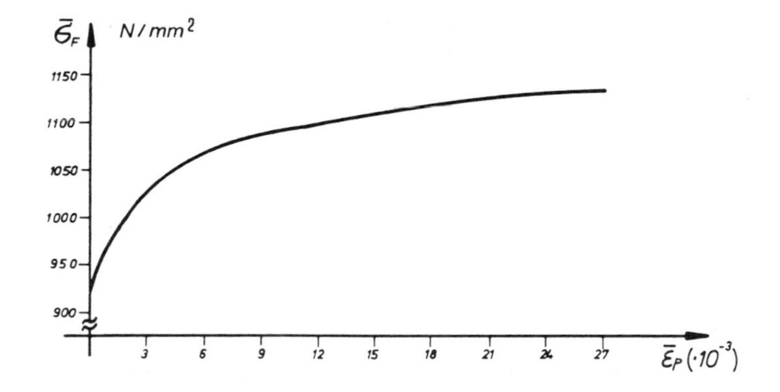

FIG. 3. Stress strain curve (uniaxial) for von Mises flow condition.

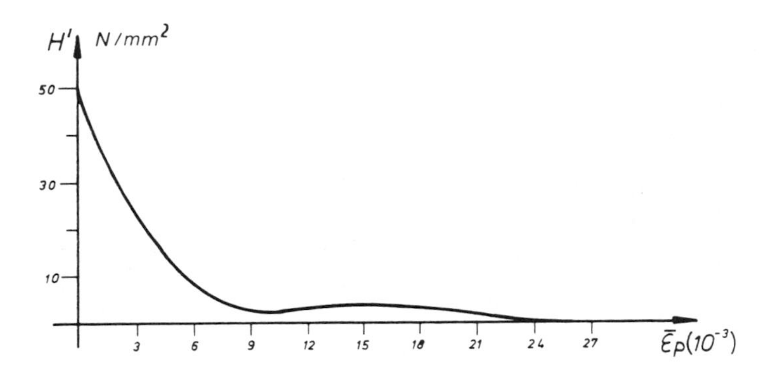

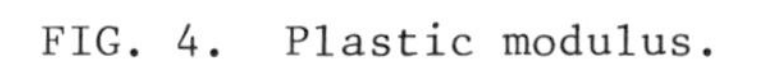
FIG. 4. Plastic modulus.

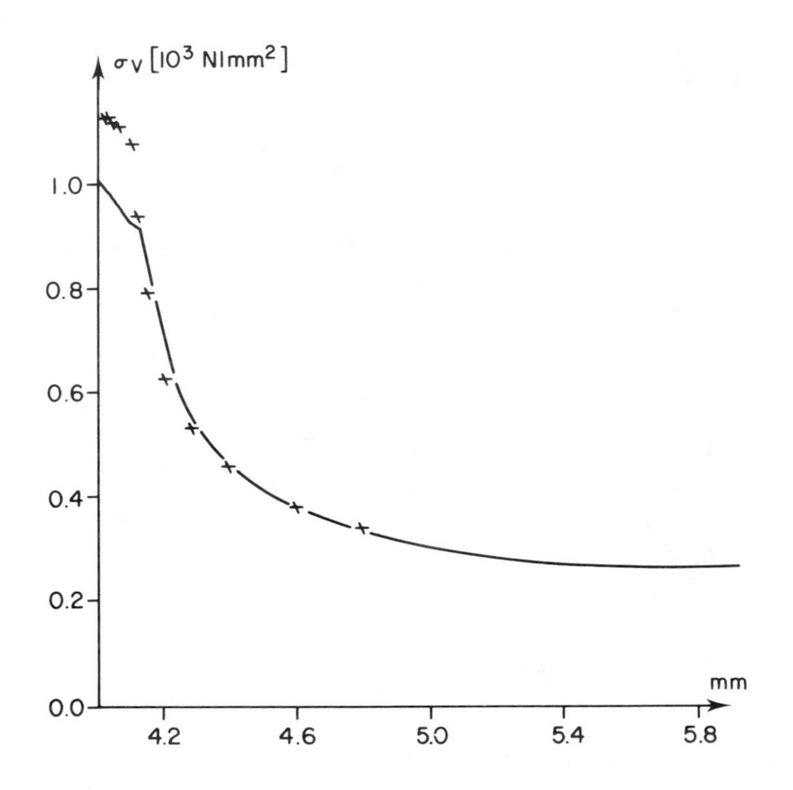

FIG. 5. Von Mises stress along the smallest cross section of the double notch tension strip.

Due to Fig. 3. the von Mises stress should reach the value of about 1100 N/mm^2 at the notch tip.

ACKNOWLEDGEMENTS

The author thanks Dipl.-Ing. P. Bischet for doing the computational work of the numerical examples.

REFERENCES

1. COURANT, R., Über ein konvergenzerzeugendes Prinzip in der in der Variationsrechnung, *Nachrichten von der Königl. Gesellschaft der Wissensch. zu Göttingen* (1922).
2. COURANT, R., Variational Methods for the Solution of Problems of Equilibrium and Vibrations, *Bull. Amer. Math. Soc.* 49 (1943).
3. ZIENKIEWICZ, O.C., Constrained Variational Principles and Penalty Function Methods in Finite Element Analysis. Proc. Conference on *Solution of Differential Equations* in Dundee (1973).
4. SCHOMBURG, U., Anwendung der Penalty-Methode bei finiten Elementen, *ZAMM 56*, T327-329 (1976).
5. SCHOMBURG, U., Die Penalty-Methode mit finiten Elementen in der Technischen Mechanik, Habilitation-Thesis, Techn. University of Aachen (1979).
6. ZIENKIEWICZ, O.C., TAYLOR, R.L., Complementary Energy in Finite Element Analysis, in: R. GLOWINSKI et al. (eds.) *Energy Methods in Finite Element Analysis*, Chichester (1979).
7. SCHOMBURG, U., DEEKEN, G., Postimproving Penalty Method for Equilibrium Constraints in Finite Element Analysis of Elasto-Plastic Structures, *ZAMM 61* (1981).
8. POWELL, M.J.D., A Method for Nonlinear Constraints in Minimization Problems in: R. FLETCHER (ed.), *Optimization* Academic Press, London (1969).
9. HESTENES, M.R., Multiplier and Gradient Methods, J. *Optimization Theory and Applications*, 4, (1969).
10. DEEKEN, G., Unpublished work (1981).
11. RIEDER, G., Iterationsverfahren und Operatorgleichungen in der Elastizitätstheory. *Abhandlungen der Braunschweigischen Wissenschaftlichen Gesellschaft. Band XIV, Zweiter Halbband* (1962).
12. ZIENKIEWICZ, O.C., *The Finite Element Method*, London (1977).

13. FRAEJS DE VEUBEUKE, B., Displacement and equilibrium models in the finite element method, Chapter 9 of *Stress Analysis*, O.C. ZIENKIEWICZ et al. (eds.) J. WILEY & SON (1965).

14. HUGHES, T.J.R., Equivalence of Finite Elements for Nearly-incompressible elasticity , *J. App. Mech.* 44 (1977).

15. DEEKEN, G., SCHOMBURG, U., Verhalten verschiedener FEM-Typen bei Variationsproblemen mit Nebenbedingungen. *ZAMM* 61 (1981).

16. NAGTEGAAL, J.C., PARKS, D.M., RICE, J.R., On Numerically Accurate Finite Element Solutions in the Fully Plastic Range. *Comp. Meths. Appl. Mech. Eng.*, 4, (1974).

17. MALKUS, D.S., HUGHES, T.J.R., Mixed Finite Element Methods Reduced and Selective Integration Techniques: A Unification of Concepts, *Comp. Math. in Appl. Mech. and Eng.* 15, (1978).

18. MALKUS, D.D., Penalty Methods in Finite Element Analysis of Fluids and Structures, *Proc. 5th Conf. Struct. Mech. in Reactor Technology* M6/1, Berlin (1979).

19. FRIED, I., Finite Element Analysis of Incompressible Material by Residual Energy Balancing, *Int. J. Solids Structures* 10 (1974).

20. STEIN, E., Die Kombination des modifizierten Trefftzschen Verfahrens mit der Methode der finiten Elemente, in: K.E. BUCK et al. (eds.) *Finite Elemente in der Statik*, ed. by K.E. Buck et al., Berlin (1973).

ON THE FINITE ELEMENT SOLUTION OF OPTIMAL DESIGN PROBLEMS USING TRANSFORMATION TECHNIQUES

*S. Lyle and †N.K. Nichols

**Kingston Polytechnic, †University of Reading*

1. INTRODUCTION

We consider the application of the finite element method to optimal design problems which involve the computation of an unknown boundary curve. The problem requires the minimization of a functional subject to a set of differential equations on a region D which is closed and bounded by the curve $\Gamma = \Gamma_1 \cup \Gamma_2$. It is assumed that Γ_2 is known and that Γ_1 is to be determined. In cases where the geometry of D is not too complicated, a set of transformations which map D onto a known region D' can be found. After transformation the corresponding system depends explicitly upon the shape of Γ_2, and a non-linear two-dimensional variational problem over the fixed region D' is obtained by the method of Lagrange. It can be shown that the necessary conditions for the solution of the transformed problem are equivalent to those for the solution of the original system. Approximate solutions to the transformed problem are obtained by the finite element method, and the convergence of the approximations to the solution of the optimal design problem is demonstrated. Application of the technique is described for the problem of torsion of a prismatic bar with a doubly-connected cross-section. The region D is of annular form and the problem is to determine the outer boundary to maximize torsional stiffness. Numerical results are given for various choices of the fixed inner boundary.

2. STATEMENT OF THE PROBLEM

The optimal design problem is described by:

$$\min_{\underline{u} \in U_{ad}} \quad I(\underline{u}) \equiv \iint_D g(r, \theta, \underline{u})\, dr d\theta , \tag{2.1}$$

subject to

$$\underline{f}(r, \theta, \underline{u}, \underline{u}_r, \underline{u}_\theta) = 0 , \qquad (r, \theta) \in D, \tag{2.2}$$

on a closed region D with piecewise smooth boundary $\Gamma = \Gamma_1 \cup \Gamma_2$, where Γ_1 is known and Γ_2 is to be determined. The admissable space U_{ad} of functions $\underline{u} = \underline{u}(r, \theta)$ is such that $\underline{u}$ and its first and second partial derivatives are square-integrable n-vector functions, and $\underline{u}$ satisfies certain given boundary conditions on Γ. It is assumed that g is a scalar-valued function and $\underline{f}$ is a p-vector valued function such that g and $\underline{f}$ are twice continuously differentiable with respect to all their arguments.

The problem may be treated by the method of Lagrange. The Lagrangian $L_D[\underline{u}, \underline{\lambda}]$ is defined by

$$L_D[\underline{u}, \underline{\lambda}] = \iint_D (g + \underline{f}^T\underline{\lambda})drd\theta , \tag{2.3}$$

where $\underline{\lambda} = \underline{\lambda}(r, \theta)$ is a square-integrable p-vector function with square-integrable first and second derivatives. At the optimal solution of (2.1) (2.2), it is necessary that the first variation of L_D vanishes and therefore that the optimal solution satisfies

$$\frac{\partial}{\partial \underline{u}} (g + \underline{f}^T\underline{\lambda}) - \frac{\partial}{\partial r} \left(\frac{\partial \underline{f}}{\partial \underline{u}_r} \underline{\lambda}\right) - \frac{\partial}{\partial \theta} \left(\frac{\partial \underline{f}}{\partial \underline{u}} \underline{\lambda}\right) = 0 , \tag{2.4}$$

$$\underline{f}(r, \theta, \underline{u}, \underline{u}_r, \underline{u}_\theta) = 0 , \tag{2.5}$$

and that the transversality condition

$$\int_\Gamma \{\underline{\delta u}^T(\partial\underline{f}/\partial\underline{u}_r)\underline{\lambda} + \delta r[- \underline{u}_r^T(\partial\underline{f}/\partial\underline{u}_r)\underline{\lambda} + g + \underline{f}^T\lambda]$$
$$+ \delta\theta[- \underline{u}_\theta^T(\partial\underline{f}/\partial\underline{u}_\theta)\underline{\lambda}]\}d\theta - \{\underline{\delta u}^T(\partial\underline{f}/\partial\underline{u}_\theta)\underline{\lambda}$$
$$+ \delta r[- \underline{u}_r^T(\partial\underline{f}/\partial\underline{u}_\theta)\underline{\lambda}] + \delta\theta[- \underline{u}_\theta^T(\partial\underline{f}/\partial\underline{u}_\theta)\underline{\lambda} + g + \underline{f}^T\underline{\lambda}]\}dr = 0 \tag{2.6}$$

holds for all admissable variations δr, $\delta\theta$, $\underline{\delta u}$ of r, θ and $\underline{u}$.

A sufficient condition for an extremal of L satisfying (2.4)-(2.6) to give a weak (local) minimum to (2.1)-(2.2) is that the second variation of L is strongly positive in a bounded convex neighbourhood of the extremal [2]. We assume that a minimum exists satisfying these conditions.

3. TRANSFORMATION TO A KNOWN REGION

To obtain solutions to problems of form (2.1)-(2.2) where the boundary segment Γ_2 is unknown, we transform D onto a fixed region D'. The transformation yields a non-linear variational problem which depends explicitly on the shape of the curve Γ_2. The finite element method, as described in section 4, is then applied. This technique may be used for a variety of regions D.

In this section we assume that D is an annular region with known inner boundary Γ_1, given in polar co-ordinates by $r = c_1(\theta)$, and outer boundary Γ_2, with equation $r = c(\theta)$ to be determined. The fixed region D' is chosen to be the annulus bounded by circles of radius $R = 1$, $R = 2$ (see Fig.1). The transformation from (r, θ) co-ordinates to (R, Θ) co-ordinates is given by

$$\Theta = \theta,\ R = (r - c_1(\theta))/(c(\theta) - c_1(\theta)) + 1,\ c \neq c_1,\ \theta \in [0, 2\pi]\ . \qquad (3.1)$$

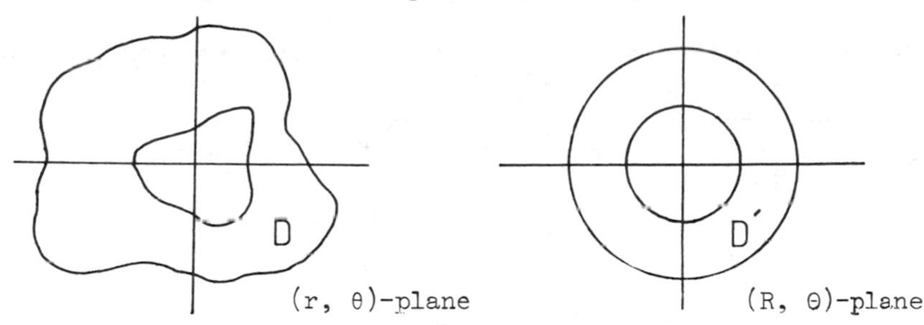

Fig.1.

The problem (2.1)-(2.2) then becomes

$$\min_{(c,\underline{U})\in V} \iint_{D'} G(R, \Theta, c, \underline{U})\,dR\,d\Theta, \qquad (3.2)$$

subject to

$$\underline{F}(R, \Theta, c, c', \underline{U}, \underline{U}_R, \underline{U}_\Theta) = 0, \quad (R, \Theta) \in D', \qquad (3.3)$$

where

$$G(R, \Theta, c, \underline{U}) \equiv g((R - 1)(c - c_1) + c_1, \Theta, \underline{U})(c - c_1),$$

$$\underline{F}(R, \Theta, c, c', \underline{U}, \underline{U}_R, \underline{U}_\Theta) \equiv$$

$$\underline{f}((R - 1)(c - c_1) + c_1, \Theta, \underline{U}, (c - c_1)^{-1}\underline{U}_R, \underline{U}_\Theta -$$

$$((R - 1)(c' - c_1') + c_1')(c - c_1)^{-1}\underline{U}_R)\ .$$

Transformation of the boundary conditions for $\underline{u}$ on Γ yields corresponding conditions for $\underline{U}$ on the boundary Γ' of D'. The admissable space V of functions $\underline{v} = (c, \underline{U})$ is defined such that $c = c(\theta)$, $\underline{U} = \underline{U}(R, \Theta)$ and their first and second derivatives are square integrable, $c(0) = c(2\pi)$, and $\underline{U}$ satisfies the required boundary conditions on Γ'.

The problem (3.2)-(3.3) may also be treated by the method of Lagrange. The Lagrangian $L[\underline{v}, \underline{\mu}]$ is now defined by

$$L[\underline{v}, \underline{\mu}] \equiv \iint_{D'} (G + \underline{F}^T \underline{\mu}) dR d\Theta, \tag{3.4}$$

and (3.1)-(3.2) is equivalent to the problem of finding $v^* = (\underline{c}^*, \underline{U}^*)$ and $\underline{\mu}^*$ such that

$$L^* \equiv L[\underline{v}^*, \underline{\mu}^*] = \sup_{\underline{\mu} \in M} \inf_{\underline{v} \in V} L[\underline{v}, \underline{\mu}], \tag{3.5}$$

where M is the space of square-integrable p-vector functions $\underline{\mu} = \underline{\mu}(R, \Theta)$ with square-integrable first and second derivatives.

At an extremal $(\underline{v}^*, \underline{\mu}^*) \in V \times M$ of the Lagrangian, it is necessary that the first variation of L vanishes. We therefore require that

$$\rho_c(\underline{v}, \underline{\mu}) \equiv \frac{\partial G}{\partial c} + \frac{\partial \underline{F}}{\partial c}^T \underline{\mu} - \frac{\partial}{\partial \Theta}\left(\left(\frac{\partial \underline{F}}{\partial c'}\right)^T \underline{\mu}\right) = 0, \tag{3.6}$$

$$\rho_{\underline{v}}(\underline{v}, \underline{\mu}) \equiv \frac{\partial G}{\partial \underline{U}} + \frac{\partial \underline{F}}{\partial \underline{U}} \underline{\mu} - \frac{\partial}{\partial R}\left(\frac{\partial \underline{F}}{\partial \underline{U}_R} \underline{\mu}\right) - \frac{\partial}{\partial \Theta}\left(\frac{\partial \underline{F}}{\partial \underline{U}_\Theta} \underline{\mu}\right) = 0, \tag{3.7}$$

$$F(R, \Theta, c, c', \underline{U}, \underline{U}_R, \underline{U}_\Theta) = 0, \tag{3.8}$$

$$\int_{\Gamma'} \delta c (d\underline{F}/\partial c')^T \underline{\mu} dR = 0, \tag{3.9}$$

$$\int_{\Gamma'} \delta \underline{U}^T (\partial \underline{F}/\partial \underline{U}_R) \underline{\mu} d\Theta - \delta \underline{U}^T (\partial \underline{F}/\partial \underline{U}_\Theta) \underline{\mu} dR = 0, \tag{3.10}$$

be satisfied by $\underline{v}^* = (c^*, \underline{U}^*)$ and $\underline{\mu}^*$ where $(\delta c, \delta \underline{U})$ is any admissable variation of $(c^*, \underline{U}^*)$.

It can be shown that (3.1)-(3.2) is equivalent to (2.1)-(2.2) in the sense that both systems yield the same set of necessary conditions. In particular we can deduce that under the transformation (3.1) the conditions (2.4)-(2.5) correspond to the conditions (3.7)-(3.8), and for the specified region D, the condition (2.6) is equivalent to (3.6), (3.9) and (3.10). Thus the transversality condition arising from the variation of the region in the original problem corresponds to the conditions ensuring that the variation of the Lagrangian (3.5) of the transformed system, with respect to the boundary curve c, vanishes at the optimum.

The transformed problem (3.2)-(3.3) attains a weak (local) minimum at a solution $\underline{v}^* = (c^*, \underline{U}^*)$, $\underline{\mu}^*$ of the equations (3.6) -(3.10) provided the second variation of L is strongly positive in a closed convex neighbourhood $N(\underline{v}^*) \times N(\underline{\mu}^*)$ of the solution. The existence of such a solution is assumed.

4. FINITE ELEMENT METHOD FOR PROBLEMS ON A FIXED REGION

Numerical solutions to problems of form (3.2)-(3.3) are obtained by the finite element method using the Lagrangian formulation (3.4)-(3.5). Approximations $(\overline{\underline{v}}, \overline{\underline{\mu}})$ belonging to finite dimensional subspaces $V_m \times M_m \subset V \times M$ are determined such that

$$\overline{L} \equiv L[\overline{\underline{v}}, \overline{\underline{\mu}}] = \sup_{\underline{\mu}\in M_m} \inf_{\underline{v}\in V_m} L[\underline{v}, \underline{\mu}], \tag{4.1}$$

where

$$V_m = \{(c, \underline{U}) : c = \sum_{j=1}^{n(m)} \gamma_j \psi_{j,m}, \ \underline{U} = \sum_{j=1}^{m} \underline{\alpha}_j w_{j,m}, \ (c, \underline{U}) \in \overline{V \cap N(\underline{v}^*)}\},$$

$$M_m = \{\underline{\mu} : \underline{\mu} = \sum_{j=1}^{m} \underline{\beta}_j w_{j,m}, \qquad \mu \in \overline{M \cap N(\underline{\mu}^*)}\}.$$

The basis functions $\psi_j = \psi_j(\Theta)$ and $w_{j,m} = (R, \Theta)$ are taken to have local compact support. The spaces V_m, M_m are closed and convex and therefore possess unique best approximations $\hat{\underline{v}}$, $\hat{\underline{\mu}}$ to $\underline{v}^*$, $\underline{\mu}^*$, and we assume that V_m, M_m are good approximating spaces in the sense that $\hat{\underline{v}}$, $\hat{\underline{\mu}}$ converge to $\underline{v}^*$, $\underline{\mu}^*$ as $m \to \infty$.

The solution of the finite dimensional problem (4.1) can be found by direct minimization. Let $\underline{v} = (c, \underline{U}) \in V_m$, $\overline{\underline{\mu}} \in M_m$ satisfy the Ritz-Galerkin equations

$$\iint_{D'} \rho_c(\underline{v}, \underline{\mu})\psi_{j,m} \, dR d\Theta = 0, \quad j = 1, 2, \ldots, n(m), \tag{4.2}$$

$$\iint_{D'} \rho_{\underline{v}}(\underline{v}, \underline{\mu}) w_{j,m} \, dR d\Theta = 0, \quad j = 1, 2, \ldots, m, \tag{4.3}$$

$$\iint_{D'} F(R, \Theta, c, c', \underline{U}, \underline{U}_R, \underline{U}_\Theta) w_{j,m} dR d\Theta = 0, \quad j = 1, 2, \ldots, m, \tag{4.4}$$

$$\int_{\Gamma'} (\partial \underline{F}/\partial c')^T \underline{\mu} \psi_{j,m} dR = 0, \qquad j = 1, 2, \ldots, n, \tag{4.5}$$

$$\int_{\Gamma'} (\partial \underline{F}/\partial \underline{U}_R) \underline{\mu} w_{j,m} d\Theta - (\partial F/\partial \underline{U}_\Theta) \underline{\mu} w_{j,m} dR = 0, \ j=1,\ldots m. \tag{4.6}$$

Then L has a degenerate saddle point at $\overline{\underline{v}}$, $\overline{\underline{\mu}}$ and equations (4.2)-(4.6) give a finite set of algebraic equations for determining the extremal.

Convergence of the approximations $(\overline{\underline{v}}, \overline{\underline{\mu}})$ to $(\underline{v}^*, \underline{\mu}^*)$ is established by the following theorem.

<u>Theorem</u> There exist constants $c_1, c_2, c_3 > 0$ such that

$$||\bar{\underline{v}} - \underline{v}^*||_{2,1} < c_1 \eta_m,$$

$$L^* - c_2 \, ||\varepsilon(\hat{\underline{\mu}})||_2^2 \leq \bar{L} \leq L^* + c_3 \zeta_m,$$

where

$$\eta_m = ||\varepsilon(\hat{\underline{\mu}})||_2 + ||\varepsilon(\hat{\underline{v}})||_{2,1} + ||\varepsilon(\hat{\underline{v}})||_{2,1}^{\frac{1}{2}},$$

$$\zeta_m = ||\varepsilon(\hat{\underline{v}})||_{2,1} + ||\varepsilon(\hat{\underline{v}})||_{2,1}^2,$$

and $\varepsilon(\hat{\underline{v}}) = \hat{\underline{v}} - \underline{v}^*$, $\varepsilon(\hat{\underline{\mu}}) = \hat{\underline{\mu}} - \underline{\mu}^*$.

Here $||\cdot||_2$, $||\cdot||_{2,1}$ are the natural L_2 and Sobolev norms. Proof of the theorem is given in [3]. From the assumptions on the approximation spaces V_m, M_m, it follows that as $m \to \infty$

$$||\bar{\underline{v}} - \underline{v}^*||_{2,1} \to 0 \quad \text{as} \quad |\bar{L} - L^*| \to 0,$$

and convergence is established.

We note that if F is quasilinear, i.e. F is linear in $\underline{U}_x$, $\underline{U}_y$ and c', then the conditions on the admissable spaces V, M may be slightly relaxed. In that case the results of the theorem hold provided c and $\underline{U}$ are square-integrable with square integrable first (partial) derivatives.

5. APPLICATION-TORSION OF A PRISMATIC BAR.

The method of sections 3-4 has been applied to the problem of torsion of a homogeneous isotropic bar with a doubly-connected cross-section. The cross-sectional domain D is of annular form with given inner boundary Γ_1. The problem is to determine the outer boundary curve Γ_2 and constant α such that the torsional stiffness

$$K \equiv 2 \iint_D \phi r dr d\theta + 2\alpha\Omega \tag{5.1}$$

is maximized and the stress function ϕ satisfies

$$\nabla^2 \phi = -2 \quad \text{on } D, \tag{5.2}$$

$$\phi = 0 \quad \text{on } \Gamma_2, \quad \phi = \alpha \quad \text{on } \Gamma_1, \tag{5.3}$$

$$\int_{\Gamma_1} (\partial\phi/\partial n) ds = 2\Omega, \qquad \iint_D r dr d\theta = S_0, \tag{5.4}$$

where Ω is the area of the domain bounded by the inner curve Γ_1 and S_0 is a given constant.

We consider the case where the inner boundary curve is symmetric and solve only on one quadrant $D_0 = \{(r, \theta) : \theta \in [0,\pi/2], c_1(\theta) \leq r \leq c(\theta)\}$ of region D. The differential equation (5.2) is written as a first order system and symmetry conditions are assumed at $\theta = 0$ and $\theta = \pi/2$.

In the case where the inner boundary Γ_1 is a circle of radius c_1, the optimal shape of Γ_2 is known to be a circle of radius c [1]. The solution is given explicitly by

$$c^* = \sqrt{S_0/\pi - c_1^2}\ , \quad \phi^* = \tfrac{1}{2}(S_0/\pi + c_1^2 - r^2), \quad \alpha^* = S_0/2\pi.$$

Numerical results for various choices of Γ_1 have been obtained by the finite element method and the transformation technique using 4×3 and 6×3 elements with piecewise linear and bilinear basis functions $\psi_{j,m}$ and $w_{j,m}$. Solutions have also been obtained using piecewise quadratic basis functions $\psi_{j,m}$ with bilinear basis $\{w_{j,m}\}$ on 4×3 elements. Details of the implementation are given in [3]. For the circular inner boundary the errors in the solutions are shown in Table 1. Convergence is indicated as the number of elements increases. The solutions ϕ also improve when the boundary curve is approximated in the higher order basis.

Table 1. *Errors*

	$\|\|\bar{c}-c^*\|\|_2/\|\|c^*\|\|_2$	$\|\|\bar{\phi}-\phi^*\|\|_2/\|\|\phi^*\|\|_2$	$\|\bar{\alpha}-\alpha^*\|/\|\alpha^*\|$
4 × 3 linear	0.014	0.039	0.014
6 × 3 linear	0.010	0.019	0.014
4 × 3 quadratic-bilinear	0.017	0.028	0.014

Results for an elliptical inner boundary and a star-shaped inner boundary are shown in Figs. 2 and 3. The computed values of the unknown boundary curve are presented and show good agreement for the three solutions obtained.

Problems with other geometries have also been investigated [3], and we observe that a wide variety of problems in optimal design may be treated using the finite element method and transformation technique described here.

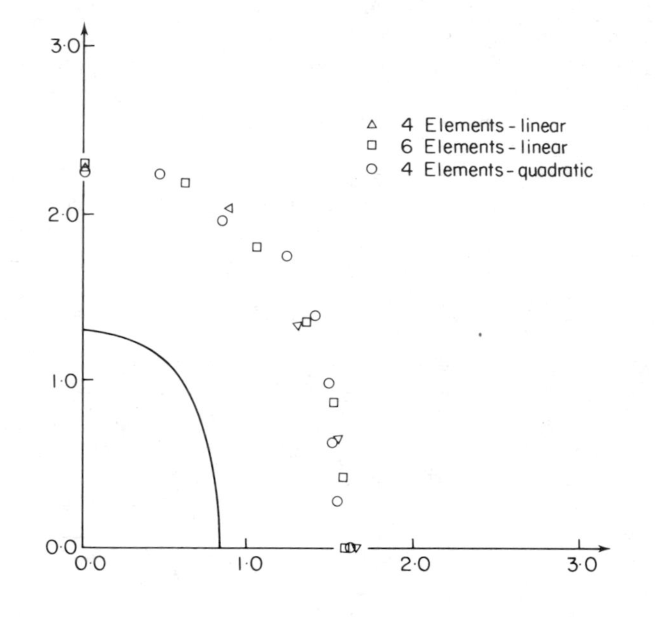
4 Elements - linear
6 Elements - linear
4 Elements - quadratic
3·0
2·0
1·0
0·0
0·0
1·0
2·0
3·0

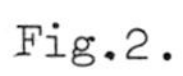
Fig.2.

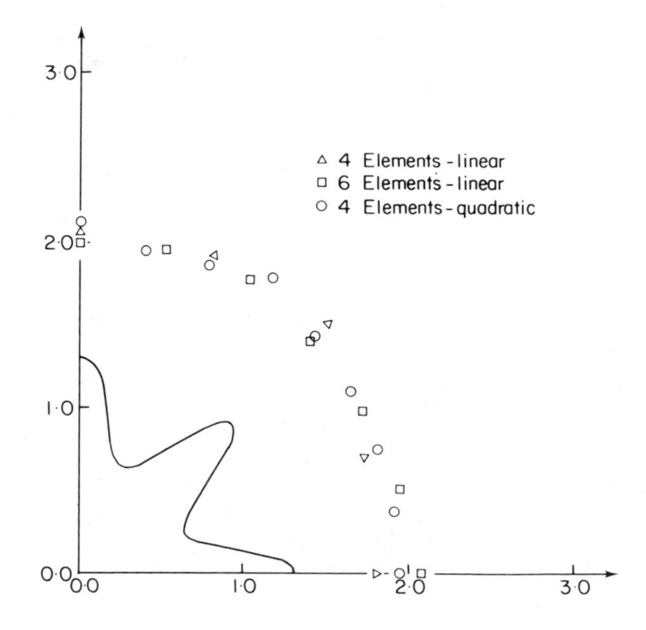
4 Elements - linear
6 Elements - linear
4 Elements - quadratic
3·0
2·0
1·0
0·0
0·0
1·0
2·0
3·0

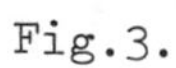
Fig.3.

REFERENCES

1. BANICHUK, N. V., On a variational problem with unknown boundaries and the determination of optimal shapes of elastic bodies, *J. Appl. Math. Mech.* 39, 1037-1047 (1975).

2. GELFAND, I. M., and FOMIN, S. V. *Calculus of Variations*, Prentice-Hall Inc., Englewood, N.J. (1963)

3. LYLE, S. Numerical methods for optimal control problems, *PhD Thesis*, Department of Mathematics, University of Reading (1981).

ANALYSIS OF ECCENTRICALLY STIFFENED WIDE PANELS OF NEARLY OPTIMUM DESIGN USING 3D-DEGENERATED CUBIC ISOPARAMETRIC SHELL ELEMENTS

G. ISE

Rheinisch-Westfalische Technische Hochschule, Aachen

1. INTRODUCTION

A nonlinear stability analysis aims at the complete description of the static response of the structure. So one has to use suitable numerical methods to trace nonlinear pre- and postcritical equilibrium paths and to obtain detailed information about the location and nature of instability phenomena. This paper presents a geometrically nonlinear stability analysis by the finite element method [7]. The algorithms used may be summarized as follows:

2. NUMERICAL METHODS

The static equilibrium of some structure that is discretized by m *load parameters* $y_1 \ldots y_m$ and n *system degrees of freedom* $x_1 \ldots x_n$ is described by the nonlinear algebraic equation

$$G(y,x) = 0 \tag{1}$$

In view of the implicit function theorem the criterium for *static instability* at a solution $(\bar{y},\bar{x})$ of eqn.(1) is

$$\det(G_x(\bar{y},\bar{x}))=0 \quad \text{or} \quad G_x(\bar{y},\bar{x}).\bar{u}=0 \quad \text{with} \quad ||\bar{u}|| = 1 \tag{2}$$

As there is a natural splitting [7],[9] of the stiffness matrix G_x, eqn.(2) is rewritten as a *linear eigenvalue problem*

$$(K_e(\bar{x})+\lambda.K_g(\bar{x})).\bar{u}=0 \quad \text{or} \quad ({}_oK_o+\lambda.({}_oK_1(\bar{x})+{}_oK_2(\bar{x})+K_g(\bar{x}))).\bar{u}=0 \tag{3}$$

with *unit* eigenvalue. The last equation is valid only in Total Lagrangian description, where ${}_oK_1$ is a linear and ${}_oK_2$ is a quadratic function of x. As (y,x) approximates the critical equilibrium state $(\bar{y},\bar{x})$, the solution (λ,u) of the eigenproblem (3) being established in (y,x) approximates $(1,\bar{u})$. Thus for structures with nonlinear subcritical elastic behaviour only a sequence of *supplementary eigenvalue analyses* determines the critical equilibrium state and the eigenvector. In order to solve eqn.(1) the following iterative process is employed:

$$(y^{p+1},x^{p+1}) = \Psi(y^p,x^p) = (y^p,x^p) + (\Delta y^p,\Delta x^p) \tag{4}$$

$$\tilde{K}(y^p,x^p)\begin{Bmatrix}\Delta y^p\\ \Delta x^p\end{Bmatrix} = \begin{Bmatrix}0\\ -G(y^p,x^p)\end{Bmatrix}, \quad \tilde{K}(y,x) = \begin{bmatrix}Q(y,x), & T(y,x)\\ G_y(y,x), & G_x(y,x)\end{bmatrix} \tag{5}$$

As long as [Q,T] is a constant $m\times(m+n)$-matrix, eqn.(4) is exactly *Newton's method* to solve the simultaneous equations

$$H(y,x) = [Q,T].(y-\hat{y},x-\hat{x}) = 0 \tag{6}$$

$$G(y,x) = 0 \tag{7}$$

where [Q,T] and $(\hat{y},\hat{x})$ define the intersection of m hyperplanes in (m+n)-space. After the first "incremental" step equation (6) is satisfied and the Newton process is then given by eqns. (4) and (5). Well known examples are:

a) *Load incrementation* : Q=id, T=0, $\hat{y}$=prescribed loads
b) *Displacement incrementation*: Q=0 , T=[id,0] e.g. $pr_1(\hat{x})$= prescribed displacements
c) *Arc-length incrementation* : Q superdiagonal, and each row of [Q,T] is normalized and perpendicular to all rows of $\tilde{K}$ below that particular row. Such a *bundle of tangents* exists at every point that is a limit point (see [10]) with respect to the load $\partial G(.)/\partial y_m$. $\tilde{K}$ has maximum determinant at such points and all noncritical points, thus extending the result of [10] to the case of m load parameters. See also [1] and [8] .

Although these three iterative procedures have the favourable properties of Newton's method near limit points and in tracing equilibrium paths that branch from bifurcation displacement and arc-length incrementation must be used instead of load increments, since those methods show very sharp convergence due to the fact that $\tilde{K}$ is well conditioned all the time. Using load increments limit points may be surrounded with the help of the *current stiffness parameter*, a scalar quantity [2],[3] that describes stability properties of the structure and is used to adjust load steps and to suppress dangerous iterations.
As an independent concept we have (see [3] :

d) *Automated load corrections* : Q=0, $T(y,x)= -G_y(y,x)^t G_x(y,x)$

This method does not result in a Newton process, since T is not constant, and $\tilde{K}$ is singular at critical points. On the other hand one has grad $f.(\Delta y,\Delta x)=-2.f$ with $f(y,x)=(1/2)G(y,x)^t G(y,x)$.

3. FINITE ELEMENT MODEL

In the analysis the 3D-degenerated cubic isoparametric plate/shell element [9] being part of the NISA program system [4] was used to model both plates and eccentric stiffeners. At each of its 16 nodes the element has 5 degrees of freedom to describe the displacements inside the shell volume: three translations u, v,w and two movements of the shell normal P, e.g. the change in angle between P and the x-axis and the rotation of P around the x-axis. In order to combine a stiffener (built from shell elements) to the plate a slightly incorrect but efficient model was used: The line of attachment is supposed to be parallel to the x-axis and to be a connection line for two elements from the plate model. The ground line of the stiffener consists of nodes located in the nodes of the plate model, whose u,v,w-displacements are coupled to those nodes. Also the rotations around the x-axis are coupled to enforce bending and twisting of the stiffener in sympathy with the deflections of the plate. The remaining degrees of freedom are left uncoupled and may be eliminated on the stiffener side.

4. NUMERICAL EXAMPLE

A longitudinally stiffened wide panel (see [11] and Fig.1, E = 210000 N/mm^2,υ =0.3) is loaded in x-direction (see Fig.2) the loaded edges being simply supported. The wide panel is supposed to be properly described by one section shown in Fig.2 and Fig.3 which is discretized with up to 30 elements assuming suitable

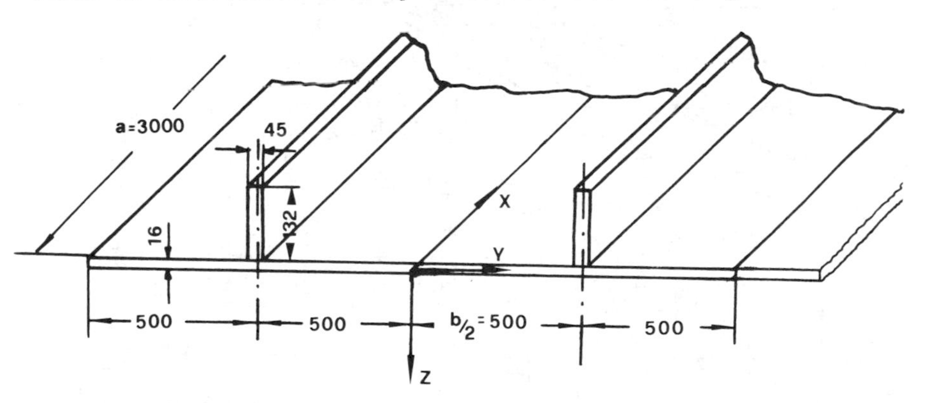

FIG. 1 Dimensions of wide panel in mm

symmetry conditions. Under subcritical compression the panel remains straight as long as no further restrictions or imperfections are present. So critical loads and buckling modes result from a linear stability analysis (see Fig.3a,b,c). Overall buckling occurs first at less than 88% of the failure load of a beam

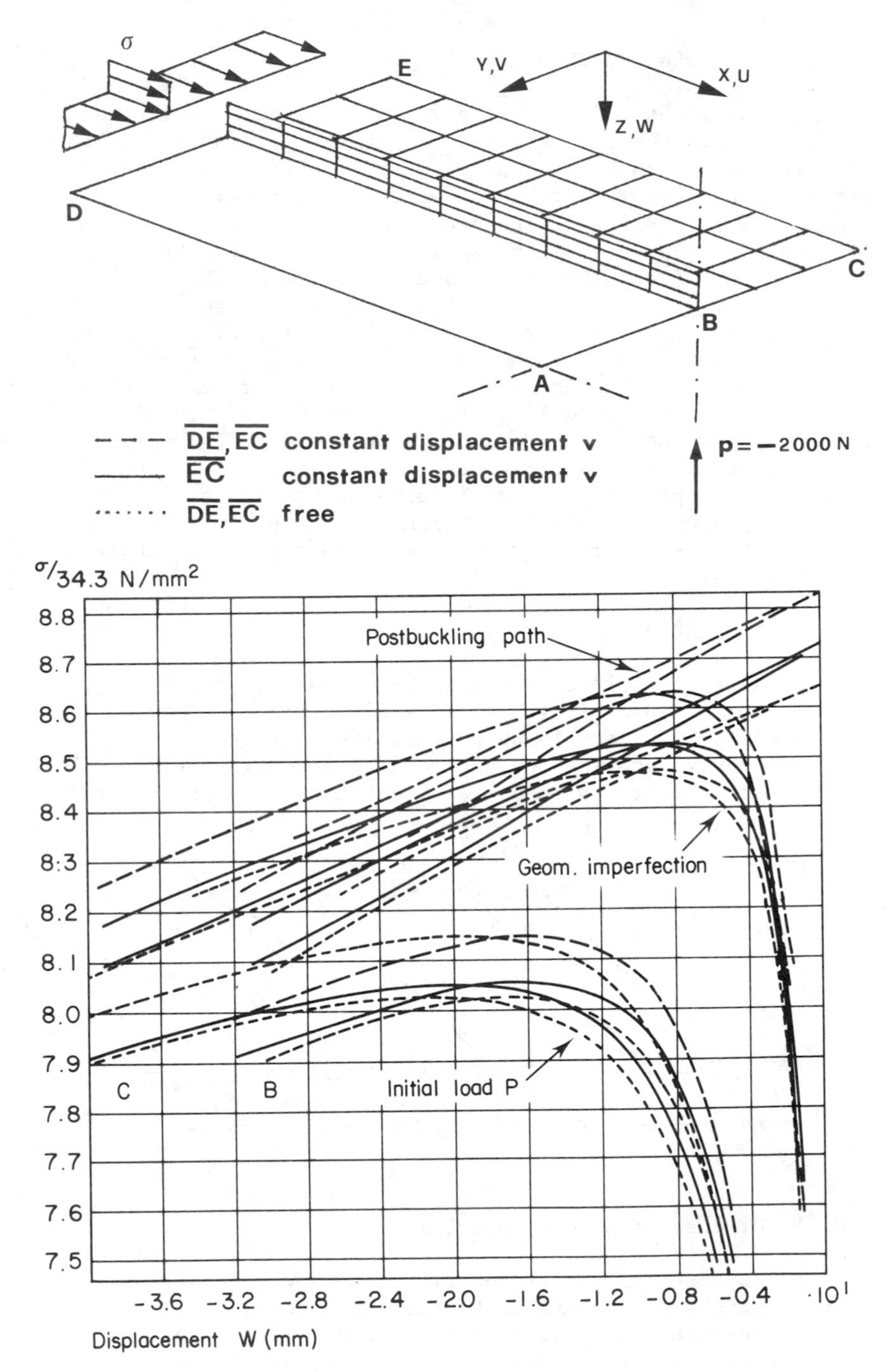
σ
E
Y,V
X,U
Z,W
D
C
B
A
DE, EC constant displacement v
EC constant displacement v
DE, EC free
p = −2000 N
σ/34.3 N/mm²
8.8
8.7
8.6
8.5
8.4
8:3
8.2
8.1
8.0
7.9
7.8
7.7
7.6
7.5
Postbuckling path
Geom. imperfection
Initial load P
C
B
-3.6 -3.2 -2.8 -2.4 -2.0 -1.6 -1.2 -0.8 -0.4 ·10¹
Displacement W (mm)

FIG. 2

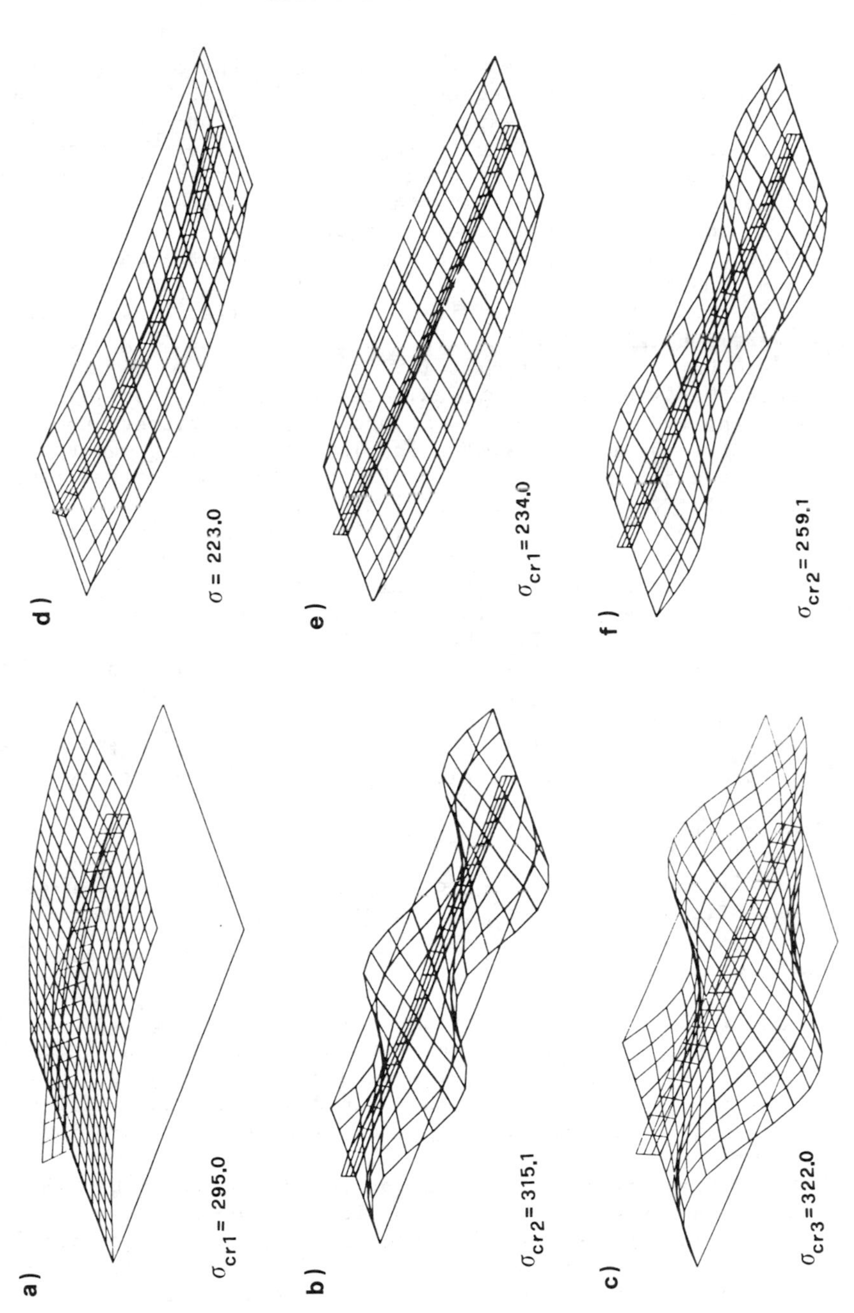
a)
σcr1 = 295.0
b)
σcr2 = 315.1
c)
σcr3 = 322.0
d)
σ = 223.0
e)
σcr1 = 234.0
f)
σcr2 = 259.1

FIG. 3

column with full effective width which is $343.N/mm^2$. The plate with clamped longitudinal edges posseses the same buckling load of $343.N/mm^2$, so one can expect some imperfection-sensitivity [5] Indeed, postcritical analyses of perfect and imperfect models show a decrease in load in connection with increasing deflections (see Fig.2). In our opinion this is due to the growing deformation of the cross-section as is typically shown in Fig.3a. The same phenomena occurs if in Fig.1 the plate thickness is 18.64 mm and the stiffener thickness is 25 mm at same weight. So it is not a necessary condition for this effect that local and overall buckling "coincide". Local buckling (Fig.3b,c) proved to be stable and symmetric.
The model bends under subcritical compression (see Fig.3d) as soon as the v-displacement in $\overline{EC}$ and $\overline{DA}$ is restricted. Local buckling (Fig.3e,f) occurs at a low load level since membrane stresses are quite different now. Fig.4 shows as a dotted line the estimates for the lowest critical load gained by three eigenvalue analyses. The branching equilibrium path remains stable and approximates the primary path again which turns out to be stable again at a higher load level. In an imperfect approach the buckling mode (Fig.3e) is constrained by a couple of forces (F=±400 N at A and C). The equilibrium path is stable and approaches the other paths after some maximum deviation again at a high load level. Obviously the buckles decrease when the overall deflection grows.
A "classical" analytical linear stability analysis confirms the bifurcation loads. To do so plate buckling is described by

$$D.\Delta\Delta w_c = \lambda.(N_x^o.w_{c,xx} + \beta.N_x^o.w_{c,yy}) \qquad 0 \leq \beta \leq \upsilon=0.3. \tag{8}$$

The conditions of elastic support by the eccentric stiffeners are established - with a minor change - as in [2].

5. REMARKS

a) Equilibrium paths that branch from bifurcation points are simply calculated as RESTART-jobs with displacement incrementation. The start position is a guess for the critical equilibrium state which is disturbed by a suitably scaled eigenmode. This information is obtained from the previous stability analysis.
b) A reduction of 25% CPU costs for the method described in [6] is worth reporting. Without any loss of convergence properties in comparison with the Newton method, the algorithm itself determines the minimum number of $L.D.L^t$-factorisations of the stiffness matrix by checking the convergence of the "secondary iterations".
c) In practical application arc-length incrementation and automated load corrections are restricted to the case m=1 .

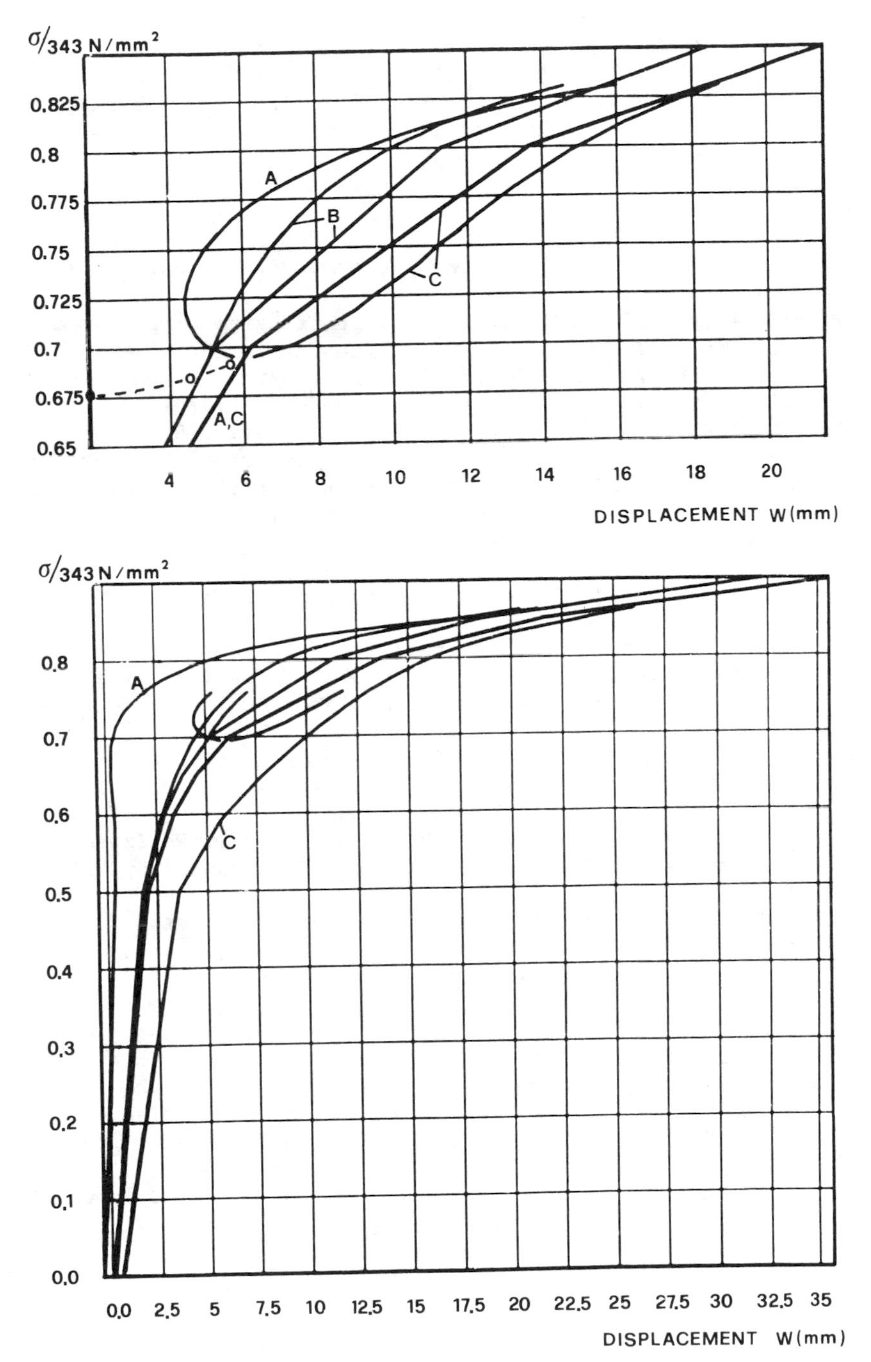
σ/343 N/mm²
0.825
0.8
0.775
0.75
0.725
0.7
0.675
0.65
A
B
C
A,C
4
6
8
10
12
14
16
18
20
DISPLACEMENT W(mm)
σ/343 N/mm²
0.8
0.7
0.6
0.5
0.4
0.3
0.2
0.1
0.0
A
C
0.0
2.5
5
7.5
10
12.5
15
17.5
20
22.5
25
27.5
30
32.5
35
DISPLACEMENT W(mm)

FIG. 4

ACKNOWLEDGEMENTS

Financial support from the Deutsche Forschungsgemeinschaft under grant Schu 173/12 is gratefully acknowledged.

REFERENCES

1. BATOZ,J.-L.,DHATT,G., Incremental Displacement Algorithms for Nonlinear Problems, *Int. J. Num. Meth. Engng. 14,1262-1267(1979)*
2. BERGAN,P.G.,SÖREIDE,T.H., Solution of Large Displacement and Instability Problems Using the Current Stiffness Parameter, in: BERGAN,P.G. et al. (ed.),*Finite Elements in Nonlinear Mechanics,* Tapir Press, Trondheim (1978)
3. BERGAN,P.G., Automated Incremental Iterative Solution Schemes, in: TAYLOR,C. et al.(ed.), *Numerical Methods for Nonlinear Problems,* Pineridge Press, Swansea (1980)
4. BRENDEL,B.,HÄFNER,L.,RAMM,E.,SÄTTELE,J.M., Programmdokumentation Programmsystem NISA, Bericht, Institut für Baustatik, Universität Stuttgart (1977)
5. BUDIANSKY,B.(ed.), *Buckling of Structures*, Springer (1976) Section IV: Mode interaction
6. ISE,G., On Convergence Properties of the Extended Alpha-Constant Stiffness Method, in:WHITEMAN,J.R.(ed.) *The Mathematics of Finite Elements and Applications,*Academic Press (1978)
7. RAMM,E.,*Geometrisch nichtlineare Elastostatik und finite Elemente,* Habilitationsschrift, Universität Stuttgart(1975)
8. RAMM,E.,Strategies for Tracing Nonlinear Response Near Limit Points, in: WUNDERLICH,E. et al.(ed.), *Nonlinear Finite Element Analysis in Structural Mechanics,*Springer (1981)
9. RAMM,E.,A Plate/Shell Element for Large Deflections and Rotations, in: ODEN,J.T. et al.(ed.),*Formulations and Computational Algorithms in Finite Element Analysis,*MIT press, Cambridge (1977)
10. RIKS,E., The Application of Newtons Method to the Problem of Elastic Stability, *J. Appl. Mech. 39,1060-1066 (1972)*
11. SÖREIDE,T.H.,*Collapse Behaviour of Stiffened Plates Using Alternative Finite Element Formulations,* Report No. 77-3, 1977, The Norw. Inst. of Techn. The University of Trondheim Norway
12. TVERGAARD,V., Imperfection-Sensitivity of a Wide Integrally Stiffened Panel under Compression, *Int. J. Solids Struct. 9, 177-192 (1973)*

A QUASI VISCOUS METHOD FOR ELASTIC AND PLASTIC POST BUCKLING ANALYSIS OF PLATES AND SHELLS

B.-H. Kröplin, D. Dinkler

Technische Universität Braunschweig

1. INTRODUCTION

In the analysis of imperfect plates and shells with or without stiffeners local snap throughs or limit points may occur at very low load level, very much lower than the elastic or elastoplastic limit load. In such cases the structure switches over from one buckling mode to another. Snap throughs with or without bifurcations may occur. In the common static incremental procedure difficulties arise often at the point of static instability. The tangential stiffness matrix displays singularities and the iteration with the out of balance forces diverges. These drawbacks are caused by the reduction of the dynamic process to a static one.

Various methods have been developed to overcome the problem in a static manner, e.g. artificial spring method [5], displacement control method [4], constant arc length method [6] or superimposing of parts of the eigenmode related to the lowest eigenvalue in the critical point [1].

In this paper the instabilities are controlled by assuming the structure to be surrounded by a viscous medium. The static equations are augmented by a damping force, which carries a part of the external loads. The damping force is used for increasing convergence as well as for governing the approximation of the load path. The classical incremental procedure can be replaced by a creep type formulation where the external loads can be applied simultaneously.

2. BASIC EQUATIONS

The analysis is governed by a viscous equation (2.1) where $K(x)$ represents the tangent structural matrix, $\underline{x}$ the vector of unknowns and $\underline{p}$ the external load. $D\dot{\underline{x}}$ describes the damping force. A dot indicates derivatives with respect to the time. For time integration a backward difference scheme is used (2.2). Substituting $\underline{x}$ by the relationship (2.3) the governing equation (2.4) for the analysis is obtained. The right hand side of (2.4), $\underline{p}^u$ is interpreted as unbalanced force. The fictitious viscosity D^*, which includes a viscosity and a time step (2.5) serves as step size device and load path control and avoids singularities of the effective structural matrix $(K(x) + D^*)$. In order to perform the above tasks it is sufficient to introduce D^* as diagonal matrix.

$$K(x)\ \underline{x} + D\ \dot{\underline{x}} = \underline{p} \tag{2.1}$$

$$\underline{x}(t) = \frac{1}{2}(1+\xi,\ 1-\xi)(\underline{x}_{n+1},\ \underline{x}_n)^T,\ \xi = 1. \tag{2.2}$$

$$\underline{x}_{n+1} = \underline{x}_n + \Delta\underline{x} \tag{2.3}$$

$$(\ K(\underline{x}) + D^*)\ \Delta\underline{x} = \underline{p}^u \quad ,\ \underline{p}^u = \underline{p} - \underline{p}_i \tag{2.4}$$

$$D^* = \frac{D}{\Delta t} \tag{2.5}$$

Every iteration cycle consists of the following steps:

1. unbalanced forces $\underline{p}^u$,

2. damping D^*,

3. current structural matrix $K(\underline{x})$,

4. increment of the unknowns $\Delta\underline{x}$.

Depending on the strength of the nonlinearity the effective stiffness can be updated in every step or after a few steps. During constant stiffness iteration the change of the effective stiffness times the incremental unknown $\Delta\underline{x}$ has to be considered on the right hand side of equation (2.4).

3. SEARCH OF STABLE EQUILIBRIUM STATES

Equation (2.4) is stable for a wide range of values of D*. However a constant D* is not economical for all parts of the solution process. Depending on $K(\underline{x})$ it may cause over or under damping. To diminish this effect D* is introduced as a function of the unbalanced forces (3.1). Equation (3.1) can be used for direct search of stable equilibrium states. The scalar λ raises the convergence rate and can be chosen within a wide range without affecting the stability of the iteration. ε is a predetermined value, which prevents D* from becoming too large, when x_i tends to zero.

$$D^*_{ii} = \left|\lambda \frac{p^u_i}{x_i}\right| \qquad |x_i| \geq \varepsilon \qquad (3.1)$$

An example is shown of a snap through and a snap back of a thin steel plate loaded by an end shortening $\bar{u}$, see fig. 1. The imperfect plate is simply supported along all edges. It is free to move in plane in the longitudinal direction and fixed in plane in transvers direction. The calculations are carried out with a four node mixed element with the displacements u_i, the axial forces $n^{\alpha\beta}$ and the bending moments $^{1}m^{\alpha\beta}$ as nodal unknowns [3].

When loaded incrementally it buckles first to the imperfection shape, later it tends to the third buckling mode, see fig. 1. Before the third eigenmode is reached, the Newton Raphson iteration breaks down despite careful increment adjustment. The same performance could be observed with displacement and with mixed finite element formulation. In contrast to the incremental analysis stable equilibrium states were calculated with the damping before the snap through (points 1, 2,3) and after the snap through (points 4 to 8) without difficulties with the load applied either in large increments or in one step. For the example 20 iterations were necessary to jump over the snap from point 3 to point 4, where the Newton Raphson iteration failed. After settling on the path behind the snap through the further behaviour is investigated by loading and unloading. In the load case the third mode is stable (point 5). During unloading further snaps occur, first to the second buckling mode (points 6 and 7), later to the first mode (point 8). For reloading from point 8 the structure takes again the load path to the third mode and the snap through. The se-

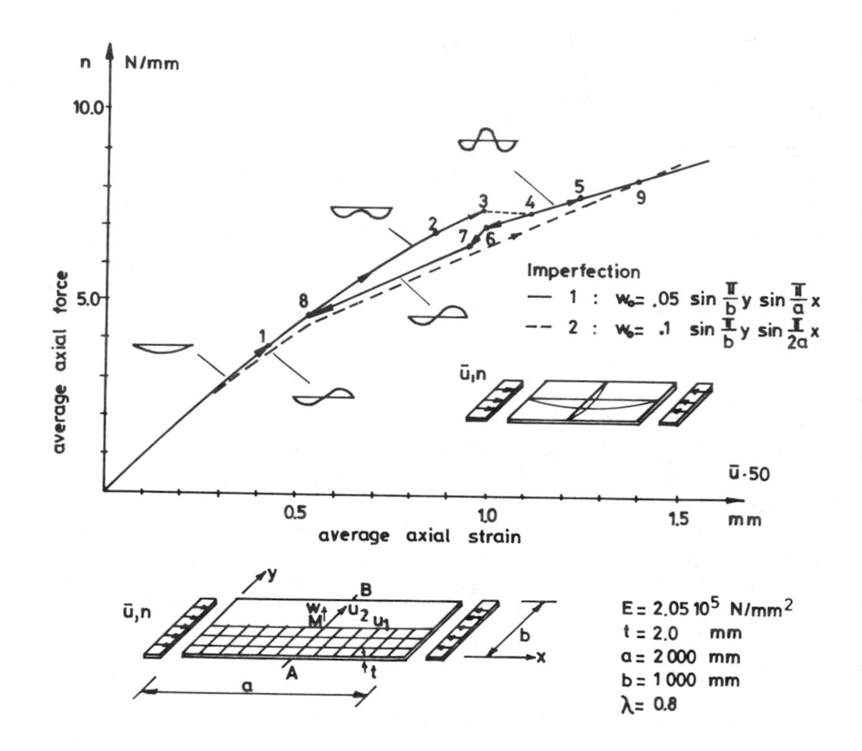

FIG. 1 Snapping plate under axial compression

cond mode, which could from classical buckling theory be expected to be related to plate geometry a/b = 2, occurs only in the unloading case. This indicates that 8 is not a bifurcation point in the classical sense.

In a different run a path associated with the second mode was calculated for a second-mode initial imperfection. This is, in the first part of the load-deflection curve, a lower path, whereas at point 9 there occurs the same phenomenon as at point 8. The switch to the lower third-mode-path did not take place. It should be noted that a switch back from point 4 to point 3 was possible for a constrained situation where the line A-B was taken as a line of symmetry.

4. TRACING OF THE LOAD PATH

After the calculation of a stable equilibrium state it is in many cases necessary to trace the path in stress and strain space in more detail. This is achieved by decomposing the damping into components tangential and orthogonal to the load vector $\underline{p}$, see fig. 2. $\underline{p}$ consists of all nodal components of the external load.

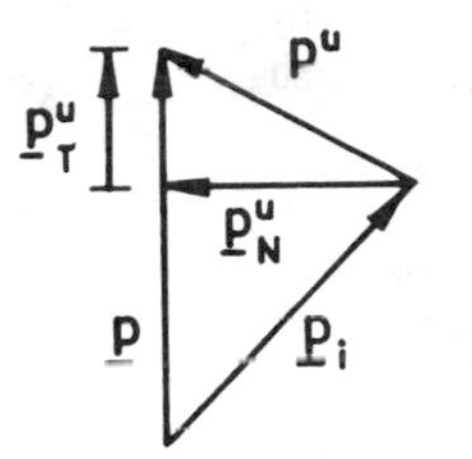

FIG. 2. Decomposition of the load vector

Due to the nonlinearity $\underline{p}$ and $\underline{p}_i$ differ in size and direction. The load path is traced better the more $\underline{p}$ and $\underline{p}_i$ coincide [5]. An approximate tracing of the load path is achieved, if the distance between the internal forces $\underline{p}_i$ and the load direction $\underline{p}$ stays within a predetermined margin. A measure for the accuracy of the approximation is the absolute value of $\underline{p}_N^u$, (4.1). Thus $\underline{p}_N^u$ has to vanish much earlier than $\underline{p}_T^u$ (4.2) during the fictitious creep process. This is easily realised by decomposing the damping D^* into the weighted parts $\alpha \underline{p}_T^u$ and $\beta \underline{p}_N^u$ (4.3). The ratio α/β governes the deviation from the load path, while the factor λ again rules the step size.

$$\underline{p}_T^u = \frac{\underline{p}_T^T \underline{p}^u}{\underline{p}^T \underline{p}} \underline{p} \tag{4.1}$$

$$\underline{p}_N^u = \underline{p}^u - \underline{p}_T^u \tag{4.2}$$

$$D_{ii}^* = \left| \frac{\lambda}{x_i} (\alpha p_{Ti}^u + \beta p_{Ni}^u) \right| \tag{4.3}$$

The strategy of weighted damping is independent of the nonlinearities considered in the problem and is therefore applicable to geometric nonlinear problems as well as to material nonlinearities.

An example is given for the snap through of a shallow arch, where the load path is traced in the elastic and plastic range. The plastic flow is described by a yield function expressed in terms of axial forces and bending moments [3].

The shallow arch, see fig. 3, is subjected to a point load at the centre. The mesh beneath the point of load is refined, in order to improve the approximation of the plastic behaviour at the centre, where eventually a plastic hinge occurs.

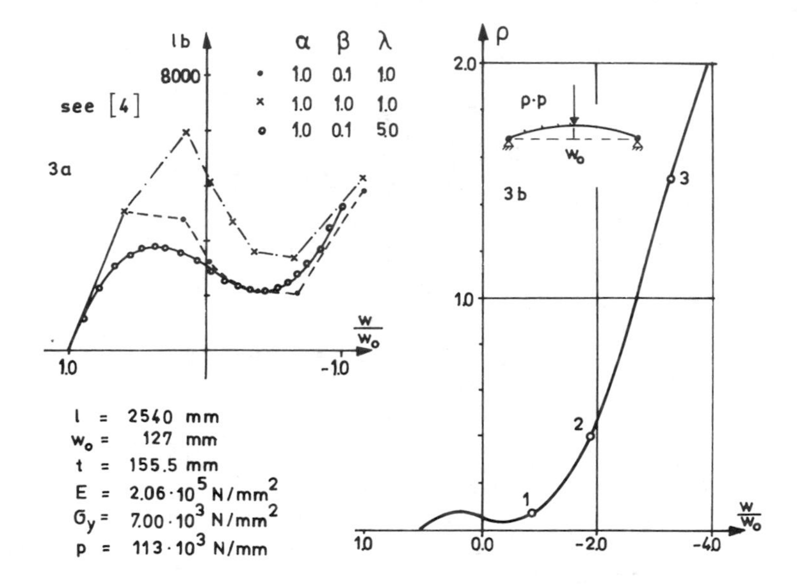

FIG. 3 Elastic and plastic load deflection curve of a shallow arch.

Various approximations of the elastic solution in the snap through region are given for different α and β. These are compared with the solution given in [4]. For $\beta/\alpha = 1/10$ and $\lambda = 5$ the approximation has an error of less than 1%.

These values were taken in the elastic-plasiic calculation of a similar example, see right hand diagram in fig. 3. At point 1 the first fibre in the central cross section starts to yield, and at point 2 this cross section is completely plastified so that a plastic hinge occurs. Further increase of the load causes a shift of the stresses in the cross section from bending moments to axial forces. The bending moments decrease, while the axial forces increase, see fig. 4. The whole structure yields at point 3.

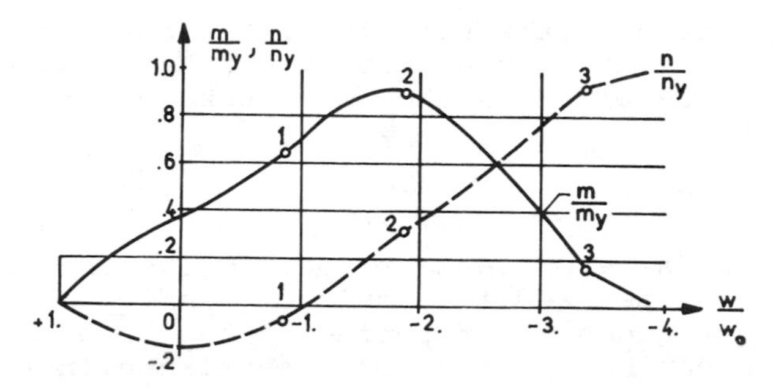

FIG. 4 Axial force and bending moment at the centre

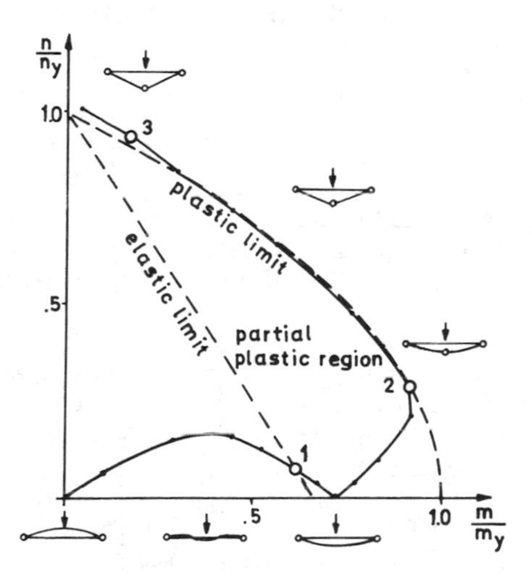

FIG. 5 Path of the central mode in the moment-force space

An increase of the load is still possible on account of the large deflection effect. Fig. 5 shows the post-creep path in the space of bending moments and axial forces. The different states are shown, together with the calculated deflection shapes of the structure. After the increase of the axial forces during the snap through, the path enters the partial plastic zone (point 1), reaches the plastic limit state, where the plastic hinge forms (point 2) and creeps very close to the plastic limit state towards a pure membrane state. As plasticity spreads out over the structure, the curvature straightens out and ultimately the structure behaves like a cable. The split damping was the sole measure for producing the path.

5. CONCLUSION

Numerical experience to date shows that artificial damping can be used as a suitable remedy in static unstable regions and for tracing the load path. This approach can be applied for general nonlinear cases. Classical load incrementation is not necessary.

ACKNOWLEDGEMENT

The authors would like to acknowledge the support provided by the "Deutsche Forschungsgemeinschaft" for this work.

6. REFERENCES

1. ARGYRIS, I.A., HILPERT, O., HINDENLANG, K., MALE-JANNAKIS, G.A., SCHELKE, E., *Flächentragwerke im Konstruktiven Ingenieurbau.* ISD-Bericht Nr. 263 (1979)

2. BERGAN, P.G., Solution Algorithmus for Nonlinear Structural Problems, *Proceedings Int. Conf. on Engineering Application of the Finite Element Method.* A.S. Computas, Hovik, 9-11 May 1979

3. EGGERS, H., KRÖPLIN, B., Yielding of Plates with Hardening and Large Deformations.*Int. J. num.Meth.* Engng. 12, 737-750 (1978)

4. HAISLER, W., STRICKLIN, J., KEY, J., Displacement Incrementation in Nonlinear Structural Analysis by the Self-Correcting Methods. *Int. J. Num. Meth. Engng.* 11 3-10 (1977)

5. SHARIFI, P., POPOV, E.P., Nonlinear Buckling Analysis of Sandwich Arches. *Proc. ASCE, J. Engng. Div.*, 97 1397-1312 (1971)

6. WEMPNER, G.A., Discrete Approximations Related to Nonlinear Theorie of Solids. *Int. J. Solids Struct.* 7, 1581-1599 (1971)

ON FINITE ELEMENT DISCRETIZATION OF A HIGHER-ORDER SHELL THEORY

Tarun Kant

University of Wales, Swansea

1. INTRODUCTION

A structural shell is a body bounded by two curved surfaces. In a shell theory the behaviour of the shell is governed by the behaviour of an appropriate reference surface. The transformation of the three-dimensional elasticity equations into an approximate system of two-dimensional shell equations is an essential feature of any shell theory. In shell theories the membrane and bending behaviour are coupled. The coupled deformations in the form of stretching and curvature change of the reference surface are required in predicting the strains existing throughout the shell space. Shells are extremely important structural elements in applications such as nuclear reactors, pressure vessels,space-craft, missiles, etc. Both their theory and analysis have always presented challenge to the researchers in the area. The so-called 'thin shell theory' is firmly established in the literature and is used extensively for analytical solutions and numerical analysis [6]. However, its reliability in many practical applications involving complex geometries and loadings, cut-outs, branches, intersections, etc. is questioned because of the inherent assumptions implicit in such a theory. The validity of this theory is further questioned when it is used in conjunction with the finite element discretization in the numerical treatment.

The finite element analysis of shell structures has developed in three distinct directions -the faceted form using flat elements, the curved shell elements formulated directly from an appropriate shell theory and the solid elements specialized to tackle shells by introducing the constraints of straight and inextensional normals. A fundamental difficulty in attempting to achieve interelement compatibility is always encountered when Kirchhoff's assumptions are introduced in the first two of these approaches. This is due to the occurrence of the second derivative of the normal displacements of the reference surface in the energy expression corresponding to flexural behaviour. A comprehensive review of all the developments can be found in the books of Cook [2] and Zienkiewicz [12]. In the first approach where an arbitrary shell is assumed to be an assemblage of flat elements, the coupling effects of membrane and bending strains in the energy expressions are absent at the element level and such a coupling occurs only on the boundary. In the second approach, general shell theories are used to derive doubly curved elements. The essential use of curvilinear coordinates in such theories makes geometrical representation of the reference surface rather complex. Because of this, only special elements in the form of cylindrical and axisymmetric shell forms have been developed. A good review of such an effort is presented by Key and Beisinger [5] in 1970. Ironically, not much has been achieved in this direction since then. The third approach involves penalty function formulation of the degenerate solid element with transverse shear energy and is used extensively for the analysis of general shells [1]. Indeed, the use of selective/reduced integration techniques in this formulation is very handy in dealing with thin shells [7-10]. However, it should be realized that there are certain limitations in the use of these elements. For example, the effects of transverse normal strain and transverse normal stress are not accounted for; the transverse shearing strains are assumed constant through the thickness and a fictitious shear correction coefficient is introduced to account for warping of the cross-section.

2. A HIGHER-ORDER SHELL THEORY

In the previous section we have presented a brief overview of the finite element analysis of shell structures. We note that there are two possibilities for errors to enter into the numerical solutions - errors due to use of inconsistent shell theory and discretization errors which occur in dealing with shells in a piecewise manner. To ensure that any errors which do arise come from the numerical discretization process and not from an unsuitable shell theory - we outline here the basis of a higher-order general shell theory in the context of the finite element method. The derivation of such a theory is available in a separate report [4]. In deriving this theory the three-dimensional system of equations of the linear elasticity theory is transformed into a two-dimensional system of shell equations in orthogonal curvilinear coordinates by deploying only the geometrical assumptions of $(h/R)^2 \ll 1$. The theory is free from the usual contradictions present in the Love's thin shell theory and is applicable to geometrically thin as well as thick shells. The derivation starts by expanding the components of the displacement vector in power series of the coordinate normal to the reference surface. The displacement model,

$$U_1 = z\,\theta_1 + z^3\theta_1^*\,,\quad U_2 = z\,\theta_2 + z^3\,\theta_2^*,\quad U_3 = u_3 + z^2\,u_3^* \qquad (2.1)$$

is assumed to represent the lowest order correction for the bending deformation effects to the Love-Kirchhoff theory embodied in the first terms of Eqn.(2.1). The membrane modes of deformation are based on the displacement model,

$$U_1 = u_1 + z^2 u_1^*, \quad U_2 = u_2 + z^2 u_2^*, \quad U_3 = z\,\theta_3 \tag{2.2}$$

which again represents the lowest order correction over the classical theories retaining only the first terms in the first two relations and U_3 equalling to zero in Eqn.(2.2). Since both membrane and bending modes of deformation are coupled in shells, the displacement forms appropriate to the theory developed here are:

$$U_i = u_i + z\,\theta_i + z^2 u_i^* + z^3 \theta_i^*, \quad (i = 1,2)$$

$$U_3 = u_3 + z\,\theta_3 + z^2 u_3^* \tag{2.3}$$

The expansions in Eqn.(2.3) contain minimum number of terms to include the effects of transverse shear deformation with warping of the cross-section and transverse normal strain. Furthermore, it is consistent in the sense that the transverse shear strains due to tangential displacement components U_1 and U_2 are of the same order in z as those determined by the transverse displacement component U_3. In Eqns.(2.3) the functions $U_i (i=1,3)$ are defined in space at a distance z with reference to a curvilinear surface described by the curvilinear coordinates $\alpha_i (i=1,2)$. The remaining functions u_i, u_i^*, θ_i and θ_i^* are defined in the curvilinear surface only and thus are two-dimensional quantities.

The shell equations are obtained by carrying out analytic integration along the thickness of the shell during the application of a variational principle [6]. This makes it possible for the usual governing equations in elasticity theory, which are in terms of stresses, strains and displacements to be expressed in terms of stress-resultants, strains and displacements. As we have pointed out earlier, the equations of this general theory are presented elsewhere [4]. Its use, as such, in the finite element context is, however, complicated because of the need to represent motion of the reference surface in curvilinear coordinates. The relevant theory essential for the finite element method and particularized for an axisymmetric shell is presented in the next section.

3. THEORY FOR AN AXISYMMETRIC SHELL

The mathematical model is based on the displacement representation

$$U_1(s,z) = u_1(s) + z\,\theta_1(s) + z^2 u_1^*(s) + z^3 \theta_1^*(s)$$

$$U_3(s,z) = u_3(s) + z\,\theta_3(s) + z^2 u_3^*(s) \tag{3.1}$$

where the functions U_1 and U_3 define the general displacement components in meridional (s or 1) and normal (z or 3) directions respectively. The parameters u_1, u_3 and θ_1 are the usual displacement components of the reference surface of the shell. The additional functions u_1^*, θ_1^*, θ_3 and u_3^* correspond to the higher-order terms in the Taylor's series expansion in the thickness coordinate z. Thus the generalized displacement vector of the reference surface in shell coordinates, $\underset{\sim}{\delta}^s$ consists of,

$$\underset{\sim}{\delta}^s = (u_1, u_3, \theta_1, u_1^*, u_3^*, \theta_1^*, \theta_3)^T \tag{3.2}$$

The physical strain components are then derived as [4,6],

$$\begin{aligned}
\varepsilon_1^z &= (\varepsilon_1 + z\chi_1 + z^2\varepsilon_1^* + z^3\chi_1^*)/(1+k_1 z) \\
\varepsilon_2^z &= (\varepsilon_2 + z\chi_2 + z^2\varepsilon_2^* + z^3\chi_2^*)/(1+k_2 z) \\
\gamma_{13}^z &= (\varepsilon_{13} + z\chi_{13} + z^2\varepsilon_{13}^* + z^3\chi_{13}^*)/(1+k_1 z) \\
\varepsilon_3^z &= \varepsilon_3 + z\chi_3
\end{aligned} \tag{3.3}$$

The 14 new functions appearing in the above Eqn.(3.3) form the generalized strain vector $\underset{\sim}{\varepsilon}$ of the reference surface and are related to the generalized displacement vector $\underset{\sim}{\delta}^s$ by the following matrix relation,

$$\underset{\sim}{\varepsilon} = \begin{bmatrix} \varepsilon_1 \\ \varepsilon_2 \\ \chi_1 \\ \chi_2 \\ \varepsilon_{13} \\ \chi_{13} \\ \varepsilon_1^* \\ \varepsilon_2^* \\ \chi_1^* \\ \chi_2^* \\ \varepsilon_{13}^* \\ \chi_{13}^* \\ \varepsilon_3 \\ \chi_3 \end{bmatrix} = \begin{bmatrix}
\frac{d}{ds} & k_1 & 0 & 0 & 0 & 0 & 0 \\
\frac{\cos\phi}{r} & \frac{\sin\phi}{r} & 0 & 0 & 0 & 0 & 0 \\
0 & 0 & \frac{d}{ds} & 0 & 0 & 0 & k_1 \\
0 & 0 & \frac{\cos\phi}{r} & 0 & 0 & 0 & \frac{\sin\phi}{r} \\
-k_1 & \frac{d}{ds} & 1 & 0 & 0 & 0 & 0 \\
0 & 0 & 0 & 2 & 0 & 0 & \frac{d}{ds} \\
0 & 0 & 0 & \frac{d}{ds} & k_1 & 0 & 0 \\
0 & 0 & 0 & \frac{\cos\phi}{r} & \frac{\sin\phi}{r} & 0 & 0 \\
0 & 0 & 0 & 0 & 0 & \frac{d}{ds} & 0 \\
0 & 0 & 0 & 0 & 0 & \frac{\cos\phi}{r} & 0 \\
0 & 0 & 0 & k_1 & \frac{d}{ds} & 3 & 0 \\
0 & 0 & 0 & 0 & 0 & 2k_1 & 0 \\
0 & 0 & 0 & 0 & 0 & 0 & 1 \\
0 & 0 & 0 & 0 & 2 & 0 & 0
\end{bmatrix} \begin{bmatrix} u_1 \\ u_3 \\ \theta_1 \\ u_1^* \\ u_3^* \\ \theta_1^* \\ \theta_3 \end{bmatrix} = \underline{L}^s \underset{\sim}{\delta}^s \tag{3.4}$$

The total potential energy Π of the system could be written as,

$$\Pi = \frac{1}{2}\int \underset{\sim}{\varepsilon}^T \underset{\sim}{\sigma}.\, 2\pi r\, ds - \int \underset{\sim}{q}^T \underset{\sim}{\delta}^s . 2\pi\, r ds \tag{3.5}$$

Here the generalized stress vector $\underset{\sim}{\sigma}$ is integral of the physical stress components through the shell thickness and is related to the generalized strain vector ε by the relation,

$$\underset{\sim}{\sigma} = \underline{D}\, \underset{\sim}{\varepsilon} \tag{3.6}$$

and is defined as,

$$\underset{\sim}{\sigma} = (N_1, N_2, M_1, M_2, Q_1, S_1, N_1^*, N_2^*, M_1^*, M_2^*, Q_1^*, S_1^*, N_3, M_3)^T \tag{3.7}$$

The derivation of the elasticity matrix $\underline{D}$ (14 x 14) is rather cumbersome and is, therefore, omitted here due to space restrictions. Interested readers may consult Ref.[4] for the definition of the coefficients of this matrix.

The Eqn.(3.6) is substituted in Eqn.(3.5) and the energy of the system is now written as,

$$\Pi = \frac{1}{2}\int \underset{\sim}{\varepsilon}^T\, \underline{D}\, \underset{\sim}{\varepsilon}\; 2\,\pi\; r ds - \int \underset{\sim}{q}^T \underset{\sim}{\delta}^S\, 2\pi\; r ds \tag{3.8}$$

We note that the above Eqn.(3.8) could be written as a penalty type functional, such that,

$$\Pi = \Pi_1 + \alpha\Pi_2 - \int \underset{\sim}{q}^T\, \underset{\sim}{\delta}^s\; 2\pi\; r\;\; ds \tag{3.9}$$

where Π_1 defines the strain energy due to normal and flexural strain terms only, while Π_2 defines the strain energy due to the remaining transverse shearing strain terms. The penalty number/s α can be recognized as half of the corresponding transverse shearing rigidity.

4. FINITE ELEMENT DISCRETIZATION

It is seen that the potential energy expression of Eqn.(3.8) contains only the first derivatives of the components of the generalized displacement vector $\underset{\sim}{\delta}^s$ and thus only C^o continuity is required for the shape functions to be used in the element formulation. Indeed, any of the numerous isoparametric formulations are possible. If the same shape function is used to define all the components of the generalized displacement vector $\underset{\sim}{\delta}^s$, we can write,

$$\underset{\sim}{\delta}^s = \sum_{i=1}^{m} N_i\, \underset{\sim}{\delta}_i^s \tag{4.1}$$

where N_i is the shape function associated with node i, $\underset{\sim}{\delta}_i^s$ is the value of $\underset{\sim}{\delta}^s$ corresponding to node i, and m is the number of nodes in the element. The vector $\underset{\sim}{\delta}^s$ corresponds to shell coordinates and thus the use of Eqn.(4.1) is helpful only for smooth shells. For composite shells with discontinuous meridional curve, it is necessary to effect a simple coordinate transformation at element level, such that

$$\underset{\sim}{\delta}^s = \underline{T}\,\underset{\sim}{\delta} \tag{4.2}$$

where $\underset{\sim}{\delta}$ is the generalized displacement vector with reference to the global coordinate system (r,z) and $\underline{T}$ is a 7x7 transformation matrix involving direction cosines of the angles between the shell (s,z) and the global (r,z) coordinate systems. The strain displacement relations given by Eqn.(3.4) can now be expressed in terms of the vector $\underset{\sim}{\delta}$ as follows:

$$\underset{\sim}{\varepsilon} = \underline{L}^s\,\underset{\sim}{\delta}^s = (\underline{L}^s\,\underline{T})\underset{\sim}{\delta} = \underline{L}\,\underset{\sim}{\delta} \tag{4.3}$$

In Eqn.(4.3) the new matrix $\underline{L}$ and the vector δ are given by,

$$\underline{L} = \underline{L}^s\,\underline{T} \quad \text{and} \quad \underset{\sim}{\delta} = (u_r,\, u_z,\, \theta_1, u_r^*, u_z^*, \theta_1^*, \theta_3)^T \tag{4.4}$$

We can now express both the geometry and the displacements in terms of an isoparametric coordinate ξ in the form,

$$r = \sum_{i=1}^{m} N_i(\xi).r_i\,, \qquad Z = \sum_{i=1}^{m} N_i(\xi).Z_i$$

$$\underset{\sim}{\delta} = \sum_{i=1}^{m} N_i(\xi)\,\underset{\sim}{\delta}_i \tag{4.5}$$

Following the standard procedure [2,12] we obtain the element matrices in the form,

$$\underline{K}_{ij} = \int_{-1}^{1} \underline{B}_i^T\,\underline{D}\,\underline{B}_j\,|\underline{J}|\,d\,\xi\,, \qquad f_i = \int_{-1}^{1} N_i\,\underset{\sim}{q}\,|\underline{J}|\,d\,\xi \tag{4.6}$$

where $\underset{\sim}{q}$ is the vector of forces corresponding to $\underset{\sim}{\delta}$ and the Jacobian $|\underline{J}|$ here is simply given by,

$$|\underline{J}| = \frac{ds}{d\xi} = \sqrt{\left(\frac{dr}{d\xi}\right)^2 + \left(\frac{dz}{d\xi}\right)^2} \tag{4.7}$$

where ds defines an arc length along the meridional curve of the shell.

The strain matrix $\underline{B}_i$ in Eqn.(4.6) is obtained as,

$$\underline{B}_i = \underline{L}\,N_i \tag{4.8}$$

with the help of Eqns.(4.3) and (4.5) such that,

$$\underset{\sim}{\varepsilon} = \sum_{i=1}^{m} \underline{B}_i\,\underset{\sim}{\delta}_i \tag{4.9}$$

The s-derivatives appearing in $\underline{B}_i$ are expressed in terms of the local ξ-derivatives as,

$$\frac{d}{ds} = \frac{d}{d\xi}\,\frac{d\xi}{ds} = \frac{1}{|\underline{J}|}\cdot\frac{d}{d\xi} \tag{4.10}$$

The evaluation of the stiffness coefficients in Eqns.(4.6) involves the strain matrix $\underline{B}_i$ containing terms like r, $\sin\phi$, $\cos\phi$ and $k_1\,(ds/d\phi)$ and all of these are functions of s. Before we proceed further it is necessary to express these parameters in terms of the ξ-coordinate. We have

$$\tan\phi = dz/dr = (dz/d\xi)\,/\,(dr/d\xi) \tag{4.11}$$

which is specified and hence,

$$\sin\phi = \tan\phi\,/\sqrt{1+\tan^2\phi} \quad \text{and} \quad \cos\phi = 1/\sqrt{1+\tan^2\phi} \tag{4.12}$$

can be evaluated. Further, with r and z coordinates of the meridional curve specified in terms of ξ, the principal curvature k_1 is given by the expression

$$k_1 = -\,(d^2r\,/\,dz^2)\,/\,[1+(dr/dz)^2]^{3/2}$$

$$= \left(\frac{dr}{d\xi}\cdot\frac{d^2z}{d\xi^2} - \frac{dz}{d\xi}\cdot\frac{d^2r}{d\xi^2}\right)/\,[(dr/d\xi)^2+(dz/d\xi)^2]^{3/2} \tag{4.13}$$

The foregoing completes the isoparametric element formulation and the usual numerical integration is carried out to evaluate the integral expression of Eqn.(4.6).

5. CONCLUDING REMARKS

A displacement-based isoparametric finite element formulation with C^0 continuous shape functions and using a higher-order shell theory is developed as a variable-number-nodes element. The problems associated with the geometrical representation of the shell's reference surface in the finite element method using a general shell theory are well known. This is so because all the shell theories use curvilinear coordinates. Further, these are assumed to be principal surface coordinates coinciding with the directions of minimum and maximum curvatures. For a general shell surface this condition is highly unlikely to be met in practice. The alternative use of non-orthogonal coordinates complicates the formulation further for any practical analyses. These difficulties have prompted the use of degenerate solid elements for general shell analyses. However, this situation should not discourage one to formulate good shell elements for specific applications and the present work is an effort in this direction.

A penalty type isoparametric formulation is presented for an axisymmetric shell structure. This differs from the generally available methods which use the classical generalized coordinate finite element formulations where rigid body and constant strain modes are not necessarily available (see [2],[5] and [12]) for original references). To the author's knowledge, there are two notable works which use isoparametric formulation. The first one, described in Ref.[12], uses C^1 continuous shape functions. In the second one [11], a linear penaltv type element is presented. A particular form of the higher-order theory discussed here is applied to plates recently, and the results obtained are very encouraging [3]. Numerical evaluation of the present formulation is presently under investigation. Extension of the present work to deal with non-symmetric loads is easily accomplished with the use of the Fourier series expansion.

ACKNOWLEDGEMENTS

The author, who is on leave from the Indian Institute of Technology, Bombay, gratefully acknowledges the 1979 Scholarship award of the Jawaharlal Nehru Memorial Trust (U.K.) and the hospitality of the Civil Engineering Department of the University of Wales, Swansea.

REFERENCES

1. AHMAD,S., IRONS,B.M. and ZIENKIEWICZ,O.C., Analysis of Thick and Thin Shell and Plate Structures by Curved Finite Elements. *Int.J.of Num.Meth. in Engng.*, 2, 419-451 (1970).

2. COOK,R.D., *Concepts and Applications of Finite Element Analysis.* Wiley, New York (1974).

3. KANT,T., OWEN,D.R.J. and ZIENKIEWICZ,O.C., A Refined Higher-Order C^0 Plate Bending Element. *Computers and Structures*, to appear (1981).

4. KANT,T., A Higher-Order General Shell Theory. *Res. Report, Civil Engg.Dept., University of Wales,* Swansea (1981).

5. KEY,S.W. and BEISINGER,Z.E., The Analysis of Thin Shells by the Finite Element Method. pp.209-252 of B.F.Veubeke (Ed.), *High Speed Computing of Elastic Structures.* Université de Liege (1971).

6. KRAUS,H., *Thin Elastic Shells.* Wiley, New York (1967).

7. MALKUS,D.S. and HUGHES,T.J.R., Mixed Finite Element Methods - Reduced and Selective Integration Techniques : A Unification of Concepts. *Comp.Meth.in Applied Mech. and Engng.*,15,63-81 (1978).

8. PAWSEY, S.F. and CLOUGH, R.W., Improved Numerical Intergration of Thick Shell Finite Elements. *Int. J. of Num. Meth. in Engng.*, 3, 575-586 (1971).

9. ZIENKIEWICZ, O.C., TAYLOR, R.L. and TOO, J.M., Reduced Integration Techniques in General Analysis of Plates and Shells. *Int. J. of Num. Meth. in Engng.*, 3, 275-290 (1970).

10. ZIENKIEWICZ, O.C. and HINTON, E., Reduced Integration, Function Smoothing and Non-Conformity in Finite Element Analysis. *J. of the Franklin Institute*, 302, 443-461 (1976).

11. ZIENKIEWICZ, O.C., BAUER, J., MORGAN, K. and ONATE, E., A Simple and Efficient Element for Axisymmetric Shells. *Int. J. of Num. Meth. in Engng.*, 11, 1545-1558 (1977).

12. ZIENKIEWICZ, O.C., *The Finite Element Method*. McGraw-Hill, New York (1977).

ON THE NUMERICAL ANALYSIS OF THIN SHELL PROBLEMS

M. BERNADOU

I.N.R.I.A., Lé Chesnay

1. INTRODUCTION

There are numerous general models of thin shells, especially that of Koiter [21,22] which is the last improvement of the well-known theories of Kirchhoff [20] and Love [24], and that of Naghdi [25,26] which originates from the surface theory of Cosserat [14] and which is more general than the Koiter model. Nevertheless, for effective computations, it seems that the Koiter model gives satisfaction to the engineers. In particular, Argyris-Haase-Malejannakis [1] and Dupuis [18] use this theory.

This paper is essentially based upon our works on *numerical analysis* of thin shell problems according to Koiter's equations [21,22].

The second Section gives a brief description of *linear equations* as well as *nonlinear shallow shell* equations. In both cases we give an existence theorem for a solution. For linear equations the same kind of method allows study of Naghdi's model-see [15,16].

In the third Section, we consider the approximation of the linear problem, by using *conforming finite element methods* in conjunction with *numerical integration techniques*. We record the *error estimate* results as well as *sufficient conditions* on the choice of the *numerical integration schemes* according to [2,3].

Finally, in a fourth Section, we consider the application of previous results to the computation of *an arch dam*. Corresponding numerical experiments are not expensive and are in agreement with experimental results for a similar arch dam.

2. THE CONTINUOUS PROBLEMS

2.1 Notations

Let Ω be a bounded open subset in a plane $\mathcal{E}^2$, with boundary Γ. Then the *middle surface* S of the shell is the image of the set $\bar{\Omega}$ by a mapping $\underline{\phi} : \bar{\Omega} \subset \mathcal{E}^2 \to \mathcal{E}^3$, where $\mathcal{E}^3$ is the usual Euclidean space. Subsequently, we shall assume that $\underline{\phi} \in (\mathcal{C}^3(\bar{\Omega}))^3$ and that all points of $S = \underline{\phi}(\bar{\Omega})$ are regular, in the sense that the two vectors $\underline{a}_\alpha = \underline{\phi}_{,\alpha}$, $\alpha = 1,2$, are linearly independent for all points $\xi = (\xi^1, \xi^2) \in \bar{\Omega}$. With the *covariant basis* $(\underline{a}_\alpha)$ of the tangent plane, we associate the *contravariant basis* $(\underline{a}^\alpha)$, which is defined through the relations $\underline{a}^\alpha . \underline{a}_\beta = \delta^\alpha_\beta$, where δ^α_β is the Kronecker's symbol. We also introduce the *normal vector* $\underline{a}_3 = \underline{a}^3 = \underline{a}_1 \times \underline{a}_2 \,/\, |\underline{a}_1 \times \underline{a}_2|$.

The *unknowns* are the three functions

$$u_i : \xi \in \bar{\Omega} \to u_i(\xi) \in \mathbb{R}, \ i = 1, 2, 3$$

which represent the covariant components of the *displacement* $\underline{u} = \underline{u}(\xi)$ of the point $\underline{\phi}(\xi)$, i.e.,

$$\underline{u} = u_i \underline{a}^i \tag{2.1}$$

In [21], Koiter shows that the evaluation of the *strain tensor of the shell* can be approximated by the evaluation of the two following surface tensors :

(i) the *middle surface strain tensor* $\gamma_{\alpha\beta}(\underline{u})$,

(ii) the *tensor of changes of curvature* $\bar{\rho}_{\alpha\beta}(\underline{u})$,

associated to a displacement field $\underline{u}$.

We assume that (*) the material of the shell is elastic, homogeneous and isotropic ; (**) the strains are small everywhere in the shell ; (***) the shell is in a state of stress in which all nonzero stress components are developed on surfaces parallel to the middle surface.

Then, we have the following relations between the tensors $\gamma_{\lambda\mu}(\underline{u})$ et $\bar{\rho}_{\lambda\mu}(\underline{u})$, on the one hand, and the symmetric tensors of tangential (membrane) stress resultants $n^{\alpha\beta}(\underline{u})$ and stress couples $m^{\alpha\beta}(\underline{u})$, on the other hand :

$$n^{\alpha\beta}(\underline{u}) = eE^{\alpha\beta\lambda\mu}\gamma_{\lambda\mu}(\underline{u}), \quad m^{\alpha\beta}(\underline{u}) = \frac{e^3}{12}E^{\alpha\beta\lambda\mu}\bar{\rho}_{\lambda\mu}(\underline{u}), \tag{2.2}$$

where e denotes the thickness of the shell and $E^{\alpha\beta\lambda\mu}$ denotes the tensor of elastic moduli for plane stress at the middle surface.

In the following, we consider for simplicity the case of a *clamped* shell. Let $\underline{p} = p^i \underline{a}_i$ be the resultant of external applied forces per unit surface area defined on the middle surface of the shell. Then, according to [21], the equations of equilibrium are given in section 2.2 for linear case and in section 2.3 for non-linear case.

2.2 The System of Linear Equations

$$n^{\alpha\beta}(\underline{u})|_{\beta} + p^{\alpha} = 0, \tag{2.3}$$
$$m^{\alpha\beta}(\underline{u})|_{\alpha\beta} - b_{\alpha\beta}n^{\alpha\beta}(\underline{u}) - p^{3} = 0, \tag{2.4}$$

with the usual conditions of a clamped boundary, i.e.,

$$\underline{u} = \underline{o},\ \partial u_3/\partial n = 0 \quad \text{on } \partial S \tag{2.5}$$

Subsequently , we shall work on the corresponding variational formulation of the problem : *Find* $\underline{u} \in \underline{V}$ *such that*

$$a(\underline{u},\underline{v}) = \int_{\Omega} \underline{p}\underline{v}\sqrt{a}d\xi^1 d\xi^2,\ \forall \underline{v} \in \underline{V} \tag{2.6}$$

where

$$a(\underline{u},\underline{v}) = \int_{\Omega} eE^{\alpha\beta\lambda\mu}[\gamma_{\alpha\beta}(\underline{u})\gamma_{\lambda\mu}(\underline{v}) + \frac{e^2}{12}\bar{\rho}_{\alpha\beta}(\underline{u})\bar{\rho}_{\lambda\mu}(\underline{v})]\sqrt{a}d\xi^1 d\xi^2 \tag{2.7}$$

$$\gamma_{\alpha\beta}(\underline{u}) = \frac{1}{2}(u_{\alpha|\beta} + u_{\beta|\alpha}) - b_{\alpha\beta}u_3 \tag{2.8}$$

$$\bar{\rho}_{\alpha\beta}(\underline{u}) = u_{3|\alpha\beta} - b_{\alpha}^{\lambda}b_{\lambda\beta}u_3 + b_{\beta|\alpha}^{\lambda}u_{\lambda} + b_{\beta}^{\lambda}u_{\lambda|\alpha} + b_{\alpha}^{\lambda}u_{\lambda|\beta} \tag{2.9}$$

$b_{\alpha\beta}$ = second fundamental form of the surface

$$\underline{V} = H_o^1(\Omega) \times H_o^1(\Omega) \times H_o^2(\Omega) \tag{2.10}$$

In [8], we have proved the following result :

Theorem 2.1 : The bilinear form (2.7) is $\underline{V}$ *- elliptic and the problem (2.6) has one and only one solution.*

Essentially, we show that the application

$$\Phi(\underline{v}) = \left\{ \sum_{\alpha,\beta=1}^{2} |\gamma_{\alpha\beta}(\underline{v})|^2_{L^2(\Omega)} + |\bar{\rho}_{\alpha\beta}(\underline{v})|^2_{L^2(\Omega)} \right\}^{1/2}$$

is an equivalent norm to the usual norm of space $\underline{V}$, and hence a(.,.) is $\underline{V}$ - elliptic. We conclude with the help of Lax-Milgram Lemma.

Remark 2.1. : For more general boundary conditions, we refer to the analysis of [2,8].

2.3 The System of Nonlinear Equations

For a classical *nonlinear shallow shell theory*, Koiter [21,§ 11] propose the following set of equations :

$$n^{\alpha\beta}(\underline{u})|_{\beta} + p^{\alpha} = 0, \qquad (2.11)$$

$$m^{\alpha\beta}(\underline{u})|_{\alpha\beta} - b_{\alpha\beta}\, n^{\alpha\beta}(\underline{u}) - (u_{3|\alpha} n^{\alpha\beta}(\underline{u}))|_{\beta} - p^{3} = 0, \qquad (2.12)$$

with the boundary conditions (2.5). Relations (2.2) are still available, but now, instead of relations (2.8)(2.9) we have the expressions

$$\gamma_{\alpha\beta}(\underline{u}) = \frac{1}{2}(u_{\alpha|\beta} + u_{\beta|\alpha}) - b_{\alpha\beta}u_3 + \frac{1}{2}u_{3,\alpha}u_{3,\beta}\,, \qquad (2.13)$$

$$\bar{\rho}_{\alpha\beta}(\underline{u}) = u_{3|\alpha\beta}\,. \qquad (2.14)$$

In [11], we have associated to the system of nonlinear equations (2.11)(2.12) and (2.5), a variational formulation stated on the space $\underline{V}$ defined in (2.10), and next, we have proved the following result

Theorem 2.2 : For a large class of nonlinear shallow shell equations, the associated variational formulation has at least one solution whenever tangential components p^{α} *of the loads are sufficiently small.*

The proof requires essentially two steps. First, we fix u_3 and we solve a linear equation with respect to u_1 and u_2 by using Lax-Milgram lemma. Then, it remains a nonlinear equation with respect to u_3. We prove that the corresponding operator is *pseudo-monotone* and *coercive*. We conclude according to Lions [23, chapter 2, theorem 2.7].

In addition, we prove that solutions are unique whenever the loads are sufficiently small.

3. THE APPROXIMATED PROBLEMS

3.1 The Discrete Space $\underline{V}_h$

From now on, we restrict our attention to linear problems and we assume that the set $\bar{\Omega}$ is a polygon. Then, with the terminology of [12], we may cover the set $\bar{\Omega}$ by an affine regular family of triangulations $\mathcal{T}_h$. Next, to every triangle $K \in \mathcal{T}_h$, we associate two finite elements so that we define two spaces of finite elements V_{h1} and V_{h2} such that $V_{h1} \subset H_0^1(\Omega)$, $V_{h2} \subset H_0^2(\Omega)$, and

$$\underline{V}_h \subset \underline{V} \qquad (3.1)$$

Hence the name : *conforming methods*.

3.2 *The Discrete Problems*

We could define an approximating solution of problem (2.6) by restricting $\underline{u}$ and $\underline{v}$ to belong to the subspace $\underline{V}_h$ of $\underline{V}$. But, in the equation (2.6), there are many variable coefficients so that we have to use numerical integration. Thus, consider the following *numerical quadrature scheme* over the set K

$$\int_K \psi(x)dx \sim \sum_{l=1}^{L} \omega_{l,K}\psi(b_{l,K}) \tag{3.2}$$

So, in the expressions (2.6),(2.7), we write $\int_\Omega (\ ) = \sum_{K\in\mathscr{C}_h} \int_K (\ )$

and we approach every integral upon K with the help of (3.2). That amounts to substitute to the bilinear form a(.,.) a new bilinear form $a_h(.,.)$. Then the corresponding *discrete problem* can be stated as follows : Find $\underline{u}_h \in \underline{V}_h$ such that

$$a_h(\underline{u}_h,\underline{v}_h) = \sum_{K\in\mathscr{C}_h} \sum_{l=1}^{L} \omega_{l,K}(\underline{p}\underline{v}_h\sqrt{a})(b_{l,K}) \tag{3.3}$$

In [2,3] we have (i) shown that the problem (3.3) has a *unique solution*. This has been achieved by showing that, under mild assumptions, the bilinear form $a_h(.,.)$ is $\underline{V}_h$ - elliptic, uniformly with respect to h ; (ii) proved the *convergence* ; (iii) obtained *sufficient conditions* on the quadrature schemes in order to preserve the error estimate of the exact integration case.

The previous results are illustrated in Fig. 1 for five different spaces $\underline{V}_h = V_{h1} \times V_{h1} \times V_{h2}$ associated to different choices of finite elements. In order to obtain the inclusion (3.1), let us note that we use $\mathscr{C}^1$- finite elements to construct the space V_{h_2}.

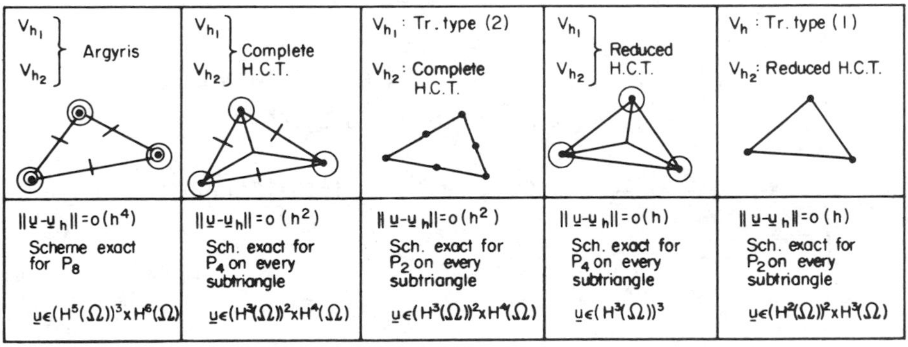

V_{h_1}, V_{h_2} : Argyris	V_{h_1}, V_{h_2} : Complete H.C.T.	V_{h_1} : Tr. type (2); V_{h_2} : Complete H.C.T.	V_{h_1}, V_{h_2} : Reduced H.C.T.	V_h : Tr. type (1); V_{h_2} : Reduced H.C.T.
$\|\underline{u}-\underline{u}_h\| = o(h^4)$ Scheme exact for P_8; $\underline{u}\in(H^5(\Omega))^3 \times H^6(\Omega)$	$\|\underline{u}-\underline{u}_h\| = o(h^2)$ Sch. exact for P_4 on every subtriangle; $\underline{u}\in(H^3(\Omega))^2 \times H^4(\Omega)$	$\|\underline{u}-\underline{u}_h\| = o(h^2)$ Sch. exact for P_2 on every subtriangle; $\underline{u}\in(H^3(\Omega))^2 \times H^4(\Omega)$	$\|\underline{u}-\underline{u}_h\| = o(h)$ Sch. exact for P_4 on every subtriangle; $\underline{u}\in(H^3(\Omega))^3$	$\|\underline{u}-\underline{u}_h\| = o(h)$ Sch. exact for P_2 on every subtriangle; $\underline{u}\in(H^2(\Omega))^2 \times H^3(\Omega)$

FIG. 1 Examples of error estimates, sufficient conditions on the quadrature schemes and regularity required for solution $\underline{u}$(we refer to [12] for notations).

Remark 3.1 For simplicity, we have assumed that the boundary Γ of the reference domain Ω is *polygonal*. This situation occurs frequently. However, the case of a *curved* boundary Γ is analyzed in [2].

Remark 3.2 When the middle surface of the shell is *plane* and when the loads are normal to this plane, i.e., $p^1 = p^2 = 0$, Koiter's model gives the usual equations of the deflection of a plate. In this case, sufficient conditions on the quadrature schemes have to be exact for polynomials of degree 6 (resp. 2) for an approximation using Argyris (resp. complete or reduced H.C.T.) - elements (see [3,9]).

Remark 3.3 In this analysis, the geometry of the shell appears only through variable coefficients which are not approximated, but only evaluated at the nodes of the numerical integration scheme. For an approximation of the geometry of the shell we refer to (i) [13] in case of conforming methods ; (ii) [10,13] in case of flat plate elements, i.e., nonconforming methods.

4. APPLICATION TO THE COMPUTATION OF AN ARCH DAM

4.1 Geometrical Definition of the Dam

In this paragraph, we give the geometrical definition of the arch dam of the project of GRAND'MAISON studied by Coyne et Bellier [17]. We consider the two following steps :

Step 1 : Definition of the middle surface of the dam :

With notations of Fig. 2, the coordinates x^i of any point M of the middle surface S of the dam are given as the components of the mapping

$$\underline{\phi} : (\xi^1,\xi^2) \in \bar{\Omega} \to \underline{OM} = \underline{\phi}(\xi^1,\xi^2) = x^i(\xi^1,\xi^2)\underline{e}_i ,$$

by the relations

$$\begin{cases} x^1(\xi^1,\xi^2) = \rho_o(\xi^2)\left[e^{\alpha\theta_o|\xi^1|}\cos(\theta_o|\xi^1|+40°) - \cos 40°\right] \\ \qquad + 0.269\, z_o\xi^2 - 0.0000085\, z_o^3(\xi^2)^3 \\ x^2(\xi^1,\xi^2) = \dfrac{|\xi^1|}{\xi^1}\,\rho_o(\xi^2)\left[e^{\alpha\theta_o|\xi^1|}\sin(\theta_o|\xi^1|+40°) - \sin 40°\right] \\ x^3(\xi^1,\xi^2) = z_o\xi^2 \end{cases}$$

where

$$\begin{cases} \alpha = \mathrm{tg}\ 40° ,\ \theta_o = 48°\ 178 ,\ z_o = 157\ \mathrm{m} \\ \rho_o(\xi^2) = 200 - 0.008233(z_o)^2(\xi^2)^2 + 0.000029(z_o)^3\ (\xi^2)^3 \end{cases}$$

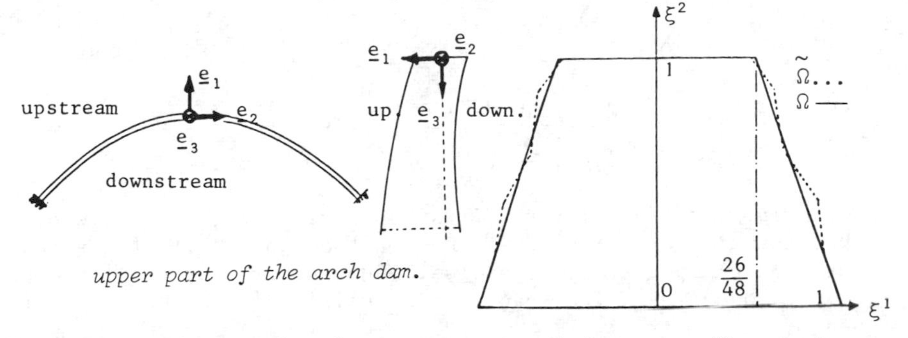

FIG.2 The orthonormal reference system of the space ξ^3 and the reference domains $\tilde{\Omega}$ (approximated) and Ω (real).

Step 2 : Definition of the thickness of the dam :

$$\begin{cases} e(\xi^1,\xi^2) = 8 + 0.248\, z_o\xi^2 - 0.000003\,(z_o\xi^2)^3 \\ \qquad + 2.10^{-8}(z_o\xi^2)^2\,[1 + 0.003\, z_o\xi^2]\left[\dfrac{e^{\alpha\theta o|\xi^1|}-1}{\sin 40°}\,\rho_o(\xi^2)\right]^2 \end{cases}$$

4.2. - Variational formulation of the problem :

We consider here the effects of *thermal loads, water pressure* and *self weight* on the behaviour of the dam. To apply Koiter's theory, we need to obtain an approximation of the work done by these loads through an integration on the thickness. In [6], we have derived the following expressions associated to a displacement $\underline{v}$ of the middle surface of the shell :

$$f(\underline{v}) = \frac{E\bar{\alpha}}{2(1-\nu)}\int_\Omega [eT_1\gamma^\lambda_\lambda + \frac{e^3}{12}T_2(2b^\lambda_\eta\gamma^\eta_\lambda - b^\lambda_\lambda\gamma^\eta_\eta - \bar{\rho}^\lambda_\lambda)]\sqrt{a}\, d\xi^1\, d\xi^2$$

$$+ \int_\Omega p\,[\frac{1}{2}\, e_{,\beta}v_\lambda a^{\lambda\beta} - (1-\frac{1}{2}\, eb^\beta_\beta)v_3]\,\sqrt{a}\, d\xi^1\, d\xi^2$$

$$+ \int_\Omega \rho_1 g_o\, e[(a^{12}v_1 + a^{22}v_2)\, z_o + (\underline{a}_3.\underline{e}_3)\, v_3]\,\sqrt{a}\, d\xi^1\, d\xi^2,$$

where the three integrals take respectively into account the effects of

(i) the *thermal loads* : On the middle surface S, we denote T_1 and T_2 the moments of order 0 and 1 through the thickness (in a mathematical sense) of the steady-state temperature distribution ; here

$$E = 2\ 10^6\ \text{ton/m}^2\ ;\ \nu = 0.2\ ;\ \bar{\alpha} = 10^{-5}\ \text{by centigrade degree}$$

(ii) the *hydrostatic pressure* p: If $\xi^2 = \bar{\xi}^2$ refer to the level of water in the reservoir, then

$$p = 0 \text{ if } 0 \le \xi^2 \le \bar{\xi}^2,\ p = \rho_2 g_o z_o(\xi^2 - \bar{\xi}^2) \text{ if } \bar{\xi}^2 \le \xi^2 \le 1,$$

where $\rho_2 = 10^3$ Kg/m^3, $g_o = 9{,}81$ m/s^2, $z_o = 157$ m ;

(iii) *the self weight*. We take $\rho_1 = 2500$ Kg/m^3.

Boundary conditions : for a first approximation, we assume here that the middle surface S is *free* on the upper part of its boundary and clamped everywhere else. Denoting by Γ_o the clamped part of the boundary Γ, the space of admissible displacement is defined by

$$\underline{V} = \{\underline{v} | \underline{v} \in (H^1(\Omega))^2 \times H^2(\Omega),\ \underline{v}|_{\Gamma_o} = \underline{0}\ ,\ \frac{\partial v_3}{\partial n}|_{\Gamma_o} = 0\}.$$

Then, the problem can be stated : *For any* $T_\alpha \in L^2(\Omega)$, *find* $\underline{u} \in \underline{V}$ *such that*

$$a(\underline{u}, \underline{v}) = f(\underline{v})\ ,\ \forall \underline{v} \in \underline{V}.$$

An extension of theorem 2.1 gives the existence and the uniqueness for a solution.

4.3 Implementation

The implementation is similar to those described in [4,5,7] The system is solved by a Choleski method using the sky-line bandwidth factorization [19]. From the solution of the system, we obtain an approximation of the displacement, strain tensor and change of curvature tensor at any point of the middle surface S. Then, applying the basic hypothesis of [21,22], we derive an approximation of the displacement and mixed stress tensor σ^{*i}_{j} everywhere in the dam. Finally, in order to get components having natural physical dimensions, we introduce the so-called *right physical components* of the stress tensors, as in [28].

4.4 Numerical Experiments

In this section, we present some results obtained by using *Argyris' triangle* to approximate the three components of the displacement.

Figs. 3 and 4 show the distribution of stresses on the *upstream* and *downstream* faces of the arch dam subjected to the hydrostatic pressure. The level of water in the reservoir is assumed to be 152 m so that $\bar{\xi}^2 = 0.032$. Figs. 3 and 4 are associated to a triangulation with 32 triangles. Other computations using triangulations with 8, 18, 50 triangles have been performed : the corresponding results - see Table 1 - show the excellent approximation properties of the Argyris element. Particularly, the results obtained from a *coarse* triangulation with *only* 8 triangles are really acceptable

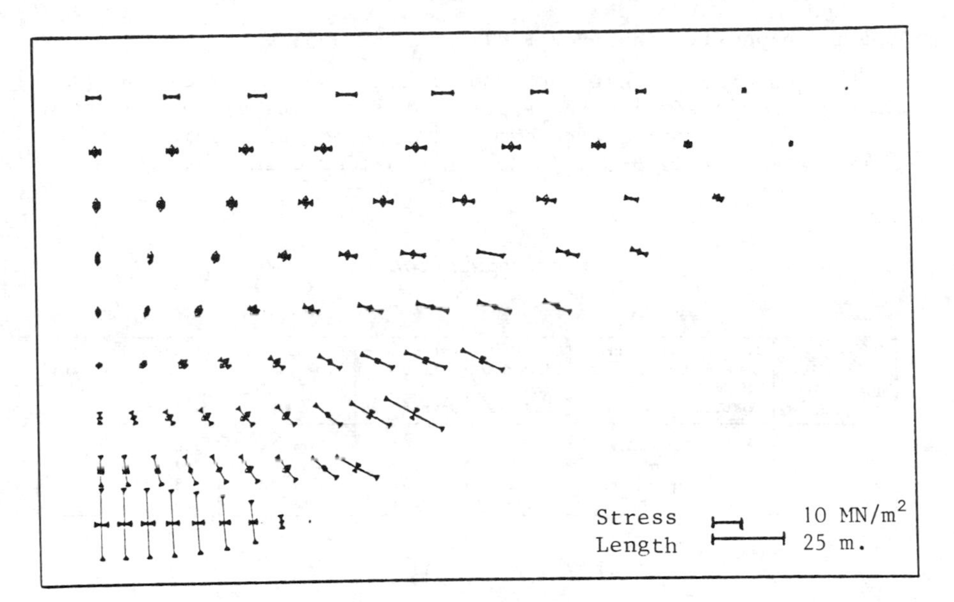

FIG. 3 Stress distribution on downstream face
(32 triangles, $\nu = 0.2$)

Stress
Length
10 MN/m^2
25 m.

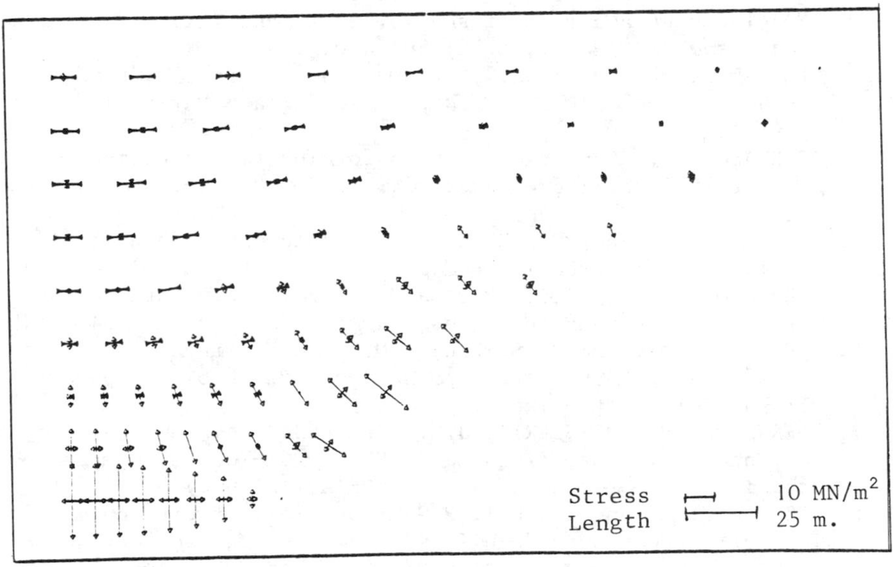

FIG. 4 Stress distribution on upstream face
(32 triangles, $\nu = 0.2$)

and non expensive (about F.F. 115 on IBM 3033).

Let us observe that these numerical experiments can be qualitatively compared with the experimental results given in [27, p. 639] for a similar arch dam. Other kinds of loadings such that *thermal* or *weight effects* are considered in [6].

TABLE 1

Effect of triangulation refinements

Number of triangles	Top of crown cantilever-*Arch Stresses* (MN/m^2)		Base of crown cantilever-*Arch Stresses* (MN/m^2)		Computing time (on IBM 3033)	Nb. of degrees of freedom	Nb. of unknowns
	Upstream	Downstream	Upstream	Downstream			
8	- 4.46	- 2.81	2.35	- 2.43	0 mn 17	210	123
18	- 4.60	- 2.77	2.50	- 2.59	0 mn 41	387	266
32	- 4.61	- 2.80	2.53	- 2.62	1 mn 20	618	463
50	- 4.62	- 2.81	2.55	- 2.64	2 mn 20	903	714

REFERENCES

1. ARGYRIS, J.H., HAASE, M. and MALEJANNAKIS, G.A. Natural geometry of surfaces with specific reference to the matrix displacement analysis of shells, I, II and III. *Proc. Kon. Ned. Akad. Wetensch.*, B.76, 361-410, (1973).
2. BERNADOU, M. *Sur l'analyse numérique du modèle linéaire de coques minces de W.T. KOITER*, Thèse d'Etat, Paris VI (1978).
3. BERNADOU, M. Convergence of conforming finite element methods for general shell problems. *Int. J. Engng. Sc.* 18, 249-276, (1980).
4. BERNADOU, M. and BOISSERIE, J.M. Implémentation de l'élément fini d'ARGYRIS-Exemples - *Rapport* IRIA-LABORIA 301, (1978)
5. BERNADOU, M. and BOISSERIE, J.M. Sur l'implémentation de problèmes généraux de coques, *Rapport* IRIA-LABORIA 317, (1978).
6. BERNADOU, M., and BOISSERIE, J.M. *The Finite Element in Thin Shell Theory; Application to an Arch Dam.* Book to appear, Birkhäuser, Boston Inc.
7. BERNADOU, M., BOISSERIE, J.M. and HASSAN, K. Sur l'implémentation des éléments finis de HSIEH-CLOUGH-TOCHER, complet et réduit. *Rapports de Recherche* INRIA, 4, (1980).
8. BERNADOU, M. and CIARLET, P.G. Sur l'ellipticité du modèle linéarite de coques de W.T. KOITER. *"Computing Methods in Applied Sciences and Engineering"*. Lectures Notes in Economics and Mathematical Systems, Vol. 134, pp. 89-136, Springer-Verlag, Berlin (1976).

9. BERNADOU, M. and DUCATEL, Y. Méthodes d'éléments finis avec intégration numérique pour des problèmes elliptiques du quatrième ordre, R.A.I.R.O., *Analyse Numérique*, 12, No. 1, 3-26, (1978).
10. BERNADOU, M. and DUCATEL, Y. Approximation of general shell problems by flat plate elements. (To appear)
11. BERNADOU, M. and ODEN, J.T. Existence theorem for general nonlinear shallow shell problems. To appear in *J. Math. Pures Appl.*
12. CIARLET, P.G. *The Finite Element Method for Elliptic Problems.* North-Holland (1978).
13. CIARLET, P.G. Conforming finite element methods for the shell problem. *The Mathematics of Finite Elements and Applications II.* J.R. Whiteman, Editor, Academic Press, pp 105-124, (1976).
14. COSSERAT, E. and COSSERAT, F. *Théorie des corps déformables.* Hermann, Paris, (1909).
15. COUTRIS, N. Théorème d'existence et d'unicité pour un problème de flexion élastique de coques dans le cadre de la modélisation de P.M. Naghdi. *C.R. Acad. Sci. Paris, Sér. A*, 283, 951-953, (1976).
16. COUTRIS, N. Théorème d'existence et d'unicité pour un problème de coque élastique dans le cas d'un modèle linéaire de P.M. NAGHDI. R.A.I.R.O., *Analyse Numérique*, 12, No.1, 51-58. (1978).
17. COYNE and BELLIER, Barrage de GRAND'MAISON, Dossier préliminaire (1977).
18. DUPOIS, G. Application of Ritz's method to thin elastic shell analysis. *Journal of Applied Mechanics* 71-APM-32, 1-9, (1971).
19. JENNINGS, A. *Matrix Computation for Engineers and Scientists*, J. Wiley and Sons, Chichester, (1977).
20. KIRCHHOFF, G. *Vorlesungen über Mathematische Physik*, Mechanik, Leipzig, (1876).
21. KOITER, W.T. On the nonlinear theory of thin elastic shells. *Proc. Kon. Ned. Akad. Wetensch.* B 69, 1-54, (1966).
22. KOITER, W.T. On the foundations of the linear theory of thin elastic shells. *Proc. Kon. Ned. Akad. Wetensch*, B 73, 169-195. (1970).
23. LIONS, J.L. *Quelques méthodes de Résolution des Problèmes aux Limites Non-linéaires*, Dunod, Gauthier-Villars, Paris (1969).
24. LOVE, A.E.H. *The Mathematical Theory of Elasticity*, Cambridge University Press, (1934).
25. NAGHDI, P.M. Foundations of elastic shell theory. *Progress in Solid Mechanics*, Vol. 4, pp 1-90, North-Holland, Amsterdam, (1963).
26. NAGHDI, P.M. The Theory of Shell and Plates. *Handbuch der Phys.* Vol. VI a-2, pp. 425-640, Springer-Verlag, Berlin, (1972).
27. RYDZEWSKI, J.R. *Theory of Arch Dams*, Pergamon Press, Oxford, (1965).
28. TRUESDELL, C. The physical components of vectors and tensors, *Z. Angew. Math. Mech.*, 33, No.10-11, 345-356, (1953).

CURVED RECTANGULAR AND GENERAL QUADRILATERAL SHELL ELEMENTS FOR CYLINDRICAL SHELLS

A. B. Sabir and T. A. Charchafchi

University College, Cardiff

1. INTRODUCTION

The present paper investigates possible improvements to the performance of a new class of curved cylindrical finite elements. The cylindrical finite element developed by Ashwell and Sabir [1] was based on simple generalised strain functions satisfying the requirements of exact representation of strain free-rigid body displacements, of constant strains and of independent strains rather than independent displacements (insofar as this is allowed by the compatibility equations). The resulting element was made to have only external "geometric" nodal degrees of freedom and thus avoiding the difficulties associated with internal degrees of freedom. The formulation of the element was based on the use of Timoshenko's [2] strain-displacement relationships. This element was tested by applying it to the solution of the pinched cylinder and roof-type barrel vault problems [1]. It was later [3] applied to the problems of diffusion of concentrated generalised forces into walls of cylinders. This was carried out to test the performance of this element in regions of rapidly varying stress distribution. In the present paper we shall confine attention to possible improvements to this element by

1 - the use of Sanders-Koiter [4] shell equations instead of Timoshenko's equations
2 - the more close satisfaction of equilibrium as well as compatibility by the inclusion of Poisson's ratio terms in the assumed strains.

Furthermore to approximate more closely boundaries not coinciding with lines of curvature and in order to analyse cylinders with curved cut-outs a general quadrilateral element based again on strain assumptions is developed and tested.

2. SANDERS-KOITER STRAIN-DISPLACEMENT EQUATIONS AND EQUILIBRIUM EQUATIONS FOR CYLINDRICAL SHELLS

If the displacements u,v and w are taken as given in Fig. 1, the Sanders-Koiter strain-displacement equations can be written as

$$\varepsilon_x = \frac{\partial u}{\partial x}, \quad \varepsilon_\phi = \frac{1}{R}\frac{\partial v}{\partial \phi} + \frac{w}{R}, \quad \gamma_{x\phi} = \frac{1}{R}\frac{\partial u}{\partial \phi} + \frac{\partial v}{\partial x}$$

$$\kappa_x = -\frac{\partial^2 w}{\partial x^2}, \quad \kappa_\phi = \frac{1}{R^2}\frac{\partial v}{\partial \phi} - \frac{1}{R^2}\frac{\partial^2 w}{\partial \phi^2}$$

$$\kappa_{x\phi} = -\frac{1}{R}\frac{\partial^2 w}{\partial x \partial \phi} + \frac{3}{4}\frac{1}{R}\frac{\partial v}{\partial x} - \frac{1}{4}\frac{1}{R^2}\frac{\partial u}{\partial \phi} \tag{2.1}$$

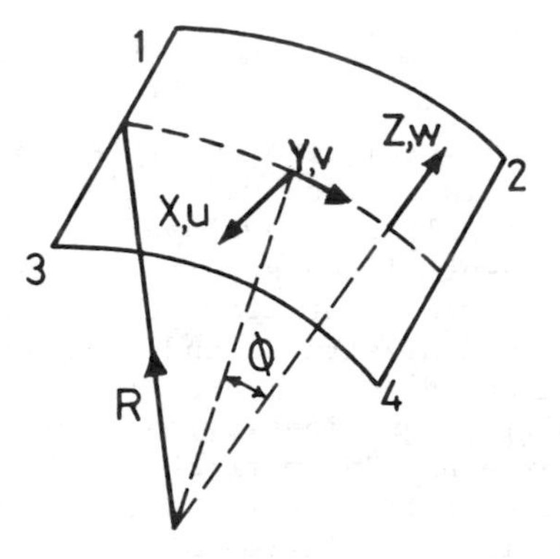

FIG. 1. Coordinate system for rectangular shell element

where $\varepsilon_x, \varepsilon_\phi$ and $\gamma_{x\phi}$ are the direct and shearing membrane strains of the mid-surface; κ_x, κ_ϕ and $\kappa_{x\phi}$ are the changes in bending and twisting curvatures of the mid-surface.

By eliminating the three displacements u,v and w from the six equations of (2.1) we obtain the three compatibility equations

$$\frac{\partial \kappa_\phi}{\partial x} - \frac{1}{R}\frac{\partial \kappa_{x\phi}}{\partial \phi} - \frac{1}{4R^2}\frac{\partial \gamma_{x\phi}}{\partial \phi} = 0$$

$$-\frac{\partial \kappa_{x\phi}}{\partial x} + \frac{1}{R}\frac{\partial \kappa_x}{\partial \phi} + \frac{3}{4R}\frac{\partial \gamma_{x\phi}}{\partial x} - \frac{1}{R^2}\frac{\partial \varepsilon_x}{\partial \phi} = 0$$

$$\frac{\partial^2 \varepsilon_\phi}{\partial x^2} + \frac{1}{R}\frac{\partial^2 \gamma_{x\phi}}{\partial x \partial \phi} - \frac{1}{R^2}\frac{\partial^2 \varepsilon_x}{\partial \phi^2} - \frac{1}{R}\kappa_x = 0 \tag{2.2}$$

Since we shall later attempt to improve the element by satisfying more closely the equilibrium equations we first write these for the case of a cylindrical shell in terms of the stress resultants and then express them in terms of strain components (2.1) by the use of the stress-strain relationships. Such consideration will lead to the three equilibrium equations

$$\frac{\partial \varepsilon_x}{\partial x} + \frac{\nu \partial \varepsilon_\phi}{\partial x} + \frac{1}{2R}(1-\nu)\frac{\partial \gamma_{x\phi}}{\partial \phi} - \frac{h^2}{24R^2}\left[(1-\nu)\frac{\partial \kappa_{x\phi}}{\partial \phi}\right] = 0$$

$$\frac{1-\nu}{2}\frac{\partial \gamma_{x\phi}}{\partial x} + \frac{1}{R}\left(\frac{\partial \varepsilon_\phi}{\partial \phi} + \frac{\nu \partial \varepsilon_x}{\partial \phi}\right) + \frac{h^2}{12R^2}\left[\frac{3R}{2}(1-\nu)\frac{\partial \kappa_{x\phi}}{\partial \phi} + \frac{\partial \kappa_\phi}{\partial \phi} + \frac{\nu \partial \kappa_x}{\partial \phi}\right] = 0$$

$$-\frac{1}{R}(\varepsilon_\phi + \nu\varepsilon_x) + \frac{h^2}{12}\left(\frac{\partial^2 \kappa_x}{\partial x^2} + \frac{\nu \partial^2 \kappa_\phi}{\partial x^2}\right) + \frac{h^2}{12R^2}\left[\frac{\partial^2 \kappa_\phi}{\partial \phi^2} + \frac{\nu \partial^2 \kappa_x}{\partial \phi^2}\right]$$

$$+ \frac{h^2}{6R}(1-\nu)\frac{\partial^2 \kappa_{x\phi}}{\partial x \partial \phi} = 0 \tag{2.3}$$

where h, is the shell thickness and ν is Poisson's ratio.

3. DERIVATION OF THE NEW ELEMENT

We first integrate equations (2.1) with all six strains equal to zero to obtain

$$\begin{aligned} u &= R\,\alpha_2 \cos\phi + R\,\alpha_4 \sin\phi + \alpha_5 \\ v &= (\alpha_1 + \alpha_2 x)\sin\phi - (\alpha_3 + \alpha_4 x)\cos\phi + \alpha_6 \\ w &= -(\alpha_1 + \alpha_2 x)\cos\phi - (\alpha_3 + \alpha_4 x)\sin\phi \end{aligned} \tag{3.1}$$

These equations give the shape function of the element corresponding to the six components of strain free-rigid body displacement, and require six of the displacement function constants $\alpha_1, \alpha_2, \ldots, \alpha_6$.

To keep the element as simple as possible, we choose five degrees of freedom at each of the four corner nodes Fig. 1, namely u,v,w $\partial w/\partial x$ and $\partial w/\partial y - v/R$. Thus the shape function for the rectangular element contains 20 constants $\alpha_1, \alpha_2, \ldots \alpha_{20}$. Having used six of these to represent the rigid body displacements, the remaining 14 constants are available for expressing the deformation of the element. These are apportioned among the strains as follows

$$\varepsilon_x = \alpha_7 + \alpha_8\phi$$

$$\varepsilon_\phi = \alpha_9 + \alpha_{10}x + \left(-\frac{\alpha_{12}}{2R}x^2 - \frac{\alpha_{13}}{6R}x^3 - \frac{\alpha_{14}}{2R}x^2\phi - \frac{\alpha_{15}}{6R}x^3\phi\right)$$

$$\gamma_{x\phi} = \alpha_{11} + \left(\frac{4}{3}\,\frac{\alpha_8}{R}\,x\right)$$

$$\kappa_x = \alpha_{12} + \alpha_{13}x + \alpha_{14}\phi + \alpha_{15}x\phi$$

$$\kappa_\phi = \alpha_{16} + \alpha_{17}x + \alpha_{18}\phi + \alpha_{19}x\phi$$

$$\kappa_{x\phi} = \alpha_{20} + \left(\frac{\alpha_{14}}{R}x + \frac{1}{2}\frac{\alpha_{15}}{R}x^2 + R\,\alpha_{17}\phi + \frac{1}{2}R\,\alpha_{19}\phi^2\right) \qquad (3.2)$$

The unbracketted terms are first assumed and the terms in the brackets are then added to satisfy the compatibility equations (2.2).

Equations (3.2) are equated to the corresponding expressions in terms of u,v and w from (2.1) and the resulting equations are integrated to obtain

$$u = \alpha_7 x + \left(\frac{3}{4}R\,\alpha_{11} + R^3\,\alpha_{19} - R^2\,\alpha_{20}\right)\phi - \frac{1}{2}R^3\,\alpha_{17}\phi^2 + \alpha_8 x\phi - \frac{1}{6}R^3\,\alpha_{19}\phi^3$$

$$v = \left(\frac{1}{4}\alpha_{11} - R^2\,\alpha_{19} + R\,\alpha_{20}\right)x + R^2\,\alpha_{16}\phi + \frac{1}{6}\frac{\alpha_8}{R}x^2 + R^2\,\alpha_{17}x\phi + \frac{1}{2}R^2\,\alpha_{18}\phi^2 + \frac{1}{2}R^2\,\alpha_{19}x\phi^2$$

$$w = \left(R\,\alpha_9 - R^2\,\alpha_{16}\right) + \left(R\,\alpha_{10} - R^2\,\alpha_{17}\right)x - R^2\,\alpha_{18}\phi - \frac{1}{2}\alpha_{12}x^2 - R^2\,\alpha_{19}x\phi - \frac{1}{6}\alpha_{13}x^3 - \frac{1}{2}\alpha_{14}x^2\phi - \frac{1}{6}\alpha_{15}x^3\phi \qquad (3.3)$$

The complete shape function is the sum of corresponding expressions from equations (3.1) and (3.3).

4. PROBLEMS SOLVED

To determine the effect of using Sanders-Koiter shell equations instead of Timoshenko's, the present element was tested by applying it to the analysis of the problems of the thin and thick pinched cylinders and the eight types of barrel vault problems. Details of these problems are given in reference [1]. The Nodal Solution Routine [5] was used and it was found that in the case of the pinched cylinder problems not only the results converged to the same answer but all the results obtained when the shell is divided into small numbers of elements were identical to three significant figures. For the barrel vault type of problems the results converged to within 1.5%.

Table 1 gives a typical comparison of the results obtained from the two elements for the case of the thin-deep barrel vault with free longitudinal straight edges.

TABLE 1

radial deflection (ft) at centre of the free edge of the thin-deep barrel vault

No. of elements in $\frac{1}{4}$ of cylinder	No. of degrees of freedom	Elements		% difference
		Ref. [1]	present	
4 x 4	125	0.12469	0.11679	6.76
6 x 6	245	0.25879	0.24421	5.97
8 x 8	405	0.34875	0.33457	4.24
10 x 10	605	0.40768	0.39562	3.05
12 x 12	845	0.44648	0.43663	2.26
14 x 14	1125	0.47272	0.46467	1.73
16 x 16	1445	0.49105	0.48446	1.36

5. THE INTRODUCTION OF POISSON'S RATIO TERMS IN THE STRAIN FUNCTIONS

When the strain element of reference [1] was used to analyse cylindrical shells loaded by rotationally symmetric uniform radial load or bending moment as shown in Fig. 2 it was noticed that small axial membrane stresses are obtained.

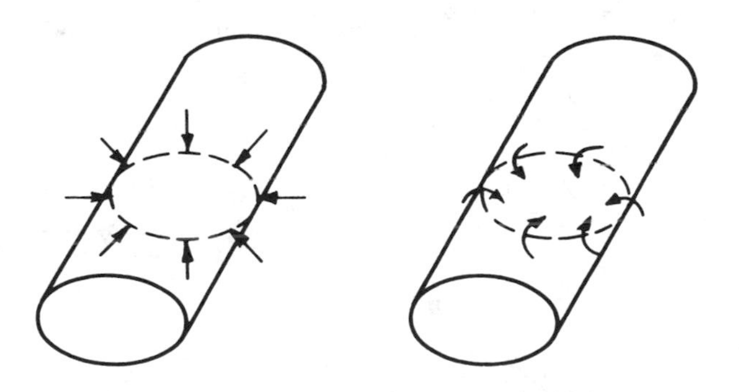

FIG. 2. Rotationally symmetric uniform loadings

Since such structures are analogous to beams on elastic foundations [6], in which these stresses do not exist, it is desirable to obtain finite element solutions giving zero values to these stresses. A close examination of the expressions for the strains ε_x and ε_ϕ as given in equations (3.2) shows that for example the axial stresses arise because the term $\alpha_{10}x$ allows a variation in ε_ϕ, which in the absence of an axial stress requires a term $-\nu\alpha_{10}x$ in ε_x. A preliminary test was therefore carried out by introducing the term $-\nu\alpha_{10}x$ in the expression ε_x

and the resulting element was tested to show that zero axial stress are obtained.

The inclusion of Poisson's ratio terms is required to satisfy equilibrium within an element and close satisfaction of equilibrium should increase the rate of convergence to the exact solution. The equilibrium equations (2.3) can be simplified by neglecting the terms that are of the same order as those disregarded in developing thin shell theories i.e. of order $(h/R)^2$. The terms appearing in [ ] brackets are therefore ignored. A further simplification is also made when the strains as given in equations (2.1) are substituted into the resulting approximate equilibrium equation by retaining first order terms only. Such consideration will show that the following additional terms for the strain may be introduced

$$\varepsilon_x = -\nu\,\alpha_{10}x + \left(\frac{\nu}{2}\frac{\alpha_{12}}{R}x^2 + \frac{\nu}{6}\frac{\alpha_{13}}{R}x^3 + \frac{\nu}{2}\frac{\alpha_{14}}{R}x^2\phi + \frac{\nu}{6}\frac{\alpha_{15}}{R}x^3\right)$$

$$\varepsilon_\phi = -\,\nu\,\alpha_8\phi$$

$$\gamma_{x\phi} = \frac{2\,\nu\,\alpha_{14}}{9\;R^2}x^3 + \frac{\nu\,\alpha_{15}}{18\;R^2}x^4 \tag{5.1}$$

Equations (5.1) when integrated will give the following additional terms to the shape function of the element.

$$u = \nu\left(\frac{-\alpha_{10}}{2}\,x^2 + \frac{\alpha_{12}}{R}\,x^3 + \frac{\alpha_{13}}{24R}\,x^4 + \frac{\alpha_{14}}{6R}\,x^3\phi + \frac{\alpha_{15}}{24R}\,x^4\phi\right)$$

$$v = \nu\left(\frac{\alpha_{14}}{72R^2}\,x^4 + \frac{\alpha_{15}}{360R^2}\,x^5\right) \tag{5.2}$$

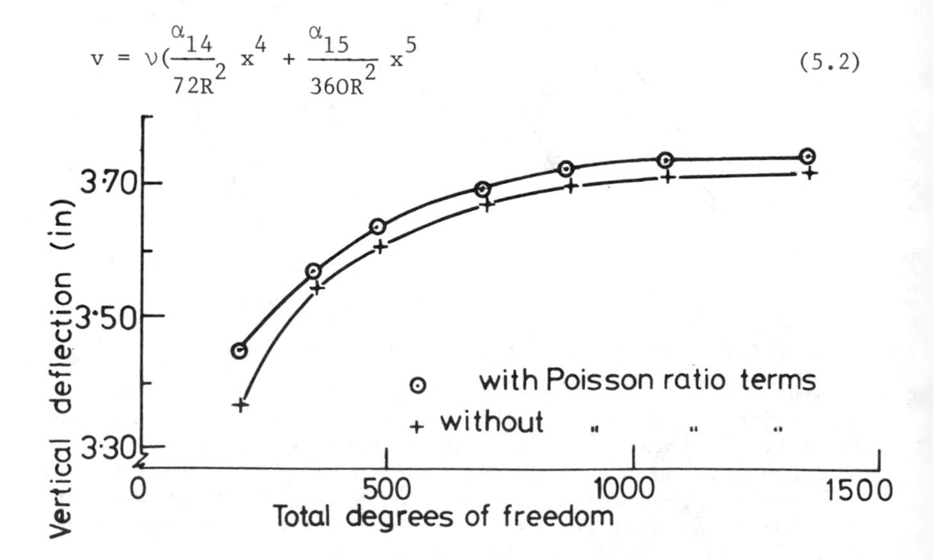

FIG.3 Vertical deflection at centre of free edge of the thick-moderate barrel vault

We note that the changes are confined to inplane membrane strains and displacements u and v, leaving the curvature strains and displacement w unchanged. We should therefore expect improvement to the rate of convergence to occur in problems where bending is not dominant. This was confirmed when the problems of the pinched cylinder were analysed. Since these are almost pure bending problems, the inclusion of the Poisson's ratio terms hardly effected the results. Figure 3 shows the improvement obtained when the free-edged, thick moderate barrel vault problem is analysed.

6. QUADRILATERAL ELEMENT

A quadrilateral element based on the strain functions given in Section 3 and having the axes x and ϕ coinciding with the principal axes of the cylinder is developed. The stiffness matrix is calculated by the use of the 7-point numerical integration formula of Zienkiewicz [8]. Since this formula is used for triangles, the quadrilateral element is divided into four triangles. This quadrilateral element was tested by analysing the pinched cylinder problems. A typical mesh (2x2) refinement is shown in Fig. 4 for one-eighth of cylinder.

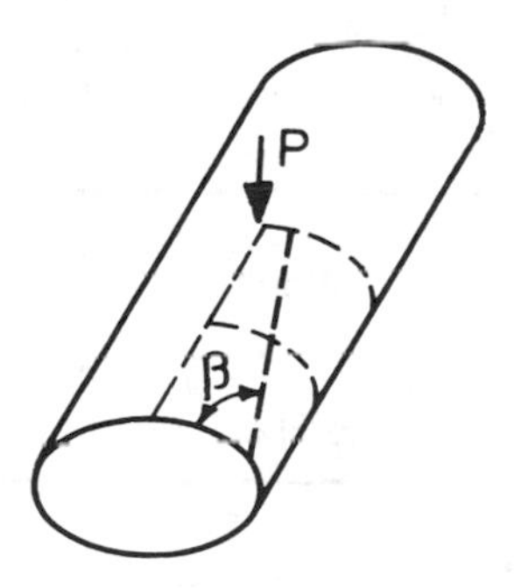

FIG. 4 pinched cylinder modelled by 2x2 quadrilateral elements

Tables 2 and 3 give the results for the thick and thin cases for two mesh arrangements. Arrangement A was for the case when the angle β has a constant value of 86.5^o while arrangement B was for β varying between 77.5^o and 86.5^o.

When this element is used to analyse the thick-moderate barrel vault problem the results again compare favourably with those of the rectangular element as shown in Table 4.

In the course of investigation of the general behaviour of this quadrilateral element it was noticed that if the centre line of the element made an angle of 22.5^o, 45° or 67.5^o with the cylinder's generator, the transformation matrix is singular.

TABLE 2

deflection under the load for the thick pinched cylinder

Mesh	Rectangular Element	Quadrilateral Arrangement A	Quadrilateral Arrangement B
2x2	0.1103	0.1106	0.1124
4x4	0.1129	0.1132	0.1141
6x6	0.1135	0.1137	0.1140
8x8	0.1137	0.1138	0.1139
10x10	0.1137	0.1138	0.1138

TABLE 3

deflection under load for the thin pinched cylinder

Mesh	Rectangular Element	Quadrilateral Arrangement A	Quadrilateral Arrangement B
2x2	0.02429	0.02431	0.02435
4x4	0.02447	0.02449	0.02456
6x6	0.02455	0.02458	0.02462
8x8	0.02460	0.02462	0.02464
10x10	0.02461	0.02463	0.02463

TABLE 4

radial deflection at centre of the free edge for the thick-moderate barrel vault

Mesh	Quadrilateral Element	Rectangular Element
2x2	0.22500	0.21401
4x4	0.29181	0.29685
6x6	0.31825	0.32438
8x8	0.33021	0.33560
10x10	0.33675	0.34112

This difficulty made it essential to avoid elements with these orientations. A full discussion of the suitability of different orientations is given in reference [9].

REFERENCES

1. ASHWELL, D.G. and SABIR, A.B., A new cylindrical shell finite element based on single independent strain functions. *Int. J. Mech. Sci.* 14, 171-183 (1972).
2. TIMOSHENKO, S., and WOINOWSKY-KRIEGER, S., *Theory of plates and shells*, McGraw-Hill, New York, (1959).
3. SABIR, A.B. and ASHWELL, D.G., Diffusion of concentrated loads into thin cylindrical shells, pp. 381-389 of J.R. Whiteman (Ed.) *The Mathematics of Finite Elements and Applications III*. Academic Press, London (1978).
4. KOITER, W.T., A Consistent First Approximation in the General Theory of Thin Elastic Shells, of Koiter, W.T. (Ed.), *Theory of Thin Elastic Shells*, Delft, (1959).
5. SABIR, A.B., The Nodal Solution Routine for the Large Number of Linear Simultaneous Equations in the Finite Element Analysis of Plates and Shells. pp.63-89 of D.G. Ashwell and R.H. Gallagher (Ed.). *Finite Elements for Thin Shells and Curved Members*. John Wiley and Sons, London, (1976).
6. SABIR, A.B. and ASHWELL, D.G., A Stiffness Matrix for Shallow Shell Finite Elements. *Int. J. Mech. Sci.* 11, 269-279 (1968).
7. ZIENKIEWICZ, O.C., *The Finite Element Method*. McGraw-Hill, New York, (1977).
8. CHARCHAFCHI, T.A. *Finite Element Analysis of Thin Shell Structures*, Ph.D. Thesis, University of Wales, (1980).

A FINITE ELEMENT ALGORITHM FOR FLOW IN A GAS CENTRIFUGE

*J. E. Akin, +M. H. Berger

**University of Tennessee*, *+Union Carbide Corporation*

1. INTRODUCTION

Flow of a viscous, heat conducting gas in a cylindrical container is an intensely studied problem which, of course, is due to the world-wide development of high-speed gas centrifugation for uranium enrichment. Many investigators have reported approximate solutions to the hydrodynamics of the flow field with varying degrees of detail. While the linearized theories invariably result in enormous simplification of the governing partial differential equations (Navier-Stokes equations) they usually, but not always, lead to a sixth order elliptic boundary value problem for either streamfunction or master potential. On the other hand, Lahargue [6] has solved the linearized equations in primitive variables using finite elements on rectangles.

The requisite mathematics to solve the resultant equations with boundary conditions and Ekman layer matching is quite formidable and still results in a considerable numerical calculation since the solution cannot, in general, be written in terms of a simple function or functions, but is expressed as a double Fourier Series. Specifically, the associated eigenfunction - eigenvalue problem is difficult, requiring asymptotic methods for the higher numbered eigenvalues and the determination of the expansion coefficients requires numerical integration or solution of a least-squares problem [10]. With this setting then, one may pose the problem of calculating the flow field using some numerical method like finite difference, finite element, collocation, discrete element, spectral technique, etc. The approach described here is a Galerkin finite element approximation using standard finite element techniques; that is, only shape functions with nearest neighbor support are considered, thereby ruling out such things as B-splines [8]. There are a great number of fine texts and monographs describing the method in detail, [3], [7], [8], [9], [11].

Albeit this linear boundary value problem has been solved by eigen-function expansion, the finite element method may prove computationally superior, while without doubt it is computationally simpler.

2. GOVERNING DIFFERENTIAL EQUATION AND BOUNDARY CONDITIONS

Linearized gas flow in a centrifuge may be described by Lars Onsager's so-called "Pancake" equation for the master potential χ away from the ends $y = 0$ and $y = y_o$ [10],

$$L\chi - F(x,y) = 0, \quad \forall\, x,y \in \Gamma \tag{2.1}$$

where F is a source term, L is the linear operator

$$L\chi \equiv L_6\chi + B^2\chi_{yy} \tag{2.2}$$

and

$$L_6\chi \equiv [e^x(e^x\chi_{xx})_{xx}]_{xx} \tag{2.3}$$

$$= e^{2x}(\chi_{xxxxxx} + 6\chi_{xxxxx} + 13\chi_{xxxx} + 12\chi_{xxx} + 4\chi_{xx}). \tag{2.4}$$

This equation is subject to the eight boundary conditions,

$$\chi_x(0,y) = \chi_{xx}(0,y) = 0 \tag{2.5}$$

and

$$L_5\chi(0,y) = \frac{R_e}{32A^{10}}\,\overline{\Theta}_y(y), \quad \forall\, x,y \in \partial\Gamma_1 \tag{2.6}$$

$$\chi_y(A^2,y) = \chi_x(A^2,y) = L_3\chi(A^2,y) = 0, \quad \forall\, x,y \in \partial\Gamma_2 \tag{2.7}$$

$$\chi_y(x,0) = -4S^{-\frac{1}{4}}R_e^{-\frac{1}{2}}A^4[e^{x/2}\chi_x(x,0)]_x, \quad \forall\, x,y \in \partial\Gamma_3 \tag{2.8}$$

$$\chi_y(x,y_o) = 4S^{-\frac{1}{4}}R_e^{-\frac{1}{2}}A^4[e^{x/2}\chi_x(x,y_o)]_x, \quad \forall\, x,y \in \partial\Gamma_4 \tag{2.9}$$

where $\overline{\Theta}_y$ is the thermal gradient, R_e is Reynolds number,

$$L_3\chi = (e^x\chi_{xx})_x, \tag{2.10}$$

$$L_5\chi = [e^x(e^x\chi_{xx})_{xx}]_x\,. \tag{2.11}$$

The rotor speed, V_w, and gas properties define constants S and

$$B = R_e S^{\frac{1}{2}}/4A^6 >> 1, \tag{2.12}$$

$$A^2 = \frac{V_w^{\,2}}{2RT_o} >> 1. \tag{2.13}$$

Here χ is a potential that defines a streamfunction, ψ, as

$$\psi = -2A^2\chi_x. \tag{2.14}$$

Domain Γ and closure $\partial\Gamma$ are illustrated schematically in Fig. 1 below. Intrinsic to this are the assumptions of infinitesimal velocity, pressure and density perturbations about rigid body rotation. The rigid body rotation leads to a density distribution that is exponential in -x.

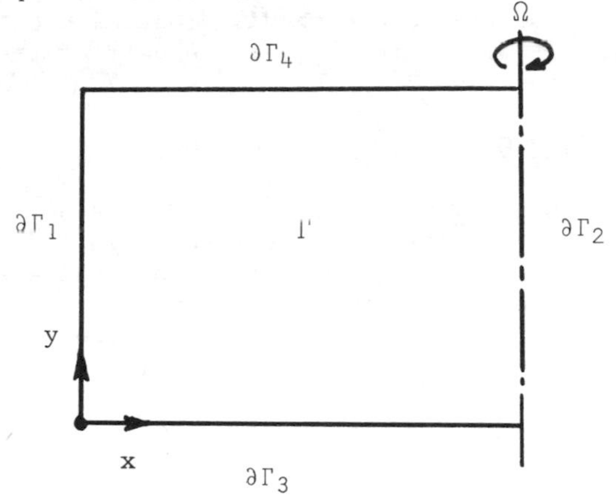

FIG. 1. Domain Γ and closure $\partial\Gamma$ of rotating fluid.

The homogeneous, source-free variant of this equation has been solved by Wood and Morton [10] using a generalized Fourier Series, while similar equations have been derived from asymptotic expansions and also solved using eigenfunction methods by Brouwers [2]. Brouwers' main contribution was the identification of domain aspect ratio (length:radius) as an important parameter. To prove that Onsager's equation is elliptic we know from the theory of higher order partial differential equations [4] that it is only necessary to show that it has no real characteristics. An expression like (2.1) requres that $\chi \in C^{6,2}$ for $F \in C^{o}$. And since it involves eight partial derivatives there needs to be eight boundary conditions. If the eight boundary conditions are all of the Neumann type, as in this case, then the solution will be determined to within an additive constant.

3. WEAK FORMULATION

Consider a restatement of the given differential or "strong" formulation equivalently as an integral or "weak" statement. Let

$$\chi = \chi^e = \underline{N}^{eT}\underline{u}^e, \ \forall\ x,y \in \Gamma^e \tag{3.1}$$

where the local basis function is $\underline{N}^e$ and $\underline{u}^e$ is the discretization of χ over an element subdomain Γ^e involving nodal values of χ as well as derivatives. Presumably χ can be approximated by a finite dimensional subspace spanned by a basis that is sufficiently differentiable at interelement boundaries such that no additional contributions occur there.

The method of weighted residuals requires that one satisfy the vanishing of the weighted residual in lieu of the original differential equation where the weighted residual is an inner product over Γ of the governing differential equation and an arbitrary weight ω. In this way it is part of the general class of projective methods [7]. Thus

$$-(\omega, L\chi - F) = 0 \tag{3.2}$$

over the x,y domain Γ with boundary $\partial\Gamma$. The negative sign appearing here is to assure a positive definite form. Certain boundary integrals arise in executing the indicated integrations that are important to the method. Formally integrating by parts thrice throws off three x-derivatives and one y-derivative onto ω and simultaneously reduces the continuity requirements on χ. The result is

$$\iint_{\Gamma}\left[L_3\omega\, L_3\chi + B^2\omega_y\chi_y + \omega F\right]dxdy$$

$$+ \int_{\partial\Gamma}\left[e^x\omega_x L_4\chi - e^x\omega_{xx}L_3\chi - \omega L_5\chi\right]\Bigg|_{x=0}^{x_T} dy$$

$$- B^2\int_{\partial\Gamma}\omega\chi_y\Bigg|_{y=0}^{y_o} dx = 0 \tag{3.3}$$

where

$$L_4\chi = (e^x\chi_{xx})_{xx}. \tag{3.4}$$

In the absence of boundary integrals this reduces to the bilinear and linear form

$$(L_3\omega,\ L_3\chi) + B^2(\omega_y, \chi_y) + (\omega, F) = 0. \tag{3.5}$$

Fortunately, for a basis with compact support it is necessary to consider only the boundary integrals for the subset of elements where $\Gamma^e \cap \partial\Gamma \neq 0$.

Discretizing χ and choosing $\underline{w}^e$ the same as $\underline{N}^e$ we have the Galerkin variant of the more general method of weighted residuals. The resulting equations are

$$\sum_e \iint_{\Gamma^e} \left[\left(\underline{L_3 N^e} \; \underline{L_3 N^{eT}} + B^2 \, \underline{N^e_y} \; \underline{N^{eT}_y} \right) \underline{u}^e + F^e \, \underline{N}^e \right] dxdy = \underline{0}, \qquad (3.6)$$

omitting for the time being those non-vanishing boundary integrals. The summation sign represents the assembly of individual element contributions. Here sufficient interelement continuity is assumed in order to write the equation as a sum over e. In general, to satisfy the given boundary conditions one may have to choose the space of functions $\underline{w}^e$ different from $\underline{N}^e$ on the boundary. This will be made clear when the boundary conditions are treated. In doing the integration by parts above, numerous boundary integrals have appeared, some of which will be satisfied naturally and others will be treated by constraining the space of admissible functions. The general rule on boundary conditions [9] is that conditions which involve only derivatives below order s,t where $\underline{w}^e \varepsilon C^{s,t}$ will make sense. Those involving derivatives of order s,t or higher, will not apply; thus, the distinction between essential and natural boundary conditions. Specifically for L of order 2m = 6, 2n = 2, and s = m = 3, t = n = 1. That is x boundary conditions of order zero, one and two, and conditions on y of zeroth order are essential. All the higher order constraints are non-essential and will be satisfied naturally. Assembly of the elemental equations into the system and correct incorporation of boundary conditions results in the usual matrix problem

$$K \, \underline{u} = \underline{f}, \qquad (3.7)$$

where K is sparse, symmetric, positive definite and invertible.

4. MODEL ORDINARY DIFFERENTIAL EQUATION

Perhaps these formalisms are best illustrated by consideration of an idealized flow. In the limit $y \to \infty$ such that $\chi_y = 0$, this problem degenerates to the so-called long bowl approximation. In addition, let $\overline{\Theta}_y\,(0,y)$ = constant which further simplifies the situation. Then

$$L_6 \chi(x) = 0, \ \forall \ x \ \varepsilon \, \Gamma \qquad (4.1)$$

subject to the boundary conditions

$$\chi'(0) = \chi''(0) = 0, \forall\, x \in \partial\Gamma \tag{4.2}$$

$$L_5\chi(0) = \text{constant (e.g., 1)} \tag{4.3}$$

$$\chi(\infty) = \chi'(\infty) = L_3\chi(\infty) = 0. \tag{4.4}$$

This has a simple exact solution [10] which can be used for comparison with a numerical calculation. The solution is

$$\chi = \frac{L_5\chi(0)}{4}\left(-\,2e^{-x} + \frac{3}{2}e^{-2x} + xe^{-2x}\right), \tag{4.5}$$

where

$$L_5\chi(0) = \frac{R_e}{32A^{10}}\,\overline{\Theta}_y(0). \tag{4.6}$$

The finite element discretization simplifies to the bilinear form

$$\sum_e \int_{\Gamma^e} \underline{L_3N^e}\;\underline{L_3N^{eT}}\,dx\;\underline{u}^e = \underline{0} \tag{4.7}$$

subject to appropriate boundary conditions. As $L_3\chi = e^x(\chi''' + \chi'')$, the variational equation requires $\underline{N}^e \in C^2$ for admissibility. An arbitrary number of derivatives may be approximated using the family of generalized Hermite polynomials which give rise to subparametric elements. Consider the members of the one-dimensional piecewise continuous Hermite polynomials $\in C^2$, the simplest being the quintic. One can easily verify the shape function to be, [5]

$$\underline{N}^e = \begin{Bmatrix} (1-\xi)^3\,(3\xi^2 + 9\xi + 8)/16 \\ (1+\xi)\,(1-\xi)^3\,(3\xi + 5)\ell^e/32 \\ (1+\xi)^2\,(1-\xi)^3\ell^{e\,2}/64 \\ (1+\xi)^3\,(3\xi^2 - 9\xi + 8)/16 \\ (1+\xi)^3\,(1-\xi)\,(3\xi - 5)\ell^e/32 \\ (1+\xi)^3\,(1-\xi)^2\ell^{e\,2}/64 \end{Bmatrix} \tag{4.8}$$

where ℓ^e is the element length and

$$x = \frac{\ell^e}{2}\xi + \frac{x^e_1 + x^e_2}{2}, \quad -1 \leq \xi \leq 1. \tag{4.9}$$

The resulting $\underline{L_3N^e}$ is (4.10)

$$\underline{L_3N^e} = e^x \begin{Bmatrix} 30(1 - 3\xi^2)/\ell^{e^3} + 15(1 - \xi)\xi(\xi + 1)/\ell^{e^2} \\ 3(-15\xi^2 + 2\xi + 5)/\ell^{e^2} - 3(1 - \xi)(-5\xi^2 - 4\xi + 1)/2\ell^e \\ 3(-5\xi^2 + 2\xi + 1)/2\ell^e - (1 - \xi)(-5\xi^2 - 2\xi + 1)/4 \\ 30(3\xi^2 - 1)/\ell^{e^3} + 15(1 + \xi)\xi(\xi - 1)/\ell^{e^2} \\ 3(-15\xi^2 - 2\xi + 5)/\ell^{e^2} + 3(1 + \xi)(-5\xi^2 + 4\xi + 1)/2\ell^e \\ 3(5\xi^2 + 2\xi - 1)/2\ell^e + (1 + \xi)(5\xi^2 - 2\xi - 1)/4 \end{Bmatrix}.$$

As the integrations are obviously quite laborious to do analytically, we resort to numerical quadrature using Gauss' rule.

Essential boundary conditions are those involving derivatives of second order or less, and natural boundary conditions are those involving derivatives of third order or higher. Here the essential conditions are homogeneous, so it is necessary to constrain the shape functions to zero on the boundaries $x = 0$ and x_T, i.e., $\chi'(0) = \chi''(0) = \chi(x_T) = \chi'(x_T) = 0$. Treatment of one natural boundary condition requires specification of $L_5\chi(0)$, while the other at $x = x_T$, $L_3\chi(x_T) = 0$, is automatically satisfied. Here x_T is chosen to be an approximation to infinity.

The element load vector vanishes in the absence of any internal generation except for the contribution from a single boundary integral. For the element with $\Gamma^e \cap \partial\Gamma^1 \neq 0$, we have

$$\underline{f}^T = L_5\chi(0)[1\ 0\ 0\ 0\ 0\ 0]\,. \tag{4.11}$$

5. RESULTS FOR MODEL EQUATION

Primary to the utility of the finite element methodology in the engineering sciences is the convenience with which an analyst may set up and solve a new problem with a modest amount of effort. This depends in part on the availability of well documented, modular user oriented codes like MODEL [1] and the existence and availability of local basis functions having nearest neighbor support. Then one need only derive the elemental stiffness and column matrices on a single typical element.

The solution of the simplified one-dimensional problem was obtained using the MODEL code described in detail by Akin [1]. Element matrices were formulated as described herein and the calculations employed the standard input, numerical integration, assembly, solution, and output program libraries of MODEL.

The computed results presented here were obtained for a uniform grid of 32 elements. Values of χ, χ', and χ'' were compared with the exact solution. The first variable is the potential, while the second and third represent the streamfunction and axial mass velocity. Figure 2 shows the comparison of the values of χ

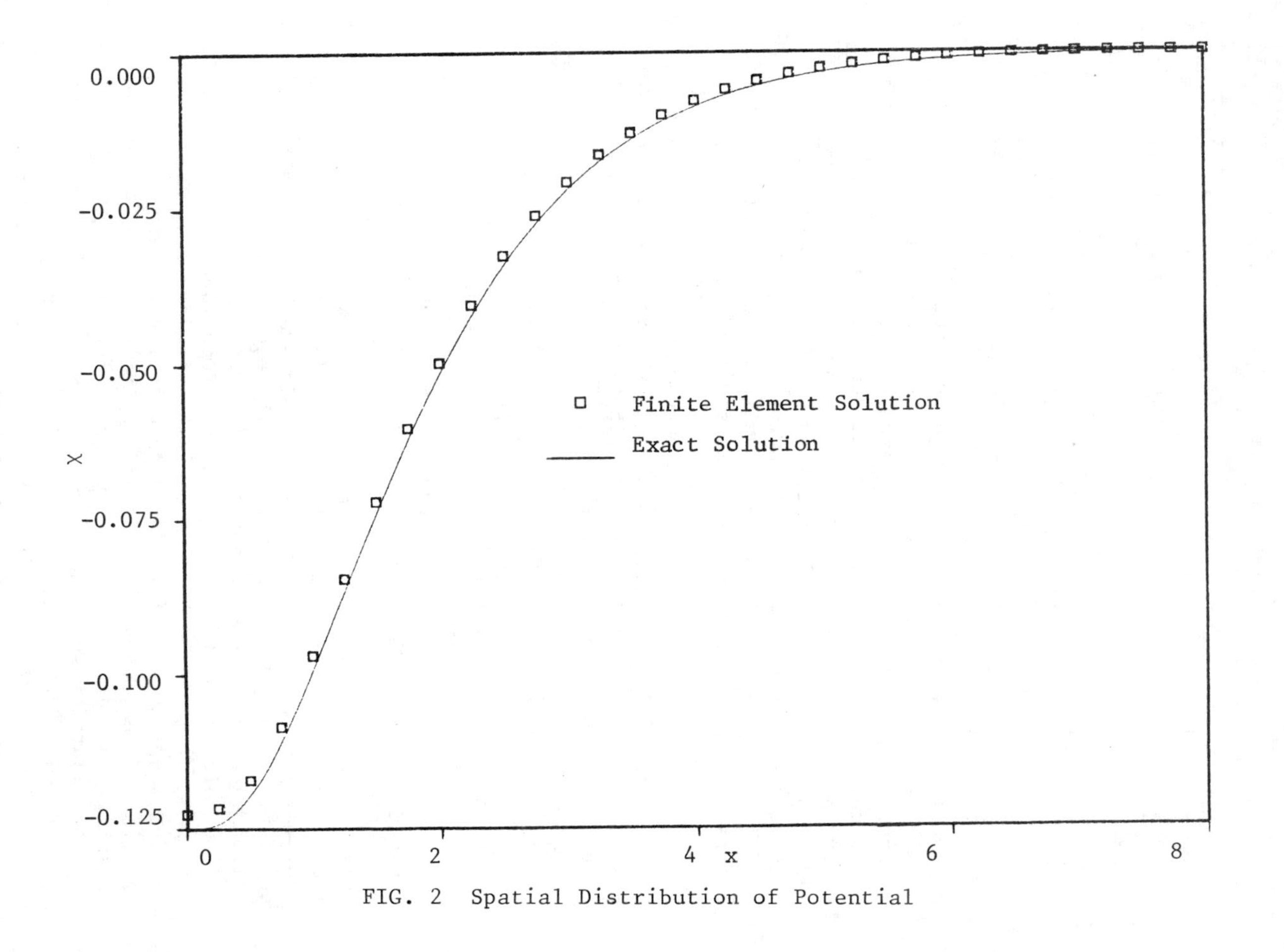

FIG. 2 Spatial Distribution of Potential

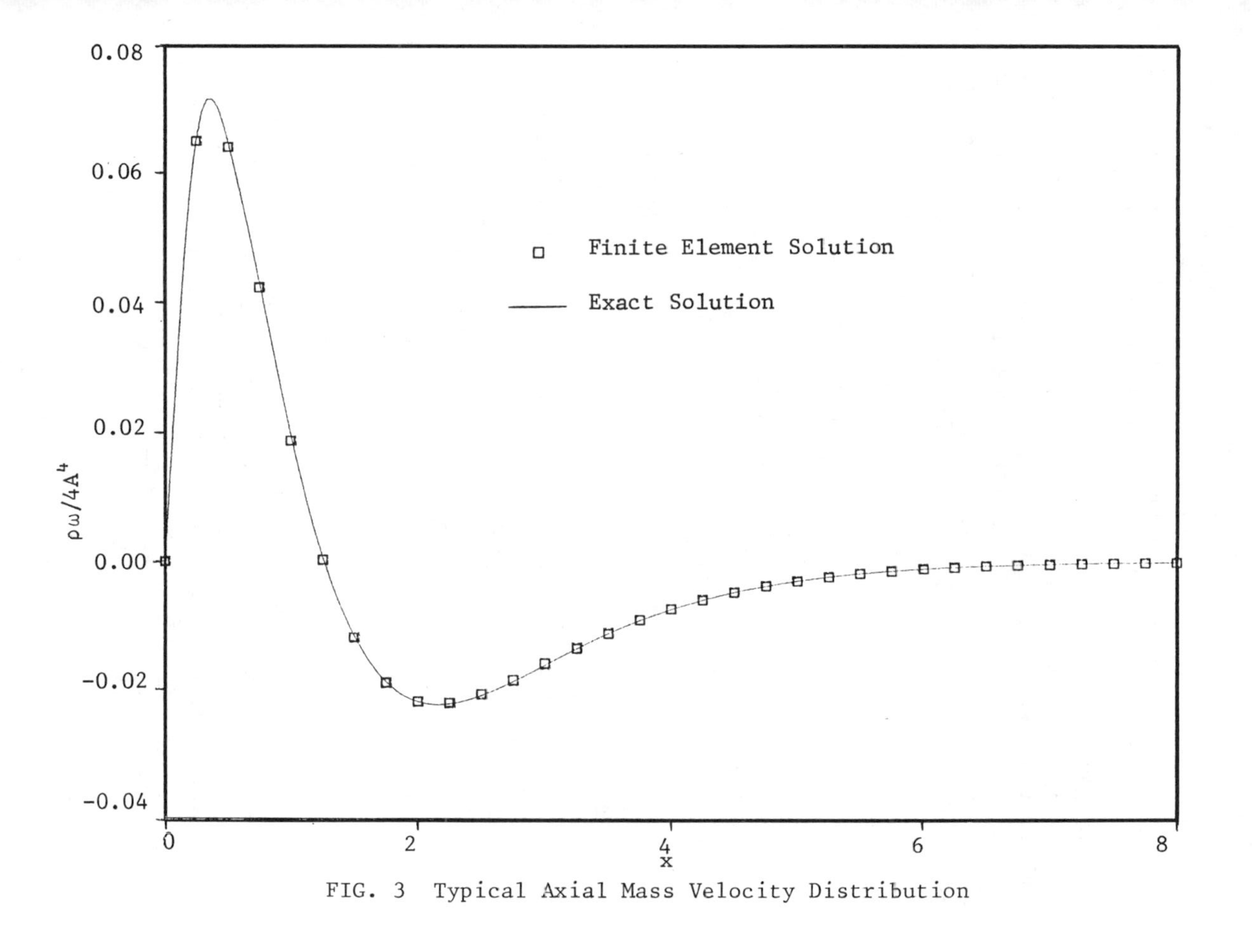

FIG. 3 Typical Axial Mass Velocity Distribution

and Fig. 3 shows those for χ''. The accuracy seems quite good, especially near the essential boundary conditions at $x = x_T = 8$, and the largest error occurs near the natural boundary conditions at $x = 0$. This is not unexpected for a uniform grid. The major source of error is due to selecting the small value for the top of the atmosphere, x_T Clearly from the above, calculating in two dimensions is a fairly straightforward extension.

ACKNOWLEDGEMENT

Portions of this work were supported by the Union Carbide Corporation-Nuclear Division, Oak Ridge Gaseous Diffusion Plant under contract W-7405 eng 26. The Oak Ridge Gaseous Diffusion Plant is operated by UCC-ND under contract to the U.S. Department of Energy.

REFERENCES

1. Akin, J. E., *Applications and Implementation of Finite Element Methods*, Academic Press, London, (1981).
2. Brouwers, J. J. H., *On the Motion of a Compressible Fluid in a Rotating Cylinder*, Ph.D. Dissertation, Twente University of Technology, Enschede, Netherlands, (1976).
3. Gallagher, R. H., *Finite Element Analysis Fundamentals*, Prentice Hall, Inc., Englewood, (1975).
4. John, Fritz, *Partial Differential Equations*, Springer-Verlag, Berlin, (1978).
5. Kawai, T., and M. Watanabe, *Analysis of a Solitary Water Wave Problem by the Method of Weighted Residuals*, Second International Symposium on Finite Element Methods in Flow Problems, S. Margherita Ligure (Italy), (1976).
6. Lahargue, J. P., and Soubbaramayer, A Numerical Model for the Investigation of the Flow and Isotope Concentration Field in an Ultracentrifuge, *Computer Methods in Applied Mechanics and Engineering*, 15, (1978).
7. Oden, J. T., *Finite Elements of Nonlinear Continua*, McGraw-Hill Book Company, New York, (1972).
8. Prenter, P. M., *Splines and Variational Methods*, John Wiley and Sons, Englewood, (1975).
9. Strang, G., and G. Fix, *An Analysis of the Finite Element Method*, Prentice Hall, Inc., (1973).
10. Wood, H. G. III, and J. B. Morton, Onsager's Pancake Approximation for the Fluid Dynamics of a Gas Centrifuge, *Journal of Fluid Mechanics*, 101, Part 1, 1-31, (1980).
11. Zienkiewicz, O. C., *The Finite Element Method*, McGraw-Hill Book Company, New York, (1977).

AN ALTERNATING ITERATION SCHEME FOR CERTAIN FREE BOUNDARY VALUE PROBLEMS

J.C. Bruch, Jr., J.M. Sloss, and J. Remar

University of California, Santa Barbara

1. INTRODUCTION

The Baiocchi method and transformation has been used on a large number of seepage problems over the past ten years. (See Baiocchi and Capelo [1], Kinderlehrer and Stampacchia [6], Oden and Kikuchi [7], and Bruch [3]). In this approach the a priori unknown solution region is extended across the free surface into a known region. Then a new dependent variable is defined using Baiocchi's transformation within this latter region. The resulting problem formulation leads to the "complementarity system" associated with its respective variational or quasi-variational inequality formulation. This method has proven effective not only from the purely theoretical point of view, that is for proving existence and uniqueness of the solution, but also from the point of view of yielding new, simple, and efficient numerical solution schemes.

Herein is discussed an extension of the Baiocchi method which is simpler and more generally applicable. This extension yields straightforward existence proofs. The method is based on an approach used by the authors in Remar et al. [8] where it was used initially to avoid a singularity in an axisymmetric flowfield at $r = 0$. The approach will be demonstrated on the steady, two-dimensional seepage through a variable shaped dam with a toe drain.

2. DIFFERENTIAL EQUATION FORMULATION

Considered here is the steady, two-dimensional seepage through a variable shaped dam with a toe drain (see Fig.). In Baiocchi [2] the problem of a dam with arbitrary shape on an impermeable, horizontal base separating two reservoirs is formulated using quasi-variational inequalities and efficient and rigorous algorithms are given for obtaining numerical results. In order to simplify the problem, several common assumptions are made. First, the soil in the flowfield is taken to be homogeneous and isotropic. Second, capillary and evaporation effects are neglected. Third, the flow is assumed laminar and follows Darcy's Law.

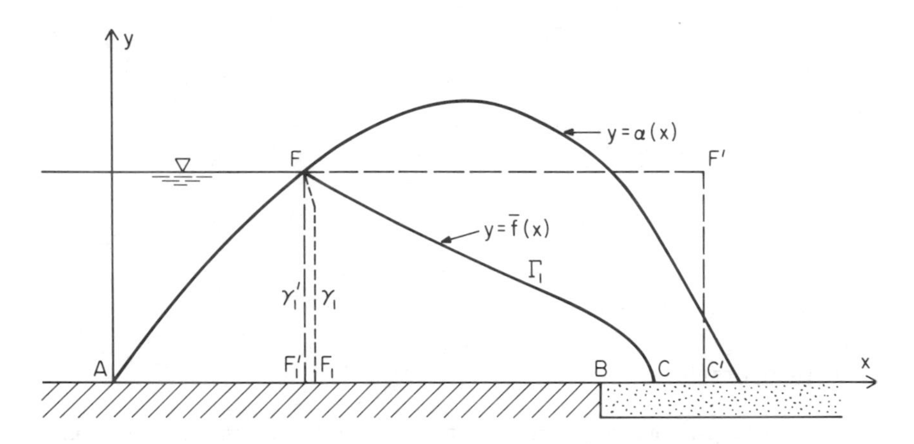

Fig. Flow problem and solution domains

Since a free surface problem is being considered, the free surface location is not known a priori. However, on the free surface two boundary conditions hold: (1) $\phi = ky$; and (2) $\psi =$ constant; or equivalently for (2) $\phi_\eta = 0$, where η is the outward normal direction, ϕ is the velocity potential, and ψ is the stream function. On the remainder of the boundaries, either Neumann or Dirichlet data is given. Let $\Gamma_1 = \{(x,y) \mid x_F < x < x_C,\ y = \bar{f}(x)\}$ be the free surface which is the boundary line between wet and dry soil. It should be noted that x's and y's having letter subscripts on them designate the x-distance and y-distance to that lettered point, respectively. Furthermore, k (the permeability) will be set equal to 1 without loss of generality.

Thus the following quantities need to be found: $y = \bar{f}(x)$, (the function describing the shape of the free surface Γ_1); Ω, (the seepage domain); $\phi(x,y) = \frac{p}{\rho g} + y$; $\psi(x,y)$; and q (the flowrate through the flowfield). The function $\bar{f}(x)$ is to be defined and smooth on its interval of existence. Furthermore, it is strictly decreasing on this interval and must satisfy

$$\bar{f}(x_F) = y_F,\ \bar{f}(x_C) = 0,\ \frac{d\bar{f}(x_F)}{dx} = -\frac{1}{\alpha'(x_F)} \text{ and } \frac{d\bar{f}(x_C)}{dx} = -\infty.$$

The functions $\phi(x,y)$ and $\psi(x,y)$ are defined on $\bar{\Omega}$ and are to be in $H^1(\Omega) \cap C^0(\bar{\Omega})$. Further, for the problem shown in the Fig.:

$$\Omega = \{(x,y) \mid 0 < x \leqq x_F,\ 0 < y < \alpha(x)\ ;\ x_F < x < x_C,\ 0 < y < \bar{f}(x)\},$$

$$\phi_x - \psi_y = 0 \text{ and } \phi_y + \psi_x = 0 \text{ in } \Omega,$$

$$\phi = y_F \text{ on } \widehat{AF},$$

$$\phi = 0 \text{ on } [BC], \tag{2.1}$$

$$\psi = q \text{ on } [AB],$$

$$\psi = 0 \text{ on } \Gamma_1 \text{ [i.e., curve } (x,\bar{f}(x))],$$

and $\phi = y$ on Γ_1.

Let the solution domain Ω be extended to the known region D $= \{(x,y)|0 < x \leq x_F,\ 0 < y < \alpha(x);\ x_F < x < x_{C'},\ 0 < y < y_F\}$ in the Fig. Then extend ϕ and ψ continuously to be defined on $\bar{D}$ by setting

$$\tilde{\phi}(x,y) = \phi(x,y) \text{ in } \bar{\Omega}, \text{ and } \tilde{\psi}(x,y) = \psi(x,y) \text{ in } \bar{\Omega},$$
$$= y \text{ in } \bar{D} - \bar{\Omega} \qquad\qquad = 0 \text{ in } \bar{D} - \bar{\Omega}. \tag{2.2}$$

This yields

$$\tilde{\phi}_x - \tilde{\psi}_y = 0 \text{ in D, and } \tilde{\phi}_y + \tilde{\psi}_x = \chi_{D-\bar{\Omega}} \text{ in D}, \tag{2.3}$$

in the sense of distributions where $\chi_{D-\bar{\Omega}}=1$ in $D-\bar{\Omega}$, $\chi_{D-\bar{\Omega}}=0$ in Ω.

Next define a new dependent variable w using the Baiocchi transformation

$$w(P) = \int_{\widehat{FP}} - \tilde{\psi}\, dx + (y - \tilde{\phi})\, dy, \tag{2.4}$$

where $\widehat{FP}$ is a smooth path in D joining F to P in D in the Fig. The integration is independent of the path due to (2.3). If (2.1) hold, then for all w in $H^2(D) \cap C^1(\bar{D})$:

$$\Delta w = \chi_\Omega \text{ in D}, \tag{2.5a}$$

$$w_y = y - y_F \text{ on } \widehat{AF},$$

$$w = (q^2/6) + q(x_B - x) \text{ on } [AB],$$

$$w_y = 0 \text{ on }]BC[, \tag{2.5b}$$

$$w = 0 \text{ in } \bar{D} - \bar{\Omega} \text{ (in particular on } \Gamma_1);$$

and $w > 0$ in Ω ($w \geqq 0$ in D).

Hence, $w(x,y) \geqq 0$, $1 - \Delta w(x,y) \geqq 0$, $w(1 - \Delta w) = 0$ in D. (2.6)

If a w is found satisfying (2.5a) subject to the conditions, (2.5b), then the following quantities can be obtained:

$$\Omega = \{(x,y) | (x,y) \text{ in } D, \ w > 0\};$$

$$\text{graph } \bar{f} = \partial\Omega - \partial D \equiv \text{points of } \partial\Omega \text{ not in } \partial D; \tag{2.7}$$

$$\phi = y - w_y \text{ in } \Omega; \ \psi = -w_x \text{ in } \Omega; \text{ and } q = \psi(x,0) \text{ on } [AB].$$

3. MODIFIED PROBLEM

Now instead of proceeding to obtain theoretical and numerical results for this problem formulation, a modified form of the problem will be considered. Domain D will be split into two regions having a common overlap. Let (see Fig.)

$$D_\phi = \{(x,y) |$$

$$0 < x \leqq x_F, \ 0 < y < \alpha(x); \ x_F < x < \bar{g}(y), \ 0 < y < y_F\}$$

$$D_w = \{(x,y) \mid x_F < x < x_{C'}, \ 0 < y < y_F\}, \text{ and}$$

$$\Omega_{f.s.} = \{(x,y) \mid x_F < x < x_C, \ 0 < y < \bar{f}(x)\},$$

where $x = \bar{g}(y)$ is the equation of the curve γ_1, i.e.

$$\gamma_1 = \{(x,y) \mid x = \bar{g}(y), \ 0 < y < y_F\}.$$

In domain D_ϕ the following boundary value problem is solved:

$$\Delta\phi = 0 \qquad \text{in } D_\phi,$$

$$\begin{aligned} &\phi = y_F && \text{on } \widehat{AF}, \\ &\phi_y = 0 && \text{on }]AF_1[, \\ &\phi = y - w_y && \text{on } \gamma_1. \end{aligned} \tag{3.1}$$

The boundary value problem to be solved in D_w is:

$$\Delta w = \chi_{\Omega_{f.s.}} \text{ in } D_w (\chi_{\Omega_{f.s.}} = 1 \text{ in } \Omega_{f.s.}, \ \chi_{\Omega_{f.s.}} = 0 \text{ in } D - \bar{\Omega}_{f.s.})$$

$$w = (q^2/6) + q(x_B - x) \text{ on } [F_1'B],$$

$$w_y = 0 \text{ on }]BC[,$$

$$w = \int_{y_F}^{y} (\bar{y} - \phi) d\bar{y} \text{ on } [FF_1'], \tag{3.2}$$

$$w = 0 \text{ in } \bar{D}_w - \bar{\Omega}_{f.s.} \text{ (in particular on } \Gamma_1),$$

$$w \geqq 0 \text{ in } D_w \text{ and } w > 0 \text{ in } \Omega_{f.s.}.$$

Hence in D_w,

$$w(x,y) \geqq 0, \ 1 - \Delta w(x,y) \geqq 0, \text{ and } w(1-\Delta w) = 0. \tag{3.3}$$

It can be seen that information is interchanged between the two domains along the line γ_1' and the curve γ_1. The approach used to solve this coupled problem is similar to that presented by Bruch and Sloss [4]. It is an alternating iteration scheme which for the complementarity system given in (3.3) yields for each iteration the associated variational inequality for $w \in K_q$

$$\iint_{D_w} \nabla(v-w) \cdot \nabla w \, dxdy \geqq - \iint_{D_w} (v-w) \, dxdy, \ \forall \, v \in K_q, \tag{3.4}$$

with $K_q = \{v \mid v \in H^1(D_w),\ v = (q^2/6) + q(x_B - x) \text{ on } [F_1'B],$

$v = 0$ on $[FF']$ and $[F'C']$, $v = \int_{y_F}^{y} (\bar{y} - \phi) d\bar{y}$ on $[FF_1']$,

$v \geqq 0$ a.e. on $D_w\}$.

In this form uniqueness and existence of the solution w can be shown (Bruch and Sloss [4]).

4. NUMERICAL SCHEME AND RESULTS

A modified alternating iteration scheme similar to that given in Remar et al. [8] will be used to solve the free surface problem as modified. The D_ϕ region will be discretized and solved using a finite element mesh and a successive over-relaxation type scheme. A finite difference successive over-relaxation scheme with projection will be used for the D_w region.

The flowrate through the flowfield, Ω, is also an unknown a priori. Thus, in addition to the inner iterations to solve for ϕ and w for a given q, there is also an outer iteration on the q to determine the flowrate. A compatibility condition is necessary for this outer iteration. The condition used herein is similar to that given by Sloss and Bruch [9], i.e.,

$$f_h(q^{(r)}) = w_{q^{(r)}}(x_F, y_F - \Delta y) - \frac{\Delta y^2}{2} \quad . \tag{4.1}$$

Then $f_h(q^{(r)}) = 0$ represents a compatibility condition, which if imposed on the set of solutions $\phi_{q,h}$ and $w_{q,h}$ (one for every q), permits the determination of an unique $\bar{q}$ such that $\phi_{\bar{q},h}$ and $w_{\bar{q},h}$ will be a solution of (3.1) and (3.2), respectively, as well as satisfy (4.1). Due to the monotonicity of $f_h(q^{(r)})$ in (4.1), the secant method is used to approximate the root.

The example to be investigated (one solved by Sloss and Bruch [9]) using the previously given approach has the following data: y_F = 30 ft; x_F = 30 ft; x_B = 60 ft; $x_{C'}$, which should be greater than $(x_B + \frac{q}{2})$, (see Harr [5] and Sloss and Bruch [9]); $q^{(0)}$ = 15.0 ft^3/sec per foot depth normal to the plane of the flow; $q^{(1)}$ = 16.0 ft^3/sec per foot depth normal to the plane of the flow; stopping error estimates

$$\max_{i,j} \left| \phi^{(n+1)}_{q^{(r)},i,j} - \phi^{(n)}_{q^{(r)},i,j} \right| < \varepsilon_1,$$

$$\max_{i,j} \left| w^{(n+1)}_{q^{(r)},i,j} - w^{(n)}_{q^{(r)},i,j} \right| < \varepsilon_1, \text{ and } \left| f_h(q^{(r)}) \right| < \varepsilon_2;$$

and a relaxation factor equal to 1.85. A triangular finite element mesh with linear shape functions and $\Delta x = \Delta y$ was used in the D_ϕ region. The Table lists the computational and iteration information for the example. The computational times listed do not include the compilation or linkage editing time. Also shown in the Table are the results obtained by Sloss and Bruch [9] solving for w throughout the whole problem. In a comparison between the results for the free surface location obtained by Sloss and Bruch [9] and those of the present method, they both gave exactly the same results for the free surface location with $\Delta x = \Delta y = 2.5$ ft. In comparing these two for the case with $\Delta x = \Delta y = 1.0$ ft there was only one point that was different. This was the point at the free surface-drain intersection. The present method gave a more accurate result.

TABLE

Computational and Iteration Information for Example

	$q^{(r)}$		$f_h(q^{(r)})$		n		
r	Sloss and Bruch [9]	Present	Sloss and Bruch [9]	Present	Sloss and Bruch [9]	Present	
0	15.00000	15.00000	47.13327	73.08655	134	79	$\Delta x = \Delta y = 2.5$ ft
1	16.00000	16.00000	59.45287	91.45297	95	67	$\varepsilon_1 = \varepsilon_2 = 0.01$
2	11.17412	11.02064	2.23226	3.33601	114	80	$x_{C'} = 70.0$ ft
3	10.98586	10.83213	0.27792	0.42894	74	60	Present: Itel AS/6--2.5 secs
4	10.95909	10.80431	-0.00152	0.00061	51	39	Sloss and Bruch [9]: IBM 360/75--16.32 secs
					458	326	
0	15.00000	15.00000	35.44714	53.93027	118	82	$\Delta x = \Delta y = 1.0$ ft
1	16.00000	16.00000	44.53114	67.48894	102	79	$\varepsilon_1 = 0.01$, $\varepsilon_2 = 0.02$
2	11.09785	11.02245	1.71725	2.39455	108	77	$x_{C'} = 68.0$ ft
3	10.90123	10.83935	0.27052	0.38672	84	60	Present: Itel AS/6--16.20 secs
4	10.86446	10.80408	0.01619	0.02431	60	45	
5		10.80172		0.00098		20	Sloss and Bruch [9]: IBM 360/75--100.84 secs
					472	363	

5. SUMMARY AND CONCLUSIONS

A modified form of the Baiocchi method [1] has been presented. In the approach presented herein, the Baiocchi transformation and method was used on the part of the problem containing the free surface while the remaining part of the solution domain was solved using classical boundary value problem techniques. These two regions of solution have an overlap so that an alternating iteration scheme can be used. A modified alternating iteration scheme was used which considerably sped up the computational time. The results obtained here have shown that both from a theoretical, as well as from a numerical computational, point of view, this approach is efficient, useful, accurate, and straightforward.

ACKNOWLEDGEMENT

The numerical computations were funded by a University of California at Santa Barbara Academic Senate Computer Research Grant.

REFERENCES

1. BAIOCCHI C., and CAPELO, A., *Disequazioni variazionali e quasi variazionali.. Applicazioni a problemi di frontiera libera, Problemi Variazionali* (Vol. 1) and *Problemi Quasi-variazionali* (Vol. 2), Quaderni dell'U.M.I., Pitagora, Bologna (1978).

2. BAIOCCHI, C., Free Boundary Problems in the Theory of Fluid Flow Through Porous Media, *Proc. Int. Congr. Math.* , Vancouver, II, 237-243 (1975).

3. BRUCH, J.C., JR., A Survey of Free Boundary Value Problems in the Theory of Fluid Flow Through Porous Media: Variational Inequality Approach, *Advances in Water Resources* , Part I, 3, 65-80, Part II, 3, 115-124 (1980).

4. BRUCH, J.C., JR., and SLOSS, J.M., Coupled Variational Inequalities for Flow from a Nonsymmetric Ditch, to be published in AMS Series, *Contemporary Mathematics, Elliptic Systems in the Plane* (1981).

5. HARR, M.E., *Groundwater and Seepage* , McGraw-Hill, New York (1962).

6. KINDERLEHRER, D., and STAMPACCHIA, G., *An Introduction to Variational Inequalities and Their Applications* , Academic Press, New York (1980).

7. ODEN, J.T., and KIKUCHI, N., Theory of Variational Inequalities with Applications to Problems of Flow Through Porous Media, *Int. J. of Engng. Sci.*, 18, 1173-1284 (1980).

8. REMAR, J., BRUCH, J.C., JR., and SLOSS, J.M., Axisymmetric Free Surface Seepage, *Num. Math.*, submitted.

9. SLOSS, J.M., and BRUCH, J.C., JR., Free Surface Seepage Problems, *J. Eng. Mech. Div. Proc.* ASCE 104, EM5, 1099-1111 (1978).

CATASTROPHIC FEATURES OF COULOMB FRICTION MODEL

V. Janovský

Charles University, Prague

1. INTRODUCTION

Consider the plane deformation of a body $\Omega \subset \mathbb{R}^2$ resting on a rigid support. The material of the body is assumed to be linearly elastic and displacements and strains are assumed to be small. The boundary $\partial\Omega$ of Ω consists of three disjoint sets Γ_u, Γ_N, Γ. Deformation arises as a result of given volume forces applied to Ω and external surface forces applied on Γ_N. The displacements are fixed along Γ_u and it is assumed that meas $\Gamma_u \neq 0$. Coulomb friction boundary conditions are prescribed on the set Γ.

Let us briefly now recall the Coulomb friction rule: For each displacement field $\underline{v} = (v_1,v_2)^T$ on Ω the normal and tangential components on Γ are defined by means of the operators N and T. Thus to be precise, $N\underline{v} = v_i\nu_i$ and $T\underline{v} = v_i t_i$ on Γ, where $\underline{\nu} = (\nu_1,\nu_2)^T$ and $\underline{t} = (t_1,t_2)^T$ are respectively the outward normal and the tangential vectors to Γ. We denote by σ_ν and σ_t the normal and tangential contact stresses on Γ. If $\underline{u} = (u_1,u_2)^T$ is the displacement field on Ω then the following conditions are required on Γ.

$$\sigma_\nu \leq 0 \;, \quad N\underline{u} \leq 0 \;, \quad \sigma_\nu \cdot N\underline{u} = 0 \;, \tag{1}$$

$$\left.\begin{array}{l} |\sigma_t| \leq \xi \;, \quad (\xi - |\sigma_t|)\cdot T\underline{u} = 0 \;; \\ \text{if } \sigma_t = \xi \neq 0 \text{ and } \sigma_t = -\xi \neq 0 \text{ then} \\ \text{respectively } T\underline{u} \leq 0 \text{ and } T\underline{u} \geq 0 \;, \end{array}\right\} \tag{2}$$

$$\xi = -\kappa\sigma_\nu \;, \tag{3}$$

where κ is a given non-negative friction coefficient and ξ is a slip stress on Γ.

Consider a formal finite element model of the above mechanical problem. In this context with reduced degrees of freedom, we can represent the displacements, the normal and tangential and slip stresses respectively by the vectors $\underline{U} \in \mathbb{R}_k$ and $\underline{s}_\nu$, $\underline{s}_t$, $\underline{\eta} \in \mathbb{R}_n$. The approximations of the operators N and T are matrices $N:\mathbb{R}_k \longrightarrow \mathbb{R}_n$ and $T:\mathbb{R}_k \longrightarrow \mathbb{R}_n$; the transpositions of N and T are denoted by N^T and T^T.

The conditions for the displacements $\underline{U} \in \mathbb{R}_k$ to satisfy the usual linear constitutive law and equilibrium conditions can be expressed, via a system of linear equations, as

$$\underline{AU} = \underline{F} + N^T\underline{s}_\nu + T^T\underline{s}_t \; . \tag{4}$$

The stiffness matrix $A:\mathbb{R}_k \longrightarrow \mathbb{R}_k$ is symmetric and positive definite. The vector $\underline{F}$ denotes a sum of both distributed volume forces on Ω and surface forces on Γ_N. The vectors $N^T\underline{s}_\nu$ and $T^T\underline{s}_t$ represent distributed normal and tangential stresses on Γ.

It remains to employ the boundary conditions on Γ. Naturally, the conditions (1), (2) and (3) are imposed on the components of the vectors $\underline{s}_\nu$, $\underline{s}_t$, $\underline{\eta}$, $N\underline{U}$, $T\underline{U}$ which approximate the quantities σ_ν, σ_t, ξ, Nu, Tu at nodal points on Γ. We rewrite conditions (1) and (2) in the form of two variational inequalities.

Let us define $\mathbb{R}_n^+$ to be the cone $\mathbb{R}_n^+ \equiv \{\underline{\omega} = (\omega_1,\ldots,\omega_n)^T \in \mathbb{R}_n : \omega_i \geq 0 \;\; \forall\, i\}$. We set $K_\eta \equiv \{\underline{r} = (\underline{r}_\nu,\underline{r}_t)^T \in \mathbb{R}_n \times \mathbb{R}_n / -\underline{r}_\nu \in \mathbb{R}_n^+ ,\; \underline{\eta} + \underline{r}_t \in \mathbb{R}_n^+ ,\; \underline{\eta} - \underline{r}_t \in \mathbb{R}_n^+\}$ for each given $\underline{\eta} \in \mathbb{R}_n^+$. Then $\underline{s}_\nu$, $\underline{s}_t$, $\underline{\eta}$, $N\underline{U}$, $T\underline{U}$ satisfy (1) and (2) iff $\underline{s} = (\underline{s}_\nu,\underline{s}_t)^T \in K_\eta$ and

$$\langle \underline{r}_\nu - \underline{s}_\nu, N\underline{U}\rangle \geq 0, \quad \langle \underline{r}_t - \underline{s}_t, T\underline{U}\rangle \geq 0 \quad \forall\, \underline{r} = (\underline{r}_\nu,\underline{r}_t)^T \in K_\eta, \tag{5}$$

where the symbol $\langle .,. \rangle$ denotes the scalar product on $\mathbb{R}_n$. For the sake of completeness, we write condition (3) in terms of $\underline{\eta}$ and $\underline{s}_\nu$, namely

$$\underline{\eta} = -\kappa\,\underline{s}_\nu \; . \tag{6}$$

We can now formulate

<u>Problem (P)</u> Let $\underline{F} \in \mathbb{R}_k$ and $\kappa \geq 0$ be given. It is required to find $\underline{U} \in \mathbb{R}_k$, $\underline{\eta} \in \mathbb{R}_n^+$ and $\underline{s} = (\underline{s}_\nu,\underline{s}_t)^T \in K_\eta$ so that conditions (4), (5) and (6) are satisfied.

We have motivated Problem (P) so that it can be regarded as a finite element model of the mechanical problem. The aim of the present paper is to analyse Problem (P). In the next Section some results concerning the existence and uniqueness of a solution to Problem (P) are quoted. A simple one-dimensional example is given in Section 3.

2. ANALYSIS OF A FINITE ELEMENT MODEL

We are to investigate Problem (P). It is natural to eliminate $\underline{U}$ from (5) by making use of (4) and the inverse A^{-1} to the stiffness matrix A. Setting

$$Z \equiv \left(\begin{array}{c|c} NA^{-1}N^{\top} & NA^{-1}T^{\top} \\ \hline TA^{-1}N^{\top} & TA^{-1}T^{\top} \end{array}\right)$$

and

$$\underline{f} \equiv -(NA^{-1}\underline{F},\ TA^{-1}\underline{F})^{\top},$$

we easily reformulate (4) and (5) as follows:

$$[\underline{r} - \underline{s},\ Z\underline{s}] - [\underline{r} - \underline{s},\ \underline{f}] \geq \underline{0} \quad \forall\ \underline{r} = (\underline{r}_{\nu}, \underline{r}_{t})^{\top} \in K_{\eta}\ , \tag{7}$$

where $[.,.]$ is a scalar product on $\mathbb{R}_n \times \mathbb{R}_n$.

The block matrix $Z : \mathbb{R}_n \times \mathbb{R}_n \longrightarrow \mathbb{R}_n \times \mathbb{R}_n$ is symmetric and positive semidefinite. It is also positive definite, if it is assumed that $N^{\top}\underline{s}_{\nu} + T^{\top}\underline{s}_t = \underline{0}(\in \mathbb{R}_k)$ implies $\underline{s}_{\nu} = \underline{s}_t = \underline{0}(\in \mathbb{R}_n)$; roughly speaking this means that the number of nodal points for the contact stress has to be in a certain proportion to the number of degrees of freedom of the displacements on the contact boundary. A similar condition is in any case satisfied in the *infinitesimal* model. All these properties of Z will now be assumed.

We define an *energy* norm $\|\cdot\|$ on $\mathbb{R}_n \times \mathbb{R}_n$ by setting $\|r\| \equiv [\underline{r}, Z\underline{r}]^{\frac{1}{2}}$ $\forall\, \underline{r} = (\underline{r}_{\nu}, \underline{r}_t)^{\top} \in \mathbb{R}_n \times \mathbb{R}_n$. One can easily observe that $\underline{s} = (\underline{s}_{\nu}, \underline{s}_t)^{\top} \in K_{\eta}$ satisfies (7) iff

$$\|\underline{s} - \underline{\theta}\| \leq \|\underline{r} - \underline{\theta}\| \quad \forall\ \underline{r} \in K_{\eta} \tag{8}$$

where

$$\underline{\theta} = Z^{-1}\underline{f}\ . \tag{9}$$

The components $\underline{\theta}_{\nu}$, $\underline{\theta}_t$ of the vector $\underline{\theta} = (\underline{\theta}_{\nu}, \underline{\theta}_t)^{\top} \in \mathbb{R}_n \times \mathbb{R}_n$ can be interpreted as the normal and tangential stresses on Γ if the Coulomb friction condition on Γ is replaced by the Dirichlet condition $\underline{u} = 0$ on Γ (i.e. if the body were *bolted* along Γ). There is a clear link between a given distributed load $\underline{F} \in \mathbb{R}_k$ and $\underline{\theta} \in \mathbb{R}_n \times \mathbb{R}_n$. According to (8), Problem (P) can be controlled by $\underline{\theta}$ instead of by $\underline{F}$. We use this fact in the sequel.

For given $\underline{\theta} \in \mathbb{R}_n \times \mathbb{R}_n$ and $\underline{\eta} \in \mathbb{R}_n^+$ there exists one and only one $\underline{s} = (\underline{s}_{\nu}, \underline{s}_t)^{\top} \in K_{\eta}$ satisfying (8). We set $S\underline{\eta} \equiv -\kappa\underline{s}_{\nu}$ which defines an operator $S : \mathbb{R}_n^+ \longrightarrow \mathbb{R}_n^+$. The operator S depends on a choice of parameters $\kappa \geq 0$, $\underline{\theta} \in \mathbb{R}_n \times \mathbb{R}_n$. In order to express that dependance, we use the notation $S(\underline{\eta}\,;\kappa,\underline{\theta})$, which means the operator S at a point $\underline{\eta} \in \mathbb{R}_n^+$ assuming *controlling parameters* to be equal to κ and $\underline{\theta}$. Let us formulate the following.

Problem (P1) Let $\underline{\theta} \in \mathbb{R}_n \times \mathbb{R}_n$ and $\kappa \geq 0$ be given. Find $\underline{\eta} \in \mathbb{R}_n^+$ such that $S(\underline{\eta}\,;\kappa,\underline{\theta}) = \underline{\eta}$.

Problems (P) and (P1) are equivalent: Let $\underline{\theta}$ and $\underline{F}$ be linked via (9). If a triple $\underline{U}$, $\underline{\eta}$, $\underline{s}$ solves Problem (P) then $\underline{\eta}$ is a solution to Problem (P1). Conversely if $\underline{\eta}$ solves Problem (P1) then we construct $\underline{s} = (\underline{s}_\nu, \underline{s}_t)^T$ and $\underline{U}$ to be the (unique) solutions of (8) and (4) respectively; the triple $\underline{U}$, $\underline{\eta}$, $\underline{s}$ solves Problem (P). In Problem (P1) the *nonlinearity* of our mechanical problem is reduced to the set where the nonlinear behaviour actually occurs, i.e. on the boundary Γ. By analysing the operator S, we can give answers to some quesitons concerning the solvability of Problem (P).

Theorem 1 For any choice of $\underline{F} \in \mathbb{R}_k$ and $\kappa \geq 0$ there exists at least one solution to Problem (P).

Proof This is given in [4].

Theorem 2 Let λ_{max} and λ_{min} be the largest and the smallest eigenvalues of the matrix Z. If the friction coefficient $\kappa \geq 0$ is sufficiently small, namely

$$\kappa\sqrt{\lambda_{max}/\lambda_{min}} < 1 , \tag{10}$$

then

(a) for each $\underline{F} \in \mathbb{R}_k$ there exists a unique solution $\underline{U}$, $\underline{\eta}$, $\underline{s}$ to Problem (P).
(b) the operator $S(.;\kappa,\underline{\theta}) : \mathbb{R}_n^+ \longrightarrow \mathbb{R}_n^+$ is contractive for each $\underline{\theta} \in \mathbb{R}_n \times \mathbb{R}_n$.

Proof This is given in [4].

There is of course the practical question of how Problem (P) may be solved. The natural strategy for computation is as follows: Guess an approximation $\tilde{\underline{\eta}} \in \mathbb{R}_n^+$ to the slip stress $\underline{\eta}$. Calculate displacements $\tilde{\underline{U}} \in \mathbb{R}_k$ and contact stresses $\tilde{\underline{s}} = (\tilde{\underline{s}}_\nu, \tilde{\underline{s}}_t)^T \in K_{\tilde{\eta}}$ from either (4) and (5) or (4) and (7) or (4) and (8), where $\underline{U}$, $\underline{s}$, $\underline{s}_\nu$, $\underline{s}_t$ and K_η have been replaced by $\tilde{\underline{U}}$, $\tilde{\underline{s}}$, $\tilde{\underline{s}}_\nu$, $\tilde{\underline{s}}_t$ and $K_{\tilde{\eta}}$; this step represents quadratic programming. If $\tilde{\underline{\eta}} = -\kappa\tilde{\underline{s}}_\nu$ then the triple $\tilde{\underline{U}}$, $\tilde{\underline{\eta}}$, $\tilde{\underline{s}}$ is a solution to Problem (P). Otherwise set $\tilde{\underline{\eta}} = -\kappa\tilde{\underline{s}}_\nu$ and iterate. In fact, the *new* value of $\tilde{\eta}$ is equivalent to S being evaluated at the *old* value of $\tilde{\eta}$.

Assuming (10), the convergence of the process is guaranteed. If κ is "too large" then the algorithm may be divergent. The reason why it happens is that there may be more than one solution to Problem (P1) (and to Problem (P)); see next Section.

3. ONE-DIMENSIONAL EXAMPLE

Assume (perhaps without much mechanical meaning) that each component of $\underline{\sigma} = (\sigma_\nu, \sigma_t)^T$ and $\underline{u} = (u_1, u_2)^T$ has one degree of freedom on the *contact* boundary $\overline{\Gamma}$. We come to a finite element model which can be formulated as a Problem (P), where $k > n = 1$. The number k of degrees of freedom for the displacements $\underline{u}$ does not really matter since we can reduce the information on $\overline{\Gamma}$ by making use of the formulation via Problem (P1).

Consider the relevant Problem (P1) for the slip stress $\eta \in \mathbb{R}_1^+$. We derive a symmetric, positive definite matrix

$$Z \equiv \begin{pmatrix} z_1, z_0 \\ z_0, z_2 \end{pmatrix}, \quad \text{i.e. } z_1 > 0 \ , \quad z_2 > 0 \ , \quad z_0^2 < z_1 z_2 \ .$$

The controlling parameters are $\kappa \geq 0$ and $\underline{\theta} = (\theta_\nu, \theta_t)^T \in \mathbb{R}_2$.

One can explicitly express the operator $S(.;\kappa,\underline{\theta}) : \mathbb{R}_1^+ \longrightarrow \mathbb{R}_1^+$ and find formulas which relate the set of solutions $\{\eta \in \mathbb{R}_1^+ : S(\eta;\kappa,\underline{\theta}) = \eta\}$ to each given couple of controlling parameters $\underline{\theta} \in \mathbb{R}_2$, $\kappa \geq 0$. It comes out that the operator S is contractive for any choice of $\underline{\theta}$ iff $\kappa \cdot |z_0| z_1^{-1} < 1$. Fig. 1 shows a typical dependance of the η's on the $\underline{\theta}$'s, assuming that $\kappa \cdot |z_0| \cdot z_1^{-1} > 1$ and that κ is fixed. The controlling parameters $\underline{\theta} \in \mathbb{R}_2$ are restricted to a rectangle ABCD. The graph of the relevant η's creates one fold with a cusp at the origin. As can be seen from the figure there are exactly three solutions η to Problem (P1) if $\underline{\theta}$ belongs to the interior of the angle GOF. In order to classify the parameters $\underline{\theta}$, let us define sets M_0, M_1, M_2 and M_3 to be the interiors of the angles GOF and GOE and EOH and FOH in the plane of θ's, see Fig. 1. It is possible to calculate contact stresses $\underline{s} = (s_\nu, s_t)^T$ for each given $\underline{\theta}$ (using the relevant value of η). There exists exactly one solution $\underline{s} = (s_\nu, s_t)^T$ for $\underline{\theta} \in M_1 \cup \overline{M}_2 \cup M_3$; moreover

$$\underline{\theta} \in M_1 \Rightarrow |s_t| < -\kappa s_\nu \tag{11}$$

$$\underline{\theta} \in M_2 \Rightarrow |s_t| = \kappa s_\nu \ , \quad s_\nu \neq 0 \tag{12}$$

$$\underline{\theta} \in M_3 \Rightarrow s_\nu = s_t = 0 \ . \tag{13}$$

The cases (13), (11) and (12) can be interpreted practically as respectively a loss of contact (between the body Ω and its rigid support), a contact without tangential slip and a contact with tangential slip. Assuming $\underline{\theta} \in M_0$, there exist exactly three solutions $\underline{s} = (s_\nu, s_t)^T$; moreover, each of the cases (11) and (12) and (13) occurs for one of the solutions $\underline{s}$. For details see [4].

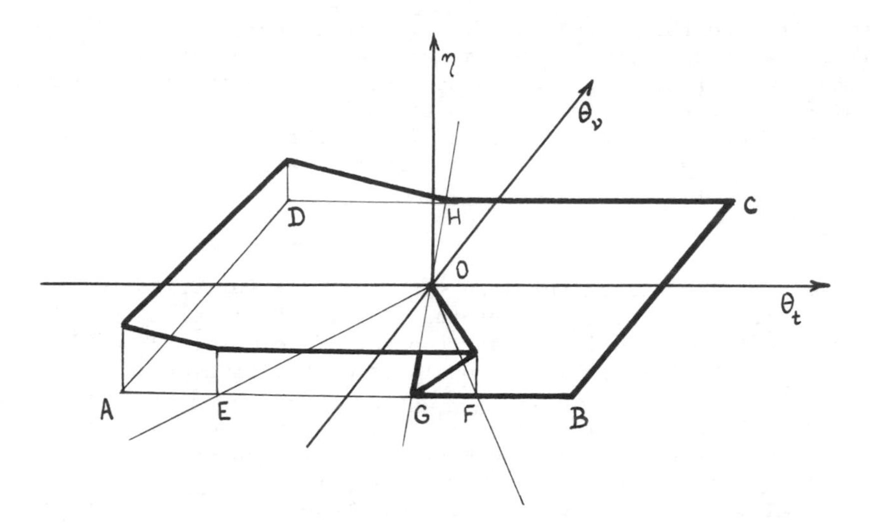

Fig. 1. Solution η to Problem (P1) as a function of θ, assuming a *large* friction coefficient κ

4. CONCLUDING REMARKS

Assuming a *small* friction coefficient, there are no difficulties from the point of view of practical computation. On the other hand, the given example suggests that large friction coefficients may be troublesome. In this case there is lack of uniqueness and a solution may not depend on the data (distributed load or $\underline{\theta}$) continuously, see the *edges* of the fold in Fig. 1. Introducting a quasistatic model, one can expect sudden jumps or branching of a solution at a particular time. This implies that one should be careful in making use of a load increment technique.

REFERENCES

1. DUVAUT, G. and LIONS, J.L., *Inequalities in Mechanics and Physics*. Springer Verlag, Berlin (1976).

2. FREDRIKSSON, B., *On Elastostatic Contact Problems with Friction*. Thesis presented at Linköping Institute of Technology, Linköping (1976).

3. JANOVSKÁ, D., *Numerical Solution of Contact Problems* (in Czech). Thesis presented at Charles University, Prague (1980).

4. JANOVSKÝ, V., Catastrophic Features of Coulomb Friction Model. *Technical Report KNM-0105044/80*, Charles University, Prague (1980).

NUMERICAL SIMULATION OF FREE BOUNDARY PROBLEMS USING PHASE FIELD METHODS

G. J. Fix

Carnegie-Mellon University

1. INTRODUCTION

We consider a material occupying the spatial region Ω. Any point in this material can be in one of two phases, say liquid or solid. We let $u = u(\underline{x},t)$ denote the temperature at the point $\underline{x}\varepsilon\Omega$ and time t, and let u_M denote the phase transition temperature. Thus if

$$u(\underline{x},t) > u_M,$$

the point $\underline{x}\varepsilon\Omega$ is in the liquid phase, and if the opposite inequality holds it is in the solid phase. The set of points $\underline{x}$ where $u(\underline{x},t) = u_M$ form the transition region.

The classical Stefan equations are normally used as a mathematical model of this situation [17]. Here the temperature field u satisfies the diffusion equation

$$\frac{\partial u}{\partial t} = \text{div}[D \text{ grad } u] \tag{1.1}$$

at points $\underline{x}\varepsilon\Omega$ which are not in the transition region and at positive time; i.e.,

$$u(\underline{x},t) \neq u_M, \; t > 0.$$

The transition region

$$\Gamma(t) = \{x\varepsilon\Omega: u(x,t) = u_M\} \tag{1.2}$$

is to be determined, and is assumed to be a surface moving with normal velocity v. Across $\Gamma(t)$ diffusion gives way to a liberation of latent heat λ, which in turn creates a jump

$$[D \text{ grad } u \cdot \underline{\nu}]_-^+$$

in the normal component $D \text{ grad } u \cdot \underline{\nu}$ of the temperature flux,

$\underline{\nu}$ being the normal to the free surface $\Gamma(t)$. The mathematical relation which models this situation is the following:

$$\lambda v + [D \text{ grad } u \cdot \underline{\nu}]_-^+ = 0 \qquad (1.3)$$

To complete the specification of the problem initial and boundary conditions must be given. For simplicity, we take

$$u(x,t) = u_\Gamma(x,t) \qquad x \varepsilon \partial\Omega,\ t > 0 \qquad (1.4)$$

and

$$u(x,t) = u_o(x,t) \qquad x \varepsilon \Omega,\ t = 0, \qquad (1.5)$$

where u_Γ and u_o are given functions.

The most common method for solving the classical Stefan problem (i.e., (1.1)-(1.5)) is by the so called H-method. Here one introduces a heat function $H = H(u)$ defined as follows:

$$H(u) = \begin{cases} u + \lambda/2 - u_M & \text{if } u > u_M \\ u - \lambda/2 - u_M & \text{if } u < u_M \end{cases} \qquad (1.6)$$

The jump conditions (1.3) and the diffusion equation (1.1) are then embodied in the single equation

$$\frac{\partial}{\partial t} H(u) = \text{div}[D \text{ grad } u], \qquad (1.7)$$

where the time derivative is taken in the distributional sense ([2]-[4]). Actually (1.7) is another way of expressing a balance of heat. Indeed, let C be any closed region in space-time with n_t being the time like normal to C, and $\underline{n}_x$ being the space like normal. Thus the balance heat in C gives

$$\int_C \{Hn_t + D \text{ grad } u \cdot \underline{n}_x\} dC = 0, \qquad (1.8)$$

which in a suitable sense is equivalent to (1.7) ([6]).

Successful numerical calculations have been done by using finite element (or finite difference) methods in conjunction with (1.7), using the standard weak formulation of the latter ([1],[3],[8]).

In many applications, most notably in crystal growth, the joining of materials, for example by welding, and in doping of semi conductor wafers ([5],[11],[19]) one has a slightly more complicated situation. First, supercooling or superheating can be present. This means that the material can possibly be in the liquid state at points where the temperature is below u_M, or in the solid state state where the temperature is above u_M. Mathematically this means that the heat function $H = H(u)$ is multivalued as shown in Figure 1.

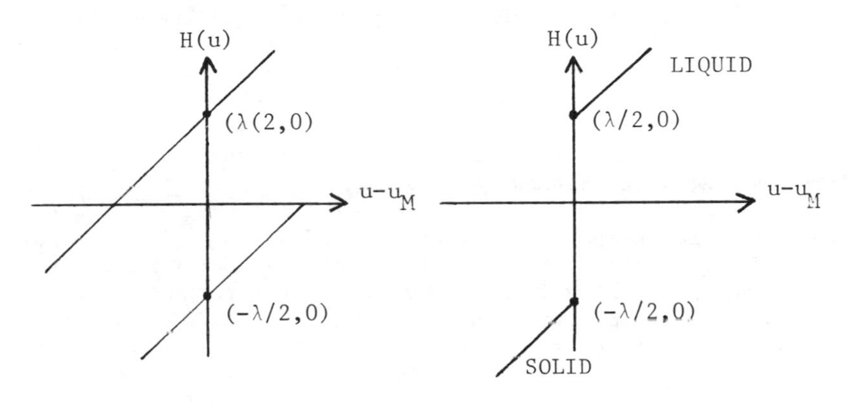

FIG. 1. Heat functions with supercooling present (left) and absent (right).

The presence of supercooling and superheating gives rise to physical instabilities not present in the classical Stefan equation. These are called dendrites [12] and are extremely important in applications. In the material science literature they are called Mullin-Sekerka instabilities [14]. Typically they occur as sharp spikes in the free surface as shown in Figure 2.

FIG. 2. Dendritic growth in transition region.

These instabilities, however, do not persist to the microscales. The effect of surface tension tends to restabilize the physical system [13]. Mathematically this is represented as a dependence of the transition temperature u_M with the curvature κ of the free surface. Namely, one has the following relation:

$$u_n = \bar{u}_M(1-\sigma\kappa), \tag{1.9}$$

where $\bar{u}_M$ is a mean transition temperature and σ is a capillary length, both assumed to be known.

There has been an interesting attempt to solve (1.7) with the transition temperature given by (1.9). [18]. The ambiguity in the heat function $H(u)$ was resolved by a mathematically ad hoc but physically intuitive rule. The remainder of the equations were discretized by a standard finite element procedure. The method was applied to a class of solidification problems with physically reasonable results. Numerical problems did occur, however, when the dendrites started to form. The transition region started to become "mushy" -- a phenomena well known to the type of methods [2] -- and this made accurate approximations to the curvature κ of the free surface difficult.

To avoid these difficulties we consider a Landau-Ginzberg model of this problem [9]. In this approach one introduces a phase field $\phi = \phi(x,t)$ which presumably is $+1$ in the solid region and -1 in the liquid region for the normal situations where supercooling is not present. The heat function H is then given by

$$H(u) = u + \lambda\phi/2, \tag{1.10}$$

where u is now the difference between the temperature and the melting temperature. One retains the balance of heat embodied in (1.7) to write the following equation

$$\frac{\partial u}{\partial t} + \frac{\lambda}{2}\frac{\partial \phi}{\partial t} = [\text{div } D \text{ grad } u] \tag{1.11}$$

To determine the phase one introduces a Helmholtz free energy functional, and requires that ϕ be a critical point. This leads to

$$\tau \frac{\partial \phi}{\partial t} = \xi^2 \Delta\phi + f(\phi) + 2u, \tag{1.12}$$

where τ is a relaxation time, ξ defines the diffusion scales for the phase field, and f is given by the following (to a first approximation):

$$f(\phi) = \tfrac{1}{2}(\phi - \phi^3) \tag{1.13}$$

Our problem is to solve (1.11)-(1.12) subject to appropriate initial and boundary conditions. In particular, we use (1.4)-(1.5) for u, and

$$\phi = \phi_\Gamma(\underline{x},t) \qquad \underline{x}\varepsilon\partial\Omega,\ t > 0 \tag{1.14}$$

$$\phi = \phi_o(\underline{x},t) \qquad \underline{x}\varepsilon\Omega,\ t = 0 \tag{1.15}$$

for the phase.

This model has all the important physical ingredients in it. For example, supercooling or superheating will be present when u and ϕ have different signs, a situation that can even be built into the initial conditions. Moreover, the stabilizing effects of surface tension are built into the derivation of (1.12).

There are some serious difficulties over and beyond the fact that the system is nonlinear. First both the relaxation time τ and the diffusion scale ξ are very small; typically

$$\tau \simeq 10^{-3},\ \xi \simeq 10^{-2} \text{ to } 10^{-3}.$$

This means that (1.12) is extremely stiff in comparison with (1.11), and hence must be discretized with great care.

2. NUMERICAL INTEGRATION OF THE FIELD EQUATIONS

The temperature field u will be treated by a standard finite element approximation. In particular, suppose for the moment that the phase field ϕ is known. Then (1.11) can be written

$$\frac{\partial u}{\partial t} - \text{div } D \text{ grad } u = -\frac{\lambda}{2\tau}(\xi^2\Delta\phi + f(\phi) + 2u) \tag{2.1}$$

Let $H_o^1(\Omega)$ denote the space of functions having square integrable gradients and vanishing on $\partial\Omega$. We replace (2.1) with the following semidiscrete variational formulation. Find a family

$$\{u(\cdot,t)\}_{t\geq 0}$$

such that

$$\int_\Omega [\frac{\partial u}{\partial t} v + D \text{ grad } u \cdot \text{grad } v + (\lambda/\tau)uv]$$
$$= (\lambda/2\tau)\int_\Omega [-\xi^2 \text{ grad } \phi \cdot \text{grad } v + f(\phi)v] \tag{2.2}$$

holds for all $v \varepsilon H_o^1(\Omega)$ and $t > 0$. Moreover, for each $t > 0$ we require that the difference $u - u_\Gamma$ between u and some extension u_Γ of the boundary data be in $H_o^1(\Omega)$. Finally, at time $t = 0$ we require that

$$\int_\Omega u(\cdot,0)v = \int_\Omega u_o v \quad \text{all} \quad v \varepsilon H_o^1(\Omega) \tag{2.3}$$

To approximate we introduce a finite dimensional space

$$S_h \subseteq H_o^1(\Omega),$$

and an approximation u_Γ^h to u_Γ. The semidiscrete approximation then takes the following form. Find a family

$$\{u_h(\cdot,t)\}_{t \geq 0} \subseteq S_h,$$

such that for each $t > 0$ we have

$$u_h - u_\Gamma^h \varepsilon S_h$$

and

$$\int_\Omega \frac{\partial u}{\partial t^h} v^h + D[\text{grad } u_h \cdot \text{grad } v^h + (\lambda/\tau) u_h v^h]$$

$$= (\lambda/2\tau) \int_\Omega [-\xi^2 \text{ grad } \phi_\delta \cdot \text{grad } v^h + f(\phi_\delta) v^h] \tag{2.4}$$

holds for all v^h in S_h. The function ϕ_δ is an approximation to ϕ that will be discussed below. Finally, at $t = 0$ we require

$$\int_\Omega u_h(\cdot,0)v^h = \int_\Omega u_o v^h \quad \text{all} \quad v^h \varepsilon S_h. \tag{2.5}$$

In the numerical examples presented in the next section S_h was taken to be piecewise linear functions on a uniform grid; i.e., no attempt was made to "track" the moving free surface by monitoring values of u. Moreover, the planar rectangular geometry of these problems permitted the use of a standard A.D.I. time integration [7].

The phase field equation (1.12) will be treated in a completely different manner. First, an explicit time differencing will be used. The well known stability condition

$$(\frac{\Delta t}{\tau})(\frac{\xi}{\Delta x})^2 < 1 \tag{2.6}$$

offers no practical difficulties because of the values of ξ and τ. (Recall that $\xi \sim 10^{-3}$, $\tau \sim 10^{-3}$). Secondly, a moving grid will be used in order to track the free surface. This grid will be considerably finer than the grid used for the temperature field u, and variable time steps will be employed.

Numerical experience with moving grids has typically been quite good (see e.g. [15]). However, a satisfactory error analysis of such schemes is apparently lacking either because the grids are assumed to be uniform ([10]), or because the estimates yield suboptimal convergence, a phenomena which has not been seen in practice. This matter is undoubtedly an important open question in finite element theory.

To give a precise specification of our approximation to the phase field we first discretize in time. In particular, we seek a family

$$\{\phi^{(n)}(\underline{x})\}_{n \geq 0}$$

defined at the times $n\Delta t$ such that

$$\int_\Omega \{\tau[\frac{\phi^{(n+1)} - \phi^{(n)}}{\Delta t}]\psi + \xi^2 \text{ grad } \phi^{(n)} \cdot \text{ grad } \psi + f(\phi^{(n)})\psi\}$$
$$= 2\int_\Omega u_h(\cdot, n\Delta t)\psi \tag{2.7}$$

holds for each $n > 0$ and $\psi \varepsilon H_o^1(\Omega)$. Also we require

$$\phi^{(n)} - \phi_\Gamma(\cdot, n\Delta t) \varepsilon H_o^1(\Omega), \tag{2.8}$$

and initially

$$\int_\Omega \phi^{(o)}\psi = \int_\Omega \phi_o\psi \quad \text{all} \quad \psi \varepsilon H_o^1(\Omega) \tag{2.9}$$

To discretize this problem we introduce a finite dimensional space

$$S_\delta(t) \subseteq H_o^1(\Omega),$$

which depends on time t. In the examples presented in the

next section these spaces consisted of piecewise linear functions defined on a grid which moves in time. The discretized variational problem seeks a family

$$\{\phi_\delta^{(n)}(\underline{x})\}_{n\geq 0}$$

with

$$\phi_\delta^{(n)} - \phi_\Gamma^\delta \in S_\delta(n\Delta t)$$

for $n > 0$, where ϕ_Γ^δ is some approximation to ϕ_Γ. In addition, for each $n > 0$ we require that (2.6) hold with $\phi^{(n)}$, $\phi^{(n+1)}$, replaced with $\phi_\delta^{(n)}, \phi_\delta^{(n+1)}$, and with ψ restricted to $S_\delta((n+1)\Delta t)$. Finally, (2.8) is required to hold with $\phi^{(o)}$ replaced with $\phi_\delta^{(o)}$, and with ψ restricted to $S_\delta(0)$.

It is of interest to note that this procedure produces a standard finite difference approximation except for the addition of a crucial drift term. In particular, consider the diffusion equation

$$\frac{\partial\phi}{\partial t} = \frac{\partial^2\phi}{\partial x^2}, \tag{2.10}$$

and let

$$x_j = x_j(t)$$

denote the nodes with

$$\phi_j = \phi(x_j(t),t)$$

denoting the nodal values of ϕ and

$$\Delta_j(t) = x_j(t) - x_{j-1}(t).$$

Then the semidiscrete version of (2.6), i.e., the version that leaves time t as a continuous parameter, reduces to the following:

$$\begin{aligned}\frac{1}{6}[\Delta_j\dot{\phi}_{j-1} + (\Delta_j+\Delta_{j+1})\dot{\phi}_j + \Delta_{j+1}\dot{\phi}_{j+1}]\\ -[\frac{1}{6}\,\dot{x}_{j-1}(\phi_j-\phi_{j-1}) + \dot{x}_j(\phi_{j+1}-\phi_{j-1}) + \dot{x}_{j+1}(\phi_{j+1}-\phi_j)]\\ = +\phi_{j-1}/\Delta_j - (1/\Delta_j + 1/\Delta_{j+1})\phi_j + \phi_{j+1}/\Delta_{j+1}\end{aligned} \tag{2.11}$$

It is the second term on the left in (2.10) which reflects the movement of the grids. Incidentally, assuming the grid moves according to

$$x_j(t) = X(jh,t),$$

where $X = X(\xi,t)$ has continuous and bounded derivatives in ξ and t, it is not difficult to show that (2.10) is second order accurate in h. Presumably this fact and the arguments used in [10] can be used to establish optimal L_2 convergence for at least this special case.

3. NUMERICAL EXAMPLES

In this section we present two numerical examples. In the first example supercooling is not present so in effect it is equivalent to a classical Stefan problem. We use this example to illustrate the type of approximations that can be obtained in a setting where one can be reasonably confident about how the solution should behave (although no closed form solution is known). In the second example we display a situation where dendrites appear and then restabilize.

In each example, the diffusion equation (2.4) was integrated by an A.D.I. using a fixed time step ΔT. The equation (2.7) for the phase ϕ used the time step

$$\Delta t = T/M.$$

The number M was chosen so the front velocity was the same order of magnitude as $\Delta/\Delta t$, where Δ is the smallest mesh length in the grid for ϕ. This is illustrated in Figure 3 for one space dimension. In addition, the grid for ϕ was moved by monitoring values of the second differences in ϕ.

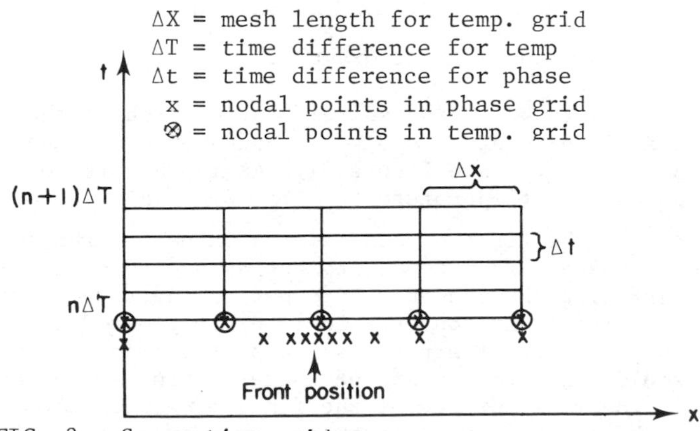

FIG. 3. Space-time grid.

The calculation started by first marching the phase equation M steps to time ΔT. In this calculation the value of u_h on the right hand side in (2.7) was frozen at its previous value, namely at $n\Delta t = 0$. Then one time step of length ΔT was used to compute u at time ΔT. This process was repeated moving from ΔT to $2\Delta T$, etc. In some cases where there were extreme variations in both u and ϕ it became necessary to iterate this process; i.e., recompute the phase based on updated values of u_h.

The first example is shown in Figure 4. It is a classical solidification problem where supercooling is not present. The interior of the rectangular domain is initially in the liquid state at the melting temperature $\bar{u}_M = 0$, while the boundary temperature u_Γ is held below $\bar{u}_M$. As time progresses a front moves and the interior starts to solidify. In Figure 5 we show sample temperature and phase profiles for $t = 0$ to $t = .5$. The plots represent functions of x for the fixed value $y = .5$. Note that the position of the front is blurred in the temperature plot on the right in Figure 5, but is quite sharp in the plot of ϕ on the left.

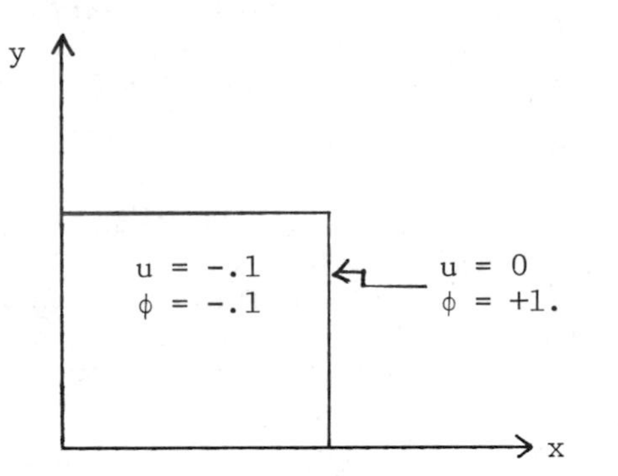

FIG. 4. Initial conditions - first example.

In the second example we have a solidification problem that is driven by seed of solid at $u = 0$ in a background of supercooled liquid at $u = -1$. (See Figure 6). As in the previous example the mean melting temperature $\bar{u}_M$ is 0. Two cases were considered. In the first a rather large value for the surface tension was chosen. This tended to act as a strong stabilizing force overriding the tendency of supercooling to promote dendritic growth. The front contours in time are plotted in x-y geometry in Figure 7. The progression of the front looks like one that would arise from a standard Stefan problem, except for the bumps in the free surface at the final time when the front neared the boundary. The latter are numerical artifacts

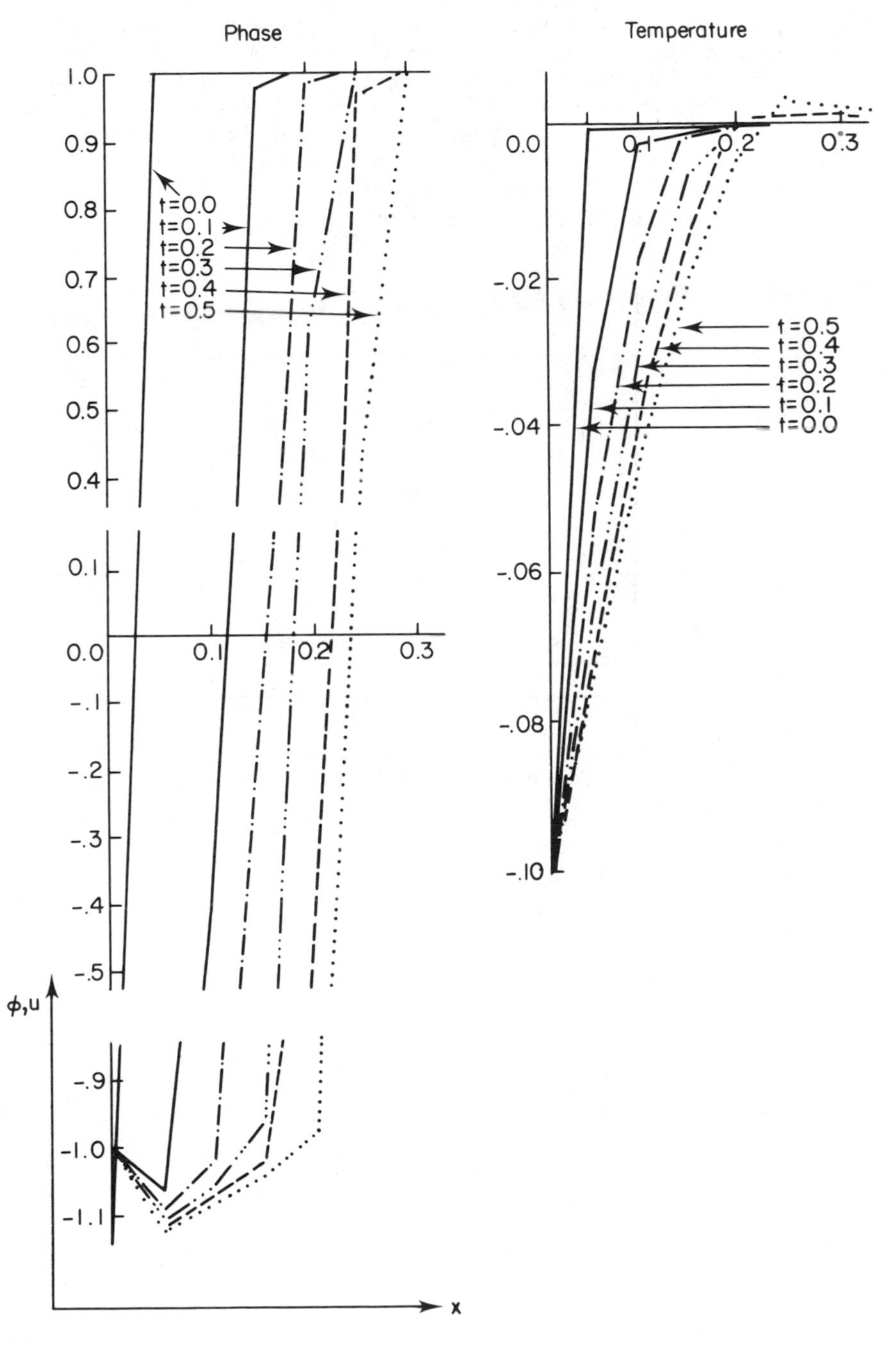

FIG. 5. Temperature and phase profiles for $y = \frac{1}{2}$.

that arose because the phase ϕ was being held at $-1.$ on the boundary and this should have been changed to $+1.$ when the point hit the boundary. In the second case the surface tension was reduced and this gave rise to dendritic growth. The front contours are displayed in Figure 8. The fact that the dendrites grew along the coordinate lines is due to the rectangular seed. If the latter were for example rotated 45°, then so would be the lines of dendritic growth.

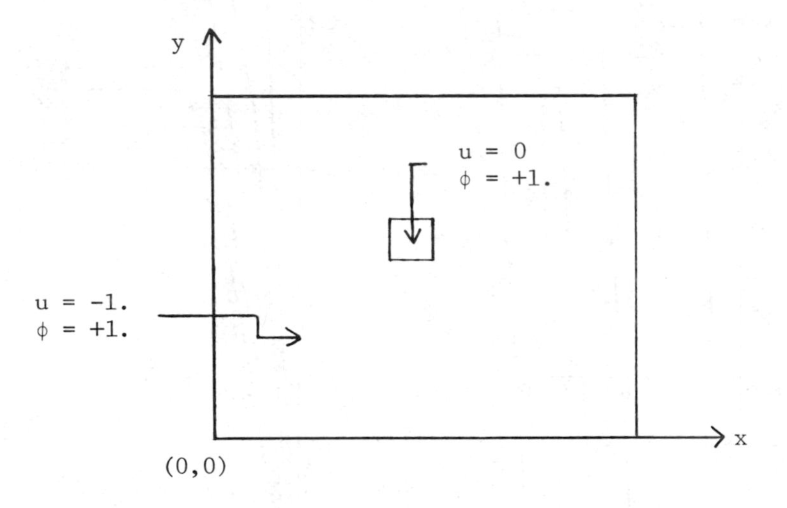

FIG. 6. Initial conditions - second example.

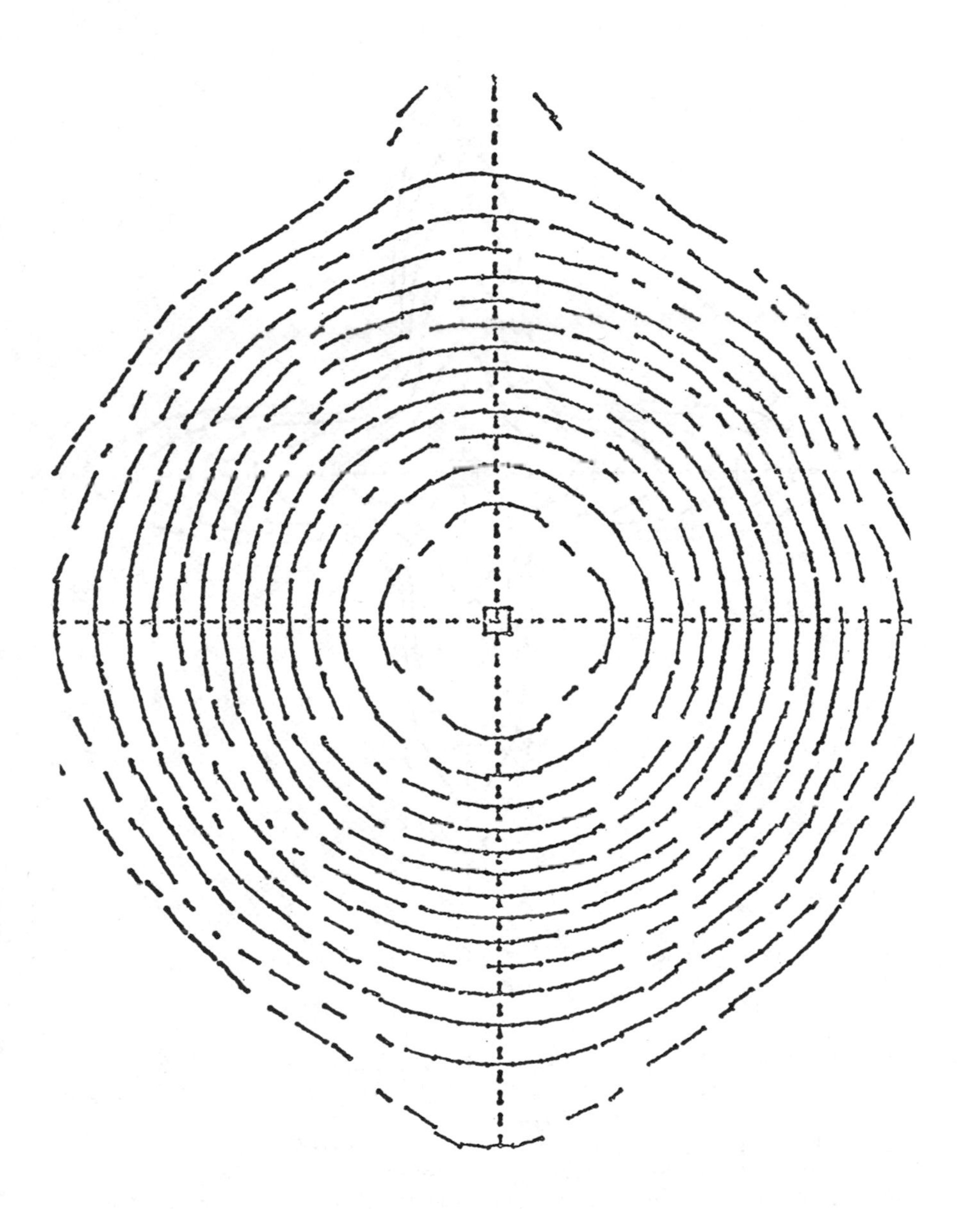

FIG. 7. First case - large surface tension.

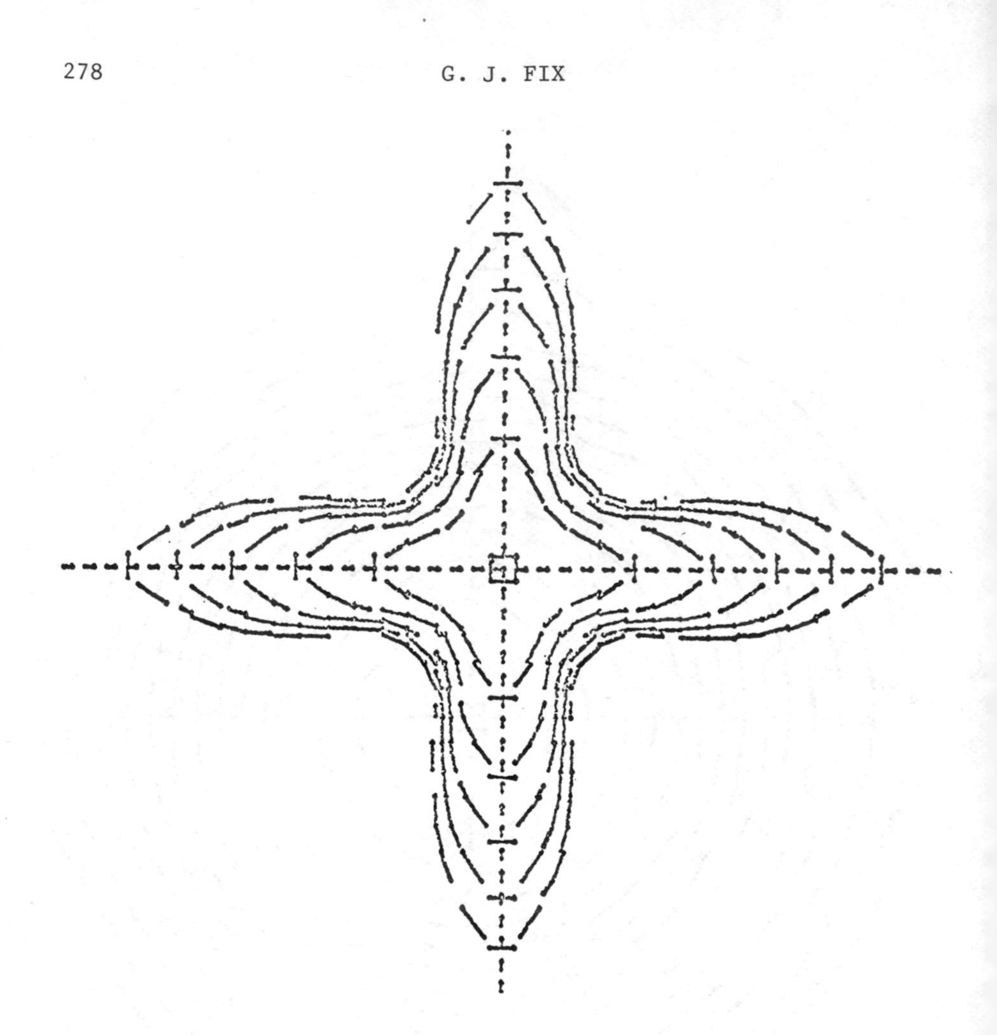

FIG. 8. Second case - small surface tension.

4. REFERENCES

1. ATTHEY, D.R., pp. 182-191 of J.R. Ockendon and W.R. Hodgkins (Eds.), *Moving Boundary Problems in Heat Flow and Diffusion*. Oxford (1974).
2. BOGGS, P.T., D.G. WILSON, and A.D. SOLOMON (Eds.), *Moving Boundary Problems*, Academic Press (1978).
3. BONNEROT, R. and P. JAMET, Numerical Computation of the Free Boundary for the Two-Dimensional Stefan Problem by Space-Time Finite Elements. *J. of Computational Physics* 25-2, 163-181 (1977).
4. BREZIS, H., pp. 101-156 of E.H. Zarantonello (Ed.), *Contributions to Non-Linear Functional Analysis*, Academic Press (1971).

5. CENTER FOR THE JOINING OF MATERIALS, Fifth Annual Technical Report of Research. Carnegie-Mellon University, Pittsburgh, Pennsylvania (1978).
6. CROWLEY, A.B., On the Weak Solution of Moving Boundary Problems. *J. Inst. Maths Applics* 24, 43-57 (1979).
7. DOUGLAS, T. and T. DUPONT, pp. 133-214 of B. Hubbard (Ed.), *Numerical Solution of Partial Differential Equations*, Academic Press (1971).
8. EPPERSON, J., Numerical Approximation of Solutions to General Stefan Problems. Doctoral Thesis, Carnegie-Mellon University (1980).
9. GINZBURG, V.L. and L.D. LANDAU, *Collected Papers of Landau*. D. ter Haar (Ed.), Gordon and Breach, New York (1965).
10. JAMET, P., Stability and Convergence of a Generalized Crank-Nicolson Scheme on a Variable Mesh for the Heat Equation. *SIAM J. Numer. Anal.* 17-4, 530-539 (1980).
11. KIRKPATRICK, J.R., G.E. GILES, JR., and R.F. WOOD, Numerical Methods for the Analysis of Laser Annealing of Doped Semiconductor Wafers. *Proceedings of the AIAA 15th Thermophysics Conference*, Colorado (1980).
12. LANGER, J.S., and L.A. TURSKI, Studies in the Theory of Interface Stability, I. Stationary Symmetric Model. To appear, *Acta. Met.*
13. LANGER, J.S., and L.A. TURSKI, Studies in the Theory of Interface Stability, II. Moving Symmetric Model. To appear, *Acta. Met.*
14. MULLINS, W.W., and R.F. SEKERKA, in *J. Appl. Phy.* 34, p. 523 (1963). See also 35, 144 (1964).
15. MURRAY, W.D., and F. LANDIS, Numerical and Machine Solutions of Transient Heat-Conduction Problems Involving Melting or Freezing. Part 1 - Method of Analysis and Sample Solutions. *Transactions of the ASME*, 106-112 (1959).
16. OCKENDON, J.R., and W.R. HODGKINS (Eds.), *Moving Boundary Problems in Heat Flow and Diffusion*. Oxford (1974).
17. RUBINSTEIN, L.I., *The Stefan Problem*, Translations of Mathematical Monographs 27, Amer. Math. Soc., Providence, Rhode Island (1971).
18. SMITH, J., Shape Instabilities and Pattern Formation in Solidification. Carnegie-Mellon Univ. Research Report, January (1980). Submitted to *Jour. of Comp. Physics*.
19. TENNEY, D.R., and J. UNNAM, Numerical Analyses for Treating Diffusion in Single-, Two-, and Three-Phase Binary Alloy Systems. National Aeronautics and Space Administration Technical Memorandum 78636 (1978).

ON THE ASYMPTOTIC CONVERGENCE OF SOME BOUNDARY ELEMENT METHODS

W. L. Wendland

Fachbereich Mathematik
Technische Hochschule Darmstadt,
Germany, Federal Republic,
temporarily University of Delaware,
Newark, Del. 19711 USA.

This work has been supported jointly by the Department of Mathematics of the University of Maryland in College Park, the Applied Mathematics Institute and the Department of Mathematical Sciences at the University of Delaware and the "Deutsche Forschungsgemeinschaft" (No. We 659/4-1).

1. INTRODUCTION

This is a brief survey on the asymptotic error analysis of boundary integral equations, in particular on the joint work of the author with M. Costabel, G. C. Hsiao, P. Kopp, U. Lamp, T. Schleicher and E. Stephan [33,34,35,37,38,39,43,44,62,63, 69,70,71,72,73].

Nowadays the most popular numerical methods for solving elliptic boundary value problems are finite differences [20], finite elements [10,19] and, more recently, boundary integral methods. The latter are numerical methods for solving integral equations (or their generalizations) on the boundary Γ of the given domain. Comparing the computational complexity of these methods it turns out that they are of the same magnitude for two-dimensional interior boundary value problems [71] namely proportional to N^3 if N denotes the number of grid points on the boundary curve Γ.

The conversion of elliptic boundary value problems into corresponding integral equations for the investigation of existence goes far back in history. For computational purposes, however, boundary integral equations of various types became more fashionable only recently. (See e.g. [40] and the proceedings [15,16,60].)

The reduction of interior or exterior boundary value problems as well as transmission problems, to equivalent boundary integral equations on Γ can be done with many different methods, since this reduction is by no means unique. The two most popular methods are called the "direct method" and the "method of potentials". In all these cases one needs a fundamental solution or fundamental matrix $\gamma(z,\zeta)$ respectively of the differential equations <u>explicitly</u> since it will be used in numerical computations.

Thus, the practical usefulness of the boundary integral methods hinges essentially upon the simple computability of a fundamental solution. This restricts these methods mainly to differential equations with constant coefficients.

Most numerical procedures for boundary integral methods are based on collocation methods whereas mathematically Galerkin's procedure and corresponding variational formulations provide a further developed error analysis. Here we shall formulate the main ideas and results for the Galerkin method with finite element approximations on the boundary Γ . Moreover we shall develop a numerical implementation of Galerkin's procedure. The resulting scheme not only provides high accuracy as Galerkin's method but also is simple to be adapted to modern computing machines. We shall term this method as the <u>Galerkin collocation method</u>.

It applies to a wide class of integral equations on the boundary manifold Γ as to integral equations of the second and the first kind, to singular integral equations with Cauchy kernels on curves and Giraud kernels on surfaces, i.e. Calderon Zygmund operators [17] and also some integrodifferential equations with finite part principal value operators. The method generalizes the Galerkin collocation in [34] that has been developed for Fredholm integral equations of the first kind with the logarithmic kernel as the principal part.

The effectiveness of the method rests on the asymptotic convergence properties of Galerkin's method. For finite element methods in the domain and for finite differences it is well known that strong ellipticity of the corresponding boundary value problem implies the asymptotic convergence. But for the boundary integral methods the strong ellipticity of the corresponding pseudodifferential operators seems not to have received the proper attention yet.

Here we shall focusing towards (i) <u>strong ellipticity</u>, (ii) <u>a priori estimates for the integral equations</u>, and for two-dimensional problems towards (iii) <u>convolution operators as the principal parts</u> and (iv) <u>smoothness of the remaining kernels</u>.

(i) <u>Strong ellipticity:</u> Since Mikhlin's fundamental work [45] and the constructive proof of the Lax-Milgram theorem by Hildebrandt and Wienholtz [31] it is well known that the Gårding inequality, i.e. coerciveness implies asymptotic convergence of Galerkin's method in the energy norm. This in turn gives optimal convergence rates in the corresponding Sobolev spaces. Using the Sobolev space norms which are equivalent to the energy norm it turns out that the strong ellipticity is even necessary for the convergence of <u>all</u> Galerkin procedures due to Vainikko [67].

As for the variational methods [19] , the use of regular finite element functions yields optimal order of convergence. (See also [63] and corresponding weaker results in [56].) Here we consider equations which are strongly elliptic with ellipticity correspoding to Agmon, Douglis and Nirenberg (see [32] p. 268) but also with pseudodifferential operators of arbitrary real orders. Strong ellipticity implies coerciveness.

It should be pointed out that strong ellipticity is a rather strong condition. Often one uses specific weaker properties of the problem for satisfying the Babuska-Brezzi conditions [9] which provide convergence.

(ii) a priori estimates: If the integral equations are interpreted as strongly elliptic pseudodifferential equations [42,59,66] then they provide a priori estimates in the whole scale of Sobolev spaces in addition to Gårding's inequality. This allows to generalize the Aubin-Nitsche Lemma [49] from differential equations to the general class of strongly elliptic pseudodifferential equations as done by Hsiao and Wendland in [38]. Nitsche's trick proves super approximation i.e. optimal order of convergence even if the error is measured in Sobolev space norms of order less than the energy norm. This superapproximation implies high convergence rates for the approximate potentials in compact subdomains away from the boundary manifold where the integral equations were solved approximately.

(iii) Convolution kernels as principal part: In any case, the principal part of a pseudodifferential operator has convolutional character [59,66]. But if it can be depicted as a simple convolution in one variable, i.e. for two-dimensional boundary value problems, then the Galerkin weights of the principal part associated with finite element functions on a regular grid form a Toeplitz matrix whose elements are given by a vector. This vector can eventually be expressed by two vectors which can be evaluated exactly up to the desired accuracy once for all independent of the boundary curve Γ as well as of the meshsize h for any fixed type of finite elements. It should be pointed out that the accuracy of the numerical results depends significantly on the computing of the approximate principal part.

(iv) Smooth remaining kernels: If the remainder of the integral operator subject to the convolutional principal part has smooth kernel then the corresponding Galerkin weights can be treated numerically by suitable quadrature formulas depending on the particular finite elements to be used and the consistency needed. This leads to simple (modified) collocation formulas.

In this way, the computation of the stiffness matrix can be done in a most efficient and simple manner. On the other hand, the solvability of the corresponding algebraic systems as well as the asymptotic convergence of the approximate solutions are assured by the strong ellipticity of the integral equations. If the consistency is of sufficiently high order then the asymptotic convergence and even the superapproximation remain valid for the fully discretized Galerkin collocation scheme as well.

Our replacement of the smooth part of the kernel is very much related to spline collocations of smooth kernels in Fredholm integral equations of the second kind [5,18,52,57] . But here we are interested in an efficient approximation of the Galerkin weights rather than of the kernel due to the much wider class of equations.

In the following we give some standard examples of boundary integral equations. E.g. the exterior Dirichlet problem for the Laplacian [71] yields the Fredholm integral equation of the first kind with side condition,

$$-\frac{1}{\pi}\int_{\Gamma} u(\zeta)\log|z-\zeta|\,ds_{\zeta} - \frac{1}{\pi}\omega = f \quad \text{for } z\in\Gamma\ , \tag{1.1}$$

$$\frac{1}{2\pi}\int_{\Gamma} u\,ds = B\ , \tag{1.2}$$

where the function f on Γ and the constant B are given and where u, the normal derivative of the desired solution on Γ and ω, the logarithmic capacity of Γ are unknowns.

However, the same exterior Dirichlet problem is also equivalent to the Fredholm integral equation of the second kind on Γ with side condition,

$$u(z) - \frac{1}{\pi}\int_{\Gamma} u(\zeta)\left(\frac{\partial}{\partial\nu_{\zeta}}\log|z-\zeta|\right)ds_{\zeta} - \frac{1}{\pi}\omega = f(z), \tag{1.3}$$

$$\frac{1}{2\pi}\int_{\Gamma} u\,ds = 0\ . \tag{1.4}$$

Here again f is given and u and ω need to be determined.

Similarly, the exterior and interior Neumann problems as well as the interior Dirichlet problem can be formulated in terms of different boundary integral equations.

The above exterior Dirichlet problem is only one very simple example leading to boundary integral equations. References to many other examples can be found in [69] , in particular from conformal mapping, electrostatics, flow problems including slow viscous flows, plate and shell problems, elasticity problems in two and three dimensions, problems of thermoelasticity, time harmonic and stationary electromagnetic fields (see also [40] and e.g. the conference proceedings [15,16,60]).

In many of these cases the integral equations become much more complicated. However, the types of integral equations are Fredholm integral equations of the second kind as in (1.3) or of the first kind as in (1.1). In addition one also finds singular integral equations on curves as

$$a(z)u(z) + \frac{1}{\pi i}b(z)\int_{\Gamma}\frac{u(\zeta)}{\zeta - z}d\zeta + \int_{\Gamma}k(z,\zeta)u(\zeta)\,ds_{\zeta} = f(z) \quad \text{for } z\in\Gamma \tag{1.5}$$

or the corresponding equations on boundary surfaces [54], or one finds operators of the form

$$-\frac{\partial}{\partial\nu_z}\frac{1}{\pi}\int_\Gamma u(\zeta)\left(\frac{\partial}{\partial\nu_\zeta}\log|z-\zeta|\right)ds_\zeta = f(z), \quad z\in\Gamma \tag{1.6}$$

and the corresponding operators in higher dimensions. (See e.g. [8,28].).

Often the above operators also appear in strongly elliptic systems of integral equations. Although these integral equations seem to be rather special, it turns out that almost all the integral equations of applications provide all these properties. In particular, the systems of integral equations of stationary and time harmonic problems of elastomechanics, thermoelsticity, of flows (viscous and inviscid) and of electromagnetics form strongly elliptic pseudodifferential equations (see examples in [69]).

A further example of a system of boundary integral equations comes from the plane mixed boundary value problem with the Laplacian [22,23,43,44,62,71,72,73]

$$\begin{aligned} &\Delta U = 0 \quad \text{in} \quad \Omega\subset\mathbb{R}^2 \ (\text{or in } \mathbb{R}^2\backslash\overline{\Omega}), \\ &U = g_1 \quad \text{on } \Gamma_1, \\ &\frac{\partial U}{\partial\nu} = g_2 \text{ on } \Gamma_2, \end{aligned} \tag{1.7}$$

(and an appropriate condition at infinity for exterior problems). Ω is a simple connected bounded domain in $\mathbb{R}^2$ with a smooth boundary curve $\Gamma = \Gamma_1\cup\Gamma_2\cup Z_1\cup Z_2$, where Γ_1 and Γ_2 are two (for simplicity) disjoint parts of Γ with endpoints Z_1 and Z_2.

We formulate the boundary integral equations via the so called "direct method". The variational solution U within the domain Ω can be represented by

$$\begin{aligned} U(z) = &\frac{1}{2\pi}\int_\Gamma U(\zeta)\frac{\partial}{\partial\nu_\zeta}(\log|z-\zeta|)ds_\zeta \\ &-\frac{1}{2\pi}\int_\Gamma\frac{\partial U}{\partial\nu}(\zeta)\log|z-\zeta|ds_\zeta - \frac{1}{2}\omega \end{aligned} \tag{1.8}$$

with $\omega = 0$ and $\int_\Gamma\frac{\partial U}{\partial\nu}ds = 0$. Here we introduce ω and impose the compatibility condition in order to avoid eigensolutions [44]. Inserting the boundary conditions and passing z to the boundary Γ, one obtains with the well known jump relations for the double layer potential the following equations on the corresponding parts of the boundary:

on Γ_2: $(I-K_{22})\, U - V_{11}(\frac{\partial U}{\partial \nu}) + \omega :=$ (1.9)

$$U(z) - \frac{1}{\pi}\int_{\Gamma_2} U(\zeta)\,(\frac{\partial}{\partial\nu_\zeta}\log|z-\zeta|)ds_\zeta + \frac{1}{\pi}\int_{\Gamma_1}(\frac{\partial U}{\partial\nu}(\zeta))\log|z-\zeta|ds_\zeta + \omega$$
$$= \frac{1}{\pi}\int_{\Gamma_1} g_1(\zeta)(\frac{\partial}{\partial\nu_\zeta}\log|z-\zeta|)ds_\zeta - \frac{1}{\pi}\int_{\Gamma_2} g_2(\zeta)\log|z-\zeta|ds_\zeta,$$

on Γ_1: $V_{11}(\frac{\partial U}{\partial \nu}) + K_{21}U - \omega :=$ (1.10)

$$-\frac{1}{\pi}\int_{\Gamma_1}(\frac{\partial U}{\partial\nu}(\zeta))\log|z-\zeta|ds_\zeta + \frac{1}{\pi}\int_{\Gamma_2} U(\zeta)(\frac{\partial}{\partial\nu_\zeta}\log|z-\zeta|)ds_\zeta - \omega$$
$$= g_1(z) - \frac{1}{\pi}\int_{\Gamma_1} g_1(\zeta)\frac{\partial}{\partial\nu_\zeta}(\log|z-\zeta|)ds_\zeta + \frac{1}{\pi}\int_{\Gamma_2} g_2(\zeta)\log|z-\zeta|ds_\zeta$$

and the side condition

$$\int_{\Gamma_1}\frac{\partial U}{\partial\nu}\,ds = -\int_{\Gamma_2} g_2 ds = : B\ .$$

These equations now serve as integral equations for the unknown boundary data $U|_{\Gamma_2}$, $\frac{\partial U}{\partial\nu}|_{\Gamma_1}$ (and $\omega = 0$) . As soon as these are known, (1.8) gives the desired solution in the whole of Ω .

The validity of the above steps must be justified and depends on the behavior and regularity of U and $\frac{\partial U}{\partial \nu}$ at the boundary [73].

The application of the boundary element method in the form of Galerkin collocation to the mixed boundary value problems requires some modifications. This is due to the singularities of the solution's gradient at the collision points in two dimensional problems or, respectively at the collision curve in three dimensional problems where the two different boundary conditions are adjoining [72].

Since Fichera's fundamental work on the Zaramba problem [26] it is well known that these singularities are unavoidable unless the data satisfy specific side conditions. These singularities generate corresponding singularities of the boundary charges in the boundary integral method. They pollute numerical computations unless they are handled separately.

In §5 we shall see how the boundary integral method can be improved by augmenting the appropriate singularity functions to the finite element scheme. This is based on a local analysis of the solution to the mixed boundary value problem due to Grisvard [29] and of the integral equations [22,23,73]. In §5 we apply Galerkin's method to the modified integral equations. A similar method but with collocation has been used by J. Blue [13] . Since our system of integral equations is strongly elliptic we find convergence in the corresponding energy norm. This estimate corresponds to [21] and [25]. In order to improve the convergence of the approximation we use local analysis for better regularity in connection with the coerciveness on one hand and a-priori estimates for the corresponding pseudodifferential operators on the other hand. We find in [43,44] improved asymptotic convergence and also super approximation which cannot be obtained by variational methods and coerciveness alone. Moreover we approximate the stress intensity factors besides the desired charges and give corresponding error estimates.

The method has been extended to polygonal domains in [22,23]. There it turns out, that equation (1.10) cannot be used anymore for Galerkin's procedure but needs to be modified. In addition, since corresponding compositions of some operators are not pseudodifferential operators anymore, Costabel and Stephan need to introduce another symbol analysis by using the Mellin-transformation. Then a nontrivial extension of [73] provides an improved boundary integral method on polygons [22,23].

The extension of our method to three dimensions raises many new difficulties. Here only preliminary results on the local behaviour of the solution are known [12,24,41,61] . The formulation of corresponding boundary integral methods can be found in [72]. The corresponding analysis will be presented in [62].

Mixed boundary value problems in two and three dimensions describe many problems of classical mathematical physics such as crack and punch problems, contact problems in thermoelasticity, heat conduction in space science, electrostatics and flow and infiltration problems - to name a few. Some of these examples can be found in [61,73].

In §6 we present some of the numerical results from [35] and [44]. It should be mentioned that the experiments in [35] for smooth problems showed that the doubling of the number of grid points on Γ improved the accuracy of the results by only one decimal digit whereas an improvement of 3 digits was obtained due to the transition from piecewise constant to piecewise quadratic trial and test functions. Moreover the latter transition required only 10% more computing time whereas the doubling of grid points also doubled the computing time.

Most numerical implementations of boundary integral methods in engineering are done with the standard collocation method. But there only a few results on its asymptotic convergence are known except in the case of Fredholm integral equations of the second kind. Here we refer to the extensive bibliography by Ben Noble [51], the surveys by K. Atkinson [6], C. Baker [11] and results on superconvergence [18,30,57,58].

For our more general equations and the collocation method there are only preliminary results available as for the special case of the Fredholm integral equation of the first kind with logarithmic kernel on the closed curve Γ [1,2,4,68] . For a singular integral equation with Cauchy's kernel S. Prössdorf and G. Schmidt [54] have proved recently that the collocation with piecewise linear functions converges if and only if the singular integral equation is strongly elliptic.

A more rigorous asymptotic analysis for the class of strongly elliptic equations is yet to be done.

It should be mentioned that the numerical computations with boundary integral equations show superconvergence where the solution is smooth. This indicates that local convergence properties also hold for the boundary integral equations.

Other open questions are uniform convergence properties and the analysis of meshrefinements as well as non uniform grids on Γ.

Although the boundary integral method is of the same complexity as the finite element method or the finite difference method there are several other properties of the boundary integral method which may be rather advantageous:

i) The experiments showed very reasonable results already for small numbers of grid points on Γ.

ii) The method is applicable to interior as well as to exterior problems without modifications.

iii) The desired potentials are given by boundary potentials integrated over Γ and hence can be differentiated analytically away from Γ.

On the other hand the boundary integral method is restricted to problems where the fundamental solution is explicitely available whereas the usual finite element procedure provides a rigorous method. Thus a sensible combination of both methods might become a most efficient numerical procedure in the future.

2. STRONGLY ELLIPTIC INTEGRAL EQUATIONS

Although all the above mentioned types of equations have very different properties in the classical theory of integral equations it turns out that if they are considered as pseudodifferential operators [66] they have a very strong, common property. Namely the equations of practical interest are all "strongly elliptic". In order to formulate this property one needs the Sobolev spaces $H^s(\Gamma)$ of generalized functions on Γ, their interpolation spaces and their dual spaces. For the definitions we refer to [3] (in particular p. 214). Then each of the above mentioned operators A defines a continuous linear mapping $A : H^s \to H^{s-2\alpha}$ for a whole scale of real s (depending on the smoothness of Γ). 2α is called the order of the pseudodifferential operator A [66]. (G. Richter calls -2α in [56] "smoothing index".) For our examples we have $2\alpha = 0$ in (1.9) and (1.11), $2\alpha = -1$ in (1.5), $2\alpha = +1$ in (1.12). The boundary integral equations we write in short

$$Au = f \quad \text{on } \Gamma . \tag{2.1}$$

The announced common property is the coerciveness in form of the Gårding inequality:

$$\mathrm{Re}(Av,v) = \mathrm{Re}\int_\Gamma vAv\,ds \geq \gamma\|v\|^2_{H^\alpha} - |k[v,v]| \qquad (2.2)$$
$$\text{for all } v \in H^\alpha(\Gamma)$$

where $\gamma>0$ is a constant independent of v and where $k[u,v]$ denotes a compact bilinear form on $H^\alpha \times H^\alpha$. In some cases k equals zero, then inequality (2.2) corresponds to strong energy estimates as in [48].

In order to characterize those equations or systems of equations providing coerciveness let us use the context of pseudodifferential operators and let us consider systems of equations in the form (2.1). Then to A there belongs a $p\times p$ matrix-valued principal symbol $a_o(x,\xi) = ((a_{q,r}(x,\xi)))_{q,r=1,\ldots,p}$ corresponding to the p equations of (2.1) for the p components $v_q, q = 1,\ldots,p$. As usual, the $a_{q,r}(x,\xi)$ are assumed to be homogeneous functions of $\xi\in\mathbb{R}^n$ for $|\xi|\geq 1$ with degrees $\alpha_{qr}\in\mathbb{R}$.

Now we define strong ellipticity (analogously to the Agmon-Douglis-Nirenberg ellipticity for differential equations) assuming that there is an index vector $\alpha = (\alpha_1,\ldots,\alpha_p) \in \mathbb{R}^p$ such that

$$\alpha_{qr} = \alpha_q + \alpha_r\,, \quad q,r = 1,\ldots,p. \qquad (2.3)$$

A is then a continuous linear pseudodifferential operator of order 2α, i.e. defining a continuous map

$$A\colon H^{s+\alpha}(\Gamma) := \prod_{q=1}^{p} W_2^{s+\alpha_q}(\Gamma) \to H^{s-\alpha}(\Gamma) := \prod_{q=1}^{p} W_2^{s-\alpha_q}(\Gamma),\ s\in\mathbb{R}, \qquad (2.4)$$

in the scale of Sobolev spaces in (2.4). (The admissible s depend also on the smoothness of Γ.)

Now for the following we assume

$$A \text{ is strongly elliptic} \qquad (2.5)$$

i.e. there exists a complex valued smooth matrix $\theta(z)$ and a constant $\gamma>0$ such that

$$\mathrm{Re}\,\zeta^T\theta(z)a_o(z,\xi)\,\overline{\zeta} \geq \gamma|\zeta|^2 \qquad (2.6)$$

for all $z\in\Gamma$, all $\xi\in\mathbb{R}^n$ with $|\xi| = 1$ and for all $\zeta\in\mathbb{C}^p$. A strongly elliptic system A satisfies the Gårding inequality [42],

$$\mathrm{Re}(Av,v)_{L^2(\Gamma)} \geq \gamma'\|v\|^2_{H^\alpha(\Gamma)} - |k[v,v]| \text{ for all } v\in H^\alpha(\Gamma) \qquad (2.7)$$

with $\gamma'>0$ and $k[v,w]$ a compact bilinear form on $H^{\alpha}\times H^{\alpha}$.

In the following we shall always consider the modified equations

$$\mathcal{A}v := \theta A = \theta f =: g \tag{2.8}$$

instead of (2.1) i.e. for $\mathcal{A}$ we have $\theta=1$. Then Gårding's inequality (2.7) for $\mathcal{A}$ implies the (non unique) decomposition

$$\mathcal{A} = D + K \tag{2.9}$$

where D is a positive definite pseudodifferential operator and $K : H^{s+\alpha} \to H^{s-\alpha}$ is compact. This is just the kind of operator that provides the convergence of Galerkin's method for <u>any</u> approximating family of finite dimensional subspaces [67].

In the following table we have collected the main properties of the simple examples of Section 1. Note that the symbols and, hence, strong ellipticity and α remain the same for equations (1.1), (1.3), (1.6) in higher dimensions if $\gamma = -\frac{1}{\pi}\log|z-\zeta|$ is replaced by the corresponding fundamental solution $\frac{1}{\omega_n}|\underline{x}-\underline{y}|^{2-n}$, $n \geq 3$ of the Laplacian. Since for D and the principal symbol we neglect compact operators, we have the same properties if the fundamental solution of the Laplacian is further replaced by that of the Helmholtz equation or by Green's functions to larger domains.

Further examples including symbols of systems can be found in [69, §5].

Equ. (1.3): $2\alpha=0$, $Du = u$	$a_o=1$, always strongly elliptic
Equ. (1.1): $2\alpha=-1$, $Du=-\frac{1}{\pi}\int_\Gamma \log\left|\frac{z-\zeta}{2\,\mathrm{diam}\,\Gamma}\right| u(\zeta)\,ds_\zeta$	$a_o=\frac{1}{\lvert\xi\rvert}$, always strongly elliptic
Equ. (1.5): $2\alpha=0$, $Du=\theta(z)\left(a(z)u(z)+\frac{b(z)}{i\pi}\int_\Gamma\frac{u(\zeta)\,d\zeta}{\zeta-z}\right)$, $\theta(z)=\bar{a}-\bar{b}\lambda_o$ with $\lambda_o := \begin{cases} \mathrm{Re}\,\bar{a}b/\lvert b\rvert^2 & \text{for } \lvert\mathrm{Re}\,\bar{a}b\rvert<\lvert b\rvert^2 \\ 1 & \text{for } \mathrm{Re}\,\bar{a}b \geq \lvert b\rvert^2 \\ -1 & \text{for } \mathrm{Re}\,\bar{a}b \leq -\lvert b\rvert^2 \end{cases}$	$a_o=a(z)+b(z)\frac{\xi}{\lvert\xi\rvert}$, strongly elliptic iff $a(z)+\lambda b(z) \neq 0$ for all $\lambda\in[-1,1]$

Equ. (1.6): $2\alpha=1$, $Du = -\frac{\partial}{\partial\nu_z}\frac{1}{\pi}\int_\Gamma u(\zeta)\cdot(\frac{\partial}{\partial\nu_\zeta}\log\|z-\zeta\|)ds_\zeta + \int_\Gamma u ds.$	$a_o = \|\xi\|$	, always strongly elliptic

Table 1: Positive definite parts, principal symbols and strong ellipticity of the simple examples in the Introduction

In order to formulate strong ellipticity also for the system (1.9) - 1.11) and to further improve our method later on we need to introduce the Sobolev spaces $H^r(\Gamma_j)$ on the corresponding boundary part and, moreover, the spaces of functions which can be extended by zero values onto the whole boundary Γ, i.e.

$$\tilde{H}^r(\Gamma_j) := \{f\in H^r(\Gamma) \text{ with } \operatorname{supp} f\subset\overline{\Gamma}_j \text{ and } \|f\|_{\tilde{H}^r(\Gamma_j)} := \|f\|_{H^r(\Gamma)}\}. \tag{2.10}$$

Then $U|_{\Gamma_2}\in L_2(\Gamma_2)$ and $\psi = \frac{\partial U}{\partial\nu}|_{\Gamma_1}\in\tilde{H}^{-1/2}(\Gamma_1)$ in (1.9) and (1.10) can be extended by zero values onto all of Γ and the operators in the system can be seen as pseudodifferential operators on Γ. Then $\alpha = (0,-1/2)$ and with respect to the mapping A : $L_2(\Gamma_2)\times\tilde{H}^{-1/2}(\Gamma_1)\times\mathbb{R} \to L_2(\Gamma_2)\times H^{1/2}(\Gamma)\times\mathbb{R}$ we find the principal symbol

$$a_o = \begin{pmatrix} 1, & 0 & , 0 \\ 0, & |\xi|^{-1} & , 0 \\ 0, & 0 & , 1 \end{pmatrix} \tag{2.11}$$

a_o is obviously strongly elliptic and, hence, we have the following coerciveness inequality:

<u>Lemma 2.1: To the system</u> (1.9 - 1.11) <u>there exists a constant</u> $\gamma_0>0$ <u>such that the Garding inequality</u>

$$(\{(I-K_{22})U - V_{11}\psi + \omega\},U)_{L_2(\Gamma_2)} + (\{V_{11}\psi+K_{21}U-\omega\},\psi)_{L_2(\Gamma_1)} \tag{2.12}$$
$$+ \int_{\Gamma_1}\psi ds\cdot\omega \geq \gamma_0\{\|U\|^2_{L_2(\Gamma_2)} + \|\psi\|^2_{\tilde{H}^{-1/2}(\Gamma_1)} + |\omega|^2\}$$
$$-|k[(U,\psi,\omega),(U,\psi,\omega)]|$$

<u>holds for all</u> $(U,\psi,\omega)\in L_2(\Gamma_2)\times\tilde{H}^{-1/2}(\Gamma_1)\times\mathbb{R}$ <u>where</u> k <u>is a suitable compact bilinear form on</u> $L_2(\Gamma_2)\times\tilde{H}^{-1/2}(\Gamma_1)\times\mathbb{R}$, [73].

Note that the coerciveness (2.12) and corresponding continuity hold in the above spaces whereas the "natural" space for the variational solution is the different space
$H^{1/2}(\Gamma_2) \times \tilde{H}^{-1/2}(\Gamma_1) \times \mathbb{R}$.

3. GALERKIN'S METHOD AND REGULAR FINITE ELEMENTS

Now let us assume that the system of linear boundary integral equations

$$Au = f \quad \text{on} \quad \Gamma \tag{3.1}$$

is strongly elliptic and has no eigensolution, i.e. the solution of (3.1) is unique, and hence, it is uniquely solvable. If (3.1) admits eigensolutions then finitely many additional side conditions determine the solution uniquely. If the side conditions are approximated as well then our following theorems remain valid (see [63].).

By H_h let us denote a family of finite dimensional approximating spaces on Γ depending on a parameter $h>0$. Let μ_j, $j=0,1,\ldots,N$ denote a basis of H_h. Then the well known Galerkin procedure for (3.1) is to find the coefficients γ_j of the approximate solution

$$\hat{v}(z) = \sum_{j=0}^{N} \gamma_j \mu_j(z) \ , \quad z \in \Gamma \tag{3.2}$$

by solving the finite system of linear equations,

$$\sum_{j=0}^{N} (A\mu_j, \mu_k)_{L_2} \gamma_j = (f, \mu_k)_{L_2} \ , \quad k=0,\ldots,N \ . \tag{3.3}$$

For the convergence of this procedure we have well known results going back to S. Mikhlin [45], S. Hildebrandt and E. Wienholtz [31]. Here we use the version known as Céa's lemma, [19, p.104].

For its formulation let P_h denote the L_2 orthogonal projection onto H_h . Then we require the approximation property

$$\lim_{h\to 0} \|P_h g - g\|_{H^\alpha} = 0 \quad \text{for any} \quad g \in H^\alpha \ . \tag{3.4}$$

This assumption implies with the Banach-Steinhaus theorem the stability

$$\|P_h\|_{H^\alpha, H^\alpha} \le c \quad \text{and} \quad \|P_h\|_{H^{-\alpha}, H^{-\alpha}} \le c \tag{3.5}$$

by duality, where c is independent of h .

These requirements are satisfied for a wide class of approximations including regular finite elements [10].

Now we are in the position to state Céa's lemma:

Theorem 3.1: Let (3.1) with A be a strongly elliptic system with unique solution $u \in H^\alpha$ to any $f \in H^{-\alpha}$. Then there exists $h_o>0$ such that (3.3) are uniquely solvable for every $0<h\le h_o$. Moreover there exists a constant c independent of h and f such that

$$\|\hat{v} - u\|_{H^\alpha} \le c \inf_{\chi\in H_h} \|u - \chi\|_{H^\alpha} \quad . \tag{3.6}$$

For convenience, in the following asymptotic error analysis we are always using $c, c', \ldots$ as generic constants which may change their size and meaning at different places.

As was mentioned above, Theorem 3.1 is not restricted to our finite element approximations but applies to a rather wide class of Galerkin methods as e.g. for the projection methods using trigonometric polynomials as in [53,55] . The proof of Theorem 3.1 is rather standard, we omit the details. (See e.g. [71].) Note that (3.6) is equivalent to the stability of the Galerkin projection

$$G_h := (P_h A P_h)^{-1} P_h A, \text{ i.e.} \tag{3.7}$$

$$\|G_h\|_{H^\alpha, H^\alpha} \le c \tag{3.8}$$

for all $0 < h \le h_o$ where c is independent of h .

Now we specify the spaces H_h to regular (ℓ,m) systems of finite element functions [10]. They have the following approximation property and satisfy an inverse assumption:

Approximation property: Let the multi-indices m,t,s satisfy componentwise $-\ell\le t\le s\le\ell$; $-m\le s$, $t\le m$. Then to any $u\in H^s(\Gamma)$ and any $h>0$ there exists a $\mu\in H_h$ such that

$$\| u_q - \mu_q \|_{H^{t_q}} \le c\, h^{s_q - t_q} \| u_q \|_{H^{s_q}} \quad \text{(see [14].)} \tag{3.9}$$

The constant c is independent of μ, h and u .

The finite element functions $\mu = (\mu_1,\ldots,\mu_p) \in H_h$ provide for $-m\le t\le s<m$ the inverse assumption:

$$\| \mu_q \|_{H^{s_q}(\Gamma)} \le c\, h^{t_q - s_q} \| u_q \|_{H^{s_q}(\Gamma)} \tag{3.10}$$

where the stability constant c is independent of μ and h, [50]. If we insert (3.9) into the right hand side of (3.6) we surely find improved asymptotic orders of convergence if $h\to 0$. Using the inverse assumption (3.10) one can also extend the estimate of the left hand side to H^t norms with $\alpha\le t\le m$ [63]. These are the results which have also been obtained with variational methods as in [19]. But as was already mentioned in the introduction, for pseudodifferential operators A one can even prove superapproximation [38]. Collecting these results we find the following improved convergence theorem.

Theorem 3.2 [38,56,63]: Let A be strongly elliptic and let (3.1) have an unique solution. Let H_h satisfy (3.9) and (3.10) and define

$$\alpha'_q := \min\{\alpha_q, 0\} , \quad q = 1,\dots,p . \tag{3.11}$$

Suppose $\alpha_q - \ell_q \le t \le s \le \ell_q - \alpha_q = \beta_2$, $\max\{o,t\} \le m_q - \alpha_q = \beta_1$ for $q = 1, \dots, p$ and $s \ge 0$. Then we have the asymptotic error estimate

$$\| u - \hat{v} \|_{H^{\alpha+t}(\Gamma)} \le c\, h^{s-t} \| u \|_{H^{\alpha+s}(\Gamma)} \tag{3.12}$$

In addition, if we consider the discrete equations (3.3) in $L_2(\Gamma)$ then we find for the stability of these equations

$$\|\hat{v}\|_{L_2(\Gamma)} \le c \sum_{q=1}^{p} h^{2\alpha'_q} \|g\|_{L_2(\Gamma)} . \tag{3.13}$$

Remarks: With a simple analysis of $P_h A P_h - A$ for $\alpha > 0$ the stability (3.13) yields the condition number of (3.3) to be of the order $\sum_{q=1}^{p} h^{-2|\alpha_q|}$.

The stability estimate (3.13) can also be used for an estimate of errors due to contamination and round off effects in the framework of ill posed problems. This can be found in [39].

The asymptotic estimate (3.12) includes the case $t<\alpha$, i.e. superapproximation. If $t=2\alpha-\ell$ then one has for sufficiently smooth data the superapproximation

$$\| u - \hat{v} \|_{H^{2\alpha-\ell}} \le c\, h^{2(\ell_q - \alpha_q)} \| u \|_{H^{\ell}} \tag{3.14}$$

That implies for the desired solution U of the boundary value problems local superconvergence

$$\| U - \hat{U} \|_{X(\tilde{\Omega})} \le \| u - \hat{v} \|_{H^{2\alpha-\ell}} \le c' h^{2(\ell_q - \alpha_q)} \| u \|_{H^{\ell}(\Gamma)} \tag{3.15}$$

where $\tilde{\Omega}$ is any compact subdomain in the interior, respecively, exterior of Γ and $X(\tilde{\Omega})$ denotes any norm. Here c, c' depend on $\tilde{\Omega}$ and $X(\tilde{\Omega})$. The proof can be found in [38].

If we apply these results to one of the examples (1.1 - 1.2), (1.3 - 1.4), (1.5) or (1.6) and if we assume that the data are sufficiently smooth then also the solution u on Γ is correspondingly smooth and the error estimates (3.12) and (3.14) are restricted only by ℓ , the biggest index of the approximation property (3.9), as long as the order of h is concerned.

For the mixed boundary value problem (1.7) and the corresponding system of integral equations (1.9 - 1.11), however, the situation changes significantly since the solution has singularities at the collision points.

Theorem 3.3 [73]: To every $g_1 \in H^{3/2+\sigma}(\Gamma_1)$, $g_2 \in H^{1/2+\sigma}(\Gamma_2)$

with $|\sigma| < \frac{1}{2}$ there exists exactly one variational solution U of (1.7). This solution has the form

$$U(z) = \sum_{i=1}^{2} \alpha_i \rho_i^{1/2} \sin \frac{1}{2} \Theta_i + v(z) \tag{3.16}$$

with a smooth function $v \in H^{2+\sigma}(\Omega)$, $|\sigma| < \frac{1}{2}$. For $g_1 \in H^2(\Gamma_1)$, $g_2 \in H^1(\Gamma_2)$ (3.16) remains valid. Here $\rho_i = |z-Z_i|$ denote the distances to the corresponding collision points Z_i and Θ_i denotes the respective angle between the tangent vector Z_i in the direction of Γ_1 and the ray $(z-Z_i)$. Because of the singularities in Z_i we find from (3.16) $U \in H^{3/2-\varepsilon}(\Omega)$ for any $\varepsilon>0$ even for C^∞-data g_1, g_2. This implies with the trace theorem the following lemma corresponding to [21] and [25].

Lemma 3.1: Let u_h, ψ_h denote the Galerkin solutions with regular finite elements to (1.9 - 1.11), $0 \le m \le \ell$. Then one finds asymptotic convergence as

$$\|U - u_h\|_{L_2(\Gamma_2)} + \|\frac{\partial U}{\partial \nu} - \psi_h\|_{\tilde{H}^{t+\varepsilon}} \tag{3.17}$$

$$\le c_\varepsilon \{h^{1-\varepsilon} \|U\|_{H^{1-\varepsilon}(\Gamma_2)} + h^{-(t+2\varepsilon)} \|\frac{\partial U}{\partial \nu}\|_{\tilde{H}^{-\varepsilon}(\Gamma_1)}\}$$

with any $\varepsilon>0$ and $-1 \le t \le -\frac{1}{2}$. The constant c_ε is independent of U, h, u_h and ψ_h but may depend on ε.

Since the singularities appear even for smooth data, this estimate cannot be improved unless one takes care of the singularities in a more specific manner. That means that our method and analysis need to be refined accordingly. These more specific methods will be presented in §5.

4. THE GALERKIN COLLOCATION METHOD

For the numerical implementation of Galerkin's procedure (3.3) the weights of the stiffness matrix,

$$a_{jk} := (A\mu_j, \mu_k) \ , \ j,k = 0,\dots,N \tag{4.1}$$

have to be evaluated. Since A is here usually given by an integral operator (in the usual of the generalized) sense, the computation of a_{jk} requires a double integration over $\Gamma\times\Gamma$. If this is done numerically, the kernels of the integral operators must be computed at all combinations of grid points on Γ. In addition, special care must be taken of the singular integrals. This often results in high computing times. In order to reduce the computing time and in order to simplify the computation of the singular integrals let us specify the further investigations to two-dimensional problems. In addition we assume that the principal parts of A are given by convolutional operators. This is the case in all of our examples. For simplicity let us here consider just one equation (3.1). The extension to systems is of simplest technical nature (see [69]). Let Γ be given by a regular parameter representation

$$\Gamma : z = z(t) \ , \quad t \in [0,1] \tag{4.2}$$

with $z(t)$ a 1-periodic sufficiently smooth vector valued function satisfying

$$\left|\frac{dz}{dt}\right| = R(t) \geq R_o > 0 \quad \text{for all} \quad t \ , \tag{4.3}$$

where R denotes the Jacobian. Then the operator A with a convolution operator as principal part has the form

$$\begin{aligned} Au|_t = \text{p.v.} \int_{|t-\tau|<\frac{1}{2}} \{ p_1(t-\tau) + \log|t-\tau| p_2(t-\tau) \} u(t)R(t)\, dt \\ + \int_{|t-\tau|<\frac{1}{2}} L(\tau,t)(u(t)R(t))dt = f(t) \ . \end{aligned} \tag{4.4}$$

Here $p_1(\zeta)$ and $p_2(\zeta)$ are homogeneous functions for $\zeta \neq 0$ of degree $\beta = -2\alpha - 1$. The principal symbol a_o and (4.4) are related by the Fourier transformation F,

$$a_o(\xi) := F(p_1(\cdot) + \log|\cdot| p_2(\cdot))|_\xi \ . \tag{4.5}$$

For singular integral equations with the Cauchy kernel, the above special form (Equation 4.4)) of A is too restrictive. Yet we leave this detail to [70].

From now on we consider strongly elliptic integral equations of the form (4.4) and we further assume that the remaining terms collected in $L(\tau,t)$ define a sufficiently smooth function of τ and t. Otherwise we again split into two terms, where the first contains the singularity and has to be treated similarly to the principal part.

Since in Equation (4.4) only R depends on Γ we consider (4.4) as an integral equation over $[0,1]$ for the 1-periodic new unknown function

$$v(t) := R(t)u(t) \ . \tag{4.6}$$

Note that the principal part in (4.4) then becomes independent of the special choice of the curve Γ. Therefore we shall adapt numerical integration to the special integrals in (4.4) and the principal part in the standard form (4.4) will yield the already mentioned Toeplitz matrix.

The Galerkin weights due to the smooth remaining parts will be treated numerically by appropriate quadrature formulas depending on the particular finite elements to be used. In them we may use only grid points in a grid connected with the finite elements such that the kernel functions are to be evaluated as seldom as necessary. This leads to simple modified collocation formulas and the computation of the corresponding stiffness matrix can be done extremely fast.

In order to utilize the convolution in the principal part we use regular finite elements on a uniform grid of [0,1] defined with shifts and stretched variables from one shape function $\mu(\eta)$. The latter we define as in [7, Chap. 4] by suitable piecewise polynomials of order m with $\mu \in C^{m-1}$. For $m = 0,1,2$ e.g. we have

$$\mu(\eta) = \left\{ \begin{array}{c|c|c|l} m=0 & m=1 & m=2 & \text{for} \\ \hline 1 & \eta & \frac{1}{2}\eta^2 & 0 \leq \eta < 1 \\ 0 & 2-\eta & -\eta^2 + 3\eta - 3/2 & 1 \leq \eta < 2 \\ 0 & 0 & \frac{1}{2}\eta^2 - 3\eta + 9/2 & 2 \leq \eta < 3 \\ 0 & 0 & 0 & \text{elsewhere} \end{array} \right. \tag{4.7}$$

With μ we define a basis of H_h by

$$\mu_j(t) := \mu\left(\frac{t}{h} - j\right)\frac{1}{R(t)} \text{ for } h_j \leq t \leq 1 + h_j \text{ , } j=0,\ldots,N,\ h=1/(N+1) \tag{4.8}$$

and their 1-periodic extensions

$$\mu_j(t+\ell) := \mu_j(t) \quad \text{for integer } \ell \text{ .}$$

For u in Equation (4.4) we use the approximation

$$u_h(t) := \sum_{j=0}^{N} \gamma_j \mu_j(t) \text{ .} \tag{4.9}$$

With this basis we define a regular $(m+1,m)$ system [10] , i.e. we choose $\ell = m+1$.

Remarks: Our boundary elements have been defined by the transplantation of a regular $(m+1,m)$ system in the parameter domain onto Γ with the local parameter representation of Γ . For calculations, the integrals will be evaluated by using the local coordinates. In those the finite elements appear as simple functions over the parameter domains. This construction of finite elements on Γ requires that the parameter representation is fully available. For the two-dimensional case this is a sensible requirement. In the space, however, the boundary surface has also to be approximated [47].

For the computations we insert (4.9), (4,8) into Equations (4.1), and we find for the terms due to the first expression in Equation (4.4)

$$d_{jk} = \int_0^1 \text{p.v.} \int_{|t-\tau|\le\frac{1}{2}} [p_1(t-\tau)+\log|t-\tau|p_2(t-\tau)]\mu_j(t)R(t)dt\mu_k(\tau)\cdot R(\tau)d\tau$$

$$= h^{2+\beta}\{\int_{\tau'=0}^{m+1} \text{p.v.} \int_{t'=0}^{m+1} [p_1(t'-\tau'+(j-k))+p_2\cdot\log|t'-\tau'+(j-k)|]\cdot\mu(t')\mu(\tau')dt'd\tau'$$

$$+ \log h \int_{\tau'=0}^{m+1} \text{p.v.} \int_{t'=0}^{m+1} p_2(t'-\tau'+(j-k))\mu(t')\mu(\tau')dt'd\tau'\},$$

$$d_{jk} = h^{2+\beta}\{W_{1\rho} + W_{2\rho} \log h\} \text{ with } \rho = j-k \in Z . \tag{4.10}$$

Here the two vectors of weights

$$W_{1\rho} = \int_{\tau'=0}^{m+1} \text{p.v.} \int_{t'=0}^{m+1} [p_1(t'-\tau'+\rho)+p_2\log|t'-\tau'+\rho|]\cdot\mu(t')\mu(\tau')dt'd\tau', \tag{4.11}$$

$$W_{2\rho} = \int_{\tau'=0}^{m+1} \text{p.v.} \int_{t'=0}^{m+1} p_2(t'-\tau'+\rho)\mu(t')\mu(\tau')dt'd\tau', \ \rho\in Z \tag{4.12}$$

can be computed once for all independent of Γ and h . For more details see [34] and [69]. For all the remaining smooth terms in the Galerkin equations to Equation (4.4) we use numerical integration.

Since in the corresponding integrals

$$\int_{\text{supp }\mu_j} f(t)\mu_j(t)R(t)dt = h\int_{\sigma=0}^{m+1} f(h(j+\sigma))\mu(\sigma)d\sigma \tag{4.13}$$

the finite element functions appear as factors, the numerical integrations are chosen accordingly to the respective reference function μ such that polynomials f up to the order $2M+1$ are <u>integrated</u> <u>exactly</u>. This leads to formulas like

$$\int_{\text{supp }\mu_j} f(t)\mu_j(t)R(t)dt = h\sum_{\ell=-M}^{M} b_\ell f(z_{j\ell}) + \tilde{R} \tag{4.14}$$

where

$$z_k := z(h(k+\frac{m+1}{2})) \quad\text{and}\quad z_{j\ell} := z(h(j+\frac{m+1}{2}+\tilde{\gamma}_\ell)) , \quad \ell = -M, \ldots, M \tag{4.15}$$

are the gridpoints subject to the boundary elements and, correspondingly subject to the integration formula. $\tilde{R}$ denotes the error term which is of order h^{2M+2}. The simplest choice $\tilde{\gamma}_\ell = \ell$ yields $z_{j\ell} = z_{j+\ell}$ and weights $b_\ell = b_{-\ell}$ as follows:

	m=0		m=1		m=2	
	b_o	b_1	b_o	b_1	b_o	b_1
M = 0:	1	0	1	0	$\frac{1}{3}$	0
M = 1:	$\frac{11}{12}$	$\frac{1}{24}$	$\frac{5}{6}$	$\frac{1}{12}$	$\frac{3}{4}$	$\frac{1}{8}$

(4.16)

For $\tilde{\gamma}_\ell = \frac{1}{2}\ell$ and $M = 2$ one has

m=1			m=2		
b_o	b_1	b_2	b_o	b_1	b_2
$\frac{13}{30}$	$\frac{4}{15}$	$\frac{1}{60}$	$\frac{2}{5}$	$\frac{7}{30}$	$\frac{1}{15}$

(4.17)

Instead of (4.17) one often uses Gaussian integration formulas, then $\tilde{\gamma}_\ell$ correspond to the Gaussian nodal points and (4.14) is modified accordingly. Using Equation (4.4) for the smooth terms of the weights in (4.4) we obtain

$$\int_{\tau=0}^{1} \int_{|\tau-t|\le\frac{1}{2}} L(\tau,t)\mu_j(t)dt\mu_k(\tau)d\tau = h^2 \sum_{\ell,i=-M}^{M} b_i b_\ell L(z_{ki},z_{j\ell}) + \tilde{R} \tag{4.18}$$

with the error term

$$|\tilde{R}| \le h^{s+2} c\{\max|\frac{\partial^s L}{\partial\tau^s}| + \max|\frac{\partial^s L}{\partial\tau^s}|\} \ , \ 0 \le s \le 2M+2 \ . \tag{4.19}$$

Now we are ready to formulate the <u>Galerkin-collocation equations</u> by using Equations (4.9), (4.10) and (4.18). They read as

$$\begin{aligned}\sum_{j=0}^{N} a_{hjk}\gamma_j := \sum_{j=0}^{N} \{h^{2+\beta}(W_{1,\rho(j,k)} + \log h\, W_{2,\rho(j,k)}) \\ + h^2 \sum_{\ell,i=-M}^{M} b_i b_\ell L(z_{ki},z_{j\ell})\}\gamma_j \\ = h \sum_{i=-M}^{M} b_i f(z_{ki}) = : F_k \quad k=0,\dots,N,\ \rho(j,k)=j-k.\end{aligned} \tag{4.20}$$

For saving computing time, the values of L and f at the grid points should be evaluated <u>only once</u> at the beginning and then be stored for further use as to build up the stiffness matrix in (2.20).

This suggests a choice $\tilde{\gamma}_\ell = \ell$ or $\frac{1}{2}\ell$ in the numerical integration formulas.

For the asymptotic error due to the Galerkin-collocation we shall use the already established error estimates for Galerkin's method. To this end we abbreviate the Equations (4.20) by

$$\sum_{j=0}^{N} a_{hjk}\gamma_j = F_k , \quad k=0,\dots,N \tag{4.21}$$

as mappings in H_h. To this end we define $\tilde{F} \in H_h$ by

$$(\tilde{F},\mu_k) = F_k \quad \text{for} \quad k=0,\dots,N . \tag{4.22}$$

Then the Galerkin Equations (3.3) and the Galerkin-collocation Equations (4.20) take the form

$$P_h A P_h \hat{v} = P_h f \quad \text{and} \quad \tilde{A}_h v_h = \tilde{F} , \quad \hat{v}, v_h \in H_h \subset L_2 \cap H^\alpha , \tag{4.23}$$

respectively. One easily obtains the estimate

$$\|v-v_h\|_{L_2} \leq \|\tilde{A}_h^{-1}\|_{L_2L_2} \{\|(\tilde{A}_h - P_h A P_h)\hat{v}\|_{L_2} + \|P_h f - \tilde{F}\|_{L_2}\} . \tag{4.24}$$

This estimate shows clearly that we need estimates for stability, i.e. $\|\tilde{A}_h^{-1}\|_{L_2L_2}$, consistency, i.e. $\|(\tilde{A}_h - P_h A P_h)\hat{v}\|_{L_2}$ and the truncation error $\|P_h f - \tilde{F}\|_{L_2}$. These estimates are collected in the following theorem.

<u>Theorem 4.1: i) Consistency: Let the weights</u> $W_{1\rho}, W_{2\rho}$ <u>be accurate to an order</u> h^a <u>and let</u> $(\frac{\partial}{\partial\tau})^{2M+2}L$ <u>and</u> $(\frac{\partial}{\partial t})^{2M+2}L$ <u>be continuous. Then we have the consistency</u>

$$|(\tilde{A}\mu,\nu) - (A\mu,\nu)| \leq \lambda(h)\|\mu\|_{L_2}\|\nu\|_{L_2} \quad \underline{\text{for all}}\ \mu,\nu\in H_h \tag{4.25}$$

where

$$\lambda(h) \leq c_1|\log h|h^{a-2\alpha-1} + c_2 h^{2M+2} . \tag{4.26}$$

<u>ii) Stability: In addition let</u> $a > 1+2(\alpha-\alpha')$ <u>and</u> $M > -1-\alpha'$. <u>Then there exists a constant</u> $h_o > 0$ <u>such that</u>

$$\|\tilde{A}_h^{-1}\|_{L_2L_2} \leq ch^{2\alpha'} \tag{4.27}$$

where c is independent of h for all $0 < h \leq h_o$.

iii) Truncation error: For $F_k = h \sum_{\ell=-m}^{M} b_\ell f(z_{k\ell})$ in Equations (4.22) there holds

$$\|P_h f - \tilde{F}\|_{L_2} \leq ch^{\sigma}\|f\|_{H^{\sigma}} \quad \text{with} \quad \frac{1}{2} < \sigma \leq 2M+2 \,. \tag{4.28}$$

iv) Collected errors: For $a > m+2+2(\alpha'-\alpha)$ and $M \geq \frac{m-1}{2} - \alpha'$ we find an error estimate

$$\|u - v_h\|_{L_2} \leq ch^s\{\|u\|_{H^s} + \|f\|_{H^{s-2\alpha'}}\} \tag{4.29}$$

with $1+2\alpha' \leq s \leq m+1$ and $0 \leq s$. For $a > 2m+3-2\alpha'$ and $M \geq m-\alpha-\alpha'$ we have even the super approximation

$$\|u - v_h\|_{H^t} \leq ch^{s-t}\{\|u\|_{H^s} + \|u\|_{L_2} + \|f\|_{H^{s-t-2\alpha'}}\} \tag{4.30}$$

provided $2\alpha-m-1 \leq t \leq s \leq m+1$, $s-t \geq 1-2\alpha'$.

For the proofs see [33] and [70].

In case of our example with the mixed boundary value problem we shall need some modifications which will be presented in §5.

5. AN IMPROVED GALERKIN-COLLOCATION FOR THE MIXED BOUNDARY VALUE PROBLEM

In order to improve the order of approximation from (3.17) to higher orders of h we need local analysis for better regularity of the desired densities on Γ. According to (3.16) the traces of U and $\frac{\partial U}{\partial \nu}$ can be written as

$$U|_{\Gamma_2} = \sum_{i=1}^{2} \alpha_i \rho_i^{1/2} \chi_i + \tilde{g}_1 + w_o \tag{5.1}$$

and

$$\frac{\partial U}{\partial \nu}\Big|_{\Gamma_1} = -\frac{1}{2} \sum_{i=1}^{2} \alpha_i \rho_i^{-1/2} \chi_i + \tilde{g}_2 + \phi_o \tag{5.2}$$

where $\tilde{g}_1 \in H^2(\Gamma_2)$ and $\tilde{g}_2 \in H^1(\Gamma_1)$ are arbitrarily chosen functions satisfying the transition conditions

$$\tilde{g}_1(Z_i) = g_1(Z_i) \quad \text{for} \quad i = 1,2 \,,$$

and

$$\tilde{g}_2(Z_i) = g_2(Z_i) \quad \text{if} \quad 0 < \sigma < \frac{1}{2} \tag{5.3}$$

and in a generalized sense for $\sigma = \frac{1}{2}$ [44]. Then the functions $w_o \in H^{3/2+\sigma}(\Gamma_2) \cap \tilde{H}^1(\Gamma_2)$ and $\phi_o \in \tilde{H}^{1/2+\sigma}(\Gamma_1)$ represent new smooth densities whereas α_i, $i=1,2$ are the unknown stress

<u>intensity factors</u>. χ_i, i=1,2 are two cut-off functions with $\chi_i \equiv 1$ in some neighbourhood of Z_i which will be specified later on. Inserting (5.1 - 5.3) into (1.9- 1.11) we find the modified system of integral equations

$$\begin{aligned}
&(I-K)_{22}w_o+R_{12}(\alpha_i,\phi_o)+\omega := \\
&:= w_o(z) - \frac{1}{\pi}\int_{\Gamma_2} w_o(\zeta)d\theta_z + \frac{1}{\pi}\int_{\Gamma_1}\phi_o(\zeta)\log|z-\zeta|ds_\zeta \\
&\quad + \sum_{i=1}^{2}\alpha_i[\rho_i^{1/2}\chi_i - \frac{1}{\pi}\int_{\Gamma_2}\rho_i^{1/2}\chi_i d\theta \\
&\quad - \frac{1}{2\pi}\int_{\Gamma_1}\rho_i^{-1/2}\chi_i \log|z-\zeta|ds_\zeta] + \omega \qquad (5.4)\\
&= \frac{1}{\pi}\int_{\Gamma_1} g_1 d\theta + \frac{1}{\pi}\int_{\Gamma_2}\tilde{g}_1 d\theta - \tilde{g}_1(z) \\
&\quad - \frac{1}{\pi}\int_{\Gamma} g_2^x \log|z-\zeta|\, ds_\zeta = F_1(z) \quad \text{for } z\in\Gamma_2 ,
\end{aligned}$$

$$\begin{aligned}
&V_{11}\phi_o - \frac{1}{2}\sum_{i=1}^{2}\alpha_i\rho_i^{-1/2}\chi_i(\zeta) + K_{21}(w_o + \sum_{i=1}^{2}\alpha_i\rho_i^{1/2}\chi_i) - \omega \\
&:= -\frac{1}{\pi}\int_{\Gamma_2}[\phi_o(\zeta) - \frac{1}{2}\sum_{i=1}^{2}\alpha_i\rho_i^{-1/2}\chi_i(\zeta)]\log|z-\zeta|ds_\zeta \\
&\quad + \frac{1}{\pi}\int_{\Gamma_2} w_o d\theta + \sum_{i=1}^{2}\alpha_i\frac{1}{\pi}\int_{\Gamma_2}\rho_i^{1/2}\chi_i\, d\theta - \omega \qquad (5.5)\\
&= g_1(z) - \frac{1}{\pi}\int_{\Gamma_1} g_1 d\theta - \frac{1}{\pi}\int_{\Gamma_2}\tilde{g}_1 d\theta + \frac{1}{\pi}\int_{\Gamma} g_2^x \log|z-\zeta|ds_\zeta \\
&= F_2(z) \quad \text{for } z\in\Gamma_1
\end{aligned}$$

and

$$\int_{\Gamma_1}[\phi_o - \frac{1}{2}\sum_{i=1}^{2}\alpha_i\rho_i^{-1/2}\chi_i]ds = -\int_{\Gamma} g_2^x ds = B . \qquad (5.6)$$

In (5.4), (5.5) we denote by $d\theta_z$ the kernel of the double layer potential,

$$d\theta_z(\zeta) := \frac{\partial}{\partial\nu_\zeta}(\log|z-\zeta|)ds_\zeta . \qquad (5.7)$$

Note that $d\theta_z$ is the total differential of the angle $\arg(\zeta-z)$.

In order to formulate the solvability properites of (5.4 - 5.6) let us introduce the new spaces of unknowns and right hand sides,

$$M_o^{1/2+\sigma} := \{ \{\alpha_i,\omega,\phi_o,w_o\} | \alpha_i,\omega\in\mathbb{R},\ \phi_o\in\tilde{H}^{1/2+\sigma}(\Gamma_1), \quad w_o\in H^{3/2+\sigma}(\Gamma_1)\ ,\ w_o(Z_i) = 0\ ,\ i = 1,2\} \tag{5.8}$$

and

$$F_o^{3/2+\sigma} := \{ \{F_1,F_2,B\} \in H^{3/2+\sigma}(\Gamma_2) \times H^{3/2+\sigma}(\Gamma_1) \times \mathbb{R} | \quad F_1(Z_i) + F_2(Z_i) = 0 \} \tag{5.9}$$

Both spaces are equipped with the corresponding norms.

A rather tedious analysis [44,73] provides the following theorem.

<u>Theorem 5.1: The mapping defined by the left hand sides of</u> (5.4 - 5.6) <u>is an isomorphism</u> $M_o^{1/2+\sigma} \to F_o^{3/2+\sigma}$ <u>for any</u> σ, $|\sigma| < \frac{1}{2}$.

Besides Theorem 5.1 one also needs a shift theorem for the Aubin-Nitsche Lemma which is also given in [44] and in [73, Theorem 2.5].

For Galerkin's procedure we use the finite elements (4.7), (4.8) with m=2 for the smooth parts w_o and ϕ_o in (5.4 - 5.6). For the collision points Z_i we require

$$Z_i \in \{z(j\cdot h)\ |j=0,1,2,\dots,N\}\ ,\ i=1,2\ . \tag{5.10}$$

For convenience let us introduce the following two sets of indices:

$$I_1 := \{j|0 \le j \le N \quad \mu_j|_{\Gamma_1} \not\equiv 0\ \}, \qquad I_2 := \{j|0 \le j \le N \quad \mu_j|_{\Gamma_2} \not\equiv 0\ \}. \tag{5.11}$$

Now we are in the position to define the subspaces on Γ_ℓ by

$$H_h(\Gamma_\ell) := \{\lambda_h = \sum_{j\ I_\ell} \gamma_j\mu_j(t)|_{\Gamma_\ell}\ \}\ \ell = 1,2\ , \tag{5.12}$$

and

$$\tilde{H}_h(\Gamma_\ell) := \{\tilde{\lambda}_h = \sum_{j\ I_\ell} \tilde{\gamma}_j\mu_j(t)|_{\Gamma_\ell} |\tilde{\lambda}_h(Z_i) = 0,\ i=1,2\}\ ,\ell=1,2. \tag{5.13}$$

In order to formulate the modified Galerkin method for (5.4 - 5.6) we first approximate the given functions g_ℓ by $g_{\ell h} \in H_h(\Gamma_\ell)$, $\ell = 1,2$, requiring

$$(g_{\ell h},\mu_j)_{L_2(\Gamma_\ell)} = (g_\ell,\mu_j)_{L_2(\Gamma_\ell)} \quad \text{for all} \quad j\in I_\ell\ . \tag{5.14}$$

Then $\tilde{g}_{\ell h} \in H_h(\Gamma_{\ell+1})$ with $\Gamma_3 := \Gamma_1$ are chosen arbitrarily satisfying

$$\tilde{g}_{\ell h}(Z_i) = g_{\ell h}(Z_i) \ , \ i=1,2 \ , \tag{5.15}$$

e.g. by linear functions of t .

For the smooth parts of the desired solutions we choose the approximations

$$w_{oh} = \sum_{j \in I_2} \tilde{\gamma}_j \mu_j(t) \text{ with } w_{oh}(Z_i) = 0 \ , \ i=1,2 \ ,$$
$$\psi_{oh} = \sum_{j \in I_1} \tilde{\beta}_j \mu_j(t) \text{ with } \phi_{oh}(Z_i) = 0 \ , \ i=1,2 \ . \tag{5.16}$$

Now the Galerkin equations for (5.4 - 5.6) read as

$$(\{w_{oh} - K_{22} w_{oh} + R_{12}(\tilde{\alpha}_i, \phi_{oh}) + \tilde{\omega}\} \ , \ \tilde{\lambda}_h)_{L_2(\Gamma_2)} = $$
$$= (F_1, \tilde{\lambda}_h)_{L_2(\Gamma_2)} \text{ for all } \tilde{\lambda}_h \in \tilde{H}_h(\Gamma_2) \ , \tag{5.17}$$

$$(V_{11}[\phi_{oh} - \frac{1}{2} \sum_{i=1}^{2} \tilde{\alpha}_i \rho_i^{-1/2} \chi_i] + K_{21}(w_{oh} + \sum_{i=1}^{2} \tilde{\alpha}_i \rho_i^{1/2} \chi_i)$$
$$- \tilde{\omega} \ , \ \tilde{\Xi}_h)_{L_2(\Gamma_1)} = (F_2, \tilde{\Xi}_h)_{L_2(\Gamma_1)} \tag{5.18}$$

for all $\Xi_h \in \tilde{H}_h(\Gamma_1)$ and $\Xi_h = \rho_i^{-1/2} \chi_i$, $i=1,2$,

$$\int_{\Gamma_1} [\phi_{oh} - \frac{1}{2} \sum_{i=1}^{2} \tilde{\alpha}_i \rho_i^{-1/2} \chi_i] ds = B \tag{5.19}$$

for $w_{oh} \in \tilde{H}_h(\Gamma_2)$ and $\phi_{oh} \in \tilde{H}_h(\Gamma_1)$.

For an asymptotic error analysis of the improved method (5.17 - 5.19) we need the approximation properties (3.9) for $u \in H^s(\Gamma_j)$, $\mu \in H_h(\Gamma_j)$, $j=1,2$, $m=2$ as well as for $u \in H^s(\Gamma_j) \cap \tilde{H}^1(\Gamma_j)$ and $\mu \in \tilde{H}_h(\Gamma_j)$ and also the corresponding inverse assumptions (3.10). In addition we need for

$$\Psi_h = \psi_{oh} + \sum_{i=1}^{2} \tilde{\alpha}_i \rho_i^{-1/2} \chi_i \ , \quad \psi_{oh} \in \tilde{H}_h(\Gamma_1)$$

the <u>generalized</u> <u>inverse</u> <u>assumption</u>

$$\|\Psi_h\|_{Z^s} \leq M h^{r-s-\varepsilon} \|\Psi_h\|_{Z^r} \tag{5.20}$$

with $-2 < r \leq s < 2$ and $\varepsilon > 0$ if $s < 0$ and $r \geq 0$, otherwise $\varepsilon = 0$. The proof in [73 , Lemma 1.5] is not complete. The complete proof can be found in [23] .

With (5.20) available, a simple modification of the results in [73] yields the following error estimates [44] :

Theorem 5.2: There exists a meshwidth $h_o > 0$ such that the Galerkin equations (5.17 - 5.19) are uniquely sovable for any h, $0<h\leq h_o$. For decreasing meshsize $h \to 0$ we have the asymptotic error estimates

$$\sum_{i=1}^{2} |\tilde{\alpha}_i-\alpha_i| + \|\phi_{oh}-\phi_o\|_{\tilde{H}^{t-1}(\Gamma_1)} + \|v_{oh} - v_o\|_{H^t(\Gamma_2)} + |\tilde{\omega}-\omega|$$
$$\leq ch^{r-t-\varepsilon}\{\|g_1\|_{H^r(\Gamma_1)} + \|g_2\|_{H^{r-1}(\Gamma_2)}\} \tag{5.21}$$

for $1<t\leq r<2$ and any $\varepsilon> 0$

and

$$\|\phi_{oh}-\phi_o - \sum_{i=1}^{2} \tfrac{1}{2}(\tilde{\alpha}_i-\alpha_i)\rho_i^{-1/2}\chi_i\|_{\tilde{H}^{t-1}} + |\tilde{\omega}-\omega|$$
$$+ \|v_{oh}-v_o + \sum_{i=1}^{2} (\tilde{\alpha}_i-\alpha_i)\rho_i^{1/2}\chi_i\|_{H^t(\Gamma_2)} \tag{5.22}$$
$$\leq ch^{r-t-\varepsilon}\{\|g_1\|_{H^r(\Gamma_1)} + \|g_2\|_{H^{r-1}(\Gamma_2)}\}$$

for $-1<t\leq r<2$, $t<1$ and any $\varepsilon> 0$ if $0<t<1$ and $\varepsilon=0$ if $-1<t\leq 0$. The constant c is independent of h , ϕ_o, v_o , α_i , ϕ_{oh}, v_{oh} and $\tilde{\alpha}_i$ but may depend on ε .

For the numerical treatment of (5.17 - 5.19) via Galerkin-collocation note that in (5.4 - 5.6) on both sides appear the same types of integrals and corresponding double integrals if g_ℓ, $\tilde{g}_\ell$ are replaced by $g_{\ell h}$, $\tilde{g}_{\ell h}$ according to (5.15) . In [44] we perform the Galerkin collocation and the corresponding error analysis carefully such that Theorem 5.2 is prevailed to the Galerkin collocation. Special care has to be taken of the singularity functions $\rho_i^{-1/2}$.

6. SOME NUMERICAL RESULTS

Numerical computations with Galerkin collocation have been made for the equations (1.1), (1.2) and corresponding systems of the more general form

$$-\int_\Gamma \log|z-\zeta|\underline{u}(\zeta)ds_\zeta + \int_\Gamma L(z,\zeta)\underline{u}(\zeta)ds_\zeta = \underline{f}(z) + \underline{\omega} \ , \quad \int_\Gamma \underline{u}(\zeta)ds_\zeta = \underline{B} \ , \ z,\zeta \in \Gamma \tag{6.1}$$

in [33,34,39,69] where $\underline{u}$ and $\underline{f}$ are vector valued functions and $\underline{\omega}$ and $\underline{B}$ are vectors. L is a given smooth matrix kernel.

We also made computations for the mixed boundary value problem (1.7) using Galerkin collocation (5.17 - 5.19) in [43,44].

The computations have been carried out on the IBM 370-168 computer at the Technische Hochschule Darmstadt.

Example 6.1: Symm's method in conformal mapping [64,65]. Interior mappings of ellipses [27 , p. 161, Table 14a] onto the unit circle with $m = 2$ and 60 grid points on Γ :

Eccentricity	0.2	0.5	0.83
Max. absolute error	4×10^{-3}	7×10^{-5}	3×10^{-6}

Interior mappings of reflected ellipses [27, p. 264, Example 2] with $m = 2$ and 36 grid points on Γ :

Eccentricity	0.25	0.6	0.65
Max. absolute error	3×10^{-3}	10^{-4}	7×10^{-5}

For further examples see [35,69] .

Example 6.2: Exterior Stokes problem [36]. The exterior Stokes problem leads to a system of the form (6.1). Numerical computations have been made for ellipses Γ with $m = 2$ (and other m) and 20 respectively 40 grid points on Γ :

Eccentricity	0.6		0.9	
Grid points	20	40	20	40
Max. absolute error	2×10^{-6}	10^{-7}	10^{-5}	10^{-7}

For further examples see [35,69,71] .

Example 6.3: Mixed boundary value problem (1.7) [43,44]. Numerical computations have been made for the unit circle Γ and $Z_\ell = (0,(-1)^\ell)$ with the harmonic function

$$U = \mathrm{Re}\,\{\sqrt{2}\,\frac{\sqrt{(x+iy)^2+1}}{x+iy+1} - \frac{1-x-iy}{1+x+iy}\}\quad (i=\sqrt{-1})\ .$$

Number of grid points N+1 :	40	80	160
Max. absolute error of $u\|_{\Gamma_2}$:	7×10^{-2}	10^{-2}	10^{-3}
Max. absolute error of $\phi_o\|_{\Gamma_1}$:	1.5×10^{-1}	7×10^{-2}	4×10^{-2}
Max. absolute error of α_i :	7×10^{-2}	3×10^{-2}	2×10^{-2}

For further examples see [43,44,71].

REFERENCES:

1. ABOU EL-SEOUD, M.S., *Numerische Behandlung von schwach singulären Integralgleichungen erster Art.* Dissertation, Technische Hochschule Darmstadt, Germany (1979).

2. ABOU EL-SEOUD, M.S., Kollokationsmethode für schwach singuläre Integralgleichungen erster Art. *ZAMM* 59, T45-T47 (1979).

3. ADAMS, R.A., *Sobolev Spaces.* Academic Press, New York (1975).

4. ALEKSIDZE, M.A., *The Solution of Boundary Value Problems with the Method of the Expansion with Respect to Non-orthonormal Functions.* Nauka, Moscow (Russian)(1978).

5. ARTHUR, D.W., The solution of Fredholm integral equations using spline functions. *J. Inst. Maths. Applics.* 11, 121-129 (1973).

6. ATKINSON, K.E., *A Survey of Numerical Methods for the Solution of Fredholm Integral Equations of the Second Kind.* Philadelphia, SIAM (1976).

7. AUBIN, J.P., *Approximation of Elliptic Boundary-Value Problems.* Wiley-Interscience, New York (1972).

8. AZIZ, A.K. and KELLOGG, R.B., Finite element analysis of a scattering problem. To appear.

9. BABUSKA, I., The finite element method with Lagrangian multipliers. *Num. Math.* 20, 179-192 (1973).

10. BABUSKA, I. and AZIZ, A.K., Survey lectures on the mathematical foundations of the finite element method. pp. 3-359 of A.K. Aziz (Ed.), *The Mathematical Foundation of the Finite Element Method with Applications to Partial Differential Equations.* Academic Press, New York (1972).

11. BAKER, C., *The Numerical Treatment of Integral Equations.* Clarendon, Oxford (1977).

12. BALDINO, P.R., *An integral equation solution of the mixed problem for the Laplacian in* $\mathbb{R}^3$. Rapport Interne No. 48, Centre de Mathématiques Appliquées, Ecole Polytechnique, Palaiseau, France (1979).

13. BLUE, J., Boundary integral solutions of Laplace's equations. *The Bell System Techn. Journal* 57, 2797-2821 (1978).

14. BRAMBLE, J. and SCHATZ, A., Rayleigh-Ritz-Galerkin methods for Dirichlet's problem using subspaces without boundary conditions. *Comm. Pure Appl. Math.* 23, 653-675 (1970).

15. BREBBIA, C.A., *The Boundary Element Method in Engineering.* Pentech Press, London (1978).

16. BREBBIA, C.A. (Ed.), *New Developments in Boundary Element Methods.* CLM Publications lim. 125 High Street Southampton SO1OAA England (1980).

17. CALDERON, A.P. and ZYGMUND, A., Singular integral operators and differential equations. *Amer. J. Math.* 79, 901-921 (1957).

18. CHANDLER, G.A., *Superconvergence of numerical solutions to second kind integral equations.* Ph.D. Thesis, Australian National University (1979).

19. CIARLET, P.G., *The Finite Element Method for Elliptic Problems.* North Holland, Amsterdam (1978).

20. COLLATZ, L., *The Numerical Treatment of Differential Equations.* Springer-Verlag, Berlin (1959).

21. COLLI-FRANZONE, P., Approssimazione mediante il metodo di penalizzazione, die problemi misti di Dirichlet-Neumann per operatori lineari ellittici del secondo ordine. *Boll. Un. Mat. Ital.* (4) 7, 229-250 (1973).

22. COSTABEL, M. and STEPHAN, E., On the boundary element method for polygonal domains. R-208H, to appear in *Proc. of the fourth IMACS - Conf.* (1981).

23. COSTABEL, M. and STEPHAN, E., *Boundary integral equations for mixed boundary value problems in polygonal domains and Galerkin approximation.* Preprint No. 593, Fachbereich Mathematik, Technische Hochschule D-61 Darmstadt, Schloßgartenstr. 7 , Germany (1981).

24. ESKIN, G.I., *Boundary Problems for Elliptic Pseudo-Differential Operators,* (Russian) Nauka, Moscow (1973).

25. ESKIN, G., BOGOMILNII, A. and ZUCHOWIZKII, S., Numerical solution of the stamp problem. *Comp. Meth. App. Mech. Eng.* 15, 149-159 (1978).

26. FICHERA, G., Analisi essistenziale per le soluzioni dei problemi al contorno misti relativi alle equazione ed ai sistemi di equazioni del secondo ordine di tipo ellittico auto aggiunti. *Ann. Scuola Norm. Sup. Pisa, Ser.* III, 1 75-100 (1949).

27. GAIER, D., *Konstruktive Methoden der konformen Abbildung.* Springer-Verlag, Berlin (1964).

28. GIROIRE, J. and NEDELEC, J.C., Numerical solution of an exterior Neumann problem using a double layer potential. *Math. of Comp.* 32, 973-990 (1978).

29. GRISVARD, P., *Boundary Value Problems in Non-Smooth Domains.* University of Maryland, Dept. Mathematics, College Park, Md. 20742, Lecture Notes 19, (1980).

30. HÄMMERLIN, G. and SCHUMAKER, L.L., *Procedures for kernel approximation and solution of Fredholm integral equations of the second kind.* CNA Report 128, Center Numerical Analysis, Univ. Texas, Austin (1977).

31. HILDEBRANDT, ST. and WIENHOLTZ, E., Constructive proofs of representation theorems in separable Hilbert space. *Comm. Pure Appl. Math.* 17, 369-373 (1964).

32. HÖRMANDER, L., *Linear Partial Differential Operators.* Springer-Verlag, Berlin (1969).

33. HSIAO, G.C., KOPP, P. and WENDLAND, W.L., A Galerkin collocation method for some integral equations of the first kind. *Computing* 25, 89-130 (1980).

34. HSIAO, G.C., KOPP, P. and WENDLAND, W.L., The synthesis of the collocation and the Galerkin method applied to some integral equations of the first kind. pp. 122-136 of C.A. Brebbia (Ed.), *New Developments in Boundary Element Methods.* CML Publ. Southampton, (1980).

35. HSIAO, G.C., KOPP, P. and WENDLAND, W.L., Some applications of a Galerkin-collocation method for integral equations of the first kind. *In preparation.*

36. HSIAO, G.C. and MACCAMY, R.C., Solution of boundary value problems by integral equations of the first kind. *SIAM Review* 15, 687-705 (1973).

37. HSIAO, G.C. and WENDLAND, W.L., A finite element method for some integral equations of the first kind. *J. Math. Anal. Appl.* 58, 449-481 (1977).

38. HSIAO, G.C. and WENDLAND, W.L., The Aubin-Nitsche lemma for integral equations. To appear in the *Journal of Integral Equations*.

39. HSIAO, G.C. and WENDLAND, W.L., Super approximation for boundary integral methods. R.-208 D, to appear in *Proc. of the fourth IMACS - Conf.* (1981).

40. JAWSON, M.A. and SYMM, G.T., *Integral Equation Methods in Potential Theory and Elastostatics*. Academic Press London (1977).

41. JOHNSON, H.L., An integral equation formulation of a mixed boundary value problem on a sphere. *SIAM J. Math. Analysis*, 6, 417-426 (1975).

42. KOHN, J.J. and NIRENBERG, L., On the algebra of pseudo-differential operators. *Comm. Pure Appl. Math.* 18, 269-305 (1965).

43. LAMP, U., SCHLEICHER, T., STEPHAN, E. and WENDLAND, W.L., The boundary integral method for a plane mixed boundary value problem. R.-208 B, to appear in *Proc. of the fourth IMACS - Conf.* (1981).

44. LAMP, U., SCHLEICHER, T., STEPHAN, E. and WENDLAND, W.L., Galerkin Collocation for an improved boundary element method for a plane mixed boundary value problem. *In preparation.*

45. MIKHLIN, S.G., *Variationsmethoden der Mathematischen Physik*. Akademie-Verlag, Berlin (1962).

46. MIKHLIN, S.G. and PRÖSSDORF, S., *Singuläre Integraloperatoren*. Akademie-Verlag, Berlin (1980).

47. NEDELEC, J.C., Curved finite element methods for the solution of singular integral equations on surfaces in $\mathbb{R}^3$. *Comp. Math. Appl. Mech. Engin.* 8, 61-80 (1976).

48. NEDELEC, J.C., *Approximation des équations intégrales en mécanique et en physique*. Lecture Notes, Centre de Mathématiques Appliquées. Ecole Polytechnique, Palaiseau, France (1977).

49. NITSCHE, J.A., Ein Kriterium für die Quasi-Optimalität des Ritzschen Verfahrens. *Numer. Math.* 11, 346-348 (1968).

50. NITSCHE, J.A., Zur Konvergenz von Näherungsverfahren bezüglich verschiedener Normen. *Num. Math.* 15, 224-228 (1970).

51. NOBLE, B., *A Bibliography on: "Methods for solving integral equations"*. Math. Res. Center Tech. Report 1176 and 1177, Madison, Wis. U.S.A. (1971).

52. PRENTER, P.M., A collocation method for the numerical solution of integral equations. *SIAM J. Numer. Anal.* 10, 570-581 (1973).

53. PRÖSSDORF, S., *Some Classes of Singular Equations.* North Holland, Amsterdam (1978).

54. PRÖSSDORF, S. and SCHMIDT, G., A finite element collocation method for singular integral equations. *To appear.*

55. PRÖSSDORF, S. and SILBERMANN, B., *Projektionsverfahren und die näherungsweise Lösung singulärer Gleichungen.* Teubner, Leipzig (1977).

56. RICHTER, G.R., Numerical solution of integral equations of the first kind with nonsmooth kernels. *SIAM J. Numer. Anal.* 17, 511-522 (1978).

57. RICHTER, G.R., Superconvergence of piecewise polynomial Galerkin approximations for Fredholm integral equations of the second kind. *Numer. Math.* 31, 63-70 (1978).

58. SCHÄFER, E., Fehlerabschätzungen für Eigenwertnäherungen nach der Ersatzkernmethode bei Integralgleichungen. *Numer. Math.* 32, 281-290 (1979).

59. SEELEY, R., Topics in pseudo-differential Operators. pp. 169-305 of L. Nirenberg (Ed.), *Pseudo-Differential Operators*, C.I.M.E., Cremonese, Roma (1969).

60. SHAW, R. et.al. (Ed.), *Innovative Numerical Analysis for the Engineering Sciences.* The University Press of Virginia, Charlottesville (1980).

61. SNEDDON, I.N., *Mixed Boundary Value Problems in Potential Theory.* North Holland, Amsterdam (1966).

62. STEPHAN, E., The boundary integral method for the three-dimensional mixed boundary value problem of the Laplacian. *In preparation.*

63. STEPHAN, E. and WENDLAND, W.L., Remarks to Galerkin and least squares methods with finite elements for general elliptic problems. *Lecture Notes Math.* 564, 461-471, Springer, Berlin (1976); *Manuscripta Geodaetica* 1, 93-123, (1976).

64. SYMM, G.T., An integral equation method in conformal mapping. *Numer. Math.* 9, 250-259 (1966).

65. SYMM, G.T., Numerical mapping of exterior domains. *Numer. Math.* 10, 437-445 (1967).

66. TREVES, F., *Introduction to Pseudodifferential and Fourier Integral Operators I*. Plenum Press, New York and London, (1980).

67. VAINIKKO, G., On the question of convergence of Galerkin's method. *Tartu Rükl. Ül. Toim.* 177, 148-152 (1965).

68. VORONIN, V.V. and CECOHO, V.A., An interpolation method for solving an integral equation of the first kind with a logarithmic singularity. *Dokl. Akad. Nauk SSR* 216; *Soviet Math. Dokl.* 15, 949-952 (1974).

69. WENDLAND, W.L. On Galerkin collocation methods for integral equations of elliptic boundary value problems. pp. 244-275 of J. Albrecht and L. Collatz (Ed.), *Numerical Treatment of Integral Equations. Intern. Ser. Num. Math.,* Birkhäuser Basel, 53, (1980).

70. WENDLAND, W.L., Asymptotic accuracy and convergence. To appear in C. Brebbia (Ed.): *Boundary Element Methods, the State of the Art*, Pentech Press, London.

71. WENDLAND, W.L., *Asymptotic Convergence of Boundary Element Methods and Integral Equation Methods for Mixed Boundary Value Problems*. Lecture Notes, University of Maryland, College Park, to appear.

72. WENDLAND, W.L. and STEPHAN, E., Boundary integral method for mixed boundary value problems. pp. 543-554 of R. Shaw et.al. (Ed.), *Innovative Numerical Analysis in the Engineering Sciences*. The University Press of Virginia, Charlottesville (1980).

73. WENDLAND, W.L., STEPHAN, E. and HSIAO, G.C., On the integral equation method for the plane mixed boundary value problem of the Laplacian. *Math. Methods in the Applied Sciences* 1, 265-321 (1979).

HIERARCHICAL FINITE ELEMENT APPROACHES, ERROR ESTIMATES AND ADAPTIVE REFINEMENT

*O.C. Zienkiewicz, *D.W. Kelly, *J. Gago and + I. Babuška

*_Department of Civil Engineering, University College of Swansea, U.K., +Institute for Physical Science and Technology, University of Maryland, U.S.A._

1. INTRODUCTION

Despite a continuing effort to identify optimal finite element grids most of the finite element computations today still rely on an a-priori mesh design based on the user's intuition and experience. Once the mesh is designed, however, there seems to be growing evidence that high order isoparametric elements provide a better refinement process than mesh subdivision.

This paper is concerned with the identification of the discretization error in finite element solution and the definition of optimal refinement processes. The advantages and limitations of the hierarchical approach [58, 36-40] will be discussed and it will be shown how the intelligent enrichment of the finite element grid can be left to the computer if a capacity for a-posteriori error estimation exists within the finite element code [1-14, 37-40, 46-49].

2. HIERARCHICAL FINITE ELEMENTS

2.1 Hierarchical Shape Functions

The concept of hierarchical finite elements dates from 1970 and these were first introduced [58] with the objective of creating elements that would allow an easy transition from a region where a finite element solution required a high degree of refinement to a region where a lower degree of refinement was sufficient. Other advantages soon become apparent and it will be shown here that the hierarchical concept is very powerful in allowing an error indication capability that can be used for adaptive mesh refinement [36-39, 46-49].

We will begin by defining hierarchical finite elements as those in which successive refinements are additive in the manner of additional terms in a Fourier series. It follows that the "stiffness" matrix corresponding to the hierarchical element at a certain level of refinement is a sub-matrix of the "stiffness" matrix corresponding to a higher level of refinement.

This leads to matrix approximation equations of the type:

$$K_{11}\, \underset{\sim}{a}_1^{(1)} + \underset{\sim}{q}_1 = \underset{\sim}{0} \tag{2.1}$$

and

$$\begin{bmatrix} K_{11} & K_{12} \\ K_{21} & K_{22} \end{bmatrix} \begin{Bmatrix} \underset{\sim}{a}_1^{(2)} \\ \underset{\sim}{a}_2^{(2)} \end{Bmatrix} + \begin{Bmatrix} \underset{\sim}{q}_1 \\ \underset{\sim}{q}_2 \end{Bmatrix} = \underset{\sim}{0} \tag{2.2}$$

where (2.1) is the finite element equilibrium equation corresponding to a certain formulation, and equation (2.2) is the same equation corresponding to a higher order or refinement. The matrices K_{11} and $\underset{\sim}{q}_1$ remain unchanged.

Consider solving the linear differential equation

$$A(\phi) \equiv L\,\phi + q = 0 \quad \text{in } \Omega \tag{2.3}$$

with boundary condition

$$B(\phi) \equiv \overline{L}\,\phi + s = 0 \quad \text{on } \Gamma \tag{2.4}$$

where Γ is boundary of Ω.

Approximate the solution ϕ of (2.3), (2.4) by the solution $\hat{\phi}$ in the form

$$\hat{\phi} = \sum_{m=1}^{M} a_m N_m \tag{2.5}$$

with a proper choice of the basis functions N_i, $i = 1, 2, \ldots, m$.

The coefficients a_m will be determined from the condition that

$$\int_\Omega W_e\,[L\hat{\phi} + q]\,d\Omega + \int_\Gamma \overline{W}_e\,[\overline{L}\hat{\phi} + s]\,d\Gamma = 0 \tag{2.6}$$

holds for all W_e and $\overline{W}_e$, where W_e and $\overline{W}_e$ are functions suitable for the problem (2.3), (2.4). In addition (2.6) is meant in a generalized way so that it can be used even if LN_i does not exist in a classical sense (see [59]).

The p-version of the finite element method is defined for a sequence of solutions as the enrichment of the trial and test function set through the introduction of shape functions N_m corresponding to higher order polynomial degree, p, while the h version is defined as the approach equivalent to the reduction of the finite element mesh size h, maintaining constant the polynomial order [8-9, 46-49].

In 1-D the p-version of the finite element method involves

the addition to the linear element of quadratic, cubic, etc. trial functions as shown in Figure 1. Considering the trial functions expressed as a function of the local element coordinates with $-1 \leq \xi \leq 1$ we may have for the p-version the following hierarchical shape functions. For the linear terms

$$N_0 = \frac{1-\xi}{2}$$

$$N_1 = \frac{1+\xi}{2} \tag{2.7}$$

For the quadratic the 'obvious' form will be a quadratic that goes to zero at points $\xi = -1$ and $\xi = 1$ so that it does not interfere with the a_o and a_1 coefficients. Thus

$$N_2 = (\xi - 1)(\xi + 1) \tag{2.8}$$

For the cubic and higher order elements the only restriction is that the shape functions will go to zero at $\xi = -1$, $\xi = 1$, so that we write for the p-th order

$$N_p = (\xi - 1)(\xi + 1)\,\xi^{p-2}, \quad p \geq 2 \tag{2.9}$$

The above set is obviously not unique and many alternatives are possible.

A convenient form of hierarchical functions is given in Peano et al. [38] as

$$N_p = \frac{1}{p!}(\xi^p - b) \qquad \begin{array}{l} b = 1 \text{ if } p \text{ even} \\ b = \xi \text{ if } p \text{ odd} \end{array} \tag{2.10}$$

where $p \geq 2$ is the polynomial order.

It is easy to observe how the associated variables have the meaning of higher derivatives of $\hat{\phi}$ i.e.,

$$a_{p+1} = \frac{d^p \hat{\phi}}{d\xi^p} \tag{2.11}$$

Although the successive importance of these hierarchical variables diminishes, the optimal form is one which gives an orthogonal set of shape functions in relation to the energy inner product. Such a set of shape functions will be given by integrals of the Legendre polynomials in the following form

$$N_p = \frac{1}{(p-1)!}\,\frac{1}{2^{p-2}}\,\frac{d^{p-2}}{d\,\xi^{p-2}}(1-\xi^2)^{p-1} \tag{2.12}$$

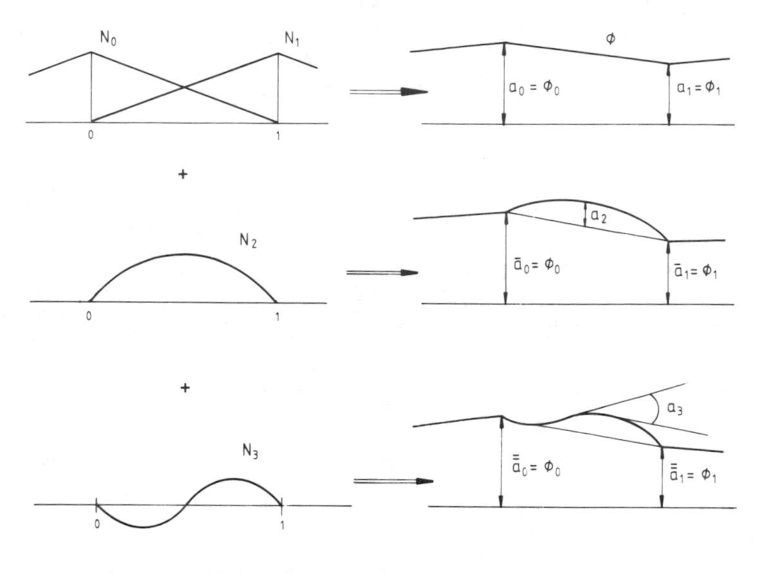

FIG. 1. One dimensional hierarchical elements for the p-version of the finite element method

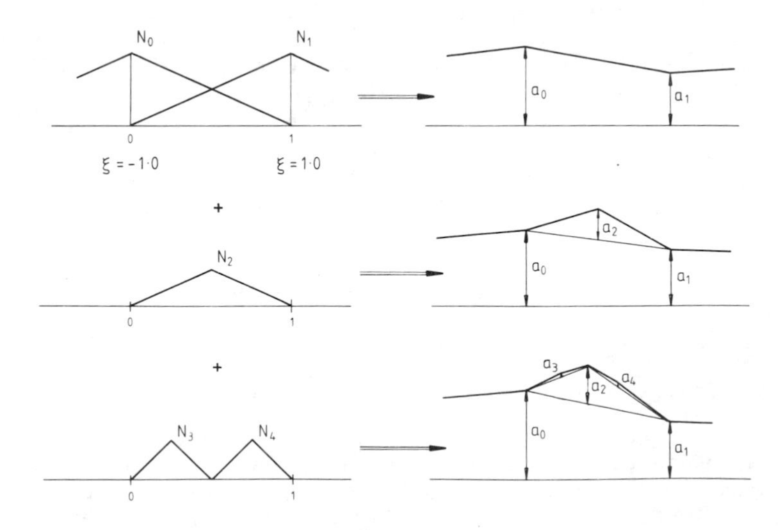

FIG. 2. One dimensional hierarchical elements for the h-version of the finite element method

if the differential equation $-\phi'' + q = 0$ is considered.

This set was introduced by Zienkiewicz et al. [58]. Again, the multipliers have the meaning of a measure of the higher derivatives of $\hat{\phi}$ at the centre of the finite element. The advantage of this set of hierarchical functions is that for 1-D problems the coupling between different higher order degrees of freedom is non-existent. We will discuss this point further in Section 2.2. We shall also see that this set of hierarchical functions will play a very important role in the error analysis study.

For the h-version there is also an infinite number of hierarchical refinement possibilities corresponding to the sub-division of the initial element in equal or unequal parts. If we consider a hierarchical refinement corresponding to Figure 2, we will have as shape functions

$$N_0 = \frac{1-\xi}{2}$$

$$N_1 = \frac{1+\xi}{2} \tag{2.13}$$

$$N_2 = \begin{cases} \xi+1 & \text{if} \quad \xi \le 0 \\ -\xi+1 & \text{if} \quad \xi \ge 0 \end{cases}$$

etc. Considering Figure 2 it is obvious that the physical meaning of the linear hierarchical variables is a relative displacement set.

Once the one-dimensional interpolation formulae have been established the generation of hierarchical shape functions for rectangular elements is almost trivial as,

a) the corner node functions are simply bi-linear products, and

b) 'hierarchical' functions of the type defined above are always zero at the corner nodes.

Polynomial shape functions of all orders in two dimensions can be obtained by simple products of the one-dimensional formulae, but in general losing the properties of the orthogonality mentioned above. The identity of hierarchical variables on any element side with those on the adjacent element then automatically guarantees the uniqueness of the polynomial along that side.

The three-dimensional case and the triangular based finite elements are just special cases of the concepts expanded above and we direct the interested reader to [36, 62].

The hierarchical and non-hierarchical shape functions for the h and p versions of the finite element method are presented in Figure 3 and Figure 4 for two-dimensional problems. It is easy to show that a direct transformation from the hierarchical to the non-hierarchical formulation and vice-versa is possible. For example, it is possible to transform a quadratic serendipity

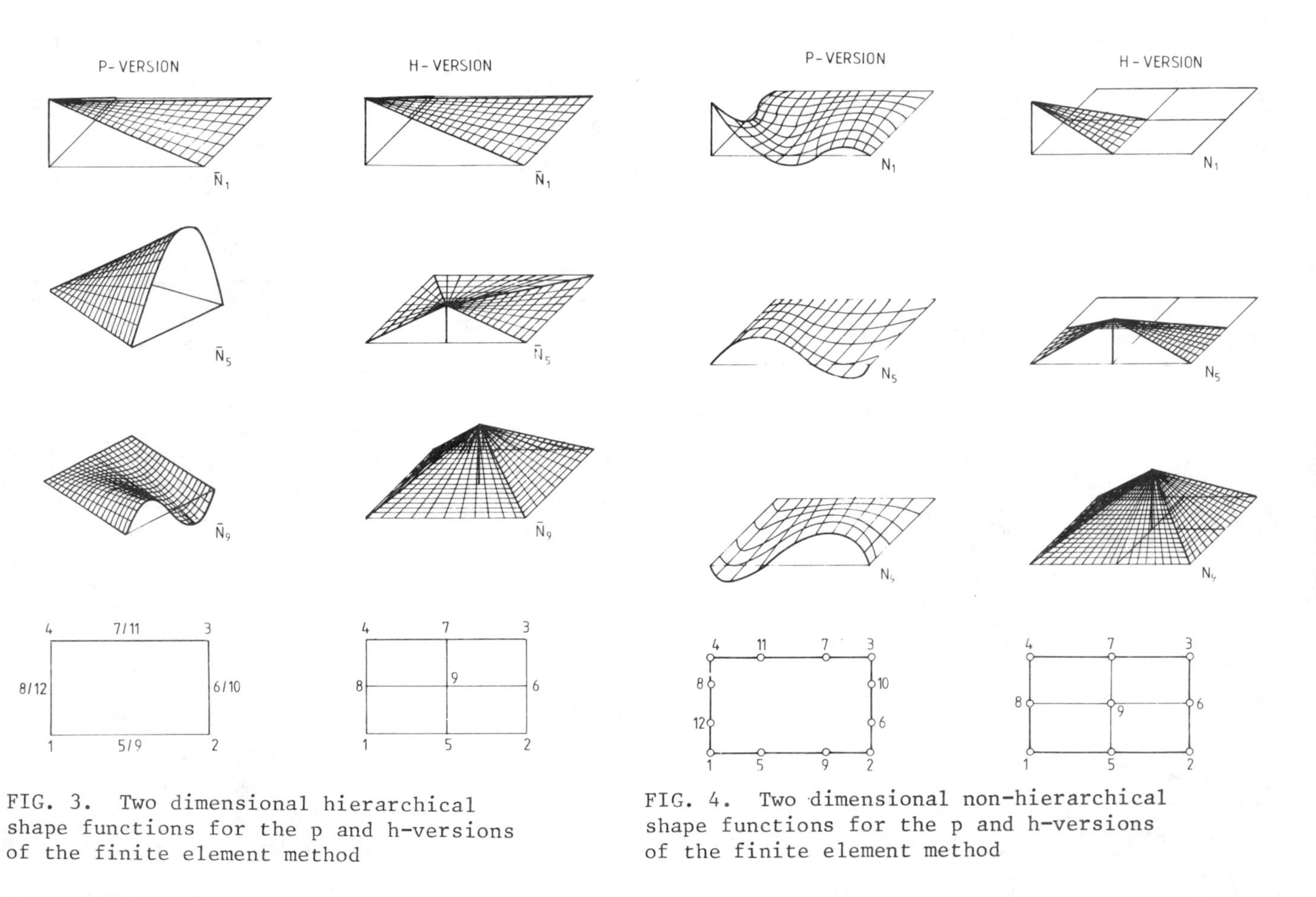

FIG. 3. Two dimensional hierarchical shape functions for the p and h-versions of the finite element method

FIG. 4. Two dimensional non-hierarchical shape functions for the p and h-versions of the finite element method

element to hierarchical form by a matrix transformation after the non-hierarchical matrix has been determined. It should also be noted that when the refinement is to the same order in Figures 1 to 4, the finite element solution $\hat{\phi}$ in (2.5) from the hierarchic and non-hierarchic formulations will be identical.

2.2 *Advantages of Hierarchical Formulations*

In 1-D the advantages of a hierarchical formulation over a non-hierarchical one are obvious because we can obtain a system of equations of the form

$$\begin{bmatrix} K_{11} & K_{12} & 0 & 0 & 0 & . & 0 \\ K_{21} & K_{22} & 0 & 0 & 0 & . & 0 \\ 0 & 0 & K_{33} & 0 & 0 & . & 0 \\ 0 & 0 & 0 & K_{44} & 0 & . & 0 \\ 0 & 0 & 0 & 0 & K_{55} & . & 0 \\ . & . & . & . & . & . & . \\ 0 & 0 & 0 & 0 & 0 & . & K_{nn} \end{bmatrix} \begin{Bmatrix} \underset{\sim}{a}_1 \\ \underset{\sim}{a}_2 \\ \underset{\sim}{a}_3 \\ \underset{\sim}{a}_4 \\ \underset{\sim}{a}_5 \\ . \\ \underset{\sim}{a}_n \end{Bmatrix} + \begin{Bmatrix} \underset{\sim}{q}_1 \\ \underset{\sim}{q}_2 \\ \underset{\sim}{q}_3 \\ \underset{\sim}{q}_4 \\ \underset{\sim}{q}_5 \\ . \\ \underset{\sim}{q}_n \end{Bmatrix} = \underset{\sim}{0} \qquad (2.14)$$

where K_{ii} are diagonal matrices if the functions are of a suitable orthogonal form. This implies an improved conditioning of the assembled system of equilibrium equations, compared to the non-hierarchical formulation which does not have the same strong diagonal character and, secondly, the possibility of a direct solution for the hierarchical variables.

In 2-D we expect the same advantages to hold although the system of equations is now not completely orthogonal and has the form

$$\begin{bmatrix} K_{11} & K_{12} \\ K_{21} & K_{22} \end{bmatrix} \begin{Bmatrix} \underset{\sim}{a}_1 \\ \underset{\sim}{a}_2 \end{Bmatrix} + \begin{Bmatrix} \underset{\sim}{q}_1 \\ \underset{\sim}{q}_2 \end{Bmatrix} = \underset{\sim}{0} \qquad (2.15)$$

The first advantage noted above is carried to the two-dimensional case by the fact that for each element the conditioning of the hierarchical stiffness matrix is better than the conditioning of the non-hierarchical one. The off-diagonal links are weakened, implying an overall better conditioning of the resulting system of equations. This is in fact observed by numerous authors [30, 53-56], who have proved that a relative displacement formulation of the finite element method yields better conditioned matrices than the classic total variable approach. Indeed, the same can be said of so-called local-global element forms.

In relation to the second advantage we can say that the hierarchical formulation is optimal because it allows for all the information to be passed from one discretization level to the second discretization level once a mesh refinement is decided. Also the implicit substructuring existent in this multi-level formulation allows for the very effective use of block iteration solution schemes. This, associated with the better conditioning of the overall system, will imply a fast rate of convergence for the iterative equation solver. Wachspress two-level elements [54] are another way of achieving these objectives although on a two-level theory, rather than a multi-level one. See also the local-global formulation of Mote [30] as another possible form of hierarchical formulation.

3. ERROR ANALYSIS (A HIERARCHICAL APPROACH)

The function $\hat{\phi}$ of the form (2.5) is not in general the exact solution of the problem. Therefore we cannot, in general, have

$$L\hat{\phi} + q = r \equiv 0 \tag{3.1}$$

$$\bar{L}\hat{\phi} + s = \rho \equiv 0 \tag{3.2}$$

(if (3.1) (3.2) would be satisfied, we would have $\hat{\phi} = \phi$).

Assume for simplicity that (3.2) (i.e. $\rho = 0$) holds; then $r \neq 0$. Function r usually (e.g. if L is a second-order equation and N_i have not continuous derivative) can be written in the form

$$r = r_1 + r_2 \tag{3.3}$$

where r_1 is the (usual) regular part of the residuum inside every element and r_2 is the singular part (Dirac) function concentrated on the interface between elements with physical interpretation of concentrated forces whose origin is indicated in Figure 5.

Considering e.g. equation

$$-\phi'' + q = 0 \quad \phi(0) = \phi(L) = 0 \tag{3.4}$$

and using piecewise linear elements (as in Figure 5) we get the regular part

$$r_1 = q$$

and the singular part r_2 are the forces (Dirac function) concentrated at the nodal point with the magnitude of the jump of the derivative of the approximate solution on the given nodal point.

For the potential problem in two dimensions the Dirac function is concentrated on the element interface and has the magnitude

$$J_s = \frac{\partial \hat{\phi}}{\partial n}_1 - \frac{\partial \hat{\phi}}{\partial n}_2$$

where $\frac{\partial \hat{\phi}}{\partial n}_i$ i=1,2 is the normal derivative on the left and right side of the interface. In the case of elasticity, J_s is the traction discontinuity between elements.

3.1 Error Definition in the Energy Norm

Denoting $e = \phi - \hat{\phi}$, the error of the finite element solution, then obviously subtracting (2.3), (3.1) and (2.4), (3.2) we get

$$Le = -r$$

$$\bar{L}e = -\rho \tag{3.5}$$

and we assume for simplicity as before that $\rho = 0$. The goal is now to measure the magnitude of e. If the set of the trial functions and test functions is the same, then for a suitable class of linear problems we can define

$$||e||_E = \left[\int e \, Le \, d\Omega \right]^{\frac{1}{2}} = \left[\int (\phi - \hat{\phi}) \, L(\phi - \hat{\phi}) d\Omega \right]^{\frac{1}{2}} \tag{3.6}$$

the so-called energy norm, and through it we can measure the magnitude of e.

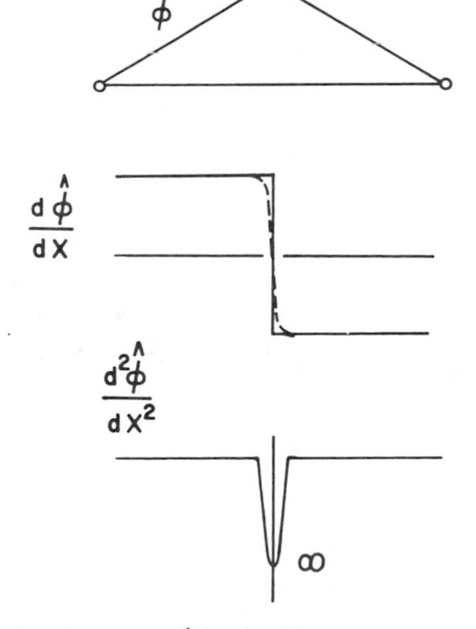

FIG.5. Interface residuals for error analysis.

Expanding $||e||_E^2$ we have

$$||e||_E^2 = \int \phi \, L\phi \, d\Omega + \int \hat{\phi} \, L\hat{\phi} \, d\Omega - \int \phi \, L\hat{\phi} \, d\Omega - \int \hat{\phi} L\phi \, d\Omega \quad (3.7)$$

Note that by equation (2.6) with $W_e = \hat{\phi}$ we can write

$$\int \hat{\phi} \, L\hat{\phi} \, d\Omega \equiv - \int \hat{\phi} \, q \, d\Omega \quad (3.8)$$

and by equation (2.3) $L\phi = - q$. Substituting in (3.7),

$$||e||_E^2 = - \int \phi (L\hat{\phi} + q) \, d\Omega \quad (3.9)$$

or $$||e||_E^2 = - \int \phi \, r \, d\Omega \quad (3.10)$$

Using once more (2.6) and (3.8) we get

$$\int \hat{\phi} \, r \, d\Omega = 0 \quad (3.11)$$

so we can write

$$||e||_E^2 = - \int (\phi - \hat{\phi}) \, r \, d\Omega \quad (3.12)$$

or $$||e||_E^2 = - \int e \, r \, d\Omega \quad \text{where } e = \phi - \hat{\phi} \quad (3.13)$$

We remark that (3.11) has a sense of a self equilibration of the residual forces.

3.2 Error Indication in One Dimension

We now come to the crux of the matter. How can we estimate the error using expressions such as (3.12) without knowledge of ϕ?

It is clear that what is needed is an approximation to ϕ (or e) on a local base because St.Venant's Principle ensures the effect of the equilibrating residuals will be local.

Employing hierarchical modes provides one possibility. On every element I_j we set

$$\phi \simeq \hat{\hat{\phi}} = \hat{\phi} + N_{i+1} \, a_{i+1} \quad (3.14)$$

where N_{i+1} is the (as yet not used) next hierarchical shape function. Now we could write

$$||e||^2_{E(I_j)} \simeq -a_{i+1} \int N_{i+1} \; r \; d\Omega \tag{3.15}$$

where $||\cdot||_{E(I_j)}$ is the norm of the error of the element I_j. This process obviously needs an estimate of a_{i+1}.

In one dimension, when we choose as shape functions polynomials whose derivatives are orthogonal, we have from (2.14)

$$a_{i+1} = - \frac{q_{i+1}}{K_{i+1,i+1}} \tag{3.16}$$

Here q_{i+1} can be determined from

$$q_{i+1} = \int N_{i+1} \; q \; d\Omega = \int N_{i+1} \; (r - L\hat{\phi}) \, d\Omega = \int N_{i+1} \; r \; d\Omega \tag{3.17}$$

because $\int N_{i+1} \; L\hat{\phi} \; d\Omega = 0$ due to the orthogonality referred to above. Now (3.11) can be written as

$$||e||^2_{E(I_j)} \simeq \frac{(q_{i+1})^2}{K_{i+1,i+1}} \tag{3.18}$$

which is the error indicator presented in [37,40].

When the shape functions do not possess a complete orthogonality a degree of approximation has to be introduced here. First, a_{i+1} has to be estimated using the previously found a_j values as

$$a_{i+1} = \frac{1}{K_{i+1,i+1}} \left(- q_{i+1} - K_{i+1,j} \; a_j \right) \tag{3.19}$$

Now

$$q_{i+1} = \int N_{i+1} \; r \; d\Omega - K_{i+1,j} \; a_j \tag{3.20}$$

because

$$\int N_{i+1} \; L\phi_i \; d\Omega \neq 0 \tag{3.21}$$

and q_{i+1} is again evaluated from the previously determined a_j. These approximations are generally tenable if near orthogonality of hierarchical functions exists, as is often the case with the p-type elements.

A more serious shortcoming of this error estimate is however immediately apparent. If the residual r is orthogonal to N_{i+1} in (3.15) the error indication will be zero, since

$$\int N_{i+1} \; r \; d\Omega = 0 \tag{3.22}$$

This implies that the proposed criterion leads to an indication of the error absorbed by a hierarchical refinement on the existing mesh rather than an accurate estimate of the error in the finite element solution. Therefore we will denote this simply as an error indicator and search further for a true error "estimate".

3.3 Error Estimation in One Dimension

Such an estimate can be obtained from a simple calculation, when $e = \phi - \hat{\phi}$ is the exact response to r.

Consider the model problem

$$-\frac{d^2\phi}{dx^2} + q = 0 \tag{3.23}$$

with boundary condition

$$\phi(0) = \phi(L) = 0$$

with e being the error involved in a linear approximation with nodal points x_i. As in (3.5), we have

$$-\frac{d^2 e}{dx^2} + r = 0 \quad \text{with} \quad e(0) = e(L) = 0 \tag{3.24}$$

In addition it can be shown that $e(x_i) = 0$ in our particular case (e.g. [59]), so e can be determined on every interval $I_j = (x_{j-1}, x_j)$ separately.

Assuming that $r = \bar{r} \sin \frac{\pi x}{L}$, then the differential equation

$$-\frac{d^2 e}{dx^2} + r = 0 \qquad e(0) = e(L) = 0 \tag{3.25}$$

is easily solvable. Substituting $e = \bar{e} \sin \frac{\pi x}{L}$ in (3.25) we get

$$\frac{\pi^2}{L^2} \bar{e} \sin \frac{\pi x}{L} = - \bar{r} \sin \frac{\pi x}{L} \tag{3.26}$$

giving $$e = -\frac{L^2}{\pi^2} r \tag{3.27}$$

Further on L

$$||e||_E^2 = -\int_0^L e\, r\, dx = \frac{L^2}{\pi^2}\int r^2\, dx \tag{3.28}$$

So far we assumed that $r = \bar{r}\sin\frac{\pi x}{L}$. In general, the residual will involve many terms and

$$e = \sum a_i \sin\frac{i\pi x}{L}.$$

It can be easily shown that we get (see Appendix)

$$||e||_E^2 \leqslant \frac{L^2}{\pi^2}\int r^2\, dx \tag{3.29}$$

i.e., we replace equality in (3.28) with inequality in (3.29). Applying now (3.28) to every interval separately [because $e(x_i)=0$, it is possible] then we get

$$||e||_E^2 \leqslant \frac{1}{\pi^2}\sum (x_{i+1}-x_i)^2 \int_{x_i}^{x_{i+1}} r^2\, dx \tag{3.30}$$

where $x_{i+1} - x_i = h_{i+1}$.

The bounding inequality in (3.30) is valid for r of all orders of polynomial variation in the one-dimensional problem (3.23) and linear elements. Note, however, that in the limit of h refinement of the linear elements we expect that r will be nearly constant on each element. In this limit only, the addition of a hierarchic quadratic term will give a valid error estimate. Here we can rewrite (3.14) on every single element I of length h

$$\phi = \hat{\phi} + x(h-x)\, a_{i+1} \tag{3.31}$$

and

$$e = a_{i+1}\, x(h-x) \tag{3.32}$$

From (3.24)

$$r = -2a_{i+1} \tag{3.33}$$

and substituting (3.33) in (3.32)

$$e = -\frac{r}{2}\, x(h-x) \tag{3.34}$$

Then

$$||e||_{E(I)}^2 = -\int e\, r\, d\Omega = \frac{r^2}{2}\int_0^h x(h-x)\, dx = \frac{r^2 h^3}{12} \tag{3.35}$$

Note that (3.32) will give the same measure of the error for constant r if the factor π^2 is changed to 12. We therefore implement

$$\|e\|_E^2 = \frac{1}{12} \sum h_i^2 \int_{x_i}^{x_{i+1}} r^2 \, dx \tag{3.36}$$

as the asymptotically correct error estimator for linear elements and one-dimensional problems. This is indeed the form presented by Babuška et al. [4-6] where a different and detailed mathematical justification is given.

3.4 Error Indication and Estimation in Two Dimensions

In two dimensions the hierarchical functions have their support on either one or two elements, as shown in Figure 6. Again we can use the hierarchical modes to make a total projection as in (3.15). We note, however, that in two dimensions the hierarchical contributions to K are not diagonal as in the 1-D equation, and the approximation of (3.19) has to be used, i.e.

$$a_{i+1} = - \frac{[q_{i+1} + K_{i+1,j} \, a_j]}{K_{i+1,i+1}} \tag{3.37}$$

(for $j \neq i+1$ and summation convention) then

$$\left(\|e\|_E^2\right)_{i+1} \simeq \frac{[q_{i+1} + K_{i+1,j} \, a_j]^2}{K_{i+1,i+1}} \tag{3.38}$$

where the subscript $i+1$ refers to the new hierarchic mode.

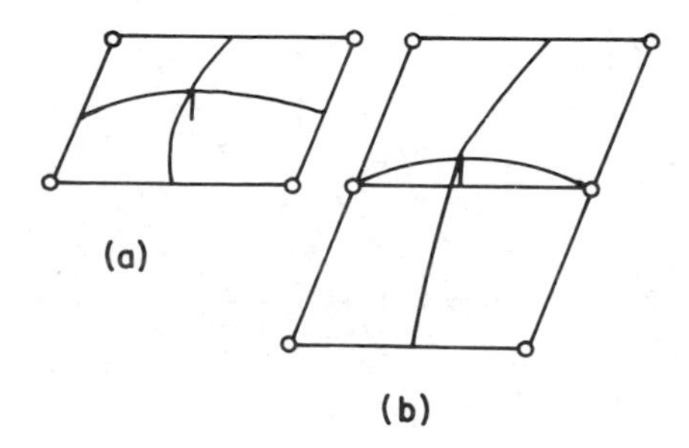

FIG.6. Hierarchical supports for two-dimensional error analysis.

Note that if we sample each hierarchical mode independently (set other hierarchical amplitudes $a_i = 0$) then the interaction of the new modes in the solution is ignored, introducing a new approximation in addition to the problem of this error indicator being a projection of the true error in the next hierarchical mode and not the total error. The effect of this new approximation cannot be evaluated a-priori.

To estimate the total error we will take some guidance from the hierarchic nodes in Figure 6 but attempt to base the analysis on the theory developed for the one-dimensional case. The hierarchic modes indicate that not only must we consider the contribution from the residual on the element (regular part of the residual) but also the contribution from the interface (singular part of the residual). In one dimension only the former was considered because the hierarchic nodes were all internal to the element.

Consider first the regular part of the residual on the element. We will over-estimate the error associated with the hierarchic node in Figure 6a by releasing boundary conditions to give two one-dimensional responses (see Figure 7a) and associating half the residual in each direction. In each direction (3.30) gives

$$||e||^2_{E(I_j)} \leqslant \frac{h_j^2}{\pi^2} \int \left(\frac{r}{2}\right)^2 d\Omega \tag{3.39}$$

so that the addition of both one-dimensional contributions gives

$$||e||^2_{E_1} \leqslant \frac{h^2}{2\pi^2} \int_{\Omega_j} r^2 \, d\Omega \tag{3.40}$$

The influence of the singular part of the residual can be treated in the same way if the residual is distributed as indicated in Figure 7b. Now we have

$$r \simeq \frac{J_s}{h} \tag{3.41}$$

and (3.30) gives

$$\begin{aligned} ||e||^2_{E_2} &\leqslant \frac{h^2}{\pi^2} \int r^2 \, d\Omega \\ &= \frac{h^2}{\pi^2} \int \left(\frac{J_s}{h}\right)^2 d\Omega \\ &= \frac{h}{\pi^2} \int J_s^2 \, dy \end{aligned} \tag{3.42}$$

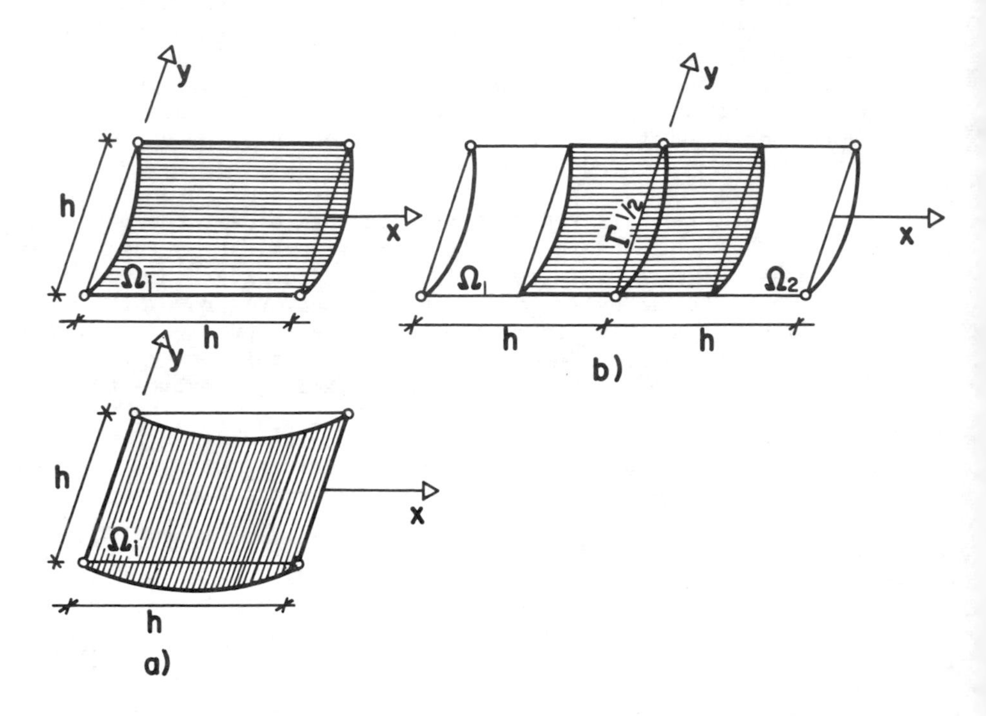

FIG.7. Deformation patterns for two-dimensional error analysis.

The error estimator will consider each element separately. We introduce a factor of 1/2 into (3.42) because each boundary will be considered twice. The error estimate becomes for the bilinear element

$$||e||^2_{E(\Omega_j)} \leq \frac{h^2}{2\pi^2}\int_{\Omega_j} r^2 \, d\Omega + \frac{h}{2\pi^2}\int_{\Gamma_j} J_s^2 \, d\Gamma \tag{3.43}$$

Since this derivation is based on a superposition of one-dimensional analyses we should obtain the one-dimensional estimate if the boundary conditions impose essentially a one-dimensional problem on the two-dimensional mesh. Using (3.43) and constant right-hand side q in (2.3) we get an estimate which is not complately identical with the one-dimensional estimate. Therefore we reduce the estimate by modifying (3.43) to

$$||e||^2_{E(\Omega_j)} \leq \frac{h^2}{2\pi^2}\int_{\Omega_j} \left(r - \bar{r}\right)^2 \, d\Omega + \frac{h}{2\pi^2}\int_{\Gamma_j} J_s^2 \, d\Gamma \tag{3.44}$$

where $\bar{r}$ is the mean value of the residual on the element. This form is mathematically justified in [2].

It follows that on the entire domain we get

$$||e||_E^2 \lesssim \sum_i \left(\frac{h^2}{2\pi^2} \int_{\Omega_i} (r - \bar{r})^2 \, d\Omega + \frac{h}{2\pi^2} \int_{\Gamma_i} J_s^2 \, d\Gamma \right) \tag{3.45}$$

where the sum is taken over all elements Ω_i and Γ_i (if J_s at the boundary Γ_i is properly defined).

We derived these estimates under various assumptions. The question arises whether they are acceptable. The answer is positive (see for example Babuska et al. [2,7]). There the term $1/\pi^2$ is replaced by $1/12$ to obtain the asymptotically correct estimate

$$||e||_E^2 \simeq \sum_i \left(\frac{h^2}{24} \int_{\Omega_i} (r - \bar{r})^2 \, d\Omega + \frac{h}{24} \int_{\Gamma_i} J_s^2 \, d\Gamma \right) \tag{3.46}$$

The estimate has essentially two parts. The one related to the regular part of residual (volume integral) and the other one to the singular part of the residual (jump of derivatives term). It can be theoretically shown that the first term (volume integral) is in the limit negligible with respect to the second one. Practical experience shows that the first term is relatively small also for coarse meshes.

The estimate is called asymptotically correct when the ratio of the right and left-hand side of (3.46) goes to one as the error goes to zero. The estimate (3.46) has this property when some mathematical assumptions are satisfied. One major one is that the element error estimators are about the same in magnitude. This can be achieved e.g. by an adaptive selection of the elements. The experience shows e.g. that the asymptotical correctness is not achieved when the solution has singular behaviour and a uniform (obviously improper) mesh is used.

3.5 *Practical Error Analysis*

Both the estimators (3.36) and (3.43) and the indicators (3.18) and (3.38) can play a fundamental role in the finite element analysis. The estimators allow for an evaluation of the total error in energy in the current finite element solution and the indicators allow a rational increase of degree of the element (made in a hierarchical way) and/or element subdivision.

The latter is a direct consequence of the hierarchical error indicators being a projection of the error in the new hierarchical modes, reflecting thus the capacity for the new modes to 'absorb' error in energy. In addition, the possibility of obtaining an accurate estimate of the error in an appropriate norm allows the program to stop automatically when a certain accuracy has been achieved or to indicate the order of accuracy when a certain pre-specified solution cost has been attained.

It seems, however, that to obtain both advantages we have not only to compute all the residuals and stress discontinuities at every stage of the interation process to evaluate the estimate (3.46), but also all the hierarchical stiffness coefficients corresponding to the possible new refinements to evaluate the indicator (3.38). This is not necessarily the case, since the error indicators can be obtained as projections of the computed residuals and stress jumps in the new hierarchical modes, i.e. in (3.38)

$$q_j + K_{ji} a_i$$

$$= \int_{\Omega_1} r_1 N_j \, d\Omega + \int_{\Omega_2} r_2 N_j \, d\Omega - \int_{\Gamma_{1/2}} J_s N_j \, d\Gamma$$

where Ω_1, Ω_2, and $\Gamma_{1/2}$ and J_s, are indicated in Figure 7b. This computation can be achieved locally.

3.5.1 Requirements for Practical Error Estimates. We consider that a practical error estimator should satisfy the following conditions:

a. Be determined a-posteriori from information defined on a local basis.

b. If we define an effectivity index $\theta = \dfrac{||e||_E}{\left(||e||_E\right)_{exact}}$ (i.e. the ratio of the predicted energy norm of the error to the exact value of this norm), then we require $\theta \geq 1$ for all meshes, and to provide reasonable bounds $1 \leq \theta \leq 2$.

c. Asymptotic convergence $\theta \to 1$.

d. A direct interpretation of errors in stresses should be available.

In the examples that follow we show that error estimates which satisfy these criteria are available from the estimators of the form (3.36) and (3.46) if the following amendments are incorporated:

1. Theorem 2 of [7] states that there exist constants $k_2 \geq k_1 > 0$ independent of the mesh, such that

$$k_1 ||e||_E \leq \left(||e||_E\right)_{exact} \leq k_2 ||e||_E$$

with $||e||_E$ given by (3.46). Experience has shown $k_1 \geq 0.5$ and $k_2 \leq 1.5$ on a large number of problems, and both are asymptotically equal to 1 for uniform meshes and smooth solutions. To prevent gross violation of the second condition required of the error estimates we seek a value of k_2 by defining a factor k_{2i} for element i as

$$k_{2i} = 1 + \alpha \frac{||e||_{E(\Omega_i)}}{||u_h||_{E(\Omega_i)}} \tag{3.47}$$

where

$$||u_h||_{E(\Omega_i)} = \left(\underset{\sim}{u}_h^T \; K^i \; \underset{\sim}{u}_h \right)^{\frac{1}{2}} \tag{3.48}$$

is the energy on the i-th finite element. The corrected error estimate becomes

$$||e||^*_{E(\Omega_i)} = k_{2i} \; ||e||_{E(\Omega_i)} \tag{3.49}$$

We have taken $\alpha = 2$ in all applications.

2. For isoparametric transformation of the element there is more than one possible choice for the length parameter h. Following [19] and because we seek an over-estimate of the error in the solution, we choose h as the length of the maximum side of the element.

3. Error estimators have not been developed for elements of higher order than linear. However, significant work is being done in this area by Szabo and his co-workers [46-49].

The sine function analysis affords the following extension.

Quadratic elements would match the predominant part of the error in the first sine function $i = 1$, used for linear elements. Assuming now the error in the form of the sine function with $i=2$ we would replace the coefficient

$$\frac{h^2}{2\pi^2} \quad \text{by} \quad \frac{h^2}{n\pi^2} \tag{3.50}$$

with $n = 8$. For the cubic element we could progress to the mode $i = 3$ to get the coefficient with $n = 18$.

Thus with changing polynomial order p on the elements the coefficient of the estimator can be replaced by

$$\frac{h^2}{2\pi^2} \left(\frac{1}{p^2} \right)$$

It is to be noted however, that this assumes that quadratic

interpolants, for example, can completely eliminate error in the form of the first sine function. This of course is not the case so the factor above would be optimistic.

Again, in the interest of producing an over-estimate of the error in the solution, we reduce the power on p to one so that a general form of (3.46) for elements of polynomial order p becomes

$$||e||^2_{E(\Omega_i)} = \frac{h^2}{24p} \int_{\Omega_i} r^2 \, d\Omega + \frac{h}{24p} \int_{\Gamma_i} J_s^2 \, d\Gamma \qquad (3.51)$$

Note we relax the requirement $\theta \to 1$ here in 2. above.

4. The analysis above strictly applies only to the Laplace operator. However, it can be generalized, for example, to problems governed by the Navier equations of elasticity. Formula (3.51) has been found adequate when the influence of Poisson's ratio has been incorporated. We take

$$||e||^2_{E(\Omega_i)} = C\left\{\frac{h^2}{24p} \int_{\Omega_i} r^2 \, d\Omega + \frac{h}{24p} \int_{\Gamma_i} J_s^2 \, d\Gamma\right\} \qquad (3.52)$$

with $C = \frac{1-\nu}{E}$ for plane stress.

5. Local estimates of the error in stress are the elusive goal of most practical error analysis. It is interesting that the error estimates advocated here are evaluated locally element-wise are justified locally by appealing to St.Venant's Principle and are backed up by a corrective factor which, in a limited number of experimental problems, ensured local bounds on the energy of each element.

We can suggest the following: a bound on some average of the stress on the element will be obtained by scaling local stresses by a factor FS_i

$$FS_i = \frac{||u_e||'_{E(\Omega_i)}}{||u_h||_{E(\Omega_i)}} \qquad (3.53)$$

where $||u_e||'_{E(\Omega_i)}$ is the prediction of the energy of the exact solution on the i-th element evaluated as

$$||u_e||'_{E(\Omega_i)} = \left(||u_h||^2_{E(\Omega_i)} + ||e||^{*2}_{E(\Omega_i)}\right)^{\frac{1}{2}} \qquad (3.54)$$

and $||u_h||_{E(\Omega_i)}$ is given by (3.48).

3.6 Applications of the Error Estimators of Section 3.5.

3.6.1 Cantilever with applied end moment, linear elements. The simple cantilever shown in Figure 8 was analysed for plane stress using the four-node, bilinear element. The exact solution has a quadratic variation of vertical displacement along the beam.

The finite elements used are compatible and fully integrated so the finite element solution u_h is known to be stiff. The error estimate $||e||_E^*$ does, however, over-estimate the error in the solution. A comparison of the results also supports the constant C in (3.52) used to include the influence of Poisson's ratio ν. Finally, the moment is constant along the beam in this simple example so that the stresses in each element are identical and $||e||_{E(\Omega_i)}$ from (3.52) is the same for each element. The error estimates in the table can therefore be interpreted locally as well as globally, and implementation of (3.53) will obviously give good estimates of the local error in stress.

The last two columns in the table give the ratios

$$\frac{||e||_E^*}{||u_h||_E} \times 100 \qquad \text{and} \qquad \frac{\left(||e||_E\right)_{exact}}{||u_h||_E} \times 100$$

respectively. This measure may provide the best qualitative indication of the accuracy of the finite element stresses.

Note that all calculations between values in these tables must be based on the formula

$$\left(||u||_E^2\right)_{exact} = ||u_h||_E^2 + ||e||_E^2$$

This formula expresses the fact that the error is orthogonal to the finite element solution.

3.6.2. Small circular hole, h and p refinement. The configuration of the problem and the finite element meshes used are shown in Figure 9. The analysis was for plane stress with $\nu = 0.3$. The results given as Case 1 in Table 2 are for the bilinear four-node element and subdivision of all elements into four, as shown in the figure. In this analysis the surface geometry of the hole was updated in the refinement. Again, the global error estimates $||e||_E^*$ are excellent.

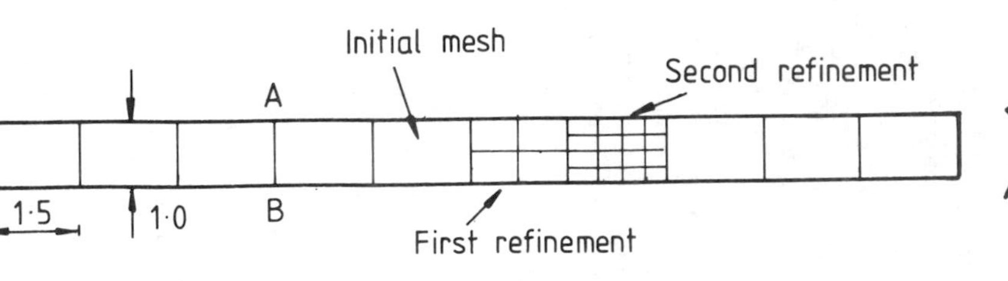

Fig. 8. Cantilever and adaptive meshes.

TABLE 1

Cantilever Applied End Moment, Linear Elements

(a) $\nu = 0.0$							
No. Elements	$\|u_h\|_E$	Error Estimate $\|e\|_E$	θ	Error Estimate $\|e\|_E^*$	θ^*	$\frac{\|e\|_E^*}{\|u_h\|_E}.100$	$\frac{(\|e\|_E)\text{exact}}{\|u_h\|_E}.100$
10	$.153 \times 10^{-2}$	$.101 \times 10^{-2}$	0.62	$.233 \times 10^{-2}$	1.43	152	107
40	$.198 \times 10^{-2}$	$.082 \times 10^{-2}$	0.78	$.150 \times 10^{-2}$	1.43	75	53
160	$.216 \times 10^{-2}$	$.049 \times 10^{-2}$	0.85	$.078 \times 10^{-2}$	1.36	36	27
"Exact" Solution	$.224 \times 10^{-2}$ (Beam Theory)						
(b) $\nu = 0.3$							
No. Elements	$\|u_h\|_E$	Error Estimate $\|e\|_E$	θ	Error Estimate $\|e\|_E^*$	θ^*	$\frac{\|e\|_E^*}{\|u_h\|_E}.100$	$\frac{(\|e\|_E)\text{exact}}{\|u_h\|_E}.100$
10	$.160 \times 10^{-2}$	$.094 \times 10^{-2}$	.60	$.229 \times 10^{-2}$	1.46	143	98
40	$.201 \times 10^{-2}$	$.076 \times 10^{-2}$	.72	$.146 \times 10^{-2}$	1.43	73	49
160	$.218 \times 10^{-2}$	$.045 \times 10^{-2}$	.84	$.077 \times 10^{-2}$	1.43	35	24
"Exact" Solution	$.224 \times 10^{-2}$ (Beam Theory)						

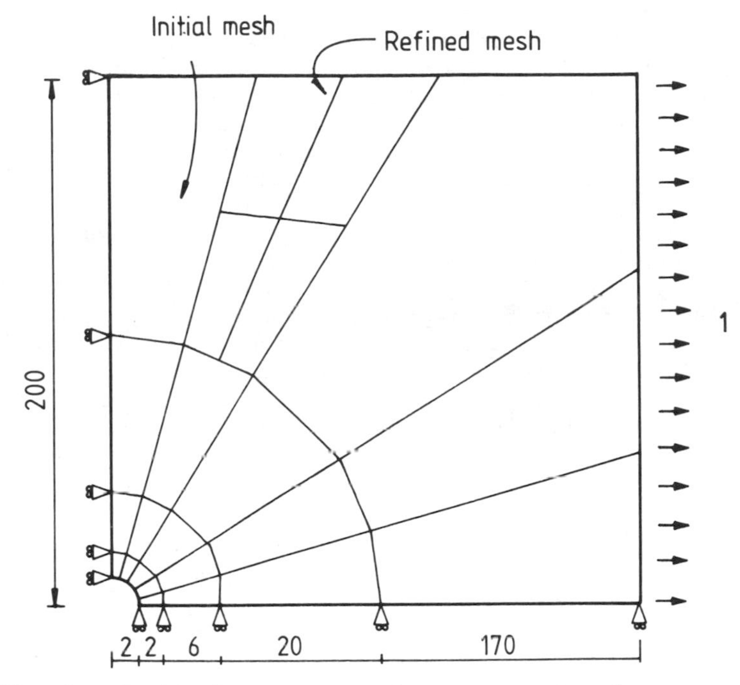

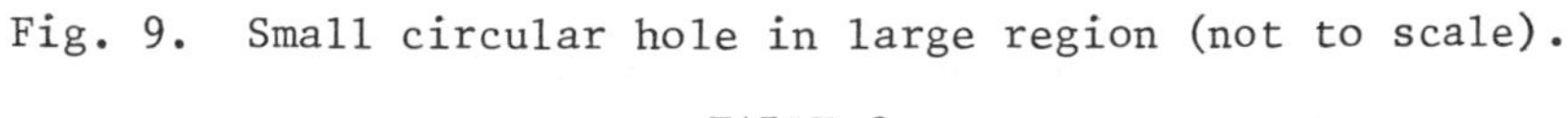

Fig. 9. Small circular hole in large region (not to scale).

TABLE 2

Large Plate with Small Circular Hole $\nu = 0.3$

Case 1. Linear Elements (Complete Subdivision)

No. Elements	$\Vert u_h \Vert_E$	Error Estimate $\Vert e \Vert_E$	θ	Error Estimate $\Vert e \Vert_E^*$	θ^*
24	1.154816	$.583 \times 10^{-2}$	.85	$.700 \times 10^{-2}$	1.02
96	1.154829	$.343 \times 10^{-2}$	.81	$.405 \times 10^{-2}$	0.96
"Exact" Solution	1.154837	(refined mesh and Richardson extrapolation)			

Case 2. Increasing Polynomal Order on 24 Element Mesh (No Update of Hole Surface) (Complete Refinement)

Element Type (p)	$\Vert u_h \Vert_E$	Error Estimate $\Vert e \Vert_E$	θ	Error Estimate $\Vert e \Vert_E^*$	θ^*
Linear (1)	1.154816	$.583 \times 10^{-2}$	.85	$.700 \times 10^{-2}$	1.06
Quadratic (2)	1.154832	$.264 \times 10^{-2}$	.93	$.297 \times 10^{-2}$	1.05
Cubic (3)	1.154834	$.162 \times 10^{-2}$	.94	$.173 \times 10^{-2}$	1.00
"Exact" Solution	1.154835	(refined mesh and Richardson extrapolation)			

The results given as Case 2 in the table are for a uniform increase of the polynomial order on all elements based on the coarse mesh indicated in Figure 9. Here the surface geometry of the hole has not been updated with refinement. Again, the error estimates $||e||_E^*$ are excellent but the example indicates the practical weakness of a global error measure. The first column in the table indicates that the significant error in stresses in the immediate region of the hole surface appears as only a small perturbation to the global energy of the region.

Obviously a local measure is required. In Figure 10 we plot the stress tangential to the hole surface in two elements adjacent to the hole. The finite element stress, plotted as a linear interpolant through the 2x 2 Gauss point values, is scaled using the factors FS_i given by (3.53). The order of the error in the stresses is accurately indicated even at this local element level.

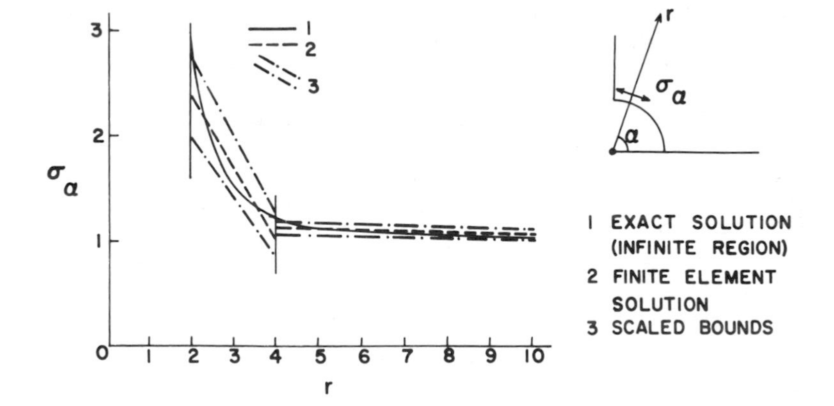

Fig.10. Local stress errors: $\alpha = 86.4^\circ$ on 24-element mesh.

4. ADAPTIVITY

The concept of adaptivity follows naturally from the previous discussion as the expansion of the trial function space S_h, hierarchically or non-hierarchically, but only where the space is shown to be deficient. It has been taken as a basis for adaptivity that the sequence of finite element solutions must follow the best rate of convergence in terms of the number of degrees of freedom of the structure. The optimal rates of convergence for both the h and p versions of the finite element method have been identified and quantified in [13].

Two programs are being developed for this research. The first is based on the h convergence process in Babuška [7], which utilizes $||e||_{E(\Omega_i)}$ as the indicator for mesh refinement.

The second p-convergence algorithm uses directly the hierarchical elements and indicators (3.38).

The strategy for selection of new degrees of freedom in adaptive processes is not uniquely resolved. Here we refine on the basis of evaluating all error indicators and including degrees of freedom whose indicator exceeds one-half of the maximum value or, in the case of the h-convergence program, subdividing elements whose indicator exceeds this value. It has been found that the path followed by the adaptive process is not greatly affected by changing the one-half factor. However, in practice this choice may affect the expense of the solution process and alternative strategies are discussed both by Babuška [7] and Peano [38].

4.1 *Examples*

4.1.1. Cantilever with applied end moment. The error estimators $||e||_{E(\Omega_i)}$ are identical on all elements in the first two meshes shown in Figure 8, so the sequence of results in Table 1 corresponds to a h-adaptive process. The accuracy of the error estimators indicates that an effective stopping criterion on the basis of the energy norm, stress or displacement, could be defined.

4.1.2. Small hole in large region. Both the h and p adaptive processes have been applied to the problem defined in Figure 9 and the results plotted on Figure 11. Notice that the convergence is expressed in terms of number of degrees of freedom and not as usual in terms of h. In the p-version h remains constant and a comparison would not be possible. For 2-D elements $O(h^2) \simeq O(1/N)$ so we expect, for example, for linear elements, slopes $\leqslant 1$.

The plots in Figure 11 can be divided in two groups: convergence using linear trial spaces (adaptive or non-adaptive), and convergence using higher order trial spaces. The first group includes solution extensions nos. 3, 12, 13, 14, and the second group extensions nos. 2, 4, 11, from meshes M1 and M3. Extensions 4, 13, 14 are adaptive and the type of meshes obtained are represented in Figures 12 and 13.

As expected, the rates of convergence are higher for the second group because of the better convergence characteristics of higher order finite elements. Within each group the adapative solutions are better because there is an intelligent criterion to select the new degrees of freedom.

The adaptive p-extension (extension no.4) tends in the limit to the cubic solution of extension no.2, due to the fact that we set a limit of complete cubic modes in the adaptive p-convergence program. This last result shows that a very accurate solution on p-convergence extensions requires the use of higher order polynomials. The disadvantage is that this leads to a loss in pointwise convergence because of the 'noise' associated with the polynomial oscillations.

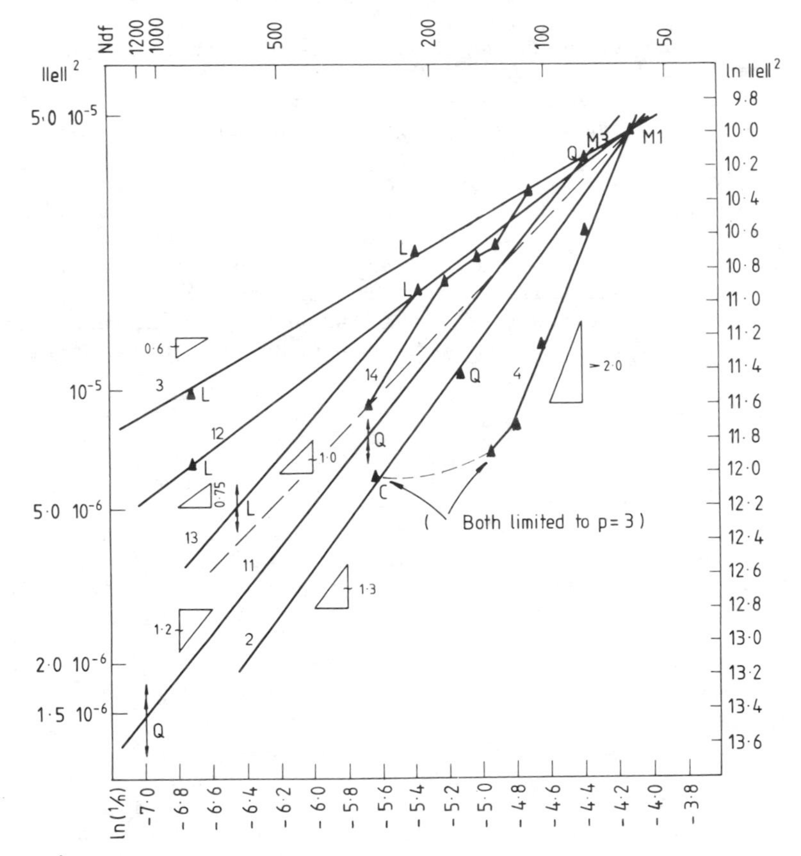

Extensions

2	p	- Convergence based on M1, complete refinement
3	h	- Convergence based on M1, complete refinement, no geometry updating
4	p	- Convergence based on M1, adaptive
11	h	- Convergence based on M3, complete refinement
12	h	- Convergence based on M1, complete refinement
13	h	- Convergence based on M1, adaptive based on local Richardson's extrapolation for error indication
14	h	- Convergence based on M1, adaptive based on refinement of top 50% of error indicators in the mesh

where M1 - mesh shown in Figure 9

M3 - a mesh of six 9-node elements based on M1

FIG.11. Experimental rates of convergence for small hole in large region

A mixed h and p convergence model could therefore be the best strategy. In [14] it is proved that the combined versions produce higher order rates of convergence than either the h or p-version by itself, indicating that research in this direction is necessary. We note, however, that the resulting program will have a very complex structure. In this context we reference the work done on the h-version [43,57] which by itself presents a highly complex situation.

The p-convergence programs, on the other hand, have a simpler structure but nevertheless more elaborate than the usual finite element codes, since we have to allow for error subroutines, automatic node generation subroutines, and multi-level finite element types.

The meshes obtained for the h and p adaptive processes of Extensions 14 and 4 are given in Figure 12 and 13 respectively. Similar refinement near the hole is seen in both cases. Finally, we note from Table 3 that the error estimators for the h adaptive process given in Figure 12 are again accurate enough to provide a stopping criterion. However, in Section 4.1.2 it was seen that the global energy norm gave little indication of the accuracy of stresses near the hole surface. The local stress error estimator discussed in that section may provide a more practical accuracy test.

Obviously very powerful solution algorithms can be based on these processes. The efficiency of the hierarchic indicators is best shown by returning to the cantilever beam of Figure 8 and considering a tip shear load. The interpolants required on the interfaces between elements such as AB, in the figure, depends on the Poisson's ratio. With $\nu = 0.0$ only cubic interpolants in the x-direction are required; with $\nu = 0.3$ the exact solution only follows if quadratic as well as cubic interpolants are added on the interfaces. The adaptive process based on the hierarchic error indicators is sensitive to exactly these requirements and the quadratics are left out of the adaptive process for the first problem.

TABLE 3

Linear Elements (Adaptive Solution - Extension 14)

	ndf	$\|\|u_h\|\|_E^2$	Error Estimate $\|\|e\|\|_E$	θ	Error Estimate $\|\|e\|\|_E^*$	θ^*
Step 1	60	1.333602	$.583 \times 10^{-2}$	.85	$.70 \times 10^{-2}$	1.02
Step 2	111	1.333616	$.428 \times 10^{-2}$	.76	$.52 \times 10^{-2}$	.92
Step 3	133	1.333625	$.382 \times 10^{-2}$	.80	$.45 \times 10^{-2}$	.94
Step 4	150	1.333626	$.355 \times 10^{-2}$	.76	$.42 \times 10^{-2}$	.90
Step 5	182	1.333629	$.323 \times 10^{-2}$	.74	$.38 \times 10^{-2}$	.87
Step 6	287	1.333639	$.295 \times 10^{-2}$	.98	$.36 \times 10^{-2}$	1.20

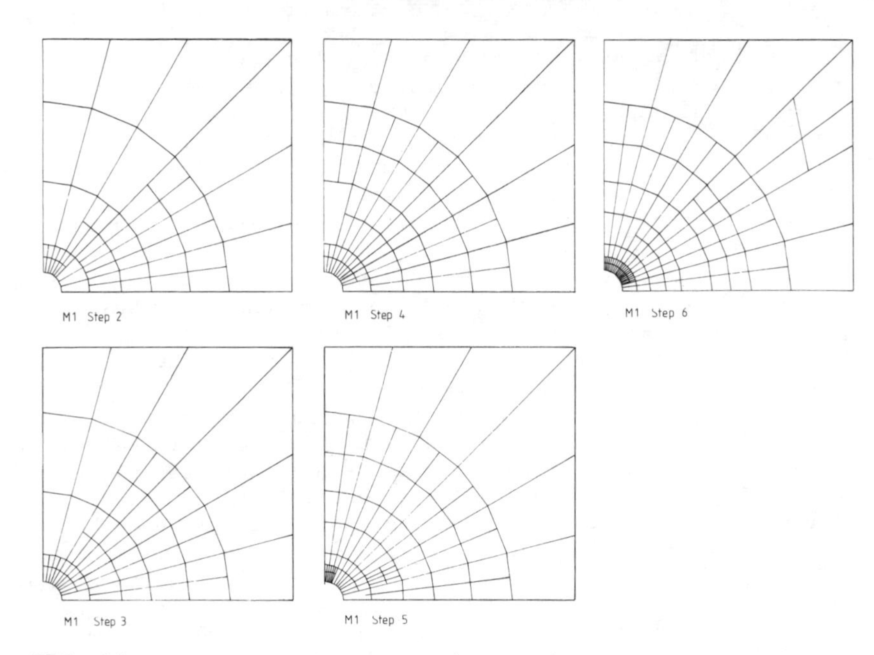

FIG. 12. H-convergence on M1 linear elements - extension 14 (figures not to scale)

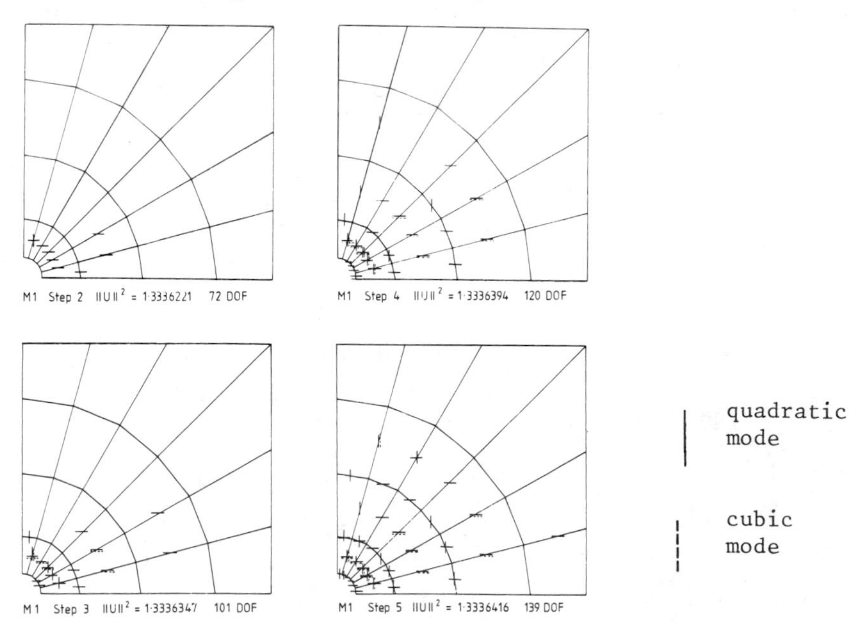

FIG. 13. P-convergence on M1, hierarchical elements - extension 4 (figures not to scale)

5. CONCLUDING REMARKS

A general description of 'hierarchical' finite elements, including both h and p versions of the finite element method has been presented. There are three main advantages of this approach. Firstly, it leads to improved conditioning of the stiffness matrix and the topology of the stiffness matrix indicates an efficient partitioning for block iteration solution procedures. Secondly, compatibility is easily enforced in meshes with a graded refinement of polynomial order or element size. Finally, in the new generation of adaptive finite element schemes higher order hierarchical modes provide an indicator for the selection of the new degrees of freedom which should be added to the finite element mesh.

We have also demonstrated in this paper that accurate error estimators are available and can be interpreted locally at least for problems with stress concentrations no greater than the circular hole. Projections of the residuals required for the evaluation of these error estimators give the hierarchical indicators which control the adaptive processes. Naturally, the program architecture becomes complex but processes allowing the accuracy required of the solution to be pre-specified are within reach.

APPENDIX

PROOF OF THE BOUND OF EQUATION (3.30)

The energy norm of the error is given by

$$||e||_E^2 = -\int_0^L r\ e\ dx \tag{A.1}$$

Take $r = \sum r_i \operatorname{Sin} \frac{i\pi x}{L}$

From (3.24) $\frac{d^2 e}{dx^2} = r$

so that

$$e = -\sum \left(\frac{L}{i\pi}\right)^2 r_i \operatorname{Sin} \frac{i\pi x}{L}$$

Substituting in (A.1)

$$||e||_E^2 = +\int_0^L \left[\sum r_i \operatorname{Sin} \frac{i\pi x}{L}\right] \left[\sum \left(\frac{L}{i\pi}\right)^2 r_i \operatorname{Sin} \frac{i\pi x}{L}\right] dx$$

but since

$$\int_0^L \left(\text{Sin}\,\frac{i\pi x}{L}\,\text{Sin}\,\frac{j\pi x}{L} \right) dx = 0, \qquad i \neq j$$

$$||e||_E^2 = \frac{L^2}{\pi^2}\int_0^L \left[\sum \frac{r_i}{i}\,\text{Sin}\,\frac{i\pi x}{L} \right]^2 dx$$

$$\leq \frac{L^2}{\pi^2}\int_0^L \left[\sum r_i\ \text{Sin}\,\frac{i\pi x}{L} \right]^2 dx$$

$$= \frac{L^2}{\pi^2}\int_0^L r^2\,dx$$

ACKNOWLEDGEMENT

J. Gago gratefully acknowledges the support of the Gulbenkian Foundation, Lisbon, Portugal, Fellowship No. 18/79/B and the Technical University of Lisbon.

The work of I. Babuška was supported in part by the Office of Naval Research under contract N00014-77-0623.

REFERENCES

1. BABUŠKA, I., The Selfadaptive Approach in the Finite Element Method, in J.R.Whiteman (Ed.), *Mathematics of Finite Elements and Applications*, Academic Press, London (1975).
2. BABUŠKA, I. and MILLER, A., A-Posteriori Error Estimates and Adaptive Techniques for the Finite Element Method, *Tech.Note BN-968, Institute for Physical Science and Technology, University of Maryland,* (June 1981).
3. BABUŠKA, I. and RHEINBOLDT, W.C., Error Estimates for Adaptive Finite Element Computations, *Siam J.Numer.Anal.*, Vol.15, No.4 (August 1978).
4. BABUŠKA, I. and RHEINBOLDT, W.C., Computational Aspects of the Finite Element Method, *Mathematical Software III,* (Ed. J.R.Rice), Academic Press, (1977), 223-253.
5. BABUŠKA, I. and RHEINBOLDT, W.C., A-Posteriori Error Estimates for the Finite Element Method, *Int.J.Numer.Meths. Engng.*,Vol.12, 1597-1615 (1978).
6. BABUŠKA, I., Analysis of Optimal Finite Element Meshes in R^1, *Math.Comput.*, 30 (1979), 435-463.
7. BABUŠKA, I. and RHEINBOLDT, W.C., Adaptive Approaches and Reliability Estimations in Finite Element Analysis, *Comput. Meths. in Applied Mechanics and Engng.*, 17/18 (1979), 519-40.
8. BABUŠKA, I. SZABO, B.A. and KATZ, I.N., The P-Version of the Finite Element Method, *Siam J.Num.Anal.*, Vol.18 (1981), 515-546.

9. BABUŠKA, I., KATZ, I.N. and SZABO, B.A., Hierarchic Families for the P-Version of the Finite Element Method, *Proc. 3rd Int.Symp. on Comp.Meths. for Partial Differential Equations*, Lehigh University (1979), 278-286.
10. BABUŠKA, I. and RHEINBOLDT, W.C., A-Posteriori Error Analysis of Finite Element Solutions for One-Dimensional Problems, *Siam J.Num.Anal.*, Vol.18 (1981), 565-589.
11. BABUŠKA, I. and RHEINBOLDT, W.C., Reliable Error Estimation and Mesh Adaptation for the Finite Element Method, in J.T. Oden (Ed.), *Comput.Meths. in Nonlinear Mechanics*, (1980), 67-108.
12. BABUŠKA, I., A-Posteriori Error Estimates and Adaptive Approaches for the F.E.M., Maryland Conference, (March 1980).
13. BABUŠKA, I. and SZABO, B.A., On the Rates of Convergence of the Finite Element Method, *Rep. WU/CCM-80/2, Centre for Comput.Mechanics, Washington University* (1980). To appear in *Int.J.Num.Meth.Engng.*
14. BABUŠKA, I. and DOOR, M.R., Error Estimates for the Combined h and p Versions of the Finite Element Method, *Tech.Note BN-95, Inst. for Physical Science and Technology* (1980). To appear in Numerische Mathematik.
15. BASU, P.K. and SZABO, B.A., Adaptive Control in p-Convergent Approximations, *Proc. 15th Annual Meeting Soc. of Engng. Science, Inc.* (1978), Gainesville, Florida.
16. BASU, P.K., SZABO, B.A. and TAYLOR, B.D., Theoretical Manual and Users' Guide for Comet - XA, *Rep. WU/CCM-79/2, Centre for Comput.Mechanics, Washington University* (1979).
17. BRANDT, A., Multi-Level Adaptive Technique (MLAT) for Fast Numerical Solution to Boundary Value Problems, *Proc. 3rd Int.Conf.Numer.Meths. in Fluid Mechanics* (Paris 1972), *Lecture Notes in Physics*, Vol.18, Springer-Verlag, Berlin and New York, (1973), pp.82-89.
18. BRANDT, A., Multi-Level Adaptive Solutions to Boundary Value Problems, *Maths. of Comput.*, Vol.31, No.138 (1977), 333-390.
19. CIARLET, P.G. and RAVIART, P.A., Interpolation Theory over Curved Elements, with Applications to Finite Element Methods, *Comp.Meths. in Appl.Mechanics and Engng.*, 1 (1972), 217-249.
20. CARROLL, W.E. and BARKER, R.M., A Theorem for Optimum Finite Element Idealizations, *Int.J.Solid Structures*, (1973), Vol.9, 883-895.
21. CARROLL, W.E., On the Reformulation of the Finite Element Method, *Int.Symp. on Innovative Numer.Anal. in Appl.Engng. Science*, Versailles-France (1977).
22. DUNAVANT, D.A., Local A-Posteriori Indicators of Error for the P-Version of the Finite Element Method, *Rep. WU/CCM-80/1, Centre for Comput.Mechanics, Washington University* (1980).
23. FELIPPA, C.A., Optimization of Finite Element Grids by Direct Energy Search, *Appl.Maths. Modelling* (1978), Vol.1.
24. FELIPPA, C.A., Numerical Experiments in Finite Element Grid Optimization by Direct Energy Search, *Appl.Maths Modelling* Vol. 1 (1977).

25. KELLY, D.W., A Bound Theorem for Reduced Integration and Error Analysis. Companion paper in this text.
26. KELLY, D.W., Bounds on Discretization Error by Special Reduced Integration of the Lagrange Family of Finite Elements, *Int.J.Num.Meths.Engng.*, Vol.15, 1489-1506, (1980).
27. MELOSH, R.J. and KILLIAN, Douglas E., Finite Element Analysis to Attain a Pre-specified Accuracy, *Proc. 3rd Nat.Congress on Computing in Structures* (1976).
28. MELOSH, R.J. and MARCAL, P.V., An Energy Basis for Mesh Refinement of Structural Continua, *Int.J.Num.Meth. in Engng.*, Vol.II, 1083-1091 (1977).
29. MELOSH, R.J., Principles for Design of Finite Element Meshes, Maryland Conference (1980).
30. MOTE, C.D., Global-Local Finite Element, *Int.J.Num.Meth. Engng.*, 3, 565-74 (1971).
31. NICOLAIDES, R.A., On Multiple Grid and Related Techniques for Solving Discrete Elliptic Systems, *J. of Comput.Physics*, 19, 418-431 (1976).
32. NICOLAIDES, R.A., On the ℓ^2 Convergence of an Algorithm for Solving Finite Element Equations, *Maths. of Comput.*, Vol.31, No.140 (1977), 892-906.
33. NICOLAIDES, R.A., On Some Theoretical and Practical Aspects of Multigrid Methods, *Inst. for Comput.Applcs. in Science and Engng. (ICASE)*, NASA Langley Research Centre, Virginia. Report No.77-19 (1977).
34. ODEN, J.T. and REDDY, J.N., *An Introduction to the Mathematical Theory of Finite Elements* (1976).
35. OLIVEIRA, E.R. de Arantes e, Optimization of Finite Element Solutions, *Proc. 3rd Conf. on Matrix Meths. in Structural Mechanics*, Wright-Patterson Air Force Base, Ohio, (1971), pp.750-769.
36. PEANO, A.G., Hierarchies of Conforming Finite Elements for Plane Elasticity and Plate Bending, *Comput. and Maths. with Appls.*, Vol.2, No.3-4 (1976).
37. PEANO, A.G., PASINI, A., RICCIONI, R. and SARDELLA, L., Self-Adapative Finite Element Analysis, *Proc. VIth Int. Finite Element Congress*, Baden Baden, (1977).
38. PEANO, A., RICCIONI, R., PASINI, A. and SARDELLA, L., Adaptive Approximations in Finite Element Structural Analysis, *ISMES*, Bergamo, Italy, (1978).
39. PEANO, A. and RICCIONI, R., Automated Discretization Error Control in Finite Element Analysis, *2nd World Congress in Finite Element Methods*, (1978).
40. PEANO, A., FANELLI, M., RICCIONI, R. and SARDELLA, L., Self-Adaptive Convergence at the Crack Tip of a Dam Buttress, *Int.Conf. on Numer.Meths. in Fracture Mechanics*, Swansea (1979).
41. PETRUSKA, G. and KATZ, I.N., Finite Element Convergence on a Fixed Grid, *Comp. and Maths. with Appls.*, Vol.4, pp.67-71.
42. RHEINBOLDT, W.C., Adaptive Mesh Refinement Processes for Finite Element Solutions, *University of Pittsburgh Report*, (March 1980).

43. RHEINBOLDT, W.C. and MESZTENYI, C.K., On a Data Structure for Adaptive Finite Element Mesh Refinements, *ACM Transaction on Maths. Software*, Vol.6, No.2, (June 1980), pp.166-187.
44. SHEPHARD, M.S., Finite Element Grid Optimization with Interactive Computer Graphics, *Program of Computer Graphics and Dept. of Structural Engng.*, Cornell University, (1980).
45. SHEPHARD, M.S., GALLAGHER, R.H. and ABEL, J.F., The Synthesis of Near-Optimum Finite Element Meshes with Interactive Computer Graphics, *Int.J.Num.Meths.Engng.*, Vol.15, 1021-1039, (1980).
46. SZABO, B.A., BASU, P.K. and ROSSOW, M.P., Adaptive Finite Element Analysis Based on P-Convergence, *NASA Conferences Pub.* 2059, pp.43-50, (1978).
47. SZABO, B.A. and MEHTA, A.U., P-Convergent Finite Element Approximations in Fracture Mechanics, *Int.J.Num.Meths.Engng.*, 12, 551-560 (1978).
48. SZABO, B.A. and KATZ, I.N., Some Recent Developments in Finite Element Analysis, *Comp. and Maths. with Appls.*, Vol.5, pp.99-115, (1979).
49. SZABO, B.A. and DUNAVANT, D.A., An Adaptive Procedure Based on the P-Version of the Finite Element Method, *Specialists' Conf. Inst. for Physical Sci. and Technology*, University of Maryland, (1980).
50. STRANG, G. and FIX, J.G., *An Analysis of the Finite Element Method*, (1973), Prentice-Hall Inc.
51. TURCKE, D.J. and MCNEICE, G.M., Guidelines for Selecting Finite Element Grids Based on an Optimization Study, *Computers and Structures*, Vol.4, pp.499-519 (1974).
52. TURCKE, D., On Optimum Finite Element Grid Configurations, *AAA Journal*, Vol.14, (Feb. 1976).
53. WACHSPRESS, E.L., *Iterative Solution of Elliptic Systems and Applications to the Neutron Diffusion Equations of Reactor Physics*, Prentice-Hall Inc. (1966).
54. WACHSPRESS, E.L., Two-Level Finite Element Computations, Ch.31, pp.877-913, *Formulations and Computational Algorithms in Finite Elements Analysis*, Ed. Bathe, Oden Wunderlich.
55. WILSON, E.L., Finite Elements for Foundations, Joints and Fluids, Ch.10, *Finite Elements in Geomechanics*. Edited by G.Gudehus (1977), pp.319-350.
56. WILSON, E.L., Special Numerical and Computer Techniques for the Analysis of Finite Element Systems, Ch.1, pp.3-25, *Formulations and Computational Algorithms in Finite Element Analysis*, Ed. Bathe, Oden Wunderlich.
57. ZAVE, P. and RHEINBOLDT, W., Design of an Adaptive Parallel Finite Element System, *ACM Transactions on Mathematical Software*, Vol.5, No.1, (March 1979), pp.1-17.
58. ZIENKIEWICZ, O.C., IRONS, B.M., SCOTT, F.E. and CAMPBELL, J.S., High Speed Computing of Elastic Structures, *Proc. of the Symposium of International Union of Theoretical and Applied Mechanics*, Liege (1970).

59. ZIENKIEWICZ, O.C., *The Finite Element Method,* Third Edition, McGraw-Hill (1977).
60. ZIENKIEWICZ, O.C., Numerical Methods in Stress Analysis - The Basis and Some Recent Paths of Development. Contribution to a volume edited by G.S. Holister in *Developments in Stress Analysis,* (1977).
61. ZIENKIEWICZ, O.C., New Paths for the Finite Element Method, *Proceedings of the Conference on Mathematics of Finite Elements and Applications III,* Brunel University (1978).
62. ZIENKIEWICZ, O.C. and MORGAN, K., Finite Elements and Approximation (To appear).

A BOUND THEOREM FOR REDUCED INTEGRATED FINITE ELEMENTS AND ERROR ANALYSIS

D.W. Kelly

University College of Swansea

1. INTRODUCTION

The use of complementary variational principles to determine bounds on the discretization error in a finite element solution has been advocated by many authors. Principal among them has been Fraeijs de Veubeke [2]. The error analysis which results differs from the error estimates described in [7] in that they are guaranteed bounds in the energy norm reliable for coarse meshes. The error estimates of [7] on the other hand, try to predict the exact error and their reliability is historically related to adaptive processes which seek optimal mesh configurations.

Reduced integration of compatible displacement finite elements can lead to a relaxation of the constraint on the finite element displacement field. Malkus and Hughes [5] showed an equivalence between certain selective/reduced integrated elements and mixed methods. In Kelly [3,4] it was noted that solutions bounding the discretization error in the compatible model could be obtained by reduced integration and certain analogies with the complementary equilibrium formulations of de Veubeke were drawn. For the Navier equations of elasticity the first of the equilibrium conditions

$$\sigma_{ij,j} = 0 \qquad (1.1)$$

is satisfied a priori in the reduced integrated models which were considered. Interelement traction continuity, however, is only forced in an average nodal sense, so that the analogy with the complementary equilibrium formulation does not prove the existence of the bound.

In this paper an attempt is made to furnish this proof for the four-node bilinear element. It is shown that a change of integration rule can change the lower bound in energy (too stiff) solution of the compatible element to an upper bound in energy (too flexible) solution, at least in a certain class of problems. Convergence to the exact solution from the opposite sides is guaranteed by preserving the constant strain modes in the element and hence satisfaction of the patch test. Local bounds in stress have also been noted and their possible existence is discussed.

2. THE ESSENCE OF THE BOUND THEOREM

Following the analysis in [1] consider the deformation modes of a single four-node bilinear Lagrangian element given by the eigenvectors of the element stiffness matrix K, i.e. the solution of

$$K\,\underline{x}_i = \lambda_i\, I\, \underline{x}_i \qquad (2.1)$$

where the eigenvalue $\lambda_i = \underline{x}_i^t\, K\, \underline{x}_i$ is a measure of the energy in the ith deformation pattern, and the eigenvector is scaled such that $\underline{x}_i^t\, \underline{x}_i = 1$.

The deformation modes are tabled in Figure 1 for full (2 x 2) and reduced (1 x 1) Gauss integration used in the evaluation of the integrals defining K. Note that the constant stress modes 4, 5 and 6 satisfy the equilibrium equation (1.1)a priori. The two bending modes 7 and 8 become mechanisms (zero energy modes) when reduced integration is used indicating the release of constraint which makes the solution more flexible.

The essence of the bound theorem is the following. We consider the case when the deformation modes 7 and 8 can be modified so that the finite element solution based on the new modes is exact. Because the compatible, fully integrated modes are an approximation, the eigenvalues for any modified modes will be less than the eigenvalues for the fully integrated modes. For example, for a constant bending moment in the x-direction the finite element solution will be exact if the

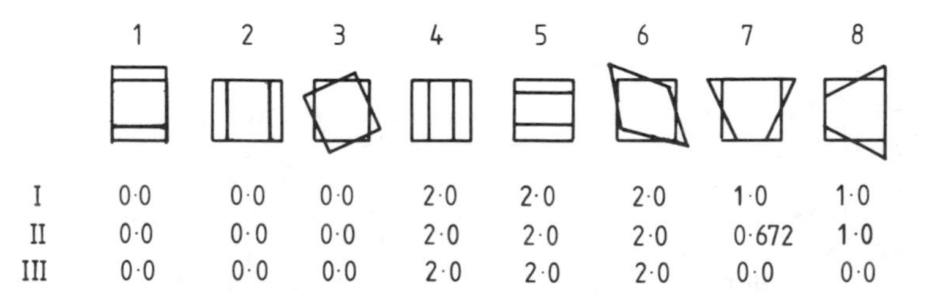

	1	2	3	4	5	6	7	8
I	0·0	0·0	0·0	2·0	2·0	2·0	1·0	1·0
II	0·0	0·0	0·0	2·0	2·0	2·0	0·672	1·0
III	0·0	0·0	0·0	2·0	2·0	2·0	0·0	0·0

FIG. 1. Deformation modes and eigenvalues of four-node element. I Eigenvalues for full integration. II Eigenvalues for full integration with new mode in Figure 2. III Eigenvalues for reduced integration.

mode shown in Figure 2 replaces mode 7. The eigenvalue for the new mode is given in II in Figure 1. If the new mode still becomes a mechanism when reduced integrated, the reduced integrated solution for the original and the exact modes will be the same. The release of constraint then ensures the reduced integrated model will provide a lower bound on the exact solution.

3. DEFINITIONS

We wish to consider any self-adjoint problem in R^n, $n \geq 2$ such that

$$Au = f \quad \text{on } \Omega \tag{3.1}$$

with boundary conditions

$$u = \tilde{u} \quad \text{on} \quad \Gamma_u \tag{3.2}$$

$$t = \tilde{t} \quad \text{on} \quad \Gamma_t \tag{3.3}$$

where $\Gamma_u \cup \Gamma_t = \Gamma$. Initially for simplicity we will restrict ourselves to the case $f = 0$ and $\tilde{u} = 0$. We also consider only the two-dimensional four-node, bilinear rectangular element. For this element define full integration as the (2 x 2) Gauss quadrature rule which integrates exactly all polynomial terms contributing to the stiffness matrix. Reduced integration uses the (1 x 1) Gauss quadrature rule.

Define $\bar{u}$ the exact solution of the problem, u_h the finite element solution using a fully integrated stiffness matrix and u_h^* the finite element solution using a reduced integrated stiffness matrix. Define $a(u,u)$ the energy inner product associated with the functional, $a_h(u,u)$ as the inner product evaluated using full numerical integration and $a_h^*(u,u)$ the inner product evaluated using reduced numerical integration.

Finally define S_h as the space containing the linear interpolating functions $N_i(\xi,\eta)$ of the compatible elements such that

$$u(\xi,\eta) = \sum_{i=1}^{4} N_i(\xi,\eta)\ \hat{u}_i \tag{3.4}$$

where the $\hat{u}_i$ are the nodal displacements. Also take the domain of the element Ω^e as $-1 \leq \xi,\eta \leq 1$.

4. THEOREMS

We aim to prove the bounding inequalities

$$a_h(u_h,u_h) \leq a(\bar{u},\bar{u}) \leq a_h^*(u_h^*,u_h^*) \tag{4.1}$$

at least for a certain class of problems. The first two

theorems given below are standard theorems proving the left-hand inequality. The third proves the right-hand inequality.

4.1 Optimality of the Finite Element Solution if $\bar{u} \varepsilon S_h$

Thereom 1: If the exact solution $\bar{u} \varepsilon S_h$ and u_h is defined by minimisation of the appropriate functional

$$a_h(u_h,u_h) = a(\bar{u},\bar{u}) \tag{4.2}$$

Proof. Define $e = \bar{u} - u_h$

then $a(e,v) = 0 \quad \forall\, v \,\varepsilon\, S_h$

but $u_h \,\varepsilon\, S_h$ and $\bar{u} \,\varepsilon\, S_h$

$\therefore u_h = \bar{u}$

and $a_h(u_h,u_h) = a(\bar{u},\bar{u})$ because the numerical quadrature is exact.

4.2 Fully Integrated Compatible Elements $\bar{u} \notin S_h$

Theorem 2: If $\bar{u} \notin S_h$

$$a_h(u_h,u_h) \leq a(\bar{u},\bar{u}) \tag{4.3}$$

Proof. u_h is defined by minimisation of a functional π and

$$\pi_{min} = -\tfrac{1}{2}\, a_h(u_h,u_h) \tag{4.4}$$

where $a_h(u_h,u_h) \geq 0$. The use of approximate deformation modes in S_h imposes constraints on the solution.

$$\therefore a_h(u_h,u_h) \leq a(\bar{u},\bar{u}) \tag{4.5}$$

4.3 Reduced Integrated Compatible Elements $\bar{u} \notin S_h$

If $\bar{u} \notin S_h$, define $\bar{S}_h$ by the addition of extra modes $\bar{N}_j$

$$u(\xi,\eta) = \sum_{i=1}^{4} N_i(\xi,\eta)\, \hat{u}_i + c_j\, \bar{N}_j(\xi,\eta) \tag{4.6}$$

such that $\dfrac{\partial \bar{N}_j}{\partial \xi}, \dfrac{\partial \bar{N}_j}{\partial \eta} = 0$ at $(0,0)$ and $\bar{u} \,\varepsilon\, \bar{S}_h$ (4.7)

$$\text{Then} \quad a_h^*(v,v) = 0 \tag{4.8}$$

for the basis vectors v of $\bar{S}_h - S_h$

Theorem 3: If the reduced integration rule ensures

$$a_h^*(u_h^*,u_h^*) = 0 \quad \forall\, u_h^* \text{ such that } A\, u_h^* \neq 0, \tag{4.9}$$

and if the additional modes $\bar{N}_j(\xi,\eta)$ satisfy (4.7) and are orthogonal to $\tilde{t}$ on Γ_t, then

$$a_h^*(u_h^*,u_h^*) \geq a(\bar{u},\bar{u}) \tag{4.10}$$

where $u_h^* \in S_h$. Here it is tacitly assumed that the stiffness matrix is nonsingular so that a solution still exists.
Proof. It is possible, with an a priori knowledge of the exact solution, to define

$$c_j = f(u_i) \tag{4.11}$$

in (4.6) and define a transformation matrix T to eliminate c_j from the solution

$$T^t\, K\, T = T^t\, \underline{P} = \underline{P}' \tag{4.12}$$

where $\underline{P}'$ is a partition of $\underline{P}$ because the additional modes are orthogonal to $\tilde{t}$ on Γ_t.

If full integration is used Theorem 1 ensures

$$a_h(u_h,u_h) = a(\bar{u},\bar{u}) \tag{4.13}$$

since $\bar{u} \in \bar{S}_h$.
If we now use reduced integration for the stiffness matrix the finite element solution u_h^* is still defined by minimisation of a functional and

$$\pi_{min} = -\tfrac{1}{2}\, a_h^*(u_h^*,u_h^*) \tag{4.14}$$

where $a_h^*(u_h^*,u_h^*) \geq 0$. Reduced integration ensures satisfaction of (4.8) and (4.9) which imply the release of constraint on the solution so that

$$a_h^*(u_h^*,u_h^*) \geq a(\bar{u},\bar{u}) \tag{4.15}$$

4.4 Alternative Boundary Conditions

If we consider the problem with boundary conditions $u = \tilde{u}$ on Γ_u, $t = 0$ on Γ_t, then

$$\pi_{min} = +\tfrac{1}{2}\, a_h(u_h,u_h) \tag{4.16}$$

and the direction of the inequalities in (4.1), (4.3) and (4.10) are reversed.

Throughout this paper it is assumed that full numerical integration is used for integrals converting the applied tractions $\underset{\sim}{t}$ to nodal loads.

5. AN EXAMPLE

Consider the cantilever beam shown in Figure 2. The exact solution is obtained from the finite element solution if we define on each element

$$u(\xi,\eta) = \sum_{i=1}^{4} N_i(\xi,\eta)\, u_i$$

$$v(\xi,\eta) = \sum_{i=1}^{4} N_i(\xi,\eta)\, v_i - 2.5(\xi^2-1)(u_1-u_2+u_3-u_4)$$

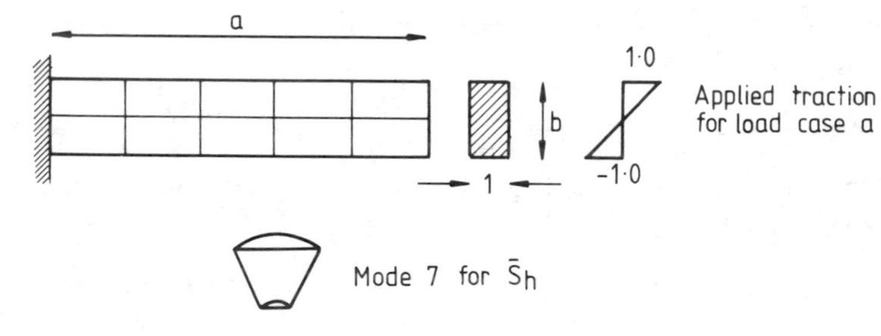

FIG. 2. Cantilever beam example a = 100, b = 2, Young's modulus = 1000, Poisson's ratio = 0.0.

TABLE 1

Strain Energy for the Cantilever Beam.
Load Case a = Tip Moment as in Figure 2,
Load Case b = Applied Tip Shear

	Description of Solution Mode	Energy Inner Product
Load Case a 4-Node Element	$u_h \in S_h$	$.0065 \times 10^{-2}$
	Beam Theory (and $u_h \in \bar{S}_h$)	$.33 \times 10^{-2}$
	$u_h^* \in S_h$	$.44 \times 10^{-2}$
Load Case b 4-Node Element	$u_h \in S_h$	1.21×10^{2}
	Beam Theory	47.61×10^{2}
	$u_h^* \in S_h$	62.87×10^{2}
Load Case b 9-Node Element (Integration rules defined in Section 6)	$u_h \in S_h$	46.94×10^{2}
	Beam Theory	47.61×10^{2}
	$u_h^* \in S_h$	52.70×10^{2}

where $N_i(\xi,\eta)$, $i = 1,4$ are the normal bilinear shape functions and the extra term in (5.1) adds a bubble to the bending mode as shown.

Since we are able to construct $\bar{S}_h$ the bound is guaranteed and the strain energy for the full and reduced integrated solutions is compared to the exact solution in the first set of results in Table 1. Monotonic convergence to the exact solution as the mesh is refined is demonstrated in [3].

The aspect ratio of the elements was deliberately chosen large to produce a strong test for the bounds as the fully integrated solution is very stiff. It is to be noted that tractions are continuous between elements in this problem. The reduced integrated model therefore satisfies both the equilibrium conditions cited in the Introduction and gives a complete complimentary formulation.

6. GENERALIZATIONS

Three generalizations are necessary. Firstly, the added modes $\bar{N}_j$ in (4.6) need not be zero at the nodes indicating it should always be possible to define the space $\bar{S}_h$. Compatibility constraints would have to take a non-standard form ensuring the continuity of the sum of contributions. This is not a problem because we do not wish to solve using $\bar{S}_h$, merely to prove its existence so that bounds are guaranteed in S_h.

Secondly, in general we wish to relax the requirement that the $\bar{N}_j$ must be orthogonal to the boundary conditions. Orthogonality ensured $\underline{P}'$ in (4.12) was a simple partition of $\underline{P}$. Note however that if the loads do not conform with this requirement we can define a statically equivalent system of loads orthogonal to $\bar{N}_j$ by a redistribution on the element. The bound then remains accurate to an order $1 + O(h)$ where h is a measure of the size of that element. In [4] special attention was paid to this point for a problem with applied displacement boundary conditions.

Extension to the case $f \neq 0$ is now apparent. The bound will be accurate to within the approximation that f can be replaced by a statically equivalent set of nodal loads which is orthogonal to the $\bar{N}_j$.

Thirdly, we wish to extend the bounds to higher order elements. Some initial work for the nine-node Lagrangian element is reported in [3]. The results given in Table 1 are for the reduced integration using a modified (2 x 2) Gauss rule sampling at $\xi,\eta = \pm 0.5$. The change from the standard Gauss rule sampling at $\xi,\eta = \pm 0.57735$ is thought necessary to ensure satisfaction of (4.9).

7. LOCAL BOUNDS

The example shown in Figure 3 is more fully described in [3]. Of significance to the present discussion is that the solution

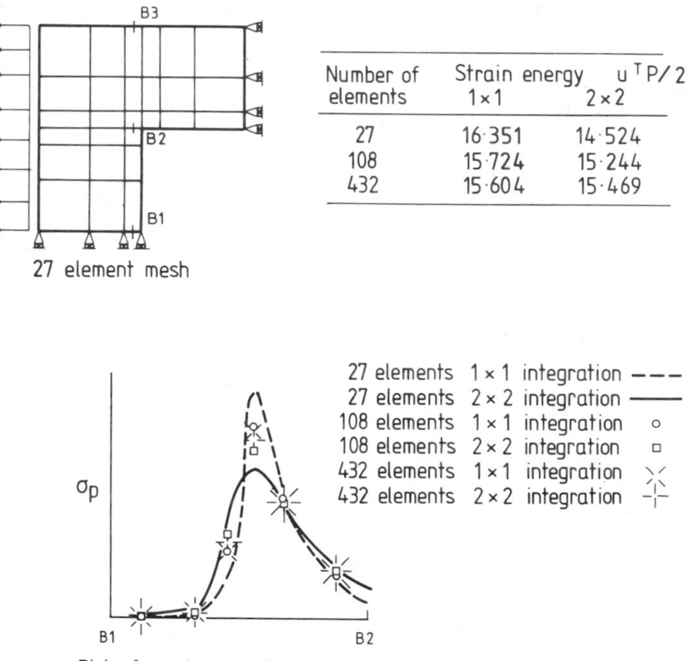

Number of elements	Strain energy $u^T P/2$ 1×1	2×2
27	16·351	14·524
108	15·724	15·244
432	15·604	15·469

FIG. 3. Global and local bounds on L-shaped domain.

exhibits a local bounding of the maximum principal stress in the region of the singularity.

The work of Moan [6] has been directed at identifying local behaviour in finite element solutions. Assymptotically local behaviour should occur if $r_p \to 0$ where

$$r_p = 1 - \frac{N_p}{\bar{N}_p} \tag{7.1}$$

where we can interpret N_p as the number of degrees of freedom in the assembled finite element model after boundary conditions have been satisfied, and $\bar{N}_p$ as the number of element deformation modes excited in the solution. The local behaviour is then characterised by a local least-square approximation for stresses.

The solution $\bar{u}_h$ which exists in $\bar{S}_h$ naturally exhibits a local (exact) fit to the exact solution. In regions of the mesh where no more than two of the constant strain modes dominate the solution both full and reduced integrated models should also exhibit a local behaviour because $N_p \sim O(2 * N_e)$ where N_e is the number of elements. We would then expect the effects of stiffening or relaxing the mesh, as identified in the theorems, to produce bounds in the neighbourhood of local perturbations. This conjecture appears to be supported by the results in Figure 3 where solutions which envelope the "exact" solution for maximum principal stress are obtained from the full and reduced integrated models.

8. CONCLUDING REMARKS

The generalizations of Section 6 ensure the existence of the bounds to an order 1 + O(h) where h is a measure of element size. However, we have not considered here the fact that the reduced integrated stiffness matrix can be singular. In some sense singularity indicates the presence of mechanisms in the mesh and hence an infinitely flexible solution. The bound theorem is therefore not violated. A more practical analysis in [3] indicates that judicious application of boundary conditions should ensure nonsingular solutions. The one special case appears to be the symmetry conditions of the example in Figure 3 and an additional constraint has to be added in that problem to obtain the solution. On the same theme a simple post-processing filter can be proposed to remove the oscillations noted in [1]. These difficulties are important for practical implementation of the bounding process indicated here and further research is needed.

ACKNOWLEDGEMENTS

The author gratefully acknowledges numerous valuable discussions with Mr. S. Nakazawa, Senior Research Assistant in the Department of Chemical Engineering, and Mr. J. Gago, Gulbenkian Foundation Fellow on leave from the Technical University of Lisbon, presently doing research in the Department of Civil Engineering, University College of Swansea.

REFERENCES

1. BICANIC, N, and HINTON, E., Spurious Modes in Two-Dimensional Isoparametric Elements. *Int. J. Num. Meth. Engng.* 14, 1545-1557 (1979).
2. FRAEIJS DE VEUBEKE, B., *Matrix Methods of Structural Analysis,* Pergamon Press (1964).
3. KELLY, D.W., Bounds on Discretization Error by Special Reduced Integration of the Lagrange Family of Finite Elements. *Int. J. Num. Meth. Engng.* 15, 1489-1506 (1980).
4. KELLY, D.W., Bounds on Discretization Error in the Finite Element Analysis of Nonlinear Problems. *Proc. First. Int. Conf. in Swansea on Numerical Methods in Nonlinear Problems,* Swansea (1980).

5. MALKUS, D.S., and HUGHES, T.J.R., Mixed Finite Element Methods - Reduced and Selective Integration Techniques: A Unification of Concepts. *Comp. Meth. Appl. Mech. Eng.* 15, 63-81 (1978).
6. MOAN, T., *Analysis of Spatial Finite Element Approximations in Structural Mechanics*. Rep. No. 76-3, The Norwegian Institute of Technology, Trondheim (1976).
7. ZIENKIEWICZ, O.C., KELLY, D.W., GAGO, J., and BABUSKA, I., 'Hierarchical' Finite Element Approaches, Error Estimates and Adaptive Refinements. Companion paper in this text.

FINITE ELEMENTS AND FINITE DIFFERENCES:
DISTINCTIONS AND SIMILARITIES

E. R. de Arantes e Oliveira

Technical University of Lisbon

1. INTRODUCTION

The finite difference method is characterized by two essential features: a local interpolation around each node and the use of the collocation technique for determining the approximate solution.

Local interpolation around each node can be used however also with finite elements if the domain is subdivided into subdomains in such a way that a unique node is located within each subdomain.

If elements of this kind are used, the finite difference and the finite element methods simply differ in that collocation is used in the first and a variational technique in the second for generating the approximate solution.

The interesting point however is that the approximate solution generated by collocation can be obtained also by a variational technique, so that parallel descriptions of both methods become possible which clarify their similarities and distinctions.

The presentation of the finite difference method as a variational technique has certainly been attempted before.

J. Cea [2], for instance, wrote on variational approximations to elliptical problems and he considered the finite difference method as providing one of such approximations. Raviart [5] did the same for certain evoltion equations.

No analogy with the finite element method has however been established in those papers. The possibility of such an analogy was later pointed out by Zienkiewicz [6]

Indices e and d will be used in the paper in connexion with finite elements and finite differences.

The following notations are adopted:

a) Real Hilbert spaces related to the domain $\Omega = \bigcup_{n=1}^{N} \Omega_n \subset \mathbb{R}^h$:

$H = L_2(\Omega)$; inner product $(u,v) = \int_\Omega uv\,d\Omega$; $\|u\| = \sqrt{(u,u)}$; $d(u,v) = \|u-v\|$.

$H_A \subset H$ — a dense linear subspace of Ω, the domain of operator A.

$H_A^e = \{u : u \in H_A; (u,v)^e = (Au,v)\}$; $\|u\|^e = \sqrt{(u,u)^e}$; $d^e(u,v) = \|u-v\|^e$.

$H_A^d = \{u : u \in H_A; (u,v)^d = (Au,Av)\}$; $\|u\|^d = \sqrt{(u,u)^d}$; $d^d(u,v) = \|u-v\|^d$.

$H \supset H^o = \{u : u = c_n$ within Ω_n; $1 \leq n \leq N$; $c_n \in \mathbb{R}\}$ (N - dimensional).

$H_A \supset H_A^o = \{u : Au \in H^o\}$ (N - dimensional).

$H_A' = \{u' : u'(\underset{\sim}{x}) = \phi_{nk}(\underset{\sim}{x}) U_k$ within Ω_n; $1 \leq n \leq N\}$ (N-dimensional).

$H_A^{e'} = \{u' : u' \in H_A'; (u',v')^e = \sum_n \int_{\Omega_n} (\underset{\sim}{R}u')^T (\underset{\sim}{R}v')\,d\Omega\}$; $\|u'\|^e = \sqrt{(u',u')^e}$.

$H_A^{d'} = \{u' : u' \in H_A'$; $(u',v')^d = \sum_n \int_{\Omega_n} (Au')(Av')\,d\Omega\}$; $\|u'\|^d = \sqrt{(u',u')^d}$.

$H \supset H' = H_A \cup H_A'$.

b) Linear spaces related to Ω_n:

$C_{An} = \{u : Au = c_n$; $c_n \in \mathbb{R}\}$.

L_n, L_{ni}^e, L_{ni}^d - the linear spaces respectively spanned by functions $\phi_{nk}(\underset{\sim}{x})$, $R_i\phi_{nk}(\underset{\sim}{x})$, $A_i\phi_{nk}(\underset{\sim}{x})$.

$L_n^d = L_{n1}^d \times L_{n2}^d \times \ldots \times L_{n,2P}^d$

P_{nq} - the space of all the polynomial functions of degree $\leq q$.

$P_{nq}^o = \{(\pi_1, \pi_2, \ldots \pi_{2p}) \in P_{nq} \times P_{nq} \times \ldots \times P_{nq} : \Pi_i = A_i p;\ p \in P_{nq} \cap C_{An}$

c) Operators:

$A : H_A \to H$ - a linear, symmetric, bounded, positive bounded below operator.

$D^q = \dfrac{\partial^q}{\partial x_1^{i_1} \ldots \partial x_h^{i_h}}$ - a qth order derivative $(q = \sum_{j=1}^{h} i_j)$; $D^o u = u$.

$A = \sum_{i=1}^{P} A_i = \sum_i^{P} a_i D_i^{2p}$ the differential operator associated with A.

$\underset{\sim}{R} = \{R_i\}$ - a differential operator such that $(Au,v) = (\underset{\sim}{R}u, \underset{\sim}{R}v) = \int_\Omega (\underset{\sim}{R}u)^T (\underset{\sim}{R}v)\,d\Omega = \int_\Omega (R_i u)(R_i v)\,d\Omega\ \forall u, v \in H_A$.

$I = \{H_A \to H_A' : u \in H_A,\ u' \in H_A',\ u'(\underset{\sim}{x}) = \phi_{nk}(\underset{\sim}{x})\, u(\underset{\sim}{x}_k)$ within $\Omega_n; 1 \leq n \leq N\}$

I^0 - the subset of I associated with $H_A^0 \subset H_A$.

$I_n = \{C_{An} \to L_n : u_n \in C_{An},\ u'_n \in L_n,\ u'_n(\underset{\sim}{x}) = \phi_{nk}(\underset{\sim}{x}) u(\underset{\sim}{x}_k)$ within $\Omega n\}$

$A^e : H'_A \to H^0$ - the approximation operator associated to the finite element approach.

$A^d : H'_A \to H^0$ - the approximation operator associated to the finite difference approach.

2. THE EQUATION

The problem consists in solving the equation

$$Au = f_o \qquad \forall f_o \in H. \tag{2.1}$$

The fact that A is positive bounded below, i.e, that a positive number γ exists such that

$$(Au, u) \geq \gamma^2 \| u \|^2 \quad \forall u \in H_A \quad , \tag{2.2}$$

ensures that the problem is well-posed.

For sake of simplicity, we assume $f^0 \in H^0$, so that $u^0 \in H_A^0$.

Equation (2.1) represents a differential equation of order 2p, together with its homogeneous principal boundary conditions, i.e., those expressed in terms of the derivatives of the function of order p or less (inclusive, p=0).

3. DISCRETIZATION

Let a set of N points selected in Ω and B points selected on $\partial\Omega$ be the internal and external nodes of Ω.

Let us subdivide Ω into N subdomains Ω_n, in such a way that each internal node becomes located within a different subdomain and not, as in the conventional finite element technique, on the subdomain boundaries.

Discretization is achieved by associating to each function $u \in H_A$ another function, u', belonging to a finite-dimensional subspace $H'_A \subset H$. We write u'=Iu and call I the interpolation operator.

Function $u' \in H'_A$ is supposed to interpolate, within each subdomain Ω_n, the values taken by u at node n and at a set of nodes located around n but external to Ω_n, and not, as in the conventional finite element technique, the values taken by u at the nodes located on the subdomain boundaries.

We use the usual summation convention and write

$$u'(\underset{\sim}{x}) = \phi_{nk}(\underset{\sim}{x})\ U_k \quad \text{within } \Omega_n \quad , \tag{3.1}$$

where $U_k = u(\underset{\sim}{x}_k)$, i.e., magnitudes U_k are the nodal values of $u\ (\underset{\sim}{x})$.

As far as the collocational approach is concerned, we assume $A\phi_{nk} = c_{nk}$ within Ω_n, c_{nk} being a non-vanishing constant. We assume thus that $L_n \subset C_{An}$.

As far as the variational approach is concerned, the functions which must not vanish are $R_i\phi_{nk}$ and not $A\phi_{nk}$. This represents of course a significant advantage with respect to the collocation technique.

The space $H'_A \subset H$ is the range of the interpolation operator.

Either d^e or d^d can be a metric on H'_A, as well as on H_A.

We can write

$$d^e(u,v) \leq \frac{1}{\gamma} d^d(u,v) \tag{3.2}$$

where γ is the positive number in (2.2).

On $H' = H_A \cup H'_A$, d^d becomes however a pseudo-metric (the approximate solution generated by collocation namely coincides with the exact solution), so that only d^e can be used on H'.

4. THE INTERPOLATION ERRORS

Let us denote with symbol Δ differences of entities associated with H_A to the same entities associated to their images in H'_A. Δu means thus $Iu-u$ and ΔF represents $F(u+\Delta u)-F(u)$, F being any function defined on H'.

The diameter of Ω_n is denoted by ℓ_n and $\ell = \max_n \ell_n$.

We assume that $\phi_{nk}(\underset{\sim}{x}) = \psi_{nk}(\underset{\sim}{y})$, where $\underset{\sim}{y} = (\underset{\sim}{x} - \underset{\sim}{x}_n)\frac{\ell_0}{\ell}$ transforms the domain Ω_n with diameter ℓ centered at $\underset{\sim}{x} = \underset{\sim}{x}_n$ into the domain Ω_{on} with diameter ℓ_o centered at $\underset{\sim}{y} = \underset{\sim}{0}$. The symbol D^α applied to functions ψ_{nk} denotes derivatives with respect to the co-ordinates y_i.

We make $S_{k\alpha} = \sup_n \sup_{\underset{\sim}{y} \in \Omega_{on}} |D^\alpha \psi_{nk}(\underset{\sim}{y})|$ and $S_\alpha = \sum_k S_{k\alpha}$. We make also $S_R = \sum_{\alpha=p} |b_\alpha|(S_\alpha + p!)$ and $S_A = \sum_{\alpha=2p} |a_\alpha|(S_\alpha + (2p)!)$.

The following theorems result which provide bounds for $\|\Delta u\|$, $\|\Delta u\|^e$, $\|\Delta u\|^d$ (see [3]).

Theorem 4.1: "If, for any n, an integer $q \geq o$ exists such that:

a) $u \in C^{q+1}(\Omega_n)$ and $\sup_{\alpha=q+1} \sup_{x \in \Omega_n} |D^\alpha u(\underset{\sim}{x})| = M_{q+1}$;

b) $P_{nq} \subset Ln$ (completeness condition);

then, $\| \Delta u \| \leq \frac{1+S_o}{(q+1)!} M_{q+1} h^{q+1} \ell^{q+1} \sqrt{(\Omega)}$".

Theorem 4.2: "If, for any n, an integer $q^e \geq o$ exists such that:

a) the condition a) of Th. 4.1 is fulfilled with $q=q^e+p$;

b) $P_{nq^e} \subset L^e_{ni}$ for each i
c) $P_{n,p-1} \subset L_n$
(completeness conditions);

then, $\| \Delta u \|^e \leq \frac{S_R}{(q^e+p+1)!} M_{q^e+p+1} h^{q^e+p+1} \ell^{q^e+1} \sqrt{(\Omega)}$".

Theorem 4.3: "If, for any n, an integer $q^d > o$ exists such that:

a) the condition a) of Th.4.1 is fulfilled with $q=q^d+2p$;

b) $P^o_{n\,q^d} \subset L^d_n$
c) $P_{n,2p1} \subset L_n$
(completeness conditions);

then, $\| \Delta u \|^d \leq \frac{S_A}{(q^d+2p+1)!} M_{q^d+2p+1} h^{q^d+2p+1} \ell^{q^d+1} \sqrt{(\Omega)}$".

5. THE APPROXIMATE SOLUTIONS

It is well-known that the solution u_o to equation (2.1) minimizes the functional $F^e(u)=(u,u)^c-2(f_o,u)$ on H_A [4].

Function u_o minimizes, also on H_A, the functional $F^d(u)=$ $=(u,u)^d-2(f_o, u)^e$.

Approximate solutions to (2.1) can be obtained by minimizing the functionals $F^e(u')=(u',u')^e-2(f_o,u')$ or $F^d(u')=(u',u')^d-$ $-2(f_o,u')^e$ on H'_A. Such minimization leads respectively to the equations:

$$A^e_{km} U_m = F^e_{ok} \quad \text{and} \quad A^d_{km} U_m = F^d_{ok}, \tag{5.1-2}$$

where

$$A^e_{km} = \sum_n (R\phi_{nk}, R\phi_{nm})_n, \; A^d_{km} = \sum_n (A\phi_{nk}, A\phi_{nm})_n ,$$

$$F^e_{ok} = \sum_n (f_o, \phi_{nk})_n \quad , \quad F^d_{ok} = \sum_n (f_o, A\phi_{nk})_n .$$

Considering that, for the collocational approach, $A\phi_{nk}$ is supposed equal to c_{nk} within Ω_n, equation (5.2) may be transformed into $\sum_n (c_{nm} U_m - f_{on}) c_{nk} \Omega_n = 0$, so that N linearly independent conditions (5.2) may be replaced by the N linearly independent conditions $c_{nm} U_m = f_{on}$ within Ω_n. This means that the approximate solution provided by the minimization of functional F^d on H'_A is the same which is generated by the collocation technique.

Operators A^e and A^d associate, to each function $u' \in H'_A$, the functions f^e and f^d of H^o which respectively take, within each subdomain Ω_n, the constant values $f^e_k = F^e_k / \Omega_k$ and $f^d_k = F^d_k / \Omega_k c_{kk}$, where $F^e_k = \sum_n (\underset{\sim}{R}\phi_{nk}, \underset{\sim}{R}u')_n$ and $F^d_k =$ $= \sum_n (A\phi_{nk}, Au')_n$.

We write therefore

$$A^e u' = f^e_o \qquad \text{and} \qquad A^d u' = f^d_o \quad , \tag{5.3-4}$$

instead of (5.1-2) and denote the approximate solutions by $u^{e'}_a$ and $u^{d'}_a$, respectively. We make also $u^e_a = (I^o)^{-1} u^{e'}_a$ and $u^d_a = (I^o)^{-1} u^{d'}_a$.

6. ON THE BOUNDEDNESS OF THE APPROXIMATION OPERATORS

The following theorems hold:

Theorem 6.1: "Let: i) $L_n \supset P_{n,2p-1}$ (completeness condition);

ii) $A^e u' = 0$ whenever u' is a polynomial of the $(2p-1)$th degree on Ω.

Then, given $u \in H^o_A$, the norm $\| A^e I^o u \|$ remains bounded, as $\ell \to 0$, iff all the derivatives of u are bounded within each Ω_n".

Theorem 6.2: "Let $L_n \supset P_{n,\, 2p-1}$ (completeness condition);

Then given $u \in H^o_A$, the norm $\| A^d I^o u \|$ remains bounded, as $\ell \to 0$, iff all the derivatives of u are bounded within each Ω_n".

As $f^e_o = A^e I u^e_a$ and $f^d_o = A^d I u^d_a$, and $\| f^e_o \|$ and $\| f^d_o \|$ are bounded, Theorems 6.1 and 6.2 ensure that magnitudes M_{q+1} of Theorems 4.1, 4.2 and 4.3 are bounded for $u = u^e_a$ and $u = u^d_a$, so that $\| u^{e'}_a - u^e_a \|$ and $\| u^{d'}_a - u^d_a \|$ are of the order of ℓ^{q+1}, $\| u^{e'}_a - u^e_a \|^e$ is of the order of ℓ^{q^e+1} and $\| u^{d'}_a - u^d_a \|$ of the order of ℓ^{q^d+1}.

Let us prove Th.6.1 and leave Th.6.2 for the reader. Assuming that all the derivatives of u are bounded within Ω_n, we prove first that the norm $\|A^e Iu\|$ must then be bounded.

Expanding indeed u within Ω_n, we find $u(\underset{\sim}{x})=p_{n,2p-1}(\underset{\sim}{x})+u^e_{n,2p}(\underset{\sim}{x})$, where $p_{n,2p-1}\in P_{n,2p-1}$, and, as all the derivatives of order 2p are bounded, $u^e_{n,2p}(\underset{\sim}{x})=O(\ell^{2p})$.

Now, by virtue of i), $P_{n,2p-1}\subset L_n$, and the function $u'=Iu$ is thus expressed, within Ω_n, by $u'(\underset{\sim}{x})=p_{n,2p-1}(\underset{\sim}{x})+\phi_{nk}(\underset{\sim}{x})U^e_{k,2p}$, where $U^e_{k,2p}=O(\ell^{2p})$.

But $\phi_{nk}(\underset{\sim}{x})=\psi_{nk}(\underset{\sim}{y})$, so that $D^p\phi_{nk}=\frac{\ell_0^p}{\ell^p}D^p\psi_{nk}$. Therefore, $R_i\phi_{nk}=O(\ell^{-p})$.

On the other hand,

$$F^e_k(u')=\sum_n(\underset{\sim}{R}\phi_{nk},\underset{\sim}{R}u')_n=\sum_n(\underset{\sim}{R}\phi_{nk},\underset{\sim}{R}\,p_{n,2p-1})_n+\sum_n(\underset{\sim}{R}\phi_{nk},\underset{\sim}{R}\phi_{n\ell})_n\,U^e_{\ell,2p},$$

and thus, by virtue of ii), $F^e_k(u')=\sum_n(\underset{\sim}{R}\phi_{nk},\underset{\sim}{R}\phi_{n\ell})_n U^e_{\ell,2p}$.

The integrand function $(\underset{\sim}{R}\phi_{nk})^T(\underset{\sim}{R}\phi_{n\ell})U^e_{\ell,2p}$ being of the order of ℓ^0, and thus bounded, $\|A^e Iu\|$ is bounded and the first part of the theorem is demonstrated.

Let us assume now that the $\|A^e Iu\|$ is bounded, and prove that all the derivatives of u are therefore bounded within each Ω_n.

Indeed, as H^0_A is a N-dimensional subspace of H, the N linearly independent functions u_n such that $Au_n=\delta_{mn}$ within Ω_m (δ_{mn}-Kronecker's symbol) form a basis for H^0_A. Function u can be expressed thus as the linear combination $u=\gamma_n u_n$. There remains to prove that the coefficients γ_n are bounded.

Applying operator I^0 to both sides, we obtain $I^0u=\gamma_n(I^0u'_n)$, and applying operator A^e, $A^e I^0u=\gamma_n f_n$, where $f_n=A^e I^0u_n$.

As both A^e and I^0 are non-singular linear transformations, their product is also a non-singular linear transformation and maps thus a basis for H^0_A into a basis for H^0. This means that functions f_n form a basis in H^0 (see [4]).

On the other hand, by virtue of the first part of the present Theorem, the norms $\|f_n\|$ are bounded, so that, as the norm of $A^e I^0u=A^e Iu$ is also bounded, the coefficients γ_n must all be bounded.

Now, the derivatives of the functions u_n are bounded, so that, as the coefficients γ_n are bounded, all the derivatives of the function $u=\gamma_n u_n$ are also bounded and the theorem is proved.

A comparison between Theorems 6.1 and 6.2 shows that no condition corresponding to ii) is needed for the second.

Now, condition ii) is associated to the so-called "patch test" [1], at least under the form it takes for two and three-dimensional Elasticity, i.e., if p=1.

For p=2, condition ii) implies that the test be satisfied by all the polynomial functions of the third degree.

This seems too severe a requirement. Let us not forget, however, that magnitudes U_k have been assumed to represent nodal values of the function (Lagrangean interpolation). Supposing that, at each node, both the nodal values of the function and of its derivatives of order up to s are interpolated (Hermitian interpolation), Theorem 6.1 has to be generalized. An adequate generalization seems (it was not proved) to consist in substituting 2p-s-1 for 2p-1 in conditions i) and ii).

7. THE ERROR INVOLVED IN THE FINITE ELEMENT TECHNIQUE

Let us assume that $\|f_o\|$ is bounded and that all the derivatives of u_o of order 2p or less are bounded within each subdomain Ω_n.

Using the inequality $d^e(u_o, u_a^{e'}) \leq \sqrt{(|\Delta_o F^e| + |\Delta_a F^e|)}$ [1] (valid if $q^e < q$), where $u_a^{e'}$ represents the approximate solution obtained from equation (5.3), $\Delta_o F^e = F^e(I u_o) - F^e(u_o)$ and $\Delta_a F^e = F^e(u_a^{e'}) - F^e((I^o)^{-1} u_a^{e'})$, the following theorem holds:

Theorem 7.1: "If a) $L_n^e \supset P^0_{nq^e}$ and $q^e < q$;

b) $L_n \supset P_{n,p-1}$;

c) the patch test is satisfied;

then, $[d^e(u_o, u_a^{e'})]^2 \leq (M_{o,q^e+p+1} \|u_o\|^e + M_{a,q^e+p+1} \|u_a^e\|^e) \cdot$

$$\cdot \frac{S_R}{(q^e+p+1)!} h^{q^e+p+1} \ell^{q^e+1} \sqrt{(\Omega)}\text{"}.$$

The discretization error is thus of the order of $\ell^{\frac{q^e+1}{2}}$.

Theorem 7.1 can easily be generalized for $q^e \geq q$.

8. THE ERROR INVOLVED IN THE FINITE DIFFERENCE TECHNIQUE

Using the inequality $d^e(u_o, u_a^{d'}) \leq \frac{1}{\gamma} \sqrt{(|\Delta_o F^d| + |\Delta_a F^d|)}$, where γ is the positive number in inequality (2.2) (remember that d^d is a pseudo-metric in H') would lead to a upper bound of the order of $\ell^{\frac{q^d+1}{2}}$.

A more refined analysis is however possible which leads to a lower upper bound.

Indeed, $[d^d(u_o, u_a^d)]^2 = (A(u_o - u_a^d), A(u_o - u_a^d))$. By virtue of collocation, Au_o coincides with $Au_a^{d'}$ within each subdomain Ω_n, so that $(A(u_o - u_a^d), A(u_o - u_a^d)) = \sum_n (A(u_a^{d'} - u_a^d), A(u_a^{d'} - u_a^d))_n = \| A \Delta u_a^d \|^2$, i.e., $d^d(u_o, u_a^d) = \| A \Delta u_a^d \|$, and, by virtue of (3.2), $d^e(u_o, u_a^d) \leq \frac{1}{\gamma} \| A \Delta u_a^d \|$.

Now, as $d^e(u_o, u_a^{d'}) \leq d^e(u_o, u_a^d) + d^e(u_a^d, u_a^{d'})$, and $d^e(u_a^d, u_a^{d'}) = \| \Delta u_a^d \|^e$, there results $d^e(u_o, u_a^d) \leq \frac{1}{\gamma} \| A \Delta u_a^d \| + \| \Delta u_a^d \|^e$.

If $q^d < q^e$, $\| \Delta u_a^d \|^e$ can be negleted once compared to $\| A \Delta u_a^d \|$, so that we obtain $d^e(u_o, u_a^{d'}) \leq \frac{1}{\gamma} \| A \Delta u_a^d \|$.

The following Theorem immediately results from this inequality and Theorem 4.3:

Theorem 8.1: "If a) $L_n^d \supset P_{nq^d}^o$

b) $L_n \supset P_{n,2p-1}$

then,

$$d^e(u_o, u_a^d) \leq \frac{1}{\gamma} \frac{S_A}{(q^d+2p+1)!} M_{q^d+2p+1} h^{q^d+2p+1} \ell^{q^d+1} \sqrt{(\Omega)}"$$

The discretization error is thus of the order of ℓ^{q^d+1}. Attention is called to the fact that only completeness conditions have to be satisfied.

9. CONCLUSIONS

The fact that the approximate solution generated by collocation can be obtained also by a variational technique make parallel descriptions of the finite difference and finite element methods possible and allows a comparison between both which clarify their relative advantages and disadvantages.

The relative advantages of the finite element method are associated with the fact that the function derivatives involved in the functional are, in the finite difference case, of the same order (2p) as those involved in the differential equation, while they are of half such order in the finite element method.

Owing to this circumstance, the co-ordinate functions are required to satisfy more severe conditions in the finite difference method than in the finite element method. In the finite difference case, functions ϕ_{nk} are indeed supposed to belong to C_{An}, i.e., to be such that their A-images take non-vanishing constant values within subdomain Ω_n, while, in the finite element case, the requirement simply is that the associated derivatives of order p involved in operator $\underset{\sim}{R}$ do not vanish everywhere on Ω_n.

The completeness conditions are also more severe in the

finite difference than in the finite element method, as they require a complete polynomial of the (2p-1)th degree to be contained in the expression of the function in the first case, and a complete polynomial of the (p-1)th degree in the second.

The relative advantages of the finite difference method stem from the fact that the associated approximate solution can be obtained using the collocation technique.

The first of such advantages is that, in the finite difference technique, the degree of accuracy depends exclusively on the degree of completeness, or, consequently, that no conditions other than the completeness conditions, are necessary for convergence.

The finite element method requires, on the contrary, the satisfaction of the completeness conditions and of some supplementary condition, like the patch test[1].

In the present paper, the patch test is considered for differential equations of order higher than the second. Confusion should not be made between this kind of patch test for higher--order equations and the so-called higher-order patch test which was not proved to be necessary [1].

The second advantage of the finite difference method consists in that an upper bound of the discretization error, lower than the one provided by the approximation theorem usually used by the author for finite element accuracy analysis[1], can be determined which shows that the error is of the order of ℓ^{q^d+1}, while in the finite element method is of the order of $\ell^{\frac{q^e+1}{2}}$.

According to such conclusion, the discretization error involved in the finite difference method can easily be lower than the error involved in the finite element method.

AKNOWLEGEMENT

The research report in this paper was supported by the Instituto Nacional de Investigação Científica (INIC) through the "Centro de Mecânica e Engenharia Estruturais da Universidade Técnica de Lisboa (CMEST)".

The author is grateful to Dr. Andrzej Karafiat of the Technical University of Cracow for his collaboration, from which Section 4 especially benefited.

REFERENCES

1. ARANTES E OLIVEIRA, E.R., The Patch Test and the General Convergence Criteria of the Finite Element Method, *Int.J. Solids and Structures* 13, 159-178 (1977).

2. CEA, J., Approximation Variationelle des Problèmes aux Limites, *Ann. Int. Fourier, Grenoble,* 14, 345-444 (1964).

3. CIARLET, P.G.; RAVIART, P.A., General Lagrange and Hermite Interpolation in R^n with Applications to Finite Element Methods, *Arch Rational Mech. Anal.,* 46, 177-199.

4. ODEN, J.T., *Applied Functional Analysis,* Prentice-Hall (1979).

5. RAVIART, P.A., Sur l'Approximation de Certaines Équations d'Évolution Linéaires et Non-Linéaires, *J. de Math. Pures et Appliquées,* 46, 11-107 (1967).

6. ZIENKIEWICZ, O.C., Finite Elements - The Background Story, pp. 1-35 of J. R. Whiteman (ed.), The Mathematics of Finite Elements and Applications. Academic Press, London (1973).

ANALYSIS OF DISCRETIZATIONS BY THE CONCEPT OF DISCRETE REGULARITY

W. Hackbusch

Ruhr-Universität Bochum

1. INTRODUCTION

The variational formulation of boundary value problems is the essential basis of all error estimates for finite element solutions. The typical form of such estimates is

$$\|u - u_h\|_{H^s(\Omega)} \le C\, h^{t-s} \|u\|_{H^t(\Omega)} \qquad (s \le t) \tag{1.1}$$

with numbers s and t varying in suitable intervals. u denotes the continuous solution, while u_h is the solution with discretization parameter h. $H^r(\Omega)$ is the usual Sobolev space of order r.

Dissimilar techniques are used to prove error estimates for *finite difference* methods. Usually, the resulting inequalities are less optimal than (1.1). In this contribution we describe a *technique that can be applied to finite element discretizations as well as to difference schemes*. This uniform approach leads to error estimates similar to (1.1). The concept is based on the '*discrete regularity*' described in §2. We discuss this property for difference schemes and finite element methods. The novel error estimates are derived in §3.

Since inequality (1.1) is well-known for finite element solutions, the concept of discrete regularity might seem to be useless for finite element methods. On that account we give some applications in §4. Super-convergence, the influence of numerical quadrature, and special upwind schemes are analysed.

The discrete regularity is also an important tool for the analysis of *nonlinear* problems as shown in §6.

2. DISCRETE REGULARITY

Discrete regularity is the counterpart of a well-known property of the continuous differential operator. It requires the introduction of discrete Sobolev norms. §2.1 contains some criteria implying discrete regularity. For finite difference schemes discrete regularity is a stronger assumption than stability. §2.3 shows that discrete regularity holds for finite element methods.

2.1 Definition and Criteria

The term 'regularity' is well-known for the (continuous) boundary value problem. Denote the problem by

$$Lu = f \qquad \text{(+ homogeneous boundary conditions)} \qquad (2.1)$$

and assume for simplicity that the differential operator is of second order. We call L *H^t-regular* if $f \in H^{t-2}(\Omega)$ implies $u \in H^t(\Omega)$:

$$\| L^{-1} \|_{H^{t-2}(\Omega) \to H^t(\Omega)} \leq C. \qquad (2.2)$$

If $t-2$ is negative, $H^{t-2}(\Omega)$ denotes the dual space of the Sobolev space $H^{2-t}(\Omega)$ or of a subspace. The variational formulation of problem (2.1) by means of a H^1-coercive bilinear form yields H^1-regularity: (2.2) holds for $t=1$. In order to ensure (2.2) for $t>1$ the boundary of Ω and the coefficients of L must be sufficiently smooth.

The discretization of problem (2.1) by finite differences or finite elements leads to the system

$$L_h u_h = f_h \qquad (2.3)$$

of linear equations. In the case of a finite difference scheme, u_h is a grid function defined on some grid Ω_h. In the case of a finite element method, u_h is the coefficient vector describing a function of the finite element subspace. For instance, the vector u_h may consist of the values at the nodal points of a triangulation.

Often, the Euklidean norm of u_h is called l_2-norm. It may be regarded as discrete analogue of the $L^2(\Omega)$-norm. We use the notation L_h^2 or H_h^o instead of l_2. As a characteristic feature of our approach we need also H_h^t-norms that are discrete analogues of the norms of the Sobolev spaces $H^t(\Omega)$ for $t>o$. For more or less natural definitions of these discrete Sobolev norms compare §§2.2, 2.3. The space H_h^{-t} ($t>o$) is equipped with the dual norm of H_h^t with respect to the l_2 scalar product.

Discrete H_h^t-regularity is the discrete counterpart of inequality (2.2):

$$\| L_h^{-1} \|_{H_h^{t-2} \to H_h^t} \leq C \qquad (C \text{ independent of } h) \qquad (2.4)$$

This is a stronger assumption than *H_h^t-stability:*

$$\| L_h^{-1} \|_{H_h^t \to H_h^t} \leq C.$$

H_h^o-stability (l_2-stability) is well-known for difference schemes. It implies $\|u_h\| \leq C\|f_h\|$ for the Euklidean norm $\|\cdot\| = \|\ \|_{H_h^o}$. From discrete H_h^2-regularity one obtains $\|u_h\|_{H_h^2} \leq C\|f_h\|_{H_h^o}$.

There are useful criteria for proving discrete regularity. The first lemma is trivial.

LEMMA 2.1 H_h^1*-coerciveness* $\langle u_h, L_h u_h\rangle \geq \varepsilon \|u_h\|^2_{H_h^1}$ *implies* H_h^1*-regularity.*

The following lemma proves discrete H_h^t-regularity by means of continuous $H^t(\Omega)$-regularity. The precise formulation of the lemmata 2.2 - 2.4 can be found in [3].

LEMMA 2.2 The essential conditions for H_h^t*-regularity of* L_h *are*
(i) $H^t(\Omega)$*-regularity of the continuous operator* L*,*
(ii) discrete H_h^s*-regularity of* L_h *for some* $s<t$*,*
(iii) a special form of $O(h^{t-s})$*-consistency of* L_h*.*

The next lemma shows that a perturbation by a lower order term does not destroy H_h^t-regularity. Since we assumed L to be of second order, a lower order term is a discrete analogue of a differential operator of order <2.

LEMMA 2.3 Let $L_h = \hat{L}_h + l_h$*.* L_h *is* H_h^t*-regular, if (i)* $\hat{L}_h$ *is* H_h^t*-regular, (ii)* l_h *is of lower order than* L_h*, (iii)* L_h *is* H_h^s*-stable for some* $s \in [t-2, t]$*.*

Condition (iii) is necessary to exclude a singular matrix L_h. But (iii) may be replaced by the invertibility of the continuous operator L.

LEMMA 2.4 Let $L_h = \hat{L}_h + l_h$ *be the discretization of* $L = \hat{L} + l$*. Assume that (i)* $\hat{L}_h$ *is* H_h^t*-regular, (ii)* l_h *is of lower order, (iii) consistency of* $\hat{L}_h$ *and* l_h*, (iv) zero is no eigenvalue of* L*, (v)* h *is sufficiently small. Then* L_h *is* H_h^t*-regular, too.*

Applications of these lemmas are mentioned below.

2.2 Discrete Regularity of Difference Schemes

For grid functions of a uniform mesh the norm of H_h^k can be defined by means of the k-th differences. Possibly, special modifications must be used near the boundary.

In [2] we proved the following theorem for general difference schemes approximating the problem (2.1) subject to homogeneous Dirichlet boundary conditions. *The difference operator* L_h *is* H_h^t*-regular for all* $t \in (1/2, 3/2)$*, if (i)* Ω *is a Lipschitzian domain, (ii)* L_h *is discrete-elliptic and* H_h^0*-stable (=* l_2*-stable), (iii) the coefficients of* L_h *are Hölder continuous.* This result is the precise analogue of a theorem of Nečas for the continuous operator L.

Discrete regularity of higher order can be proved in the following way. Applying Lemma 2.1 to the principal part $\hat{L}_h$ of the difference operator L_h, one obtains H_h^1-regularity of $\hat{L}_h$. Then, prove H_h^t-regularity of $\hat{L}_h$ by means of Lemma 2.2 (with s=o). The H_h^t-regularity of L_h is a consequence of Lemma 2.3 or Lemma 2.4.

By this technique it is possible to prove H_h^1 and H_h^2-regularity of classical difference schemes. These schemes may also use special discretizations at points near to the boundary. Examples are the Shortley-Weller scheme (cf. [2],[3]), interpolation and extrapolation techniques at points near the boundary.

2.3 Discrete Regularity of Finite Element Discretizations

Let $H_h \subset H^1(\Omega)$ be a finite element subspace of dimension N. After a suitable choice of a basis, there is a bijection (prolongation) $P_h: R^N \to H_h$. The coefficient vector u_h corresponds to the function $u = P_h u_h \in H_h$. A natural choice of the discrete norms is

$$\|u_h\|_{H_h^s} := \|P_h u_h\|_{H^s(\Omega)} \qquad (s \geq 0)$$

as long as $H_h \subset H^s(\Omega)$. If $H_h \not\subset H^2(\Omega)$ this definition fails for H_h^2. But for $0 \leq s \leq$ order of H_h, an extension is given by

$$\|u_h\|_{H_h^s} := \inf\{\ \|u\|_{H^s(\Omega)} + h^{-s}\|R_h u - u_h\|_{H_h^0}:\ u \in H^s(\Omega)\} \qquad (2.5)$$

where $R_h: L^2(\Omega) \to R^N$ is a suitable restriction. For example, $P_h R_h$ may be the L^2 projection to H_h. In case of a *uniform* triangulation, H_h^k-norms can be defined by virtue of k-th differences as discussed in §2.2.

Now we show that H_h^t-regularity ($0 \leq t \leq 2$) is a simple consequence of the usual error estimates. For given f_h let u be the solution of $Lu = f := P_h f_h$. The continuous H^2-regularity yields $\|u\|_2 = C\|f_h\|_0$. Hence, $\|R_h u - u_h\|_0 \leq C(\|P_h R_h u - u_h\|_0 + \|u - P_h u_h\|_0) \leq C' h^2 \|u\|_2 \leq C'' h^2 \|f_h\|_0$ holds for the finite element solution u_h. Definition (2.5) shows $\|u_h\|_{H_h^2} \leq C\|f_h\|_{H_h^0}$. Therefore, H_h^2-regularity is proved for L_h. Similarly, H_h^1-regularity can be shown. By duality also $L_h: H_h^{-2} \to H_h^0$ is uniformly bounded. Thus one obtains H_h^0-regularity, too. More generally, it is possible to prove

LEMMA 2.5 If the finite element subspace is of order k (i.e. $\inf\{\|u - v\|_0: v \in H_h\} \leq Ch^k \|u\|_k$) and if L and its adjoint L satisfy $H^k(\Omega)$-regularity, then L_h is H_h^s-regular for $s \in [2-k, k]$.*

Here some technical conditions (inverse assumption, stability of P_h) are omitted.

3. ERROR ESTIMATES

The continuous solution $u = L^{-1} f$ and the discrete solution $u_h = L_h^{-1} f_h \in R^N$ cannot be compared directly. We need some restriction $R_h: L^2(\Omega) \to R^N$ in order to estimate the difference $R_h u - u_h$. Also the right-hand side f_h is some (other) restriction $R_h' f$ of f. The representation

$$R_h u - u_h = L_h^{-1}\ (L_h R_h - R_h' L)\ u$$

yields the following immediate result.

LEMMA 3.1 Assume that L_h is H_h^s-regular and that L_h satisfies the consistency condition $\|L_h R_h - R_h' L\|_{H^t(\Omega) \to H_h^{s-2}} \leq Ch^{t-s}$. Then,

$$\|R_h u - u_h\|_{H_h^s} \leq C\, h^{t-s} \|u\|_{H^t(\Omega)} \qquad (3.1)$$

holds.

In particular, this formulation is suitable for difference schemes. For finite element discretizations an alternative version is more convenient. Let P_h denote the prolongation $R^N \to H_h$ as in §2.3 and define $R_h = P_h^*$ (adjoint of P_h). The splitting

$$P_h u_h - u = (P_h L_h^{-1} R_h L - I)(I - P_h R_h')u + P_h L_h^{-1}(R_h L P_h - L_h) R_h' u$$

(R_h': $L^2(\Omega) \to R^N$ arbitrary) results in the following statement.

LEMMA 3.2 The error estimate

$$\|P_h u_h - u\|_{H^s(\Omega)} \leq C h^{t-s} \|u\|_{H^t(\Omega)} \qquad (t \geq s) \qquad (3.2)$$

holds if (i) L_h is H_h^s-regular, (ii) approximation property $\|I - P_h R_h'\|_{H^t(\Omega) \to H^s(\Omega)} \leq Ch^{t-s}$ is valid for a suitable R_h', (iii) consistency condition $\|R_h L P_h - L_h\|_{H_h^t \to H_h^s}$ $\varepsilon \leq Ch^{t-s}$ holds, (iv) P_h, R_h and R_h' satisfy some stability estimates.

Note that condition (iii) is trivially satisfied for finite element methods, since $L_h = R_h L P_h$ is the definition of the stiffness matrix L_h. Condition (ii) with $0 \leq s \leq t \leq k$ holds for finite element subspaces H_h of order k.

We summarize: The usual error estimates of finite element methods yield discrete regularity of L_h. On the other hand, by virtue of Lemma 3.2 the error estimates can be regained from discrete regularity.

Finally we remark that *interior error estimates* can be obtained by 'interior regularity' estimates. Interior regularity is discussed, e.g., by Thomée and Westergren [8].

4. APPLICATIONS FOR FINITE ELEMENT DISCRETIZATIONS

In the sequel we list some problems that can be analysed very elegantly by the concept of discrete regularity.

4.1 Super-Convergence

Usually, the error estimate $O(h^k)$ with respect to $L^2(\Omega)$ corresponds to the worse estimate $O(h^{k-1})$ with respect to $H^1(\Omega)$. Nevertheless, first *differences* of the nodal values u_h may be accurate of order $O(h^k)$. In this case we are led to inequality (3.1) with H_h^1-norm defined by first differences. Since the stiffness matrix L_h is H_h^1-regular (cf. Lemma 2.5), only the consistency condition $\|L_h R_h - R_h' L\|_{H^{k+1}(\Omega) \to H_h^{-1}} = O(h^k)$ has to be verified. In particular in the case of a *uniform* triangulation, it should be possible to prove the consistency condition. In the same manner it is possible to analyse the accuracy of second differences of the nodal points.

If the triangulation is uniform in an inner part of Ω, one can make use of the interior error estimate mentioned in §3.

4.2 Upwind Finite Element Schemes

The usual finite element discretization of

$$-\kappa\Delta\Theta + \underline{u}\cdot grad\,\Theta = f \quad in \quad \Omega$$

may become instable for small values of κ. One remedy is the use of an upstream *difference* formula for the advection term $\underline{u}\cdot grad\,\Theta$. The directional derivative can be approximated by suitable differences of the nodal values u_h (cf. Thomasset [7]). The resulting scheme is no longer based on a variational formulation. Therefore error estimates cannot be obtained by classical considerations.

The concept of discrete regularity can be applied in the following way. Let $\hat{L}_h$ be the stiffness matrix corresponding to $\hat{L}=-\kappa\Delta$. Since the difference analogue l_h of $\underline{u}\cdot grad$ is of lower order than L_h, Lemma 2.3 implies H_h^t-regularity ($o\leq t\leq 2$). As shown in §3, this result yields the usual error estimates (3.2). Of course, it is a further problem to analyse the influence of the parameter κ on the constant C.

4.3 Numerical Quadrature

Consider the finite element subspace of piece-wise linear functions of a triangulation τ. For the integration of the scalar product $(\nabla u)^T A(x,y)\nabla u$ $(u,v\in H_h)$ over $T\in\tau$ use a quadrature formula of order 2. The resulting stiffness matrix can be written as $L_h = \hat{L}_h + l_h$, where L_h is the result of exact integration. Thanks to Lemma 2.5, $\hat{L}_h$ is H_h^t-regular ($o\leq t\leq 2$). Assuming that the coefficient $A(x,y)$ is twice differentiable, one obtains the estimate $O(h^2)$ for l_h as a mapping from H_h^1 into H_h^{-1}. Therefore, l_h may be regarded as a (uniformly bounded) difference operator of order zero. Lemma 2.4 ensures H_h^t-regularity ($o\leq t\leq 2$) of L_h, too. By Lemma 3.2, the usual error estimate (3.2) ($o\leq s\leq 1\leq t\leq 2$) holds also for the finite element discretization with numerical quadrature.

4.4 Saddle Point Problems

Several *systems* of differential equations are not the solution of a classical minimization problem but of a saddle point problem. An example is Stokes' equation

$$-\Delta\underline{u} + grad\; p = \underline{f}, \qquad div\;\underline{u} = o \qquad in\;\Omega, \tag{4.1}$$

where $\underline{u}=(u_1,\ldots,u_n)$ is the vector of velocities. Error estimates of such finite element solutions are much more complicated than classical estimates (cf. references in [5] and [7]).

The analysis of discrete saddle point problems by the concept of discrete regularity is described in [5]. The considerations are as follows. As other systems (e.g., the mixed formulation of the biharmonic equation) the system (4.1) is of the form

$$\begin{bmatrix} A & B \\ C & O \end{bmatrix} w = g, \text{ here: } A=-\Delta,\ B=grad,\ C=-div,\ w=\binom{u}{p},\ g=\binom{f}{o} \quad (4.2)$$

Similarly, the finite element discretization results in a system

$$\begin{bmatrix} A_h & B_h \\ C_h & O \end{bmatrix} w_h = g_h. \quad (4.3)$$

Since A_h is the usual finite element discretization of $A=-\Delta$, A_h is H_h^t-regular (cf. Lemma 2.5). The solution of system (4.3) requires the invertibility of the product

$$D_h = C_h A_h^{-1} B_h.$$

A certain form of the discrete regularity of D_h is equivalent to the well-known *Brezzi condition*. In [5] useful criteria for the estimation of D_h^{-1} can be found. The discrete regularity of D_h can be reduced to the continuous regularity of $D=CA^{-1}B$, the coerciveness of C_hB_h, and the consistency of A_h, B_h, and C_h. Optimal error estimates can be obtained immediately from discrete regularity of A_h and D_h and from consistency of A_h, B_h, C_h.

5. ANALYSIS OF MULTI-GRID METHODS BY DISCRETE REGULARITY

Multi-grid algorithms are one of the most efficient methods for the numerical solution of the discrete systems (2.3) or (4.3) of linear equations. Multi-grid methods are iterative methods. The computational work of one iteration corresponds to few relaxation steps, but the rate of convergence is a *small* number *independent* of the discretization parameter h. For a precise proof of the convergence of the multi-grid iteration an error estimate similar to (3.2) is needed. For two discretization parameters h and H $(h<H)$ the inequality

$$\| u_h - pu_H \|_{H_h^s} \le C\, H^{t-s}\, \| f_h \|_{H_h^{t-2}} \quad (5.1)$$

must hold for $u_h=L_h^{-1}f_h$, $u_H=L_H^{-1}f_H$, $f_H:=rf_h$, and for some values $t>s$. p is a prolongation from mesh size H to size h, while r is a restriction from size h to H. In the case of general difference schemes, there seems to be no other proof technique than discrete regularity (cf. [4]).

It should be emphasized that with the same efficiency multi-grid methods can be applied to *nonlinear* problems. In this case (5.1) must be valid for the derivative L_h of the nonlinear mapping $\mathcal{L}_h$ at the solution u_h of the discrete problem

$$\mathcal{L}_h(u_h) = o, \quad (5.2)$$

which is a discretization of the nonlinear boundary value problem

$$\mathcal{L}(u) = o. \quad (5.3)$$

6. ANALYSIS OF DISCRETE NONLINEAR PROBLEMS

Let u^* be one (not necessarily unique) solution of the continuous problem (5.3) and define a suitable restriction $u_h^* = R_h u^*$. The *consistency* of $\mathcal{L}_h$ can be expressed by

$$\| \mathcal{L}_h(u_h^*) \|_{H_h^{s-2}} \leq Ch^\kappa \tag{6.1}$$

since $\mathcal{L}_h(u_h^*) = (\mathcal{L}_h R_h - R_h' \mathcal{L})(u^*)$. Let $L_h(u_h) := \partial \mathcal{L}_h(u_h)/\partial u_h$ be the derivative and assume the *Lipschitz condition*

$$\| L_h(u_h) - L_h(u_h') \|_{H_h^s \to H_h^{s-2}} \leq Ch^{-\mu} \| u_h - u_h' \|_{H_h^s} \tag{6.2}$$

for all $u_h, u_h' \in \{ v_h : \| v_h - u_h^* \|_{H_h^s} \leq rh^\mu \}$ and all $\mu \in (\eta, \kappa)$, where $r > o$ and $\eta \in (o, \kappa)$ are fixed. Because of the negative power $h^{-\mu}$ condition (6.2) is very weak. H_h^s*-regularity* is required only at u_h^*:

$$\| L_h^{-1}(u_h^*) \|_{H_h^{s-2} \to H_h^s} \leq C. \tag{6.3}$$

The following result is a special form of the Newton-Kantorovič theorem. It is proved in [3].

THEOREM 6.1 Let s, κ, $\eta \in (o, \kappa)$ be fixed numbers and assume consistency (6.1), Lipschitz condition (6.2), and discrete regularity (6.3). Then, for sufficiently small h ($h < h_o$) there is a unique solution u_h of the discrete nonlinear problem (5.2) satisfying the error estimate

$$\| u_h - u_h^* \|_{H_h^s} \leq C h^\kappa .$$

Applications to Navier-Stokes equations are discussed in [5].

REFERENCES

1. D'JAKONOV, E.G., *Raznostnye metody rešenija kraevych zadač I.* Is-vo MGU, Moscow (1971).
2. HACKBUSCH, W., On the regularity of difference schemes. *Arkiv för Matematik* 19, 71-95 (1981).
3. HACKBUSCH, W., Regularity of difference schemes - Part II: Regularity estimates for linear and nonlinear problems. Report 8o-13, Mathematisches Institut, Universität zu Köln (198o).
4. HACKBUSCH, W., Convergence of multi-grid iterations applied to difference equations. *Math. Comp.* 34, 425-44o (198o).
5. HACKBUSCH, W., Analysis and multi-grid solution of mixed finite element and mixed difference equations. Preprint (198o).
6. LAPIN, A.V., Study of the W_2^2-convergence of difference schemes for quasilinear elliptic equations. *Ž. vyčhisl. Mat. mat. Fiz.* 14, 1516-1525 (1974).
7. THOMASSET, F., Finite element methods for Navier-Stokes equations. In: *Computational Fluid Dynamics, vol. 2, Lecture Series 198o-5*, von Karman Institute, Rhode St. Genese (198o).
8. THOMÉE, V. and WESTERGREN, B., Elliptic difference equations and interior regularity. *Numer. Math.* 11, 196-21o (1968).

THE APPLICATION OF ITERATED DEFECT CORRECTION TO FINITE ELEMENT METHODS

Jörg Hertling

Technical University of Vienna

1. INTRODUCTION

We consider a formally self-adjoint, uniformly elliptic differential operator

$$Lu := \sum_{j,k=1}^{2} - \frac{\partial}{\partial x_j}\left(\mu_{jk}(\underline{x})\frac{\partial u}{\partial x_k}\right) + c(\underline{x})u \qquad (1.1)$$

with coefficients μ_{jk} and c defined on a simply connected domain $\Omega \subset \mathbb{R}^2$, $\underline{x} = \{x_1,x_2\}^T$. We assumc $\mu_{jk} \equiv \mu_{kj}$, $\mu_{jk} \in \mathbb{C}^1(\Omega)$, $c \in \mathbb{C}(\Omega)$.

With this differential operator we consider the Dirichlet problem

$$\begin{aligned} Lu &= f(\underline{x},u) \\ u|_\Gamma &= g \end{aligned} \qquad (OP) \qquad (1.2)$$

where the boundary Γ of Ω is either polygonial or piecewise smooth and approximated by a polygon. Furthermore $c(\underline{x}) \geq 0$ for all $\underline{x} \in \bar{\Omega}$, $g \in \mathbb{C}(\Gamma)$, $g(\underline{x}) > 0$. We call the exact solution of this problem z.

If one defines a scalar product

$$[u,v]_L := (Lu,v) = \int_\Omega \left(\sum_{j,k=1}^{2} \mu_{jk} \frac{\partial u}{\partial x_j} \frac{\partial v}{\partial x_k} + c\,u\,v \right) d\Omega \quad (1.3)$$

the method of finite elements consists in minimizing the functional

$$F[u] = [u,u]_L + 2 \int_\Omega \left(\int_0^u f(\underline{x},v)\,dv \right) d\,\Omega \quad (1.4)$$

in a finite dimensional subspace of the space of the solution which satisfies the boundary conditions of the Dirichlet problem.

As finite dimensional subspace we choose the space of continuous, piecewise linear functions which may be generated by a triangulation of Ω. As basis functions (finite elements) of this space we choose pyramids which have their peaks in the vertices of the triangulation of Ω.

Now we will briefly sketch the idea of iterated defect correction: We assume that by means of our finite element method we have obtained approximate values

$$\eta_\nu^{[0]} \approx z(\underline{x}_\nu)$$

at the vertices $\underline{x}_\nu$ of our triangulation of Ω. We interpolate these approximate values $\eta_\nu^{[0]}$ by a bivariate interpolating function $P_h^{[0]}(\underline{x})$:

$$P_h^{[0]}(\underline{x}_\nu) = \eta_\nu^{[0]}.$$

Then we construct a new boundary value problem which has $P_h^{[0]}$ as exact solution:

$$\begin{aligned} Lu &= f(\underline{x},u) + LP_h^{[0]} - f(\underline{x},P_h^{[0]}(\underline{x})) \\ u|_\Gamma &= P_h^{[0]}|_\Gamma \end{aligned} \qquad \text{(NP)} \qquad (1.5)$$

If $\eta_\nu^{[0]}$ is "sufficiently close" to $z(\underline{x}_\nu)$ and $P_h^{[0]}(\underline{x})$ has been chosen in an "appropriate" class of interpolating functions, then the perturbation

$$d_h^{[0]}(\underline{x}) := LP_h^{[0]} - f(\underline{x},P_h^{[0]}(\underline{x}))$$

is small and therefore we call (1.5) "neighbouring problem" (NP) of the "original problem" (OP) (1.2).

In spite of our knowledge of the exact solution $P_h^{[0]}(\underline{x})$ of our NP (1.5) we solve this NP with the same finite element method as our original problem (OP) (1.2). We obtain approximate values

$$\pi_\nu^{[0]} \approx P_h^{[0]}(\underline{x}_\nu).$$

Since the NP is close to the OP we use now the well known global discretization error

$$\pi_\nu^{[0]} - P_h^{[0]}(\underline{x}_\nu)$$

of our NP as an estimate for the unknown global discretization error

$$\eta_\nu^{[0]} - z(\underline{x}_\nu)$$

of our OP: In the identity

$$z(\underline{x}_\nu) = \eta_\nu^{[0]} - (\eta_\nu^{[0]} - z(\underline{x}_\nu))$$

we substitute $\eta_\nu^{[0]} - z(\underline{x}_\nu)$ by its estimation $\pi_\nu^{[0]} - P_h^{[0]}(\underline{x}_\nu)$ and obtain a presumably better approximation

$$\eta_\nu^{[1]} := \eta_\nu^{[0]} - (\pi_\nu^{[0]} - P_h^{[0]}(\underline{x}_\nu))$$

for $z(\underline{x}_\nu)$.

Obviously this procedure can be iteratively repeated.

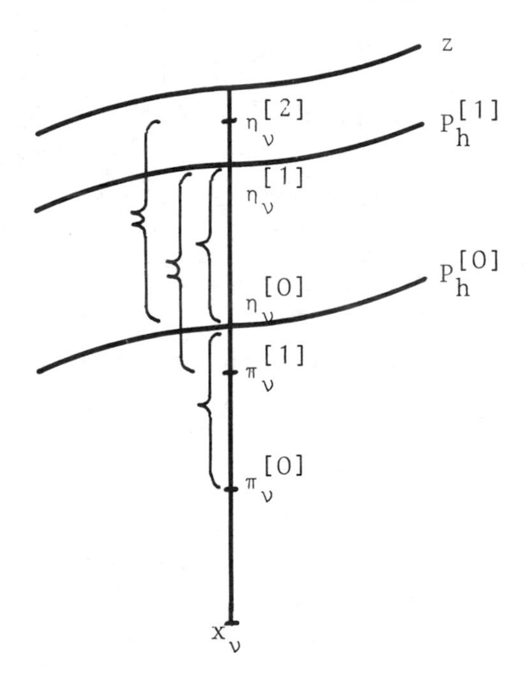

2. PRACTICAL IMPLEMENTATION

In order to be able to construct an interpolating function $P_h^{[0]}(\underline{x})$ we start with a rough triangulation of Ω.

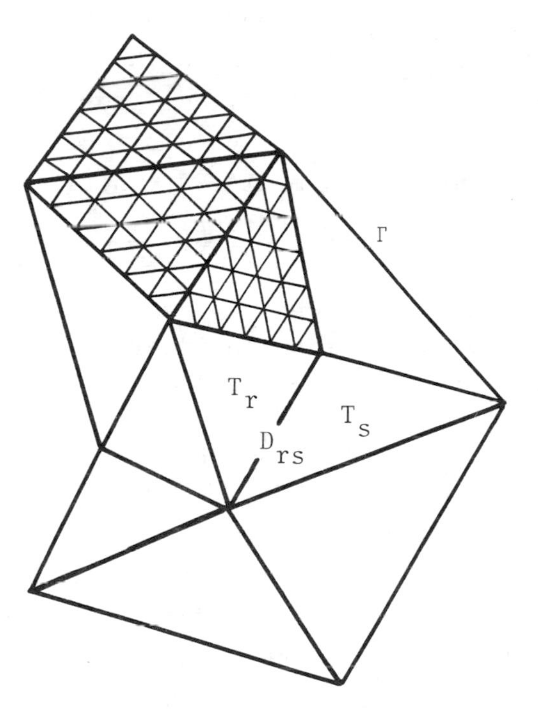

Each of these triangles is divided "equidistantly" into n^2 smaller triangles and this "finer" partition yields the vertices for our basic finite element method. If we have obtained the approximate values $\eta_\nu^{[0]}$ at these vertices we are able to construct a unique interpolating polynomial of the form

$$p^{[0]r}(\underline{x}) = \sum_{0 \leq l+m \leq n} c_{lm} \, x_1^l \, x_2^m$$

on each triangle T_r. Globally one obtains a piecewise polynomial continuous function $P_h^{[0]}$ on Ω which has jumps of the first derivatives along interior lines D_{rs}.

This implies that the NP has to be considered as a modified boundary value problem which has jumps of the first derivatives along prescribed lines and these jumps have to be added to the data set of the NP. We are forced to construct a modified functional which takes this situation into account.

One can show: These jumps can be taken into account by defining a function κ on the interior lines D_{rs} in the following way

$$\kappa\Big|_{D_{rs}} P^{[0]}\Big|_{D_{rs}} := \sum_{j,k=1}^{2} \mu_{jk}\left(\frac{\partial P^{[0]r}}{\partial x_k} - \frac{\partial P^{[0]s}}{\partial x_k}\right)\cos(\nu_{rs}, x_j) \tag{2.1}$$

where ν_{rs} is a vector orthogonal to D_{rs} and pointing from T_r to T_s. The use of the first Green's formula then enables us to reduce the finite element solution of the NP to the minimization of the functional

$$\tilde{F}[u] = \{u,u\}_L + 2\int_{\Omega}\left(\int_{0}^{u} f(\underline{x},v)\,dv\right)d\Omega \tag{2.2}$$

where

$$\{u,v\}_L := \sum_r \int_{T_r}\left(\sum_{j,k=1}^{2} \mu_{jk}\frac{\partial u}{\partial x_k}\frac{\partial v}{\partial x_j} + c\,u\,v\right)dT_r - \sum_{r,s}\int_{D_{rs}} u\,v\,\kappa\,dD_{rs}. \tag{2.3}$$

Due to the idea of iterated defect correction one has also to solve the OP with this modified functional and

$\kappa \equiv 0$.

If we compare our method with an extrapolation method, we remark, that the triangulation of Ω remains fixed and the systems of equations do not explode. This is for multidimensional problems a very important point.

The numerical example will show that the number of steps of the iterated defect correction which yield an improvement of the accuracy depends on the degree of the interpolating polynomials.

3. A NUMERICAL EXAMPLE

$$(\mu_{ij}) = \begin{pmatrix} 2 & 1.5 \\ 1.5 & 3 \end{pmatrix} \qquad c = 4$$

$$f(\underline{x},u) = 24\sin(x_1+2x_2) + 4 + \frac{1}{4} x_1 x_2 ((\sin(x_1+2x_2)+1)^2 - u^2)$$

exact solution: $u(\underline{x}) = \sin(x_1 + 2x_2) + 1$

hexagonial domain with 6 interpolating polynomials.

degree of pol.	appr. sol.	error 1 IDeC	2 IDeC	3 IDeC	4 IDeC
3	.15-02	.20.02	-	-	-
5	.59-03	.61-04	.49-04	-	-
7	.23-03	.39-05	.46-06	.17-06	.14-06
9	.13-03	.13-05	.10-06	.13-06	.13-06
11	.90-04	.76-06	.44-07	.50-07	-

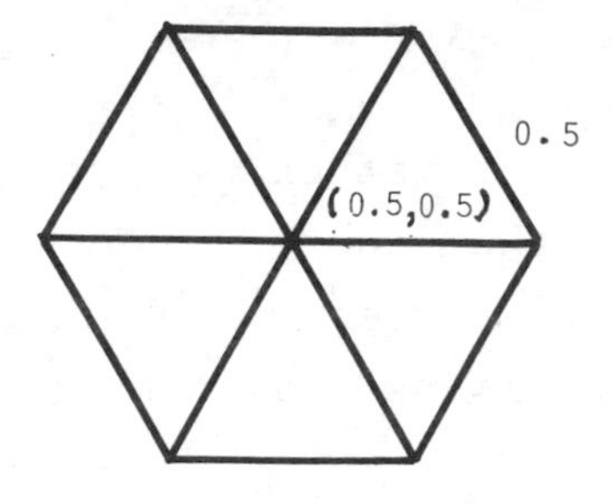

Details will appear in a joint paper with R. Frank and J.-P. Monnet. The numerical results are due to J.-P. Monnet.

REFERENCES

1. FRANK, R., The method of iterated defect-correction and its application to two-point boundary value problems, Part I: *Numer. Math.* 25 409-419; (1976), Part II: *Numer. Math.* 27 407-420, (1977).

2. FRANK, R., HERTLING, J., Die Anwendung der Iterierten Defektkorrektur auf das Dirichletproblem. *Beiträge zur Numerischen Mathematik* 7 19-31, (1979).

3. FRANK, R., HERTLING, J., UEBERHUBER, C.W., An extension of the applicability of iterated deferred corrections. *Math. Comp.* 31 907-915, (1977).

4. FRANK, R., UEBERHUBER, C.W., Iterated defect correction for differential equations. Part I: Theoretical results. *Computing* 20, 207-228, (1978).

5. MIKHLIN, S.G., *Variationsmethoden der mathematischen Physik.* Akademie-Verlag, Berlin, 1962.

6. STETTER, H.J., The defect correction principle and discretization methods. *Numer. Math.* 29 425-443, (1978).

7. ZADUNAISKY, P.E., On the estimation of errors propagated in the numerical integration of ordinary differential equations. *Numer. Math.* 27 21-39, (1976).

ON THE IMPLEMENTATION OF FINITE ELEMENTS FOR PARABOLIC EVOLUTION PROBLEMS

J.P. HENNART

Universidad Nacional Autónoma de México

1. INTRODUCTION

In a previous paper by the same author [12], a class of finite elements in time schemes has been presented, which leads to finite elements in time and space schemes when applied to the semidiscrete Galerkin equations obtained from the standard discretization of mixed initial-boundary value parabolic evolution problems by finite element techniques. In the first part of this paper, we briefly recall the general formalism presented in [12] and we propose a practical implementation of some of those schemes which seem to be the best candidates because they combine the advantages of reasonably high order (3) with good stability properties (L-stability). The practical implementation proposed here is based on the iterated defect correction idea developed in [6] for collocation methods, and consists of repeatedly solving with a simple Backward Euler (BE) scheme and with different second members. Through the values obtained at three successive mesh points, a quadratic polynomial is built up which can be shown to converge to the particular piecewise approximation considered. Theoretical arguments confirmed by numerical experiments with parabolic sample problems show that this convergence is very fast and the overall procedure turns out to be quite efficient.

In the second part of this paper, it is briefly shown how in Cartesian x, y and z coordinates, the $\nabla.p\nabla$ discretized operator can be decoupled in an easy way by the use of reduced integration techniques [20, Chap. 20] with tensor product rectangular elements. As a result, the application of fractional steps techniques leading to sequences of 1D problems is straightforward and the proposed implementation leads to an interesting alternative to the Galerkin ADI schemes proposed in [5].

2. TIME DISCRETIZATION ASPECTS

Let us consider (large) systems of ODE's of the general form

$$M\dot{\underline{U}} + K(t)\underline{U} = \underline{F}(t), \tag{2.1}$$

resulting from the application of semidiscrete Galerkin techniques [4] in conjunction with finite elements in space to mixed initial-boundary value parabolic evolution problems. In (2.1), M and K are the mass and stiffness matrices respectively.

For a given partition $\pi \equiv \{t_i, i=0,\ldots,I\}$ of $[0,T]$, integration of (2.1) over $[t_i, t_{i+1}]$ gives

$$M\underline{U}(t_{i+1}) = M\underline{U}(t_i) - \int_{t_i}^{t_{i+1}} (K(t)\underline{U}(t) - \underline{F}(t))dt, \tag{2.2}$$

which is approximated by

$$M\underline{U}_{i+1} = M\underline{U}_i - \int_{t_i}^{t_{i+1}} (K(t)\underline{U}_{pq}(t) - \underline{F}(t))dt, \tag{2.3}$$

where p,q are nonnegative integers with $p+q=\ell+1 > 0$, each component of $\underline{U}_{pq}$ belonging to a linear space S^h of dimension p+q. A generalized Hermite interpolant is then introduced in S^h with p (resp. q) interpolation conditions at $t=t_i$ (resp. t_{i+1}). By (possibly multiple) collocation at $t=t_i$ and (or) $t=t_{i+1}$, the successive derivatives are eliminated and the only parameters left are $\underline{U}_i^{(o)} = \underline{U}_i$, known from the initial condition or from previous time integration and $\underline{U}_{i+1}^{(o)} = \underline{U}_{i+1}$. The result is a class of one-step integration methods, each particular member depending on which quadrature scheme is used to approximate the righthand side of (2.3). For more details, see [12]. For nonlinear ODE's, the case where $S^h \equiv P_\ell$, the linear space of polynomials of degree less than or equal to ℓ, has been treated in [10]. Extensions to exponential basis functions leading to exponentially fitted schemes are studied in [7, 14].

Here for the sake of simplicity, we shall restrict ourselves to the case where $S^h \equiv P_\ell$. A particular scheme is thus completely defined by (p,q) and by the particular quadrature scheme (QS) used in (2.3), the general rule being that the QS must be exact for members of S^h[10]. The commonly used one-step finite difference schemes [4] are at most of second order in time, and therefore consistent with the constant parameter assumption where K(t) and $\underline{F}(t)$ are replaced by some mean value over $[t_i, t_{i+1}]$. They are in fact equivalent with some of the schemes presented in [12]: the BE scheme is a (p,q)=(0,1) scheme associated with the Radau one-point right-hand scheme (R1) while the Crank-Nicolson (CN) schemes are (1,1) schemes associated with the Gauss one-point (G1) or Lobatto two-points (L2) QS, depending on which version (midpoint or trapezoidal) is considered. It turns out that the highest order among these schemes (2) is achieved by the CN ones which are A-stable but not L-stable. In the presence of fast transients, some oscillations may appear due to their bad asymptotic behaviour. This is not the case for the BE scheme which enjoys a good asymptotic behaviour but is unfortunately of first order only.

It seems therefore that the best candidates among the class of schemes which has been proposed in [12] are the (1,2) schemes using piecewise quadratic polynomials in time. They combine the following advantages:

1/They are of reasonably high order (3) and consistent with the generally accepted use of quadratic finite elements in space, which seem to achieve a compromise between the ease of their implementation and accuracy for elliptic problems with singularities [9] and certainly also for parabolic evolution problems, due to their exponential smoothing property.

2/they do not lose accuracy betwwen meshpoints [10, 14] and are therefore finite element (rather than finite difference) oriented.

3/they are L-stable and therefore exhibit a good asymptotic behaviour: fast transients would be damped out almost immediately.

Let us give two examples, the (1,2) scheme based on the L3 QS which reads

$$[M+(\frac{2h}{3})(\frac{3}{4}K_{i+1/2}+\frac{1}{4}K_{i+1})+(\frac{h^2}{6})K_{i+1/2}M^{-1}K_{i+1}]\underline{U}_{i+1}=$$

$$=[M-(h/6)(K_i + K_{i+1/2})]\ \underline{U}_i, \tag{2.4}$$

and the (1,2) scheme based on the R2 QS which reads

$$[M+(\frac{2h}{3})(\frac{5}{8}K_{i+1/3}+\frac{3}{8}K_{i+1})+(\frac{h^2}{6})K_{i+1/3}M^{-1}K_{i+1}]\underline{U}_{i+1}=$$

$$= [M-(h/3)K_{i+1/3}]\underline{U}_i, \tag{2.5}$$

with $h \equiv t_{i+1} - t_i$ and assuming $\underline{F} \equiv 0$ for the sake of simplicity. The above schemes, as it turns out [13], are particular Implicit Runge Kutta schemes. The R2 one is moreover an ordinary collocation scheme [8] previously derived [16, 17] with collocation at $t=t_i+h/3$ and $t=t_{i+1}$, while the other one based on L3 is a "generalized collocation scheme" [19] with collocation at $t=t_{i+1}$ and in a weighted way at $t=t_i$ and $t=t_i+h/2$ simultaneously. From Eqs. (2.4) and (2.5) the R2 scheme shows possible advantages for storage since it only requires $K_{i+1/3}$ and K_{i+1}.

Numerical implementations of scheme (2.5), which seems the most attractive one, must cope with some obvious difficulties:

1/M and K_α are band matrices, however M^{-1} turns out to be normally full and as a result it is practically impossible to treat the matrix of the left-hand side of (2.5) directly. A possible remedy consists in evaluating M in a nonconsistent way, via lumping by reduced integration [20, Chap. 20], in which case the bandwidth of the left-hand side matrix in (2.5) is approximately twice the bandwidth of K_α. This is certainly of interest for 1D problems but probably not for multidimensional ones.

2/Even if M and K_α are symmetric matrices, in general $K_\alpha M^{-1} K_\beta$ is not symmetric. In [3], Descloux and Nassif derive related schemes where symmetric final expressions are obtained, basically by interpolating $K\underline{U}$ instead of $\underline{U}$, in which case no specific QS is further required. If an Hermite (p,q) interpolation scheme is used, the same convergence rates as above can be proved at the mesh points. However, they are associated with a polynomial of degree $\ell+1$ (instead of ℓ) for $\underline{U}$ since $K\underline{U}$ (and not $\underline{U}$) is interpolated. The related schemes are therefore suboptimal. Moreover, as soon as p and (or) q=2, they require that not only K but also its derivative be known at $t=t_i$ and (or) t_{i+1}.

3/In general, the left-hand side matrix is not factorizable in real factors of the form $M+\alpha h K_\beta$. Such schemes have been obtained for instance by Nørsett [18]: as a rule, they use higher degree polynomials allowing suboptimal convergence rates in order to ensure factorisability. The right-hand sides are consequently more complex and the factorisability property itself implies the presence of a pole on the real positive axis in the resulting rational approximation to $\exp(-hM^{-1}K)$ (assuming K constant over $[t_i, t_{i+1}]$). The (1,2) schemes proposed hereabove only give rise to complex poles, a situation which seems more favourable for problems where an unstable exponentially increasing solution is feasible.

The implementation we are proposing for the above schemes essentially copes with all the difficulties mentioned above. It is based on the iterated defect correction idea developed in [6, 19], to which we refer the reader for more details. In [6], Frank and Ueberhuber apply that idea to ordinary collocation techniques on uniform meshes. Here we do the same for the R2 schemes on nonuniform meshes. Extensions to "generalized collocation schemes" (L3) on uniform meshes are also feasible (see e.g. [15] where complementary aspects are considered).

The defect correction idea applied to the R2 scheme implies the following steps over each successive time interval

1/Solve (2.1) from t_i to $t_i+h/3$ and from $t_i+h/3$ to t_{i+1} using the BE scheme, namely

$$[M+(\frac{h}{3})K_{i+1/3}]\underline{U}^0_{i+1/3}=M\underline{U}^0_i+(\frac{h}{3})\underline{F}_{i+1/3}, \qquad (2.6a)$$

$$[M+(\frac{2h}{3})K_{i+1}]\underline{U}^0_{i+1} \quad =M\underline{U}_{i+1/3}+(\frac{2h}{3})\underline{F}_{i+1}, \qquad (2.6b)$$

where $\underline{U}^0_i\equiv\underline{U}_i$ is obtained from the initial conditions or from integration over the previous time step. The result of this first step is $\underline{U}^0=[\underline{U}^0_{i+1/3},\underline{U}^0_{i+1}]$.

2/Build up the vector $\underline{P}^0(t)$ of quadratic components interpolating $\underline{U}^0_i,\underline{U}^0_{i+1/3}$ and $\underline{U}^0_{i+1}$, and solve

$$M\dot{\underline{V}}^0= -K\underline{V}^0 \ +M\dot{\underline{P}}^0 \ +K\underline{P}^0 \ =-K\underline{V}^0+\underline{F}^0, \qquad (2.7)$$

with the same strategy as in 1/ and the same initial conditions. The result is $\underline{V}^0=[\underline{V}^0_{i+1/3},\underline{V}^0_{i+1}]$. Because problem (2.7) is a neighbouring problem with a known solution ($\underline{P}^0$) to the initial problem (2.1), the exact error $\underline{V}^0-\underline{P}^0$ at the nodes $t_i+h/3$ and t_{i+1} can be assumed to be a (good) approximation to the same error for problem (2.1). $\underline{U}^0$ can be corrected correspondingly

$$\underline{U}^1 = \underline{U}^0 - (\underline{V}^0-\underline{P}^0), \quad , \qquad (2.8)$$

this last step being eventually repeated in an iterative way

$$\underline{U}^{n+1}=\underline{U}^0-(\underline{V}^n-\underline{P}^n), \qquad (2.9)$$

where $\underline{P}^n$ has quadratic components interpolating the corresponding components of $\underline{U}^0_i$, $\underline{U}^n_{i+1/3}$ and $\underline{U}^n_{i+1}$. In this "local iteration strategy", the approximation over $[t_i, t_{i+1}]$ is improved possibly until convergence, before we move to the next time interval. From (2.9), a fixed point will eventually be reached if $\underline{U}^{n+1} = \underline{U}^n = \underline{P}^n$, that is if $\underline{U}^0 = \underline{V}^n$, implying from (2.7) that

$$\underline{F}^n = M\dot{\underline{P}}^n + K\underline{P}^n = \underline{F} \quad \text{at} \quad t = t_i + \frac{h}{3} \text{ and } t_{i+1} \qquad (2.10)$$

In other words, $\underline{P}^n$ *is the solution of our collocation scheme* (2.5). In general, the iterative procedure (2.9) can be rewritten as $\underline{U}^{n+1} = S\underline{U}^n + \underline{G}$, and the existence of a fixed point requires $\rho(S) < 1$. Applying the above procedure to $\dot{u} = \lambda u, \lambda \in \mathbb{R}$ it is easy to obtain the spectral radius as a function of $z = \lambda h$. As shown in Fig. 1, ρ exhibits a quite favourable behaviour when compared in particular with the corresponding spectral radius of the parabolic collocation method over a uniform mesh studied in [6] and which incidentally is only of second order.

As a simple example, we have solved a simple 1D parabolic sample problem (Test Case 3 of Ref. 12). On our Burroughs B6800 computer, a direct implementation of scheme (2.5) over a 20x20 space-time mesh required 275 secs of CP time. Essentially the same result (up to 7 significant digits) was reached in 28 secs using defect correction in an adaptive way. It is important to realize that solving (2.7) repeatedly with the same strategy as in (2.6) implies solving for new second members linear systems whose matrices have been factorized once for all in step 1. Accordingly, the above figures are only indicative: indeed for 1D problems, the solution time is proportional to the factorization time and each correction is almost as costly as the basic BE steps (4 secs. in our example where 220 corrections where performed over 20 time intervals). For multidimensional problems, it should be clear that the correction phase will only consume a small extra amount of computer time, compared with the expensive basic step where the matrix factorizations have to be carried out. At the same time, the direct implementation of (2.5) becomes prohibitively expensive and the only reasonable implementations must be based on a basic step where only matrices of the form $M + \alpha h K_\beta$ are to be factorized, as with the BN and CN schemes.

As a final comment, we would like to point out that the proposed implementation could conveniently be combined with the one proposed in [1] where the emphasis is on trying to preserve the same accuracies, while moving ahead in time with a high order scheme similar to the ones exhibited here [3, 14] but *constant in time*.

In a sense, we are showing in this paper how the basic high order scheme of [1] could be efficiently implemented.

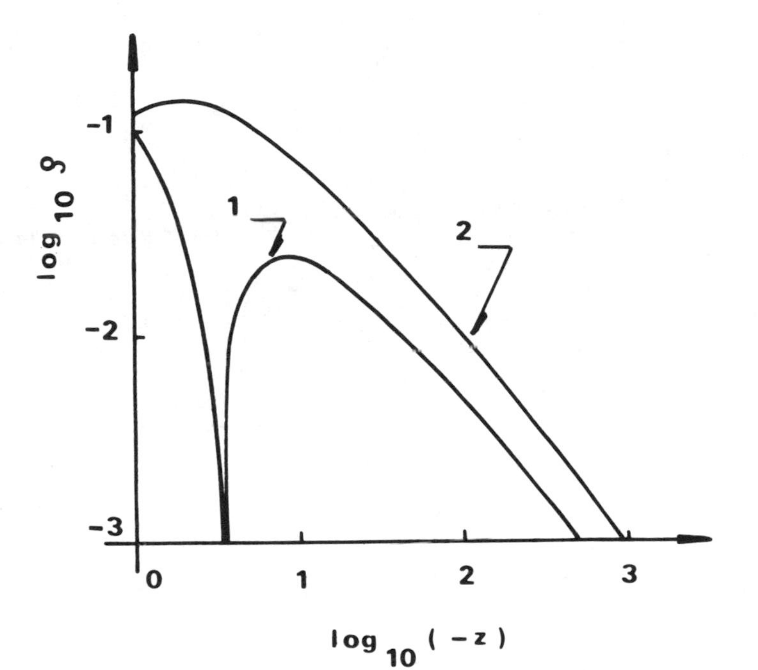

FIG. 1. Spectral radius ρ as a function of $z=\lambda h$ for the defect correction implementation of scheme (2.5) (1) and of the parabolic collocation scheme of Ref. 6 (2).

3. SPACE DISCRETIZATION ASPECTS

In the case of multidimensional domains Ω of the tensor product type, for instance a rectangle in $\mathbb{R}^2$, the natural choice is to use a tensor product finite element basis in space

$$S^k = \text{span}\{\alpha_1\beta_1,\ldots,\alpha_{N_x}\beta_{N_y}\} , \tag{3.1}$$

where the α_i's and β_j's are basis functions in the x and y direction respectively.

If, for the sake of simplicity, we consider the simple parabolic equation $u_t=\nabla.p\nabla u$ and semidiscretize it over Ω using S^k, we essentially get the system (2.1) of ODE's with $\underline{F}=\underline{0}$ and where $K=K_x+K_y$, K_x (resp. Ky) being that part of the stiffness matrix which corresponds to the derivatives in the x (resp. y) direction. In standard implementations, too much coupling results and the operator $\partial_x p \partial_x$ would for instance couple the unknowns not only in the x direction but also in the y direction, leading for instance to a 9 point coupling with bilinear elements.

When p is factorizable into functions of one coordinate only and in particular when p reduces to a constant, Douglas and Dupont [5] have shown how decoupling can be restored, leading to ADI-like implementations, which are straightforward when finite differences are used. They show moreover how to extend such a formalism when p is any function, basically by replacing K in (3.2) by $\bar{K}+(K-\bar{K})$ where $\bar{K}$ correspond to some constant value of p and by solving (2.1) implicitly in $\bar{K}$ and explicitly in $(K-\bar{K})$. For more details about this interpretation of Douglas and Dupont formalism, see [11].

Here we show how a more direct approach to decoupling is available by the use of reduced integration techniques [20, Chap. 20]. Namely, consider a matrix element $k_{x,ij}$ of K_x evaluated by numerical integration

$$
\begin{aligned}
k_{x,ij} &= \int_\Omega p \; u_{ix} \; u_{jx} \; d\bar{r} \\
&= \int_\Omega p(\alpha'_k \; \beta_\ell)(\alpha'_m \; \beta_n)\,dxdy \\
&\sim \sum_q w_q p_q (\alpha'_{kq} \; \beta_{\ell q})(\alpha'_{mq} \; \beta_{nq}) \;, \qquad (3.3)
\end{aligned}
$$

where $p_q \equiv p(\bar{r}_q)$, $\alpha'_{kq} = \alpha'_k(x_q)$ and $\beta_{\ell q} = \beta_\ell(y_q)$. Clearly if we choose the integration points $\bar{r}_q$ to be the nodes of the tensor product element considered, $\beta_{\ell q} = \delta_{\ell q}$ and $\beta_{nq} = \delta_{nq}$ so that a nonzero matrix element exists only if $q=\ell=n$, in other words the coupling due to K_x is only between those nodes which have the same y coordinate.

As a result the $\nabla.p\nabla$ operator in 2 or more dimensions is easily decoupled and lends itself to the application of fractional steps schemes (for instance ADI ones) where only 1D problems are to be solved in sequence. In [2], various ADI and LOD (Locally One Dimensional) schemes have be implemented in a 2D demonstration space-time kinetics code using biquadratic elements and the first numerical results have been quite satisfactory. More comprehensive results will be reported elsewhere.

REFERENCES

1. BRAMBLE, J.H. and SAMMON, P.H., Efficient Higher Order Single Step Methods for Parabolic Problems: Part I. *Math. Comp.* 35, 655-677 (1980).
2. DEL VALLE, E., *Master Thesis in Nuclear Engineering*, National Polytechnic Institute, Mexico (1981).
3. DESCLOUX, J. and NASSIF, N.R., Padé Approximations to Initial Value Problems. *Report EPFL*, Lausanne (1975).
4. DOUGLAS, J., Jr. and DUPONT, T., Galerkin Methods for Parabolic Equations. *SIAM J. Numer. Anal.* 7, 575-626 (1970).
5. DOUGLAS, J., Jr. and DUPONT, T., Alternating-Direction Galerkin Methods on Rectangles. pp.133-214 of B. Hubbard (Ed.), *SYNSPADE 1970*. Academic Press, New York (1971).
6. FRANK, R. and UEBERHUBER, C.W., Iterated Defect Correction for the Efficient Solution of Stiff Systems of Ordinary Differential Equations. *BIT* 17, 146-159 (1977).
7. GOURGEON, H. and HENNART, J.P., A Class of Exponentially Fitted Piecewise Continuous Methods for Initial Value Problems, To Appear in J.P. Hennart (Ed.), *Third Workshop on Numerical Analysis*. Springer Verlag, Berlin (1981).
8. GUILLOU, A. and SOULE, J.L., La Résolution Numérique des Problèmes Différentiels aux Conditions Initiales par des Méthodes de Collocation. *R.I.R.O.*, *3e Année* No. R-3, 17-44 (1969).
9. HENNART, J.P. and MUND, E., Singularities in the finite element approximation of two-dimensional diffusion problems. *Nucl. Sci. and Engng.* 62, 55-68 (1977)
10. HENNART, J.P., One Step Piecewise Polynomial Multiple Collocation Methods for Initial Value Problems. *Math. Comp.* 31, 24-36 (1977).
11. HENNART, J.P., Some Recent Numerical Methods for the Solution of Parabolic Evolution Problems. *Seminaire d'Analyse Numérique No. 298*. Université Scientifique et Médicale de Grenoble (1978).
12. HENNART, J.P., Multiple Collocation Finite Elements in Time for Parabolic Evolution Problems. pp.271-278 of J.R. Whiteman (Ed.), *The Mathematics of Finite Elements and Applications III*. Academic Press, London (1979).
13. HENNART, J.P. and ENGLAND, R., A comparison between several piecewise continuous one step integration techniques. pp.33.1-33.4 of R.D. Skeel (Ed.), *Working Papers for the 1979 Signum Meeting on*

Numerical Ordinary Differential Equations. Department of Computer Science, University of Illinois, Urbana-Champaign (1979).

14. HENNART, J.P. and GOURGEON, H., One Step Exponentitally Fitted Piecewise Continuous Methods for Initial Value Problems, *IIMAS-UNAM, Com. Tecnicas Serie NA*, To Appear (1981).
15. HENNART, J.P., Topics in Finite Element Discretization of Parabolic Evolution Problems, To Appear in J.P. Hennart (Ed.), *Third Workshop on Numerical Analisis*. Springer Verlag, Berlin (1981).
16. HULME, B.L., One-step Piecewise Polynomial Galerkin Methods for Initial Value Problems. *Math. Comp.* 26, 415-426 (1972).
17. HULME, B.L., Discrete Galerkin and Related One-step Methods for Ordinary Differential Equations. *Math. Comp.* 26, 881-891 (1972).
18. NØRSETT, S.P., Runge-Kutta Methods with a Multiple Eigenvalue Only. *BIT* 16, 388-393 (1976).
19. UEBERHUBER, C.W., Implementation of Defect Correction Methods for Stiff Differential Equations, *Computing* 23, 205-232 (1979).
20. ZIENKIEWICZ, O.C., *The Finite Element Method, 3rd Edition*. McGrawHill, London (1977).

A SURVEY OF THE FACTORISED FORMULATION OF THE DISPLACEMENT METHOD

V. Wilhelmy

Det norske Veritas, Industrial and Offshore Division, Oslo.

1. INTRODUCTION

A factorised formulation of the finite element displacement method was first introduced in 1973. Since then, some interesting developments have taken place such that dramatic improvements in accuracy of solution for displacements, forces and stresses are possible both in the natural and in the cartesian systems. New relevance of the method is evident as the popularity of minis and superminis with their short word representation increases.

2. BASIC DEVELOPMENTS

In the classical formulation of the displacement method, numerical errors are introduced due to

(a) the representation of the element stiffness matrix k in cartesian form. These matrices are singular due to redundant degrees of freedom. In practice, there are small departures from these singularities ("springs to ground", Rosanoff et al, 1968 [9]).

(b) Computation and representation of the structural stiffness matrix K resulting in truncation errors.

(c) Computation of $\bar{U}$, $\bar{U}^t \bar{U} = K$ by the Cholesky square root method.

Based on analogy with Q-R decomposition in least-square computations, Brønlund first introduced the basic idea of obtaining a factorised form of the structural stiffness matrix directly in 1973 [5]. In this rather extensive publication Brønlund showed how the building of K could be circumvented completely and he also suggested several ways of obtaining corresponding decomposed forms of element stiffnesses. A complete and consistent substructure formulation was included.

The gain in accuracy to be expected was only illustrated by a small two-spring system and in an appendix containing some basic error analysis. No element work was done in this reference.

Basically, the method proposes to replace k by a "factorised" form s such that

$$s^t s = k$$

and K by a corresponding structural factor S = sa, where a is the kinematic assembly matrix such that

$$S^t S = K = a^t s^t sa$$

S would be triangularised by an orthogonal matrix Q, $Q^t Q = I$

$$S^t S = [U^t \mid 0] Q^t Q \begin{bmatrix} U \\ \hline 0 \end{bmatrix}$$

U is an upper triangular matrix taking the place of the Cholesky factor. With it, the forward and backward substitution can be carried out as usual.

In 1975 a generally more available paper was published by Argyris and Brønlund [2], where the natural stiffness approach [1] was introduced for obtaining the element factors. Examples were given for a truss bar element and a constant strain triangle formulated in the natural system. The advantage of working in this system lies in the fact that natural stiffness matrices k_N have no redundant degrees of freedom and are therefore non-singular. The natural stiffness factor u_N is then simply the Cholesky factor of k_N. Transformation into cartesian systems is obtained through simple postmultiplication with transformation matrices a_N instead of the usual congruent transformations. In this paper, the method was first called the natural factor method because of the above reflections.

$$k_N = s_N^t s_N = u_N^t u_N$$

$$s = u_N \, a_N$$

In 1974 a detailed mathematical error analysis was published jointly by Johnsen and Roy [8] in which the error bounds to be expected in typical solutions were obtained. Essentially, this paper concludes that for typical engineering situation loading vectors the loss of digits in the solution is about half of that occuring when the conventional method is used. This implies that single precision (S.P.) computations suffice with the new method where douple precision (D.P.) computations are already needed with standard formulations.

3. A FINITE ELEMENT SYSTEM

In the same year work was begun on a new general purpose finite element system [6,7] based on the method. The two comparisons of the accuracy achievable by the new as compared to the traditional method as shown in these references and more generally in [14] are reproduced in Figs. 1 and 2.

Fig. 1 shows a statically determinate truss whose conditioning is made progressively worse through weakening of one of its diagonals. On the graph below it, the number of achievable correct digits in displacements is shown as the condition worsens. It can be seen that although a 27-bit mantissa was available in S.P. against a 60 bit mantissa in D.P., the accuracy of the method matched D.P. quality of the traditional method, especially when using residual iteration.

In Fig. 2 it is shown how the eigenvalue errors for the lowest modes, where initial truncation errors of K dominate, can be virtually eliminated by building the Rayleigh quotient with S rather than with K. The errors for higher modes, due to discretization, remained unchanged.

4. ELEMENT FACTORS

During his involvement in the development of the TOPAS system, the author derived and in 1977 published examples of the stiffness factor formulations for several types of elements [13]. It was shown in this paper that s could be obtained directly for all elements without referring first to an element stiffness matrix. For simple elements like beams, the factor is obtained in the natural system and then expanded into the global system by successive postmultiplications with transformation matrices which in this way allow such effects as unsymmetric bending and eccentricity.

For continuum elements like the six node triangle, quadrilaterals and hexahedra, factors are expressed as a matrix containing rows of integration evaluation points, premultiplied by a decomposed material matrix, square roots of thicknesses, Jacobian determinants etc. which are always positive. Element factors with too many rows are reduced to upper triangular by Householder-Givens transformations.

5. DIRECT "STRESS" COMPUTATIONS.

Scharpf published a paper in 1978 [10] where, based on the same truss example of Fig. 1, now rotated in space and with one member added, it was shown that although displacements increase dramatically in accuracy, this is not necessarily the case for stresses (forces). He derived a new, simple formula for direct computation of the "stresses" (forces and moments in the natural system, P_N):

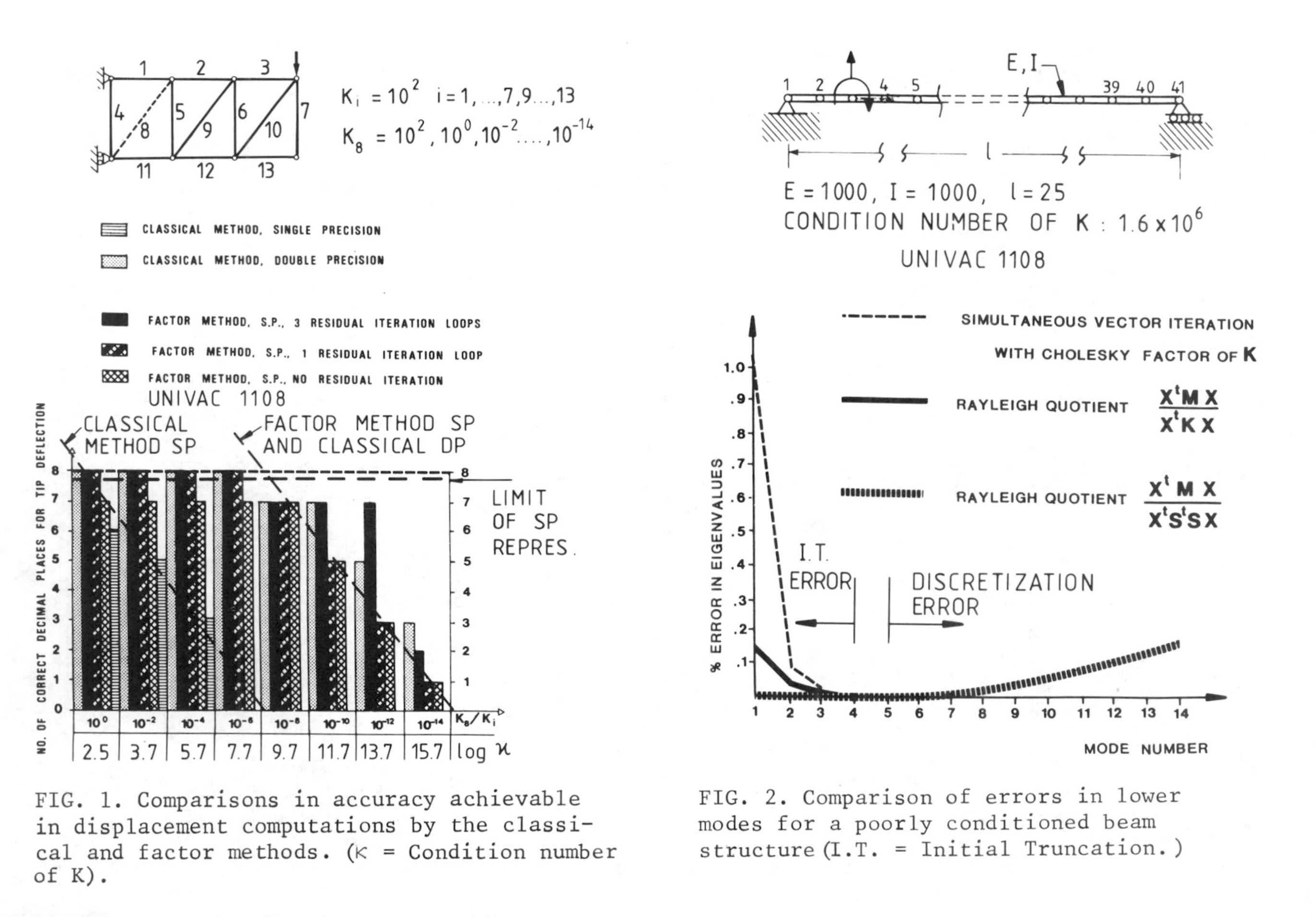

FIG. 1. Comparisons in accuracy achievable in displacement computations by the classical and factor methods. (κ = Condition number of K).

FIG. 2. Comparison of errors in lower modes for a poorly conditioned beam structure (I.T. = Initial Truncation.)

$$P_N = u_N^{\,t}\ Qt$$

where t is the result of the forward substitution such that actual computation of the structural displacements was not really longer necessary. Scharpf concluded that the relatively "poor" quality of stresses was due to lack of orthogonality in Q when S was ill conditioned. Through iterative orthogonality improvements of Q, he obtained impressive gains in accuracy, not only in displacements, but also on stresses, Fig. 3(a) and 3(b).

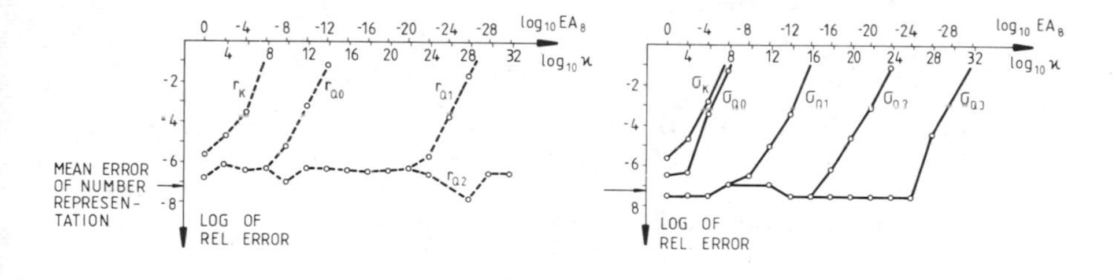

(a) Displacements (b) Stresses

FIG. 3. Numerical error for statically determinate truss using the traditional matrix (K), the factor method without orthogonality improvement (Q_o) and with 1 (Q_1) and 2 (Q_2) iterative improvements. (Machine PDP 10/11).

In the same paper, examples are also given on accuracy improvements in frames, which can be very susceptible to ill-conditioning.

6. RECENT ITERATIVE TECHNIQUES

In 1970 Argyris, Johnsen and Mlejnek [4] presented an alternative to the orthogonality improvements suggested by Scharpf, which is too costly for large systems and which cannot exploit sparsity. Based on a unified form of writing equilibrium and element elastic behaviour equations

$$\begin{bmatrix} -f_N & a_N a \\ (a_N a)^t & 0 \end{bmatrix} \begin{bmatrix} P_N \\ r \end{bmatrix} = \begin{bmatrix} 0 \\ R \end{bmatrix}$$

where f_N is the element natural flexibility and r and R the structural displacements and loads, respectively, the authors arrived at the iterative scheme

$$r^{(o)} = U^{-1}\, U^{-t} R$$

$$P_N^{(o)} = k_N a_N a r^{(o)}$$

$$\eta^{(i)} = R - a^t a_N^t\, P_N^{(i)}$$

$$\Delta r^{(i)} = U^{-1}\, U^{-t}\, \eta^{(i)}$$

$$\Delta P_N^{(i)} = k_N a_N a\, \Delta r^{(i)}$$

$$r^{(i+1)} = r^{(i)} + \Delta r^{(i)}$$

$$P_N^{(i+1)} = P_N^{(i)} + \Delta P_N^{(i)}$$

Note that $\Delta P_N^{(i)}$ is calculated from $\Delta r^{(i)}$ and not from $r^{(i+1)}$, where much information could be lost. Figures 4(c) and (b) show the examples of a cantilever modelled with many elements (this also worsens the condition number) as shown on the abscissa. The gains in accuracy, especially in "stresses" (moments) are impressive.

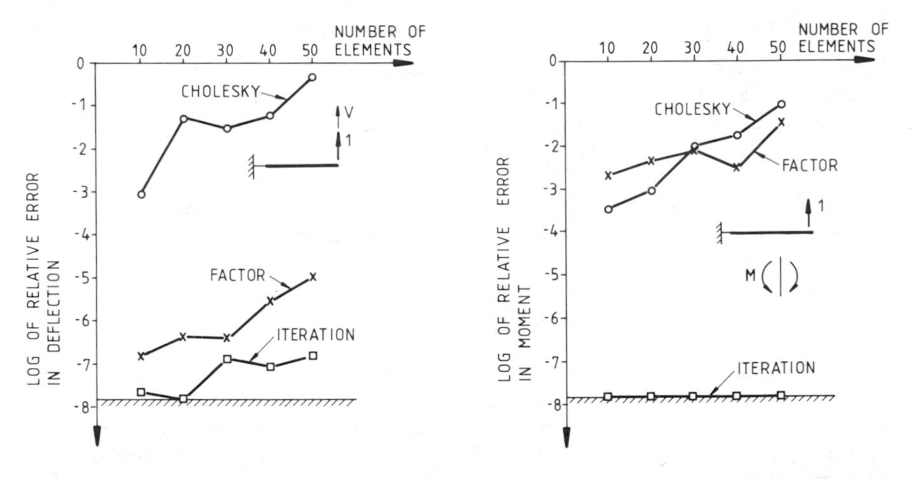

FIG. 4. Gain in accuracy for displacements and for "stresses" (moments) in a cantilever beam using iterative schemes based on the factor method. (UNIVAC 1108).

Extended to general cartesian formulation these equations can be interpreted as

$$r^{(o)} = U^{-1} U^{-t}\, R$$

$$\sigma^{(o)} = E\alpha\; a\; r^{(o)}$$

$$\eta^{(i)} = R - a^t \int_V \alpha^t \sigma^i\; dV$$

$$\Delta r^{(i)} = U^{-1}U^{-t}\,\eta^{(i)}$$

$$\Delta\sigma^{(i)} = E\alpha a\Delta r^{(i)}$$

$$r^{(i+1)} = r^{(i)} + \Delta r^{(i)}$$

$$\sigma^{(i+1)} = \sigma^{(i)} + \Delta\sigma^{(i)}$$

where σ are stresses, E is Hooke's matrix and α the matrix of strain coefficients.

Much inspired by the above work, the same authors together with P.C. Dunne published a further iterative technique for structures with very stiff or rigid parts [3]. The structural stiffness is divided into two parts, one representing the soft and the other the stiff parts. Then an iterative scheme is applied with displacement in the "soft" part and in addition with stresses in the stiff part until convergence. The method can be applied for the iterative solution improvement for any existing finite element package, provided the Cholesky factor is available. However, very accurate and well converging stresses and displacements can be obtained, also for very ill conditioned problems where the classical method fails to yield a Cholesky factor, using the factorised formulation.

To conclude this section it should be mentioned that in [4] the authors also published a rather promising technique for the solution of nonsupported structures under self-equilibrating loads.

7. OTHER DEVELOPMENTS.

In 1977 Vold [11] expanded the formulations necessary for the substructure analysis with linear constraints and in 1978 [12] published a paper on the discrete Fourier Transform Analysis of cyclically symmetric structures in connection with the factor formulation.

8. ACKNOWLEDGEMENTS

The author is indebted to Th. L. Johnsen for his support with many fruitful discussions and critical review of the paper. The author's former employer, IKOSS GmbH, is acknowledged for providing the necessary environment which led to the development of the TOPAS system.

REFERENCES

1. ARGYRIS, J.H., Continua and Discontinua, *Opening Address to the International Conference on Matrix Methods of Structural Mechanics*, Wright-Patterson Air Force Base, Ohio, (1965).

2. ARGYRIS, J.H. and BRØNLUND, O.E., The Natural Factor Formulation of the Stiffness for the Matrix Displacement Method, *Comp. Meth. Appl. Mech. Engr.* 5, 97-119 (1975).
3. ARGYRIS, J.H., DUNNE, P.C., JOHNSEN, Th.L., and MLEJNEK, H.P., A New Iterative Solution for Structures and Continua with Very Stiff or Rigid Parts, *Comp. Meth. Appl. Mech. Engr.* 24, 215-248 (1980).
4. ARGYRIS, H.H., JOHNSEN, Th. L., and MLEJNEK, H.P., On Accurate Stress Calculation in Static and Dynamic Problems Using the Natural Factor Approach, *Comp. Meth. Appl. Mech. Engr.* 19, 277-308 (1979).
5. BRØNLUND, O.E., Computation of the Cholesky Factor of a Stiffness Matrix Direct from the Factor of its Initial Quadratic Form, *ISD Report 142*. Univ. Stuttgart (1973).
6. BRØNLUND, O.E., Kiesbauer, H. and Wilhelmy, V., TOPAS,ein neues Finite Element System, pp. 23-54 of IKOSS GmbH (Ed.), *Finite Element Congress*, Baden-Baden (W.Germany) (1975).
7. BRØNLUND, O.E., Kiesbauer, H., Wilhelmy, V. and Zschau-Dombrowa, C., TOPAS Benutzerhandbuch (User's Manual),IKOSS GmbH, Stuttgart (1976).
8. JOHNSEN, Th. and ROY, J.R., On Systems of Linear Equations of the Form $A^tAx=b$; Error Analysis and Certain Consequences for Structural Applications, *Comp. Meth. Appl. Mech. Engr.* 3, 357-374 (1974).
9. ROSANOFF, R.A., GLOUDEMAN, J.F. and LEVY, S., Numerical Conditions of Stiffness Matrix Formulations for Frame Structures, *Proceedings of the Conference on Matrix Methods in Structural Mechanics*, Wright-Patterson Air Force Base, Ohio (1968).
10. SCHARPF, D.W., A Method of Stress Calculation in the Matrix Displacement Analysis, *Computers and Structures*, Vol. 8, No. 3/4, 465-477, May (1978).
11. VOLD, H., Substructure Analysis with Linear Constraints Using the Natural Factor Formulation, *Comp. Meth. Appl. Mech. Engr.* 10, 151-163 (1977).
12. VOLD, H., Efficient Implementation of the Discrete Fourier Transform Analysis of Cyclically Symmetric Structures, pp. 467-476 of J.R. Whiteman (Ed.), *The Mathematics of Finite Elements and Applications*, Academic Press, London (1979).
13. WILHELMY, V., On the Element Stiffness Factor Formulation, *Comp. Meth. Appl. Mech. Engr.* 11, 75-95, (1977).
14. WILHELMY, V., Gain in Numerical Accuracy through Direct Computation of the Cholesky Factor of a Structural Stiffness Matrix, Preprint No. 3333, *ASCE Convention and Exposition, Chicago*, Oct. 16-20 (1978).

OPTIMAL FINITE ELEMENT APPROXIMATION FOR DIFFUSION-CONVECTION PROBLEMS

*J. W. Barrett and †K. W. Morton

**Imperial College, London,*
†University of Reading

1. INTRODUCTION

In considering the diffusion-convection problem:

$$Lu \equiv - \underline{\nabla} . (a\underline{\nabla}u) + \underline{\nabla} . (\underline{b}u) = f \quad \text{in} \quad \Omega, \tag{1.1a}$$

$$u = g \quad \text{on} \quad \partial\Omega_1, \quad \partial u/\partial n = 0 \quad \text{on} \quad \partial\Omega_2, \tag{1.1b}$$

$\underline{n}$ being the outward normal on $\partial\Omega_2$, many authors have proposed Petrov-Galerkin methods, either based on a form of upwinded finite element or employing the use of rapidly varying exponential functions, in order to remove the spurious oscillations obtained with the standard Galerkin approach when the convection term is large - see [6] as a general reference. In [2] we proposed an alternative approach which is also of Petrov-Galerkin type. Its objective, however, was to approximately symmetrize the associated bilinear form of (1.1) so that many of the optimality properties associated with Galerkin methods for self-adjoint problems could be re-established and exploited.

With the standard notation H^1 for the Sobolev space of functions with first derivatives square integrable, and with basis functions ϕ_j for our trial space $S^h \subset H^1$, we define the following spaces to be used throughout this paper:

$$\begin{aligned} H^1_E &= \{v \in H^1 | v = g \text{ on } \partial\Omega_1\}, \; H^1_0 = \{v \in H^1 | v = 0 \text{ on } \partial\Omega_1\}, \\ S^h_E &= \{V \in S^h | V = g \text{ on } \partial\Omega_1\}, \; S^h_0 = \{V \in S^h | V = 0 \text{ on } \partial\Omega_1\}. \end{aligned} \tag{1.2}$$

The heart of the method in [2] was to look for a mapping $N : H^1_0 \to H^1_0$ such that

$$\begin{aligned} &\underline{\nabla}(Nw) = \rho[a\underline{\nabla}w - \underline{b}w] \quad \text{in} \quad \Omega, \\ &Nw = 0 \quad \text{on} \quad \partial\Omega_1. \quad Nw = \rho a w \quad \text{on} \quad \partial\Omega_2, \; \forall \; w \in H^1_0, \end{aligned} \tag{1.3}$$

where ρ is a positive weighting function, normalised to have unit integral. With $<\cdot, \cdot>$ denoting the standard L_2 inner product, we see that such an N symmetrizes the associated bilinear form of (1.1):

$$<Lv,Nw> + \int_{\partial\Omega_2} a(\partial v/\partial n)NwdS = <a\underline{\nabla}v-\underline{b}v, \underline{\nabla}(Nw)> + \int_{\partial\Omega_2} \underline{b}.\underline{n}v(Nw)dS \tag{1.4a}$$

$$= <a\underline{\nabla}v-\underline{b}v, \rho[a\underline{\nabla}w-\underline{b}w]> + \int_{\partial\Omega_2} \rho a\underline{b}.\underline{n}\ vwdS \tag{1.4b}$$

$$= <\rho a^2\underline{\nabla}v,\underline{\nabla}w> + <[\rho b^2 + \underline{\nabla}.(\rho a\underline{b})]v,w>$$

$$\equiv B_S(v,w),\ \forall\ v \in H^1,\ w \in H_0^1. \tag{1.4c}$$

Thus by using test functions $N\phi_j$ in a Petrov-Galerkin approach we obtain an optimal approximation to u in the norm induced by the symmetric bilinear form $B_S(\cdot, \cdot)$, denoted by $||\cdot||_S$.

Unfortunately, an exact mapping N satisfying (1.3) only exists if $\rho = e^{-\lambda}$, where λ satisfies $a\underline{\nabla}\lambda = \underline{b}$ which requires that $\underline{\nabla}\times(\underline{b}/a) = 0$. In any case, due to quadrature difficulties and the poor weighting in the resultant norm $||\cdot||_S$ with this choice, we considered in [2] the possibility of only approximately symmetrizing, which kept the choice of ρ more flexible. The technique worked extremely well in one-dimension and the next section reviews the results obtained in this case. The Petrov-Galerkin viewpoint of seeking test functions which lead to an approximately symmetric bilinear form is less appropriate in two dimensions. However, in section 3, we see that by exploring the similarities of the above technique to a least squares approach a more suitable extension of the method to higher dimensions is possible.

2. ONE-DIMENSIONAL PROBLEMS

Consider the one-dimensional problem corresponding to (1.1) on the interval $[0, 1]$, without turning points, that is $b(x) > 0$ say. We assume $\{x = 0\} \subset \partial\Omega_1$, which corresponds to Dirichlet data being specified on the inflow boundary. The approximate symmetrization is based upon looking for a mapping $N_\varepsilon : H_0^1 \to H_0^1$ which approximately satisfies (1.3). We see that by setting

$$(N_\varepsilon w)(x) = \rho aw(x) - \int_0^x [\rho b + (\rho a)']w(t)dt + \left[\int_0^1 [\rho b + (\rho a)']w(t)dt\right] \int_0^x \varepsilon(t)dt, \tag{2.1}$$

where ε is a non-negative function normalised to have unit integral, leads to

$$<Lv, N_\varepsilon w> + \int_{\partial\Omega_2} a(\partial v/\partial n)N_\varepsilon w dS = B_S(v, w) + <\rho b + (\rho a)', w> \ell(v),$$
$$\forall\ v \in H^1,\ w \in H_0^1, \quad (2.2)$$

where $B_S(\cdot,\cdot)$ is defined by (1.4c) and

$$\ell(v) = <av' - bv, \varepsilon>. \quad (2.3)$$

The corresponding Petrov-Galerkin method is: find $U^N \in S_E^h$ such that

$$B_S(U^N, \phi_j) + <\rho b + (\rho a)', \phi_j> \ell(U^N) = <f, N_\varepsilon \phi_j>,\ \forall\ \phi_j \in S_0^h. \quad (2.4)$$

Suppose that we now introduce the best fit $U^* \in S_E^h$ to u in the norm $||\cdot||_S$. Then the aim is to choose ε so as to minimise the effect of asymmetry in (2.4); that is, so that the solution U^N is close to U^* . In the case $ab > 0$ the solution of (1.1) may have a boundary layer at $x = 1$ and by choosing ε near $x = 0$, one would expect to interfere as little as possible with the best fit at the right-hand end, the opposite holding for $ab < 0$. In fact this can be rigorously established and we have the following result:

Theorem. Suppose we define $V^* \in S_0^h$ by

$$B_S(V^*, \phi_j) = <\rho b + (\rho a)', \phi_j>, \quad \forall\ \phi_j \in S_0^h. \quad (2.5)$$

Then if $1 + \ell(V^*) > 0$, there exists a unique solution U^N to (2.4) which satisfies

$$||u - U^N||_S^2 \le ||u - U^*||_S^2 + \{||V^*||_S/[1 + \ell(V^*)]\}^2[\ell(u - U^*)]^2. \quad (2.6)$$

The proof is given in [3].

A good choice is to set $\varepsilon(x) \equiv \delta(x)$, the delta function at $x = 0$. Then with a and b constant, $\rho \equiv 1$ and S^h the space of piecewise linears on a uniform mesh, it is shown in [3] that $\ell(V^*) > 0$, leading to

$$||u - U^N||_S^2 \le ||u - U^*||_S^2 + a^2[u'(0) - U^{*\prime}(0)]^2. \quad (2.7)$$

Thus when the convection term dominates, that is $b >> a$, the second term on the right-hand-side is entirely negligible resulting in U^N being indistinguishable from U^*. In fact the approach given above works well for all values of b/a and is easily generalised to cope with turning point problems. In [2] various numerical examples are given and in [3] sharp error estimates are obtained of both an a posteriori and an a priori type which demonstrate how close U^N is to the best fit U^*. One should note that the test functions $N_\varepsilon \phi_j(x)$ defined by (2.1), which are non-local, can be localised easily by taking weighted differences, leading to a sparse matrix in (2.4); or alternatively the system can be solved iteratively. In either case it is simple to program and leads to hardly any more work than solving a self-adjoint problem by the Galerkin method, especially in the case $f \equiv 0$ when the test functions $N_\varepsilon \phi_j$ do not need to be constructed explicitly.

Finally one can exploit the near optimality property of the solution U^N, to recover more detailed information about u. The main interest lies with boundary and interior layers and here the construction of effectively a least square fit to u (since $||\cdot||_S$ tends to a weighted L_2 norm as the convective term becomes more dominant) gives valuable information regarding integrals over the layer; thus the method shows to advantage over methods giving only accurate approximations to $u(jh)$. For details of recovery techniques see [2].

3. TWO-DIMENSIONAL PROBLEMS

In order to symmetrize approximately as in one dimension, we look for a mapping $N_\varepsilon : H_0^1 \to H_0^1$ satisfying

$$\begin{aligned} &\underline{\nabla}(N_\varepsilon w) = \rho[a\underline{\nabla}w - \underline{b}w] + \underline{\varepsilon} \quad \text{in} \quad \Omega , \\ &N_\varepsilon w = 0 \quad \text{on} \quad \partial\Omega_1, \quad N_\varepsilon w = \rho a w \quad \text{on} \quad \partial\Omega_2 . \end{aligned} \tag{3.1}$$

For N_ε to exist we require the right-hand-side of (3.1) to be irrotational and this implies in general that the support of $\underline{\varepsilon}$ cannot be isolated to a particular region where the underlying true solution is smooth. In fact there appear to be no obvious choices of $\underline{\varepsilon}$ which keep the effect of only approximately symmetrizing negligible. In [7] with piecewise bilinear trial functions on a uniform mesh, $\phi_{j,k}(x, y) \equiv \phi_j(x)\phi_k(y)$, we proposed using test functions generated by the product of the corresponding one-dimensional operators as the approximate symmetrizer, that is $N_\varepsilon^{(x)}\phi_j(x). N_\varepsilon^{(y)}\phi_k(y)$. Although this gives rise to considerable asymmetry in the resulting bilinear form, the numerical results obtained were encouraging. However, there is a practical disadvantage in that the test functions have to be constructed explicitly and localised by taking differences, leading to a sixteen point discrete operator, even if $f \equiv 0$. Thus there is a need for an alternative generalisation.

In [4] an H^{-1} least squares optimal control formulation was put forward for a class of non-linear problems. By applying this technique to (1.1) with $\partial\Omega \equiv \partial\Omega_1$ we see its similarity to the symmetrizing Petrov-Galerkin approach. The method basically minimises the residual in the H^{-1} norm over a trial space, that is

$$\min_{U \in S_E^h} ||LU - f||_{H^{-1}} . \tag{3.2}$$

Equipping H_0^1 with the norm $<\underline{\nabla}\cdot, \underline{\nabla}\cdot>$, one can show through the standard isometry that (3.2) is equivalent in one dimension to

$$\min_{U \in S_E^h} ||[- d/dx]^{-1} (LU - f)|| , \tag{3.3}$$

where $||\cdot||$ is the L_2 norm. Thus this can be viewed as a Petrov-Galerkin method with test functions given by $[- d^2/dx^2]^{-1} L\phi_j \in H_0^1$. For the constant coefficient problem, the test functions obtained from this viewpoint are given by $N_\varepsilon \phi_j$ in (2.1) with $\rho \equiv 1$ and $\varepsilon \equiv 1$. Therefore we can establish the existence and uniqueness of this approximation and the resulting error estimate by appealing to the theorem in Section 2. A simple calculation for piecewise linear trial functions on a uniform mesh of length h shows that $1 + \ell(V^*) \sim h$ for large values of b/a and thus one cannot guarantee the resulting approximation to be near U^*. In fact, in practice the method performs poorly on these types of problems, see [1]. So it appears that the symmetrizing Petrov-Galerkin method cannot be extended along these lines.

Another approach is to introduce auxiliary variables as in least squares and mixed methods, for instance see [5]. We start from the viewpoint that we wish to construct the best fit U^* to u from S_E^h in the norm $||\cdot||_S$; that is, find $U^* \in S_E^h$ such that

$$B_S(U^*,\phi) = B_S(u, \phi) , \quad \forall \phi \in S_0^h , \tag{3.4}$$

where $B_S(\cdot, \cdot)$ is defined by (1.4c). Using the differential equation (1.1), we can relate some of the right-hand side of (3.4) to the data: that is:

$$B_S(u, \phi) = <f,\rho a\phi> + <\underline{b}u - a\underline{\nabla}u,[\rho\underline{b} + \underline{\nabla}(\rho a)]\phi>, \ \forall \phi \in S_0^h . \tag{3.5}$$

Thus we need to know $\underline{b}u - a\underline{\nabla}u$ in order to define U^*. However, we could consider solving for $U \in S_E^h$ such that

$$B_S(U,\phi) = <f,\rho a\phi> + <\underline{V},[\rho\underline{b} + \underline{\nabla}(\rho a)]\phi>, \ \forall \phi \in S_0^h , \tag{3.6}$$

where $\underline{V}$ is an approximation to an auxiliary unknown vector $\underline{v} = \underline{b}u - a\underline{\nabla}u$, obtained by approximating $\underline{\nabla} \cdot \underline{v} = f$. This approach can be viewed as a modified least squares method applied to (1.1) in the form of a first order system.

Consider the one-dimensional case; then the auxiliary problem is $v' = f$ and this requires v to be fixed at some point in order to yield a unique solution. An obvious approximation is to set $V(0) = (bU - aU')(0)$ if $ab > 0$, i.e. specify V on the inflow boundary, and then march forward setting

$V_{j+1} = V_j + \int_{jh}^{(j+1)h} f dx$, assuming for convenience we are using a uniform mesh of length h. We can either substitute this approximation in (3.6) and solve directly for U, or we can use an iterative procedure to solve alternately for U and V : the latter has the advantage that $B_S(\cdot, \cdot)$ generates a symmetric positive definite system of equations for U, to which all the standard techniques can be applied. In practice this iteration converges extremely quickly. In the common situation where $f \equiv 0$, then v is constant and $V \equiv (bU - aU')(0)$: by comparing (2.4) and (3.6) we see that this procedure then reduces exactly to the symmetrizing Petrov-Galerkin method of section 2 with $\varepsilon(x) \equiv \delta(x)$.

In two dimensions with $f \equiv 0$, the auxiliary problem is $\underline{\nabla} \cdot \underline{v} = 0$. Again we specify $\underline{V} = \underline{b}U - a\underline{\nabla}U$ on the inflow boundary, that is where $\underline{b} \cdot \underline{n} \leq 0$, assuming it is part of $\partial\Omega_1$; but we also require an extra equation in the interior and we shall base this on $\underline{b} \times \underline{v} = - \underline{b} \times (a\underline{\nabla}u)$. In modelling diffusion-convection problems, there are two distinct properties a numerical scheme should possess: one is to model convection well, that is solve the corresponding reduced problem accurately, and the second is to capture possible boundary and interior layers on a coarse mesh. The former requirement is trivial in one dimension and the success of the symmetrizing approach is to perform the latter easily and effectively. However, in two dimensions the modelling of pure convection causes difficulties due to the notorious problem of crosswind diffusion when convecting sharp fronts, see pp 19-35 in [6]. The basis of the above scheme of introducing auxiliary variables is to split the procedure into its distinct parts. The defining equation (3.6) for U tries to capture possible boundary layers through the associated best fit property: while the modelling of convection is determined by how one approximates the auxiliary equations for $\underline{v}$. There are many ways of approximating these equations and also one has the extra flexibility of choosing the associated trial space for $\underline{V}$. However, no scheme appears to be an obvious choice at present and we have selected a procedure on the basis of its ability to track the streamlines as closely as possible.

Many numerical experiments have been performed for the case where Ω is equal to the unit square, a and the components of $\underline{b}$ are positive constants, with the flow inclined at an angle θ to the x-axis. With S^h, the space of piecewise bilinear functions on a uniform grid of length h, chosen as the trial space for $\underline{V}$ as well as U and $\rho \equiv 1$, encouraging results for various values of Θ and bh/a have been obtained. The procedure adopted for the auxiliary equations for $\underline{v}$ was to set

$$<\underline{b}\times(\underline{V} + a\underline{\nabla}U), \chi_{\Omega_e}> = 0, \qquad (3.7)$$

for all elements $\Omega_e \subset \Omega$, where χ_{Ω_e} is the associated characteristic function, and although $<\underline{\nabla} . \underline{V}, \chi_{\Omega_e}> = 0$ was an obvious candidate for the other equation, this led to excessive crosswind oscillations. A more appropriate choice can be obtained by considering the reduced problem, in which information is convected along the $\underline{b}$ streamlines. Thus we have $u(ih, jh) = u((i - 1)h, (j - \tan\Theta)h)$, if $b_1 \geq b_2$; and $u(ih, jh) = u((i - \cot\Theta)h, (j - 1)h)$, if $b_2 \geq b_1$, where $\tan\Theta = b_2/b_1$. Assuming a linear profile for U across each element boundary leads to the following:

$$U_{i,j} = (1 - \tan\Theta)U_{i-1,j} + \tan\Theta\, U_{i-1,j-1} \quad \text{if } b_1 \geq b_2 \;; \qquad (3.8a)$$

$$U_{i,j} = (1 - \cot\Theta)U_{i,j-1} + \cot\Theta\, U_{i-1,j-1} \quad \text{if } b_2 \geq b_1 \;. \qquad (3.8b)$$

In terms of the auxiliary vector $\underline{V} = [V^{(1)}, V^{(2)}]^T \equiv U[b_1, b_2]^T$, in the convection limit this reduces to

$$V^{(1)}_{i,j} = V^{(1)}_{i-1,j} - V^{(2)}_{i-1,j} + V^{(2)}_{i-1,j-1}, \quad \text{if } b_1 \geq b_2; \qquad (3.9a)$$

$$V^{(2)}_{i,j} = V^{(2)}_{i,j-1} - V^{(1)}_{i,j-1} + V^{(1)}_{i-1,j-1}, \quad \text{if } b_2 \geq b_1. \qquad (3.9b)$$

However, the scheme (3.9) is now discontinuous in the parameters b_1 and b_2. To overcome this, we compromise by taking a weighted average of (3.9a) and (3.9b), ratio $b_1 : b_2$, as our approximation to $\underline{\nabla} . \underline{v} = 0$ in the element $[(i - 1)h, ih]\times[(j - 1)h, jh]$ and use this for all values of the Péclet number b/a. Clearly this approach can be generalised to cope with non-uniform meshes. The resulting system for U and $\underline{V}$ was solved iteratively and was found to converge extremely quickly for all values of bh/a, although one might not have expected this from examining the right-hand side of (3.6).

To illustrate the technique, we consider the case $b/a = 10^6$ and $\Theta = 67.5°$ as considered by Hughes, pp 19-35 in [6]. The boundary data specified is $u = 1$ on $y = 0$ and on $x = 0$, $y \leq 0.2$, and $u = 0$ on $x = 0$, $y > 0.2$ with either (a) $\partial u/\partial n = 0$ on the outflow boundary, i.e. $x = 1$ and $y = 1$ or (b) $u = 0$ on the outflow boundary. With $h = 0.1$, the approximation U obtained along the line $x = 0.5$ compared with the true solution u is given below, together with the values of U at $x = 0.2$.

y =	0.5	0.6	0.7	0.8	0.9	1.0
(a)						
U(0.5, y)	1.02	0.96	0.98	1.05	1.08	1.05
u(0.5, y)	1.00	1.00	1.00	1.00	1.00	1.00
U(0.2, y)	0.97	0.67	0.42	0.26	0.15	0.09
(b)						
U(0.5, y)	1.03	0.96	1.00	0.98	1.37	0.00
u(0.5, y)	1.00	1.00	1.00	1.00	1.00	0.00
U(0.2, y)	0.97	0.66	0.43	0.25	0.17	0.00

Thus one can see that the scheme has worked well for case (a) when no boundary layer occurs; and in case (b) the overshoot in U at (0.5, 0.09) indicates it attempts to capture the boundary layer present. One could now try to generalise the recovery techniques used in [2] for one dimensional problems to handle the two-dimensional boundary layer.

REFERENCES

1. BARRETT, J.W., *PhD Thesis*, University of Reading (1980).

2. BARRETT, J.W. & MORTON, K.W., Optimal finite element solutions to diffusion-convection problems in one dimension. *Int. J. Num. Meth. Engng.* 15, 1457-1474 (1980).

3. BARRETT, J.W. & MORTON, K.W., Optimal Petrov-Galerkin methods through approximate symmetrization, *U. of Reading Num. Anal. Report* 4/80, (1980). To appear in *IMA J. Numer. Anal.*

4. BRISTEAU, M.O., PIRONNEAU, O., GLOWINSKI, R., PERIAUX, J., PERRIER, P. & POIRIER, G., Application of optimal control and finite element methods to the calculation of transonic flows and incompressible viscous flows. *IMA Conf. Num. Meths. in Applied Fluid Dynamics* (B.Hunt Ed.) Academic Press, London, 203-312 (1980).

5. FIX, G.J., GUNZBURGER, M.D. & NICOLAIDES, R.A., On finite element methods of the least squares type. *Comp. and Maths. with Applics.* 5, 87-98 (1979).

6. HUGHES, T.J.R. (Ed.), *Finite Element Methods for Convection Dominated Flows*, AMD-Vol.34, Amer. Soc. Mech. Eng., New York (1979).

7. MORTON, K.W. & BARRETT, J.W., Optimal finite element methods for diffusion-convection problems, *Boundary and Interior Layers - Computational and Asymptotic Methods* (J.J.H. Miller Ed.), Boole Press, Dublin, 134-148 (1980).

A PETROV-GALERKIN METHOD FOR HYPERBOLIC EQUATIONS

D. F. Griffiths

University of Dundee

1. INTRODUCTION

Petrov-Galerkin or weighted residual finite element methods generalise the standard Galerkin or Ritz finite element methods in that they allow the test and trial functions to differ. This flexibility in the choice of test and trial functions enables discrete equations with varying degrees of local accuracy and 'upwinding' to be derived. Such schemes have so far been applied successfully to many problems in diffusion-convection (see [1], [2], [4], [5] and [6]).

A similar approach is now adopted for first order hyperbolic equations, the Petrov-Galerkin formulation allowing the introduction of dissipation into otherwise conservative schemes. We shall assume that the trial functions are continuous piecewise linear functions and then determine properties of the corresponding test functions which will lead to certain orders of accuracy and dissipation. Our schemes, which we develop initially for scalar equations and then extend to linear systems, are similar in spirit to the methods described in [3], [7], [8] and [10]. Certain of the present results, together with extensions into other areas, have previously been summarized in Griffiths and Mitchell [5].

2. APPROXIMATION OF SCALAR EQUATIONS

We begin by describing the Petrov-Galerkin process for the simple scalar equation

$$\frac{\partial u}{\partial t} + a\frac{\partial u}{\partial x} = 0\ , \quad t > 0 \tag{2.1}$$

in which the speed of propagation a is constant. The real line is divided into equal elements of size h by the nodes $x_j = jh,\ j = 0,\pm1,\pm2,\ldots$ and, at each time level $t = n\Delta t$

$n = 0,1,2,\ldots$ the solution $u(x,t)$ is approximated by a piecewise linear function $U^n(x)$ which is expressed in the form

$$U^n(x) = \sum_j U_j^n \phi_j(x), \quad n = 0,1,\ldots \tag{2.2}$$

where $\{\phi_j(x) \equiv \phi(x/h - j), \text{ for all } j\}$ are the linear 'hat' functions and

$$\phi(s) = \begin{cases} 1 - |s| \,, & |s| \le 1 \\ 0 & |s| \ge 1 \end{cases} . \tag{2.3}$$

The functions $\{\phi_j\}$ span a linear space Φ_h which we refer to as the trial space. Discretizing in time by the θ-method, the Petrov-Galerkin solution of (2.1) is given by

$$(\psi_j, [U^{n+1}-U^n] + a\Delta t[\theta U_x^{n+1}+(1-\theta)U_x^n]) = 0 \,, \tag{2.4}$$

$$j = 0,\pm1,\pm2,\ldots$$
$$n = 0,1,2,\ldots$$

where $(\cdot,\cdot)$ denotes the usual L_2 linear product and the test functions $\psi_j(x)$ span a linear space, Ψ_h, of dimension equal to that of Φ_h. In what follows each test function $\psi_j(x)$ will be assumed to be a perturbation of the corresponding trial function $\phi_j(x)$; this is a convenient rather than an essential ingredient of our treatment. Accordingly, we write

$$\psi_j(x) = \phi_j(x) + \sigma_j(x) \,, \quad \text{for all } j \tag{2.5}$$

and, in order that the test functions enjoy the same translational invariance as the trial functions, we let $\sigma_j(x) \equiv \sigma(x/h - j)$ with $\sigma(s) = 0$ for $|s| \ge 1$. Note that the weak regularity of the test functions required by (2.4) allows the perturbing functions $\{\sigma_j\}$ to be discontinuous at interelement boundaries.

Properties of the discrete system (2.4) are more readily assessed if the component terms are expressed in finite difference notation. The test functions may, without loss of generality, be normalised so that

$$\int_{-\infty}^{\infty} \psi_j(x)dx = \int_{-\infty}^{\infty} \phi_j(x)dx \ , \quad \text{for all } j$$

which, through (2.5), leads to

$$\int_{-1}^{1} \sigma(s)ds = 0. \tag{2.6}$$

By substituting (2.2) and (2.5) into (2.4) we obtain a difference equation for the nodal values $\{U_j^n\}$ of the form

$$\{1-\beta\Delta_o+\tfrac{1}{2}(\alpha+\tfrac{1}{3})\delta^2\}(U_j^{n+1}-U_j^n) + p\{\Delta_o-\tfrac{1}{2}\gamma\delta^2\}(\theta U_j^{n+1}+(1-\theta)U_j^n) = 0 \tag{2.7}$$

where $p = a\Delta t/h$ is the Courant number, $0 \leq \theta \leq 1$, the central differential operators Δ_o and δ^2 are given by

$$\Delta_o U_j = \tfrac{1}{2}(U_{j+1} - U_{j-1}) \tag{2.8a}$$

$$\delta^2 U_j = U_{j+1} - 2U_j + U_{j-1} \tag{2.8b}$$

and the parameters α, β and γ are defined in terms of the function $\sigma(s)$ by

$$\alpha = \int_{-1}^{1} |s|\sigma(s)ds \ , \tag{2.9a}$$

$$\beta = -\int_{-1}^{1} s\ \sigma(s)ds \tag{2.9b}$$

and

$$\gamma = \int_{-1}^{1} \text{sgn}\ s\ \sigma(s)ds \ . \tag{2.9c}$$

Insofar as the approximation of (2.1) is concerned, the precise nature of the perturbation $\sigma(s)$ is unimportant since only the moments of σ appearing in equations (2.9) have any effect on the discrete equations. Our approach is to derive suitable values of α, β and γ by analysing (2.7) and then, by employing (2.9), to deduce a suitable form for $\sigma(s)$. The choice of σ will clearly not be unique, but an elementary example, constructed from discontinuous linear functions, is

$$\sigma(s) = \alpha(6|s|-3) + 3\beta(1-2|s|)\text{sgn } s + \gamma(3|s|-2)\text{sgn } s, \quad |s| \le 1. \tag{2.10}$$

It is evident from (2.9) and (2.10) that α is controlled by the even part of $\sigma(s)$ on $(-1,1)$ whereas both β and γ are controlled by the odd part of σ.

An alternative interpretation of the Petrov-Galerkin approximation (2.7) is that it is equivalent to the Galerkin approximation of the modified equation

$$\{1 - \tfrac{1}{2}\gamma a h \frac{\partial}{\partial x}\}\left(\frac{\partial u}{\partial t} + a\frac{\partial u}{\partial x}\right) - h(\beta-\tfrac{1}{2}\gamma)\frac{\partial^2 u}{\partial x \partial t} + \tfrac{1}{2}\alpha h^2 \frac{\partial^3 u}{\partial x^2 \partial t} = 0 \tag{2.11}$$

where both test and trial functions are from the same space of C^0 linear functions Φ_h and the θ-method is used for time discretization.

Rearranging (2.7) we find

$$\{1 - \hat{\beta}\Delta_o + \tfrac{1}{2}(\hat{\alpha}+\tfrac{1}{3})\delta^2\}(U_j^{n+1}-U_j^n) + p\{\Delta_o - \tfrac{1}{2}\gamma\delta^2\}U_j^n = 0 \tag{2.12}$$

in which there are only three independent parameters $\hat{\alpha}$, $\hat{\beta}$ and γ where

$$\hat{\alpha} = \alpha - \gamma p\theta , \qquad \hat{\beta} = \beta - p\theta . \tag{2.13}$$

The introduction of the parameter θ is therefore unnecessary for constant coefficient problems but it does allow a greater flexibility when dealing with more general hyperbolic equations.

The accuracy of the scheme (2.12) may be assessed by following the evolution of a single Fourier mode. Solving (2.1) and (2.12), respectively, for the initial data

$$u(x,0) = \exp(i\omega x), \quad -\infty < x < \infty \tag{2.14a}$$

and

$$U_j^0 = \exp(i\omega x_j) , \quad j = 0,\pm1,\pm2,\ldots \tag{2.14b}$$

leads to

$$u(x,t+\Delta t) = \exp(i\omega h)u(x,t) , \quad t = n\Delta t , \tag{2.15a}$$

and

$$U_j^{n+1} = \xi U_j^n \ , \quad j = 0,\pm 1,\pm 2,\ldots \qquad (2.15b)$$

where the amplification factor ξ is given by

$$\xi = 1 - 2ip\chi\, g(\chi), \qquad (2.16)$$

$\chi = \frac{1}{2}\omega h$ and

$$g(\chi) = \frac{\sin\chi}{\chi} \ \frac{\cos\chi - i\gamma\sin\chi}{1 - i\hat{\beta}\sin 2\chi - 2(\hat{\alpha}+\frac{1}{3})\sin^2\chi} \ . \qquad (2.17)$$

Adopting the usual definitions (see for example [9]), the family (2.12) is *accurate of order* r if

$$\xi(\chi) = \exp(-2i\chi p) + O(\chi^{r+1}) \quad \text{for} \quad \chi \to 0$$

and *dissipative of order* $2d$ if there is a positive constant C such that

$$|\xi|^2 \le 1 - C|\chi|^{2d} \ , \qquad |\chi| \le \tfrac{1}{2}\pi \ .$$

Expanding (2.17) for small values of χ leads to

$$g(\chi) = 1 + i(2\hat{\beta}-\gamma)\chi + 2(\hat{\alpha}-2\hat{\beta}^2+\gamma\hat{\beta})\chi^2 +$$
$$i[(\gamma-2\hat{\beta})/3 - 2\hat{\beta} + 8\hat{\beta}(\hat{\alpha}+\tfrac{1}{3}) - 8\hat{\beta}^3 + \gamma(4\hat{\beta}^2-2\hat{\alpha}-\tfrac{2}{3})]\chi^3 + O(\chi^4)$$

from which we can deduce the orders of accuracy and dissipation. The main features are summarized in Table 1. Note that the Galerkin approximation $(\alpha = \beta = \gamma = 0)$ is at most second order accurate at the nodes. Analogous results for the semi-discrete (i.e. continuous intime) Petrov-Galerkin approximation may be deduced by taking the limit $\Delta t/h \to 0$.

Morton and Parrott [7] introduce a useful criterion for maintaining accuracy for times steps $\Delta t = O(h)$. This is the *unit CFL condition* which requires that the discrete solution be transmitted undistorted through the grid when the Courant number p is unity. In terms of the amplification factor ξ, this requires that $\xi = \exp(-2i\chi)$ when $p = 1$. In the present context this leads to

$$\alpha = 1/6 - \gamma(\tfrac{1}{2}-\theta) \ , \quad \gamma = 2\beta + 1 - 2\theta$$
$$\text{for} \quad p = 1. \qquad (2.18)$$

TABLE 1

Orders of Accuracy and Dissipation

Order of Accuracy	Order of Dissipation	Restriction on parameters
1	2	$p[\beta - \frac{1}{2}\gamma - p(\theta-\frac{1}{2})] \le 0$ $\gamma p[\alpha - 1/6 - \gamma p(\theta-\frac{1}{2})] \le 0$
2	4	$\beta = \frac{1}{2}\gamma + p(\theta-\frac{1}{2})$ $\gamma p[\alpha - 1/6 - \gamma p(\theta-\frac{1}{2})] \le 0$
2	∞^*	$\beta = \frac{1}{2}\gamma + p(\theta-\frac{1}{2})$, $\alpha = 1/6 + \gamma p(\theta-\frac{1}{2})$ or $\gamma = 0$, $\beta = p(\theta-\frac{1}{2})$, any α
3	4	$\beta = \frac{1}{2}\gamma + p(\theta-\frac{1}{2})$, $\gamma p \ge 0$ $\alpha = p^2/6 + \gamma p(\theta-\frac{1}{2})$
4	∞^*	$\alpha = p^2/6$, $\beta = p(\theta-\frac{1}{2})$, $\gamma = 0$

* denotes a non-dissipative scheme.

Applying a similar condition for characteristics of negative slope gives

$$\alpha = 1/6 + \gamma(\tfrac{1}{2}-\theta), \quad \gamma = 2\beta - 1 + 2\theta \quad \text{for} \quad p = -1. \tag{2.19}$$

It is apparent that a particular scheme with parameters α, β, γ, θ fixed, independent of p, cannot satisfy the unit CFL condition at both $p = \pm 1$ unless $\theta = \frac{1}{2}$. Choosing $\theta = \frac{1}{2}$ gives a second order scheme when $\alpha = 1/6$, $\gamma = 2\beta$. For a higher order of accuracy the parameters must be allowed to vary with p. These are important considerations when dealing with hyperbolic systems.

The two-level schemes of Morton and Parrott appear as special cases of (2.7). Two such schemes which performed well in numerical experiments on both constant and variable coefficient problems were, in the current notation,

i) Euler-Petrov-Galerkin II (EPG II)

$$\alpha = (\theta-\tfrac{1}{3})p^2 \ , \quad \beta = \theta p \ , \quad \gamma = p \tag{2.20}$$

and ii) Crank-Nicolson-Petrov-Galerkin (CNPG)

$$\alpha = p^2/6 + \tfrac{1}{2}\gamma p(\theta-\tfrac{1}{2}) \ , \ \beta = \tfrac{1}{2}\gamma + p(\theta-\tfrac{1}{2}) \ , \ \gamma = p^2 \mathrm{sgn}\ p. \tag{2.21}$$

As implied by their names, these methods were originally derived with $\theta = 0$ (EPG II) and $\theta = \frac{1}{2}$ (CNPG); from Table 1 it may be seen that both are third order accurate and unconditionally stable (dissipative of order 4). It also is worth noting that the Lax-Wendroff scheme can be obtained by choosing $\alpha = -\frac{1}{3} + \theta p^2$, $\beta = \theta p$ and $\gamma = p$.

For initial-boundary value problems such as (2.1) with $a > 0$ on the quadrant $\{x \geq 0, t \geq 0\}$ and the boundary condition $u(0,t) = g(t)$, an artificial boundary must be imposed at $x = X$ (= Jh) say. The discrete boundary equation which results at $j = J$, derived in the usual spirit of natural conditions, is

$$(1-\alpha')U_J^{n+1} + \alpha' U_{J-1}^{n+1} = (1-p-\alpha')U_J^n + (p+\alpha')U_{J-1}^n$$

where $\alpha' = (\alpha+\frac{1}{3}+\beta)/(1+\gamma) - p\theta$. This is in general a first order approximation to (2.1), increasing to second order when $\alpha' = \frac{1}{2}(1-p)$. This is conveniently satisfied (cf Table 1) by choosing $\alpha = 1/6 + \gamma p(\theta-\frac{1}{2})$ and $\beta = \frac{1}{2}\gamma + p(\theta-\frac{1}{2})$. The stability of the discrete initial-boundary value problem for EPG II with the second order boundary scheme may be verified when $\gamma \geq 0$.

A final remark concerns the application of these schemes to scalar conservation laws of the form

$$\frac{\partial u}{\partial t} + \frac{\partial f}{\partial x} = 0 \ , \quad f \equiv f(x,u).$$

High order accuracy requires that the parameters vary with Courant number. For this purpose the Courant number is defined locally and there is a choice of whether it should vary from grid point or from element to element. It is only the latter which ensures that the difference equations are also conservative.

3. FIRST ORDER HYPERBOLIC SYSTEMS

The methods described in the preceding section are extended to the hyperbolic system

$$\underset{\sim}{u}_t + A\underset{\sim}{u}_x = 0 \ , \quad t \geq 0 \tag{3.1}$$

where $\underset{\sim}{u} \in \mathbb{R}^N$ and A is an $N \times N$ matrix with real eigenvalues.

Generalization of the earlier schemes in which the parameters α, β, γ are fixed (independent of p) are obtained by writing

$$\underset{\sim}{U}^n(x) = \sum_j \underset{\sim}{U}_j^n \phi_j(x) \qquad (3.2)$$

and taking, at the jth node of the grid, the test functions $\psi_j(x)\underset{\sim}{e}_i$, $i = 1,2,\ldots,N$ where $\{\underset{\sim}{e}_i\}$ are the canonical basis vectors for $\mathbb{R}^N$. The resulting scheme is given by (2.7) with the scalar variable U replaced by $\underset{\sim}{U}$ and with p replaced by $(\Delta t/h)A$.

Generalizations of schemes in which the parameters vary with Courant number require closer study. Suppose that A has independent left eigenvectors $\underset{\sim}{v}_1, \underset{\sim}{v}_2, \ldots, \underset{\sim}{v}_N$ with corresponding (real) eigenvalues $\lambda_1 \le \lambda_2 \le \ldots \le \lambda_m < 0 < \lambda_{m+1} \le \ldots \le \lambda_N$. Defining $z_i = \underset{\sim}{v}_i^T \underset{\sim}{u}$ we find that

$$\frac{\partial z_i}{\partial t} + \lambda_i \frac{\partial z_i}{\partial x} = 0 \ , \ i = 1,2,\ldots,N \qquad (3.3)$$

so that the appropriate test function for a particular component, z_k say, should depend on parameters defined in terms of the Courant number $p_k = \lambda_k \Delta t/h$. Denoting this test function by $\psi_j(x;\lambda_k)$, the appropriate test functions at the jth grid point should be $\psi_j(x;\lambda_k)\underset{\sim}{v}_k$, $k = 1,2,\ldots,N$. This is not a very convenient form for computational purposes. However, when the test functions are *polynomial functions* of λ_k, we can write $\psi_j(x;\lambda_k)\underset{\sim}{v}_k \equiv \psi_j(x;A^T)\underset{\sim}{v}_k$. The test space, at the jth node, is spanned by $\{\psi_j(x;A^T)\underset{\sim}{v}_k\}_{k=1}^N$ or, more conveniently, by $\{\psi_j(x;A^T)\underset{\sim}{e}_k\}_{k=1}^N$. These test functions lead to a discretized equation of the form (2.7) in which $p = (\Delta t/h)A$ and α, β, γ are matrix valued functions (where appropriate) of p, defined by Table 1. For example, the counterpart of the third order scalar scheme is obtained by choosing

$$\alpha = r^2\left(1/6 + \mu(\theta-\tfrac{1}{2})\right)A^2 \ , \quad \beta = r(\tfrac{1}{2}\mu + \theta - \tfrac{1}{2})A, \ \gamma = \mu A$$

where $r = \Delta t/h$ and μ is a scalar parameter. This method is dissipative of order four for all $\mu > 0$ and non-dissipative when $\mu = 0$. By choosing $\mu = 1$ we obtain the analogue of the EPG II method

$$\{I + \tfrac{1}{6}(I-r^2A^2)\delta^2\}(\underset{\sim}{U}_j^{n+1}-\underset{\sim}{U}_j^n) + rA\{\Delta_o I - \tfrac{1}{2}rA\delta^2\}\underset{\sim}{U}_j^{n+1} = 0. \qquad (3.4)$$

Note that the CNPG method (equation 2.21) does not fit within the framework.

The main difficulty arises when we consider initial-boundary value problems; for example, the solution of (3.1) in the strip $\{0 \le x \le 1, t \ge 0\}$. Considering only the boundary $x=1$, m boundary conditions must be supplied (since A has m negative eigenvalues) in the form $B\underset{\sim}{u} = \underset{\sim}{0}$, where B is an $m \times N$ matrix such that the $m \times m$ matrix $\{B\underset{\sim}{v}_1, B\underset{\sim}{v}_2, \ldots, B\underset{\sim}{v}_m\}$ is non-singular. In terms of the characteristic variables, this implies that $z_1, \ldots, z_m$ may be uniquely expressed in terms of $z_{m+1}, \ldots, z_N$. The remaining $N - m$ boundary conditions should ideally be constructed from approximation of equations (3.3) for $i = m+1, \ldots, N$. This would require test functions, at the boundary node $j = J$, to be given by $\{\psi_J(x;A^T)\underset{\sim}{v}_k\}_{k=m+1}^{N}$ which would, in turn, require knowledge of the eigenvectors. Where these are not available, computational experience supports the assertion that it may be sufficient to choose test functions $\{\psi_J(x;A^T)\underset{\sim}{w}_k\}_{m+1}^{N}$, where $\{\underset{\sim}{w}_k\}$ are $N - m$ vectors, satisfying $B\underset{\sim}{w}_k = 0$, $k = m+1, \ldots, N$. One advantage of using boundary test vectors which involve eigenvectors is that stability may be reduced to consideration of approximations of (3.3), i.e. stability follows by analysing the scalar system.

REFERENCES

1. CHRISTIE, I., D.F. GRIFFITHS, A.R. MITCHELL and O.C. ZIENCIEWICZ, Finite Element Methods for Second Order Differential Equations with Significant First Derivatives. *Int. J. Num. Meth. Eng.*, 10, 1389-1396 (1976).

2. CHRISTIE, I. and A.R. MITCHELL, Upwinding of High Order Galerkin Methods in Conduction-Convection Problems. *Int. J. Num. Meth. Eng.*, 12, 1764-1771 (1978).

3. DENDY, J.E., Two Methods of Galerkin Type Achieving optimal L_2-accuracy for First Order Hyperbolics. *SIAM. J. Num. Anal.*, 11, 637-653 (1974).

4. HEINRICH, J.C., P.S. HUYAKORN, A.R. MITCHELL and O.C. ZIENCIEWICZ, An Upwind Finite Element Scheme for Two Dimensional Convective Transport Equations. *Int. J. Num. Meth. Eng.*, 11, 131-143 (1977).

5. GRIFFITHS, D.F. and A.R. MITCHELL, On Generating Upwind Finite Element Methods. pp 91-104 of T.J.R. Hughes (ed.). *Finite Element Methods for Convection Dominated Flows*, Am. Soc. Mech. Eng., AMD Vol 34 (1979).

6. MITCHELL, A.R. and D.F. GRIFFITHS, *The Finite Difference Method in Partial Differential Equations*. John Wiley, New York (1980).

7. MORTON, K.W. and A.K. PARROTT, Generalized Galerkin Methods for First Order Hyperbolic Equations. *J. Comp. Phys.* 36, 249-270 (1980).

8. RAYMOND, W.H. and A. GARDER, Selective Damping in a Galerkin Method for Solving Wave Problems on Variable Grids. *Month. Weath. Rev.*, 104, 1583-1590 (1976).

9. RICHTMEYER, R.D. and K.W. MORTON, *Difference Methods for Initial Value Problems*. John Wiley, New York (1967).

10. WAHLBIN, L.B., A Dissipative Galerkin Method for the Numerical Solution of First Order Hyperbolic Equations. pp 147-170 of C. de Boor (ed.), *Mathematical Aspects of Finite Elements in Partial Differential Equations*, Academic Press, New York (1974).

GENERALISED GALERKIN METHODS FOR HYPERBOLIC EQUATIONS

K. W. Morton and A. Stokes

University of Reading

1. INTRODUCTION

The advantages to be gained from approximating hyperbolic equations by finite element methods are never likely to be as great as with equilibrium problems which are governed by extremal principles. Nevertheless, the spatial flexibility of these methods is a great help when using coarse meshes to model, for example, hydraulic flows within complicated coastlines, estuaries or river systems: and it is now well-established [3, 5, 9] that very high spatial accuracies can be obtained with linear model problems on uniform meshes. Several developments have also taken place recently [3, 4, 5] to improve both the efficiency and accuracy with which non-linear operators are approximated.

The crucial issue with any time-dependent problem, however, is whether these features can be coupled effectively with a time-marching procedure. There are basically three choices: (i) use of finite elements in both space and time variables, which seems to have no special advantages and is complicated and expensive to implement; (ii) use of a semi-discrete formulation to derive a system of ODEs in the time variable to which an ODE-solving package can be applied - this is an attractive option regarding both flexibility and accuracy but it may be unnecessarily expensive for large problems; (iii) use of finite-differencing in time with finite elements in space.

Choices (ii) and (iii) overlap to some extent but, in adopting the latter, we emphasise that the idea here is to choose the time-differencing first, and then to design a finite element method which gives accurate results more economically than comparable wholly finite difference schemes. This may be accomplished and tested in one space dimension first, but the acid test comes with real problems in two and three dimensions.

In an earlier paper [8], Morton and Parrott developed special test functions which led to Petrov-Galerkin methods with remarkably high accuracy and good stability. This was for hyperbolic systems but only for linear problems in one dimension. We here develop these ideas into two dimensions and also move away from the Petrov-Galerkin formulation in order to deal better with non-linear problems. We shall consider only linear (or multi-linear) elements, though the ideas are general, as these seem to be adequate and economical in practice. Recently, we have learned of similar test functions developed and used by Hirsch [6] in turbo-machinery problems.

2. PETROV-GALERKIN METHODS IN ONE DIMENSION

We seek to approximate the solution $u(x, t)$ of the system $\partial_t u = L(u)$, at time levels $n\Delta t$, by the finite element expansion

$$U^n(x) = \sum_{(j)} U_j^n \phi_j(x) \tag{2.1}$$

in basis functions $\phi_j(x)$. Suppose the time-discretisation is based on a linear k-step method with parameters $\{\alpha_\nu, \beta_\nu;\ \nu = 0, 1, \ldots, k\}$. Then, if test functions $\psi_i(x)$ are used, we obtain a Petrov-Galerkin method of the form

$$< \sum_{\nu=0}^{k} \{\alpha_\nu U^{n+\nu} - \Delta t \beta_\nu L(U^{n+\nu})\}, \psi_i > = 0, \quad \forall\ i, \tag{2.2}$$

where $<\cdot, \cdot>$ denotes the usual L_2 inner product over the space domain. For given trial functions ϕ_j, the objective is to choose the test function ψ_i to give greatest accuracy, with reasonable economy of effort and stability.

Suppose $U^n, U^{n+1}, \ldots, U^{n+k-1}$ were optimal approximations, in some sense, to the values of u at these time-levels. Then our viewpoint is that the choice of test functions should be directed towards making U^{n+k} an optimal approximation in the same sense. The Galerkin method, in which $\psi_i = \phi_i$, achieves this optimality in the L_2 or least squares sense in the limit of small Δt. We shall therefore normally assume that we are seeking optimality in this sense, but exploit the greater generality of the Petrov-Galerkin formulation to improve accuracy at larger values of Δt. For the simple model problem $\partial_t u = a\partial_x u$ on a uniform mesh, and when the CFL number $a\Delta t/\Delta x$ equals one, optimal approximations can be maintained by simple translation through one mesh interval Δx. Most finite difference methods are exact in this case, but not Galerkin methods. Morton and Parrott [8] therefore developed special test functions χ_i that yielded this unit CFL property in the Petrov-Galerkin case: for intermediate values of Δt they then used linear combinations of ϕ_i and χ_i to form ψ_i.

We will summarise first the main features of three of the schemes they produced, for linear elements on a uniform mesh, i.e. $\phi_i(x) = 1 - |x - x_i|/\Delta x$ for $x \in (x_{i-1}, x_{i+1})$. In the first instance, we consider only the scalar model problem

$\partial_t u = a(x)\partial_x u$ with the local CFL number $\mu_i = a(x_i)\Delta t/\Delta x$.

Euler-Petrov-Galerkin (EPG).

This is dependent on the sign of $a(x_i)$: for $\mu_i > 0$ we have

$$\langle U^{n+1} - U^n - a\Delta t\partial_x U^n, \quad (1 - \mu_i)\phi_i + \mu_i\chi_i^+\rangle = 0$$

where

$$\chi_i^+(x) = 4 - 6(x - x_i)/\Delta x, \quad x \in (x_i, x_{i+1}) .$$

The scheme is similar to but considerably more stable than Lax-Wendroff and is stable for $0 \le \mu_i \le 1$; reflection about x_i gives χ_i^- to be used when $a(x_i) < 0$. Unfortunately, because span$\{\chi_i^+\}$ deos not include the constant function, the scheme is not conservative and cannot be recommended for non-linear problems without modification.

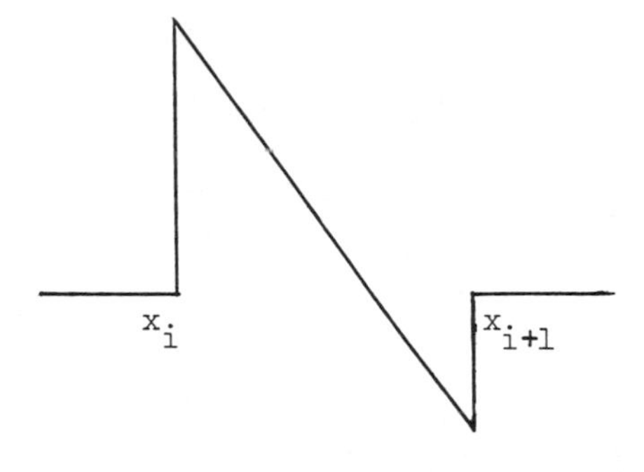

Leapfrog-Petrov-Galerkin (LPG).

As usual, this has no directional dependence and gives

$$\langle U^{n+1} - U^{n-1} - 2a\Delta t\partial_x U^n, (1 - \mu_i^2)\phi_i + \mu_i^2\chi_i\rangle = 0$$

where

$$\chi_i(x) = 2 - 3|x - x_i|/\Delta x, \quad x \in (x_{i-1}, x_{i+1}) .$$

This scheme is fourth order accurate in both Δt and Δx for constant a and is non-dissipative and stable for $|\mu| \le 1$: by comparison, the purely Galerkin scheme is only second order accurate in Δt and is stable only for $|\mu| \le 1/\sqrt{3}$; tables of phase errors show that the accuracy of

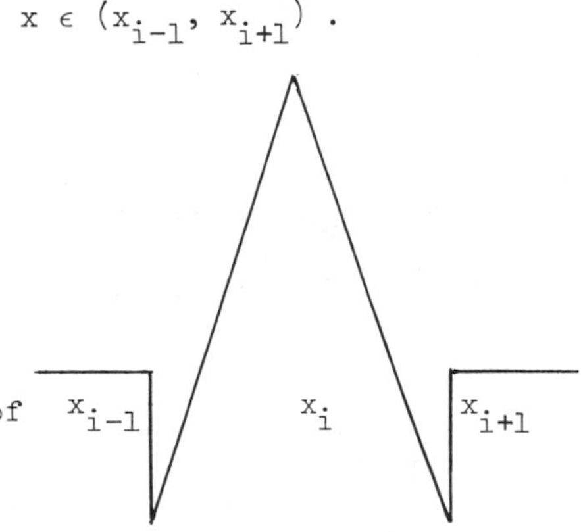

the Galerkin scheme up to $|\mu| \approx 0.3$ is maintained by LPG over the whole range of μ by the gradual modification of the test function. The symmetric form of χ_i means that the scheme is conservative and could be used for non-linear problems, though of course the leapfrog time-stepping can cause problems with spurious solutions.

Crank-Nicolson-Petrov-Galerkin (CNPG).

An implicit scheme, for $|\mu| \le 1$ and $a > 0$, this is

$$\langle U^{n+1} - U^n - \tfrac{1}{2}a\Delta t\partial_x(U^{n+1} + U^n), (1 - \mu_i^2)\phi_i + \mu_i^2\phi_i^0\rangle = 0$$

where

$$\phi_i^0(x) = 1 , \quad x \in (x_i, x_{i+1}).$$

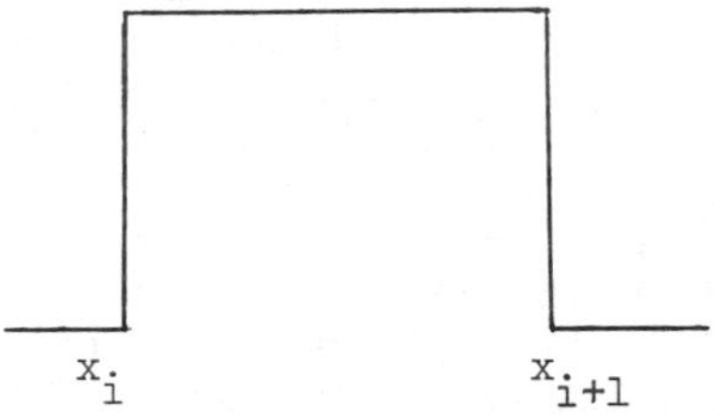

This scheme interpolates between Crank-Nicolson-Galerkin and a scheme which for constant a is identical to the Thomée-Keller box scheme. Thus, unlike these two schemes, it is slightly dissipative for $|\mu| < 1$ and is unstable for $|\mu| > 1$: in practice one would use ϕ_i^0 as the test function for $|\mu| > 1$ and so retain unconditional stability. The scheme is formally third order accurate and has very small phase errors. Apart from the implicitness, it is very simple in form and is conservative. A predictor-corrector form has been developed in which the corrector is the "conservative sub-domain" method (i.e. with ϕ_i^0 replaced by $\phi_{i-\frac{1}{2}}^0$) but the predictor is rather complicated if accuracy and stability are to be preserved. Note that for $a < 0$ one would replace ϕ_i^0 by ϕ_{i-1}^0, and $\frac{1}{2}(\phi_i^0 + \phi_{i-1}^0)$ could be used in all cases with some loss of accuracy.

The development of schemes for non-linear equations will be considered in section 4 but in [8] extensions were given for linear systems of equations. Thus for $\partial_t \underline{u} = A\partial_x \underline{u}$, applying the LPG scheme to the characteristic form and then transforming back to the original variables, gives

$$\begin{aligned} &\langle \underline{U}^{n+1} - \underline{U}^{n-1} - 2\Delta t A\partial_x \underline{U}^n, \phi_i \underline{e}_{(r)}\rangle \\ &\quad + \langle \underline{U}^{n+1} - \underline{U}^{n-1}, (\Delta t/\Delta x)^2 A^2 \sigma_i \underline{e}_{(r)}\rangle = 0 , \end{aligned} \qquad (2.3)$$

where $\sigma_i = \chi_i - \phi_i$ and $\underline{e}_{(r)}$ runs through the basis of the vector space for $\underline{u}$. Note how this form emphasises that only the mass matrix is modified by the test functions. This is a special property of the LPG scheme.

3. PETROV-GALERKIN METHODS IN TWO DIMENSIONS

In extending these schemes to two dimensions, simplicity and compactness have to be important objectives. Thus we shall always assume that the support of the special test functions is contained within that of the trial function with which they are to be combined. The simplest case to consider first is that of bilinear basis functions on rectangular elements applied to

$$\partial_t u = a\partial_x u + b\partial_y u \ . \tag{3.1}$$

As the basis elements for the trial space are just product functions $\phi_{ij}(x, y) = \phi_i(x)\phi_j(y)$, we can clearly use similar product functions as the special test functions. There is, however, some choice available as to how these should be combined. For LPG, for example, we have the following possibilities:

$$\psi_{ij}(x, y) = [(1 - \nu_i)\phi_i(x) + \nu_i\chi_i(x)][(1 - \nu_j)\phi_j(y) + \nu_j\chi_j(y)]$$
$$\text{or} \quad (1 - \nu_{ij})\phi_i(x)\phi_j(y) + \nu_{ij}\chi_i(x)\chi_j(y), \tag{3.2}$$

where ν_i, ν_j and ν_{ij} are appropriate parameters. Either will enable the unit CFL condition to be satisfied in the axial directions but there is no choice of test function which will also achieve this in the diagonal direction. Fourier analysis and test calculations have led us to use the second choice in (3.2) with

$$\nu_{ij} = (a_{ij}\Delta t/\Delta x)^2 + (b_{ij}\Delta t/\Delta y)^2 = \mu_{ij}^2 \ . \tag{3.3}$$

Similarly, for CNPG we can use this form with the special test function $\phi_i^0(x)\phi_j^0(y)$ being the characteristic function for the rectangle in the upwind direction from the node (i, j): that is, for this time-stepping algorithm we have generated a fairly typical upwind finite element method.

Figure 1 demonstrates the greatly improved accuracy obtained in practice with LPG for waves moving in any direction. The diagrams show contours obtained with the convection of a circularly symmetric Gaussian profile on a mesh with $\Delta x = \Delta y = 0.25$, taking $\mu = 0.5$ and, for LPG only, $\mu = 0.9$. There is clearly a very marked improvement over the Galerkin scheme as well as increased stability (up to $\mu = 1$). We also show, in Figure 2, cross-section profiles compared with the true solution when $b = 0$, $\mu = 0.5$. Similar but not quite such good accuracy is obtained with CNPG.

In applying these schemes to more realistic systems of equations, such as the shallow water equations, much can be gained by using the special test functions on just the convective terms $\partial_t + \underline{v} \cdot \underline{\nabla}$. To go further in general, one needs to carry out

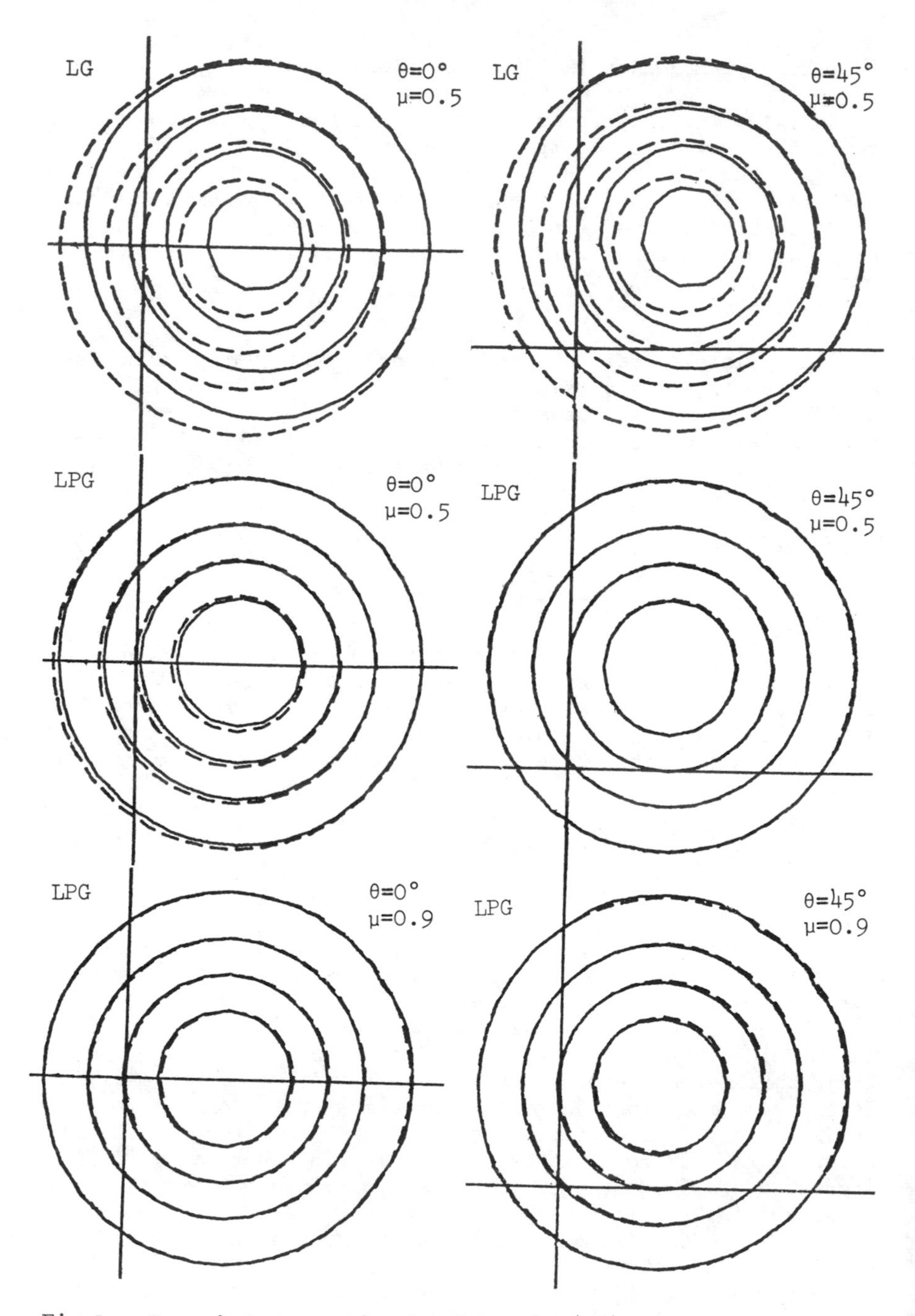

Fig.1. Gaussian convection by Galerkin (LG) and LPG: $\tan\theta = b/a$ and broken line is exact solution.

the transformation which led to the schemes for a system of equations in one dimension, such as (2.3): but this is not always possible for a system in two dimensions. Alternatively, one could exploit the special form of (2.3), namely that only the mass matrix is affected in this case, and assume that this would also be adequate for the operator $A\partial_x + B\partial_y$, perhaps replacing A^2 by $A^2 + B^2$ as in (3.3). As a last resort one can use fractional step methods, including ADI. These alternatives are currently being investigated by means of the model problem due to Grammeltvedt (see [5] for details).

Using piecewise linear trial functions on a regular mesh of equilateral triangles, one might expect to be able to satisfy the unit CFL condition in the three co-ordinate directions. Unfortunately this is not possible with $\text{supp}\chi \subset \text{supp}\phi$. For LPG one can construct test functions with this support to satisfy this condition in any two of the directions: but even then these cannot be piecewise linear and continuous in the interior of $\text{supp}\phi$, as they are in one dimension. The simplest direct generalisation of the one-dimensional function is piecewise linear with value 1 at the central node and $-1/3$ at the six surrounding neighbours. This diagonalises the mass matrix (it is the unique piecewise linear χ with this property and is that used by Hirsch [6] for this purpose) but does not satisfy the unit CFL condition in any direction. It gives some improvement in both accuracy and stability as compared with the Galerkin method but is not as effective as in the bilinear case. Further evidence will be given in section 4 that the restriction $\text{supp}\chi \subset \text{supp}\phi$ is too onerous and we have therefore not taken the development of these schemes further.

The situation is similar for CNPG although a piecewise linear test function can be constructed to satisfy the unit CFL condition in one direction. This would be constant, say 1, from the central node along the element side in that direction and falling off to -1 at the third node of the two neighbouring elements. This would not lead to a conservative scheme and does not seem a useful choice.

4. GENERALISED GALERKIN METHODS USING CHARACTERISTICS

Consider the scalar conservation law problem

$$\partial_t u + \partial_x f(u) = 0, \quad u(x, 0) = g(x) . \tag{4.1}$$

Apart from actual discontinuities, the solution u is constant along the straight-line characteristics $dx/dt = a(u)$, where $a = \partial f/\partial u$: that is,

$$u(x, t) = g(s) \quad \text{where} \quad x = s + a(g(s))t . \tag{4.2}$$

Suppose now that a finite element approximation of the form (2.1) is sought. It has been widely observed that, if the semi-discrete Galerkin equations are formed and these are integrated accurately by taking small time-steps, then the approximation to a steep wave-front exhibits oscillations which are close to those in the best L_2 fit to the time solution: one of the most detailed demonstrations of this phenomenon is given by Johnson et al. [7] for the Buckley-Leverett equations.

Consider then the problem of forming the best L_2 fit to $u(x, t + \Delta t)$ from $u(x, t)$, where Δt is reasonably large. We have from (4.2) that $u(x, t + \Delta t) = u(\sigma, t)$, where $\sigma = \sigma(x, t)$ is given by $\sigma + a(u(\sigma, t))\Delta t = x$. Thus

$$<u(t + \Delta t) - u(t), \phi_j> = \int_{-\infty}^{\infty} \phi_j(x) \int_x^{\sigma} \partial_x u(y, t)dydx$$

$$= \int_{-\infty}^{\infty} \partial_x u(y, t) \int_{y+a\Delta t}^{y} \phi_j(x)dxdy. \quad (4.3)$$

Replacing $\partial_x u$ by $a^{-1}\partial_x f$, we obtain the result

$$<u(t + \Delta t) - u(t), \phi_j> + \Delta t<\partial_x f(u(t)), \Phi_j> = 0 \quad (4.4)$$

where

$$\Phi_j(y) = \frac{1}{\Delta t a(u(y, t))} \int_y^{y+a\Delta t} \phi_j(x)dx. \quad (4.5)$$

In this sense, then, Φ_j which is an <u>upwind-average</u> of ϕ_j is the ideal test function to use on $\partial_x f$: but note that this is applied to only this term of the differential equation.

There are several ways in which we might base an approximation on these formulae. Clearly the most direct is to use the information at the foot of the characteristic and set

$$U_j^{n+1} = U^n(\sigma_j) \quad \text{where} \quad \sigma_j + a(U^n(\sigma_j))\Delta t = x_j \ : \quad (4.6)$$

many finite difference methods can be regarded as approximating this procedure and several finite element schemes have made use of it too (see, for example, [1] and [2]). But in the latter case, this fails to exploit the L_2 character of the approximation U^n and we feel that more use should be made of the form (4.4), which displays the way in which reduction to the Galerkin mwthod occurs as $\Delta t \to 0$. Going back to (4.3) and integrating by parts, a convenient form in which to approximate (4.4) is

$$<U^{n+1} - U^n, \phi_j> + <U^n, \phi_j - (1 + \Delta t\partial_x a)\phi_j(\cdot + \Delta t a)> = 0 \ ,$$

$$\text{i.e. } <U^{n+1}, \phi_j> = <U^n, \tilde{\phi}_j> \quad (4.7)$$

where

$$\tilde{\phi}_j(x) = [1 + \Delta t a'(U^n(x))\partial_x U^n(x)]\phi_j(x + \Delta t a(U^n(x))).$$

This scheme can be shown to be third order accurate for a linear equation and stable up to unit CFL number. However, non-linear instabilities presently limit its use to small Δt. A recovery procedure for estimating a and a' from U^n would in any case be needed for larger Δt and this is now being developed so as to dampen the instabilities.

Figure 3 shows results obtained with Burger's equation $\partial_t u + \partial_x(\frac{1}{2}u^2) = 0$ using the Euler time-stepping method (4.7). It is compared with CNPG, the best of the schemes given in section 2 for this problem. The initial data has the form $u(x, 0) = \cos^2\frac{1}{2}\pi x$ and the resulting steepening wave is shown at time $t = \frac{1}{2}$ using a mesh ratio $\Delta t/\Delta x = 0.1$ for $\Delta x = 1/6$.

REFERENCES

1. BENQUE, J.P., IBLER, B. & LABADIE, G. A finite element method for Navier-Stokes equations. *7th Int. Conf. on Numerical Methods in Fluid Dynamics*, (1980).

2. BREBBIA, C.A. & SMITH, S. Solution of Navier-Stokes equations for transient incompressible flow. *Proc. 1st Int. Conf. on Finite Elements in Water Resources*, Pentech Press, 4.205-4.230 (1977).

3. CHIN, R.C.Y., HEDSTROM, G.W. & KARLSSON, K.E. A simplified Galerkin method for hyperbolic equations. *Math. Comp.* 33, 647-658 (1979).

4. CHRISTIE, I., GRIFFITHS, D.F., MITCHELL, A.R. & SANZ-SERNA, J.M. Product approximation for non-linear problems in the finite element method. To appear in *IMA J. Numer. Anal.* (1981).

5. CULLEN, M.J. & MORTON, K.W. Analysis of evolutionary error in finite element and other methods. *J. Comput. Phys.* 34, 245-267 (1980).

6. HIRSCH, Ch. & WARZEE, G. An orthogonal finite element method for transonic flow calculations. *6th Int. Conf. on Numerical Methods in Fluid Dynamics* (1978).

7. JOHNSON, K.H., LEWIS, R.W. & MORGAN, K. An analysis of oscillations in the finite element modelling of Buckley-Leverett problems. *Proc. of BAIL I Conf.* (ed. J.J.H. Miller) Boole Press (Dublin), 327-331 (1980).

8. MORTON, K.W. & PARROTT, A.K. Generalised Galerkin methods for first order hyperbolic equations. *J. Comput. Phys.* 36, 249-270 (1980).

9. THOMEE, V. & WENDROFF, B. Convergence estimates for Galerkin methods for variable coefficient initial-value problems. *SIAM J. Numer. Anal.* 11, 1059-1068 (1974).

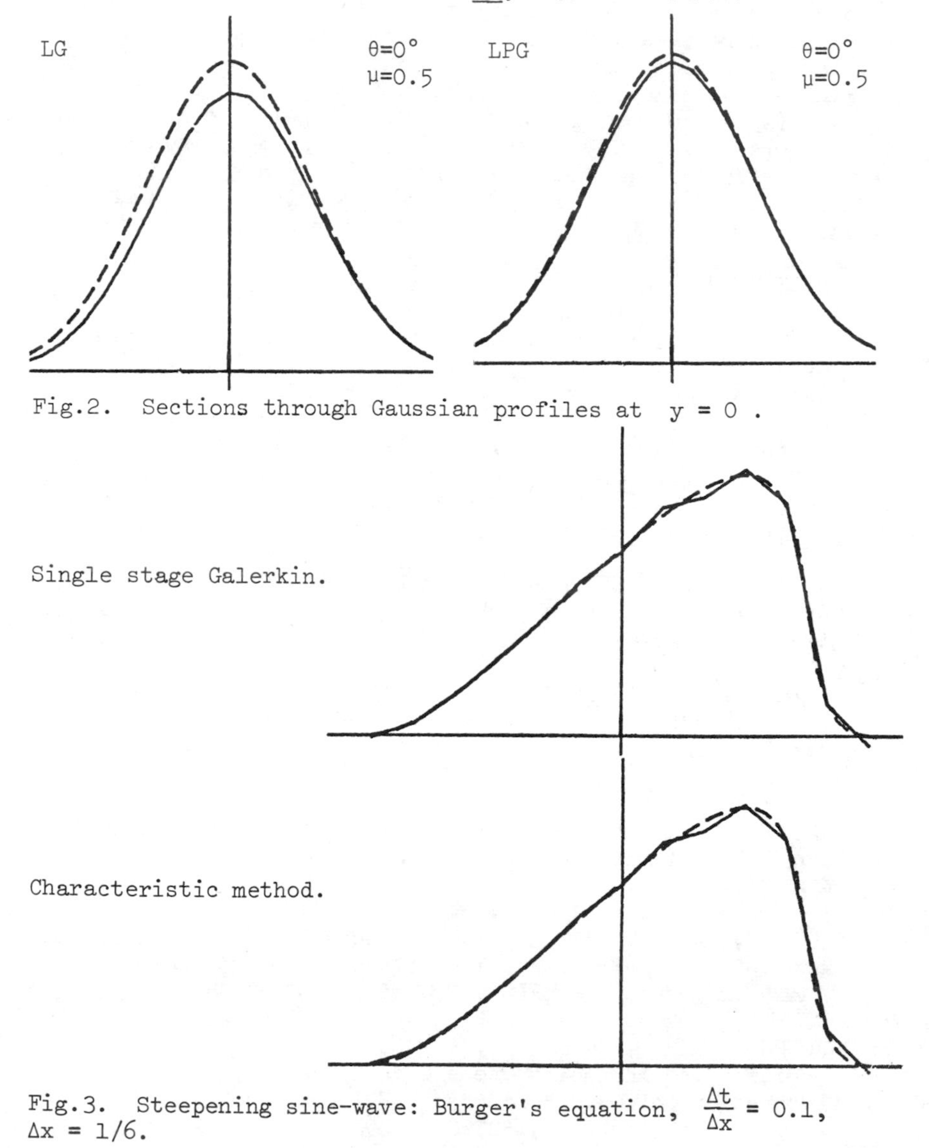

Fig.2. Sections through Gaussian profiles at $y = 0$.

Fig.3. Steepening sine-wave: Burger's equation, $\frac{\Delta t}{\Delta x} = 0.1$, $\Delta x = 1/6$.

NUMERICAL IMPERFECTIONS AND PERTURBATIONS IN THE APPROXIMATION OF NONLINEAR PROBLEMS

F. Brezzi and H. Fujii

Laboratorio di Analisi Numerica, Pavia

1. INTRODUCTION

The aim of this paper is to analyse the relationship between the imperfections introduced by the discretisation of nonlinear problems and the theory of perturbations of singularities, in particular from the point of view of Golubtsky and Schaeffer [9]. Essentially we try to answer the following questions; 1) under what conditions can one say that the discrete problem realizes "the whole bifurcation diagram" of the continuous problem? 2) What is the minimum set of perturbation parameters that should be considered, in the continuous problem, in order to interpret all the possible imperfections introduced by the discretisation? Some answers are already known for particular cases; for instant Kikuchi [8] has shown that a nondegenerated turning point is always reproduced by the numerical scheme, whilst Fujii and Yamaguti [7] have shown that, in the case of a symmetry breaking bifurcation, a *pitchfork bifurcation* is reproduced provided that the discretised equations preserve the necessary symmetry invariance. On the other hand, Brezzi et al. [3] showed that, in general, a simple bifurcation is not reproduced by the discrete equations, although one can have optimal error bounds for the distance between continuous and discretised solutions. In the present paper we show that, whenever the codimension (in the sense of Golubtsky and Schaeffer) of the problem is 1 then the introduction of a suitable parameter into the continuous problem allows us to reproduce all the possible imperfections introduced by the numerical scheme. We also show that, in that case, the perturbed discretised problem reproduces the whole bifurcation diagram. Special attention will be devoted to the problem involving the expression $x^3 - \lambda x$ which has codimension 2. We show that the introduction of two perturbation parameters, giving $x^3 - \lambda x + \mu + \alpha x^2$, allows our questions to be answered positively. On the other hand $x^3 - \lambda x + \mu + \alpha x^2$ can be considered as a one-parameter perturbation of the original two-parameter problem

$x^3 - \lambda x + \mu$. Although there is as yet no extension of the theory of Golubtsky and Schaeffer to the case of two parameters, we find that, from our point of view, $x^3 - \lambda x + \mu$ behaves as a (two-parameter) problem of codimension 1.

The paper can be outlined as follows. In Section 2 we introduce, in an abstract form, the class of nonlinear problems to be treated and the general form of discretised problems. We then recall the properties of the Liapunov-Schmidt decomposition, when applied both to continuous and to discretised problems. In Section 3 we recall the definitions of contact equivalence and of codimension in the case of one-parameter problems [9]. We briefly discuss the case of codimension 0 and we analyse systematically the case of codimension 1. At the end of Section 3 we discuss a particular problem ($x^5 - \lambda x$) which has codimension 1 with respect to Z/2-invariant functions, see [10]. This example seems to us to be particularly interesting, since its unfolding $x^5 + \mu x^3 - \lambda x$ is often present in applications as a true two-parameter problem (as in reaction diffusion problems; see e.g. Fujii [5]). Finally in Section 4 we analyse both the pitchfork $x^3 - \lambda x$ as a problem of codimension 2 and $x^3 - \lambda x + \mu$ as a two-parameter problem with codimension 1. This is also justified by applications to plate bending problems where λ and μ are both present as original physical parameters.

For the sake of simplicity, we shall consider only the case of C^∞ functions. However a limited number of continuous derivatives would be sufficient to carry out the theory in the different cases.

2. REDUCED PROBLEMS

We consider nonlinear problems of the type in which

$$F(u,\underline{\lambda}) \equiv u + TG(u,\underline{\lambda}) = 0 , \tag{2.1}$$

where G is a C^∞ mapping from $V \times R^n$ into W(V and W are assumed to be Banach spaces) and T is a linear compact operator from W into V. We also assume that we are given a sequence T_h of linear continuous operators $W \to V$ such that

$$\lim_{h\to 0} \| T - T_h \|_{L(W,V)} = 0 . \tag{2.2}$$

We now define the discretised problem as

$$F_h(u,\underline{\lambda}) \equiv u + T_h G(u,\underline{\lambda}) = 0 . \tag{2.3}$$

Remark The class of problems that can be written in the form (2.1) is quite general. For these most of the discretisations used in the practice (finite elements of various kinds, spectral methods, etc.) can be written in the form (2.3). We refer to Brezzi, Rappaz and Raviart, [1] - [3], for various examples of applications.

In that framework, the derivation of optimal error bounds for the case of the branches of regular solutions is straightforward, see for instance [7] and [1]. Let us go, therefore, directly to the case of simple singular points.

Definition *A solution* $(u_0,\underline{\lambda}_0)$ *of* (2.1) *is called a simple singular point if* $D_uF(u_0,\underline{\lambda}_0)$ *has* 0 *as a simple eigenvalue.*

If $(u_0,\underline{\lambda}_0)$ is a simple singular point of (2.1), it has been shown in [2] that the classical Liapunov-Schmidt procedure can be applied simultaneously to both problems (2.1) and (2.3). As a result we are left with two *reduced problems*,

$$f(x,\underline{\lambda}) = 0 \quad f \in C^{\infty}(R\times R^n; R) \tag{2.4}$$

with

$$f(0,\underline{0}) = f_x(0,\underline{0}) = 0 \tag{2.5}$$

and

$$f_h(x,\underline{\lambda}) = 0 \quad f_h \in C^{\infty}(R\times R^n; R) , \tag{2.6}$$

with

$$f_h \to f \quad (h \to 0) \tag{2.7}$$

uniformly in a neighbourhood of the origin (independent of h) with all the derivatives. Moreover the *distance* between f and f_h (and between their derivatives) can be estimated in terms of the *distance* between F and F_h (and between their derivatives) with no loss in the optimality of the error bounds (reference should be made to [2] for details). We also remark that, in a number of applications, the term

$$|f_h(0,\underline{0})| = |f_h(0,\underline{0}) - f(0,\underline{0})|$$

tends to zero twice as fast as the *optimal error bound.*

From now on we shall deal only with the reduced problems (2.4) and (2.6) as if they were our *original* problems, expressing the error bounds in terms of $f_h - f$ (and their derivatives). The reconstruction of the error bounds in terms of $F_h - F$, and then in terms of powers of h in the applications, is a straightforward application of the results of [2] and [3] and of the estimates for linear problems $T - T_h$.

Remark The analysis in [2] and [3] is actually carried out with $\lambda \in R$. It is easy to check that *nothing changes* when one uses instead $\underline{\lambda} \in R^n$.

3. GENERAL RESULTS FOR CODIMENSION ONE

We analyse in this section the relationships between the reduced problem

$$f(x,\underline{\lambda}) = 0 \quad (f \in C^\infty(R^{n+1}; R)) \tag{3.1}$$

and the reduced problem

$$f_h(x,\underline{\lambda}) = 0 \quad (f_h \in C^\infty(R^{n+1}; R)) \tag{3.2}$$

in the cases in which (3.1) has codimension 1. We shall always assume, from now on, that

$$f(0,0) = f_x(0,0) = 0 \tag{3.3}$$

and that

$f_h \to f$ *uniformly with all the derivatives in a neighbourhood* N_0 *of the origin, independent of* h . (3.4)

We recall first the definition of codimension in the case $x \in R$, $\lambda \in R$ (see [9]).

Definition *Let*

$$G = \{\textit{germs of } C^\infty(R^2; R)\} , \tag{3.5}$$

let $f \in G$ *and let*

$$\tilde{T}f = \{g_0 f + g_1 f_x | g_0 \in G,\ g_1 \in G\} , \tag{3.6}$$

$$Tf = \tilde{T}f \oplus \{g_2(\lambda) f_\lambda | g_2 = \textit{germ of } C^\infty(R; R)\} . \tag{3.7}$$

Then, if $G/\tilde{T}f$ *has finite dimension, we set*

$$\text{codim } f = \dim (G/Tf) . \tag{3.8}$$

Otherwise we say that f *has infinite codimension.*

Note: setting "$f_1 \simeq f_2$ iff f_1 and f_2 coincide in a neighbourhood of the origin", then $G = C^\infty(R^2; R)/\simeq$.

Let us discuss now the cases in which codim $f \leq 1$. We introduce first the definition of contact equivalence, [9].

Definition *Let* f,g *be two germs in* G. *We say that* $f \overset{c.e.}{\simeq} g$ *(*f *is contact equivalent to* g*) if there exists* $\tau(x,\lambda) \in C^\infty(R^2; R)$, $X(x,\lambda) \in C^\infty(R^2; R)$ *and* $\Lambda(\lambda) \in C^\infty(R; R)$ *with*

$$\tau(0,0) \neq 0 \quad \Lambda(0) = 0 ; \quad X(0,0) = 0 , \tag{3.9}$$

$$\frac{\partial \Lambda}{\partial \lambda}(0) > 0 \ , \quad \frac{\partial X}{\partial x}(0,0) > 0 \ , \tag{3.9}$$

such that

$$g(x,\lambda) = \tau(x,\lambda) f(X(x,\lambda),\Lambda(\lambda)) \tag{3.10}$$

(*the equality* (3.10) *holding in the sense of germs*).

Remark Obviously two functions are said to be contact equivalent if the corresponding germs are contact equivalent. In such a case relationship (3.10) holds for the two functions f and g in a neighbourhood of the origin.

From the previous results of [7] and [2] one has the following results.

Theorem 3.1 *Assume* (3.1) - (3.4). *If* f *has codimension* 0 *then there exists a neighbourhood* U *of the origin, an* $h_0 > 0$ *and a constant* c *such that for all* $h \leq h_0$ *there exists a unique point* $(x_0^h, \lambda_0^h) \in U$ *such that*

$$f_h(x + x_0^h, \ \lambda + \lambda_0^h) \overset{c.e.}{\simeq} f(x,\lambda) \ , \tag{3.11}$$

and moreover

$$\begin{aligned} |x_0^h| &\leq c(|f_h(0,0)| + |\partial_x f_h(0,0)|) \ , \\ |\lambda_0^h| &\leq c(|f_h(0,0)| + |\partial_x f_h(0,0)|^2) \ . \end{aligned} \tag{3.12}$$

Proof It is sufficient to remark that f has codimension 0 iff $f_\lambda(0,0) \neq 0$ and $f_{xx}(0,0) \neq 0$, that is if f has a nondegenerated turning point at (0,0). It is known, in that case, that f_h has a nondegenerated turning point at some (unique) (x_0^h, λ_0^h) satisfying (3.12). Thus from Lemma 3.14 of [9] one obtains (3.11).

We shall show, essentially, that a similar result holds in the case of the codimension 1, provided that an additional parameter $\mu = \lambda_2$ is considered in (3.1) and (3.2) with some suitable nondegeneracy hypotheses.

The following theorem will somehow classify the problems of codimension 1.

Theorem 3.2 *Assume that* f *satisfies* (3.3) *and*

$$f_\lambda(0,0) \cdot f_{xx}(0,0) = 0 \ , \tag{3.13}$$

then one of the following four cases holds

$$f \overset{c.e.}{\simeq} x^3 \pm \lambda \quad \textit{(nondegenerated hysteresis, codim f = 1)}, \tag{3.14}$$

$$f \overset{c.e.}{\simeq} x^2 + \lambda^2 \quad \textit{(isola, codim f = 1)} \ , \tag{3.15}$$

$$f \overset{c.e.}{\simeq} x^2 - \lambda^2 \text{ (simple transcritical bifurcation, codim } f = 1) , \tag{3.16}$$

$$\textit{codim } f > 1 . \tag{3.17}$$

Proof Assume first that $f_\lambda(0,0) \neq 0$. Hence from (3.13) $f_{xx}(0,0) = 0$. Let m be the smallest integer such that $(\partial/\partial x)^m f(0,0) \neq 0$. If $m = +\infty$ then f has infinite codimension; otherwise $f \overset{c.e.}{\simeq} x^m \pm \lambda$ (see [9] prop. 4.1) and codim $f = m-2$ (see [9] prop. 4.2). Hence (3.14) holds if $m=3$ and (3.17) holds otherwise. Assume now that $f_\lambda(0,0) = 0$ and consider det(Hf°), that is the determinant of the Hessian matrix of f at the origin. If det(Hf°) > 0 then (3.15) holds and if det(Hf°) < 0 then (3.16) holds (see [9] prop. 4.1 and 4.2). If finally det(Hf°) = 0 then $f(x,\lambda) = \xi^2 + \text{h.o.t.}$ with $\xi = ax + b\lambda$ and hence f has codimension >1.

It is an easy matter to check that a result similar to Theorem 3.1 is false in the cases (3.14) - (3.16) (and in general if codim f > 0). However, we shall prove in the following that Theorem 3.1 can be generalised to cover cases (3.14) - (3.16) provided that an additional parameter $\mu(=\lambda_2)$ is considered in (3.1) and (3.2) with suitable nondegeneracy conditions. More precisely we shall consider respectively the three cases

$$f(x,\lambda,\mu) = x^3 - \lambda + \mu x , \tag{3.18}$$

$$f(x,\lambda,\mu) = x^2 + \lambda^2 - \mu , \tag{3.19}$$

$$f(x,\lambda,\mu) = x^2 - \lambda^2 - \mu, \tag{3.20}$$

where obviously (3.1) and (3.2) should now be considered with $\lambda_1 = \lambda$ and $\lambda_2 = \mu$. Expressions (3.18) - (3.20) are the *one-parameter universal unfoldings* (see [9]) of (3.14) - (3.16) respectively.

In order to analyse (3.18) - (3.20) we first recall two basic properties of the uniform convergence of mappings.

Lemma 3.1 Let Φ *be a* C^∞ *mapping* $R^m \to R^m$ *such that* $\Phi(0) = 0$ *and* $D\Phi(0)$ *is nonsingular. Moreover let* Φ_h *be a sequence of* C^∞ *mappings which converges uniformly to* Φ *in a neighbourhood of* 0 *with all the derivatives. Then there exists a neighbourhood* U *of* 0, *an* $h_0 > 0$ *and a constant* c *such that for each* $h \leq h_0$ *there exists a unique* $x_0^h \in U$ *such that*

$$\Phi_h(x_0^h) = 0 , \tag{3.21}$$

and moreover

$$\| x_0^h \| \leq c \, \|\Phi_h(0)\| . \tag{3.22}$$

Lemma 3.2 *Let* $\Phi(x,\lambda) \in C^\infty(R^m \times R^n;\ R^m)$ *with* $\Phi(0,0) = 0$, $D_x\Phi(0,0)$ *nonsingular, and let* $g(\lambda) \in C^\infty(R^n;\ R^m)$ *be the corresponding implicit function*

$$\Phi(g(\lambda),\lambda) \equiv 0 \ . \qquad (3.23)$$

Let now $\Phi_h(x,\lambda)$ *be a sequence converging uniformly to* $\Phi(x,\lambda)$ *in a neighbourhood of the origin with all the derivatives. Then there exists a neighbourhood* $U_x(0)$ *in* R^m, *a neighbourhood* $U_\lambda(0)$ *in* R^n, *an* $h_0 > 0$ *and a sequence* c_k *such that for each* $h \le h_0$ *there exists a unique mapping* $g_h(\lambda) \in C^\infty(R^n;\ R^m)$ *from* $U_\lambda(0)$ *into* $U_x(0)$ *such that*

$$\Phi_h(g_h(\lambda),\lambda) \equiv 0 \quad \text{in } U_\lambda(0) \ , \qquad (3.24)$$

and moreover $\forall\ k > 0,\ \ \forall\ \lambda \in U_\lambda(0)$ *we have*

$$\| D^k(g(\lambda) - g_h(\lambda))\| \le c_k \sum_{r=0}^{k} \| d^r/d\lambda^r\ \Phi_h(g(\lambda),\lambda)\| \ , \qquad (3.25)$$

where $d^r/d\lambda^r$ *represents the* r-th *total derivative with respect to* λ.

The proofs of Lemmas 3.1 and 3.2 are easy. For more general results in Banach spaces, see [1] and [2].

Let us deal now with the case (3.18).

Theorem 3.3 *Let* $f(x,\lambda,\mu) = x^3 - \lambda + \mu x$ *and let* $f_h(x,\lambda,\mu)$ *be a sequence of* C^∞ *functions satisfying* (3.4). *Then there exists a neighbourhood* U *of the origin, an* $h_0 > 0$ *and a constant* c *such that for each* $h \le h_0$ *there exists a unique point* $(x_0^h,\lambda_0^h,\mu_0^h)$ *in* U *such that*

$$f_h(x + x_0^h, \lambda + \lambda_0^h, \mu_0^h) \overset{c.e.}{\simeq} f(x,\lambda,0) \ , \qquad (3.26)$$

and moreover

$$|x_0^h| + |\lambda_0^h| + |\mu_0^h| \le c(|f_h(0)| + |\partial_x f_h(0)| + |\partial_{xx} f_h(0)|); \qquad (3.27)$$

obviously here $(0) \equiv (0,0,0)$.

Proof Let us consider the mapping

$$\Phi : (x,\lambda,\mu) \to (f(x,\lambda,\mu)\ ,\ f_x(x,\lambda,\mu), f_{xx}(x,\lambda,\mu)) \qquad (3.28)$$

and the corresponding approximation

$$\Phi_h : (x,\lambda,\mu) \to (f_h, \partial_x f_h, \partial_{xx} f_h) \ . \qquad (3.29)$$

Clearly $\Phi_h \to \Phi$ uniformly with all the derivatives. Since $\Phi(0)=0$ and $\det(D\Phi(0)) = -6 \ne 0$ we can apply Lemma 3.1 which gives a unique point $(x_0^h,\lambda_0^h,\mu_0^h)$ such that

$$\left.\begin{aligned} f_h(x_0^h,\lambda_0^h,\mu_0^h) &= 0 , \\ \partial_x f_h(x_0^h,\lambda_0^h,\mu_0^h) &= 0 , \\ \partial_{xx} f_h(x_0^h,\lambda_0^h,\mu_0^h) &= 0 , \end{aligned}\right\} \tag{3.30}$$

together with (3.27). Since $f_\lambda(0) \neq 0$ and $f_{xxx}(0) \neq 0$ we will have, for $h \leq h_0$, $\partial_\lambda f_h \neq 0$ and $\partial_{xxx} f_h \neq 0$ at $(x_0^h,\lambda_0^h,\mu_0^h)$. As in the proof of Theorem 3.2 this gives the contact equivalence (3.26).

Remark A more careful analysis will show that

$$|\lambda_0^h| \leq c(|f_h(0)| + |\partial_x f_h(0)|^2 + |\partial_{xx} f_h(0)|^2) , \tag{3.31}$$

$$|\mu_0^h| \leq c(|f_h(0)| + |\partial_x f_h(0)| + |\partial_{xx} f_h(0)|^2) , \tag{3.32}$$

which is some cases could provide a better asymptotic estimate.

We shall deal now with cases (3.19) and (3.20).

Theorem 3.4 *Let* $f(x,\lambda,\mu) = x^2 \pm \lambda^2 - \mu$ *and let* $f_h(x,\lambda,\mu)$ *be a sequence of* C^∞ *functions satisfying* (3.4). *Then there exists a neighbourhood* U *of the origin, an* $h_0 > 0$ *and a constant* c *such that for each* $h \leq h_0$ *there exists a unique point* $(x_0^h,\lambda_0^h,\mu_0^h)$ *in* U *such that*

$$f_h(x + x_0^h, \lambda + \lambda_0^h, \mu_0^h) \overset{c.e.}{\simeq} f(x,\lambda,0) , \tag{3.33}$$

and moreover

$$|\mu_0^h| \leq c(|f_h(0)| + |\partial_x f_h(0)|^2 + |\partial_\lambda f_h(0)|^2) , \tag{3.34}$$

$$|\lambda_0^h| + |x_0^h| \leq c(|f_h(0)| + |\partial_x f_h(0)| + |\partial_\lambda f_h(0)|) . \tag{3.35}$$

Proof Since $f_\mu(0) \neq 0$ we have $\mu = x^2 \pm \lambda^2$ as implicit function. Lemma 3.2 gives now a unique mapping $\mu_h(x,\lambda)$ such that

$$f_h(x,\lambda,\mu_h(x,\lambda)) \equiv 0 \text{ in U} \tag{3.36}$$

and $\mu_h - \mu$, with its derivatives, is bounded in terms of $f_h - f$ as in (3.25). Let us consider now the mapping

$$\Phi : (x,\lambda) \to (\mu_x(x,\lambda), \mu_\lambda(x,\lambda)) (\equiv (2x, \pm 2\lambda)) \tag{3.37}$$

and its approximation

$$\Phi_h : (x,\lambda) \to (\partial_x \mu_h, \partial_\lambda \mu_h) . \tag{3.38}$$

Applying Lemma 3.1 one gets (x_0^h, λ_0^h) such that

$$\partial_x \mu_h(x_0^h, \lambda_0^h) = \partial_\lambda \mu_h(x_0^h, \lambda_0^h) = 0 \; . \tag{3.39}$$

Setting now

$$\mu_0^h = \mu_h(x_0^h, \lambda_0^h) \tag{3.40}$$

and differentiating (3.36) with respect to x one has

$$\partial_x f_h(x_0^h, \lambda_0^h, \mu_0^h) = 0 \tag{3.41}$$

and similarly

$$\partial_\lambda f_h(x_0^h, \lambda_0^h, \mu_0^h) = 0 \; . \tag{3.42}$$

Since $\det(Hf_h)$ converges uniformly to $\det(Hf)$, (3.33) follows as in Theorem 3.2. The estimate (3.35) is obvious from (3.22) and (3.25), whilst estimate (3.34) follows from (3.25) and (3.40) using the Taylor expansion:

$$\mu_0^h = \mu_h(0,0) + \partial_x \mu_h(0,0) x_0^h + \partial_\lambda \mu_h(0,0) \lambda_0^h + \text{h.o.t.} \tag{3.43}$$

Remark In the statements of Theorems 3.3 and 3.4 the specific form of $f(x,\lambda,\mu)$ is used only for convenience. It could obviously be replaced by means of suitable nondegeneracy conditions.

Remark Statements (3.26) and (3.33) imply that $f_h(x+x_0^h, \lambda+\lambda_0^h, \mu+\mu_0^h)$ is a universal unfolding of $f_h(x+x_0^h, \lambda+\lambda_0^h, \mu_0^h)$ equivalent to $f(x,\lambda,\mu)$ (unfolding of $f(x,\lambda,0)$); the proof is immediate.

We shall end this section with a somehow different problem in which the codimension 1 holds only in some generalised sense, namely within the space of anti-symmetric functions ($f(x,.,.) = -f(-x,.,.)$). A different definition of codimension should be introduced to deal with the problem (see [10]). Here, for the sake of simplicity, we shall limit ourselves to some elementary qualitative consideration. The reason that such a problem is of major interest is that it has applications to reaction diffusion problems,(see [6],[5]). In particular here, for the first time in our paper, the second parameter μ is a genuine parameter and not some artificial imperfection.

Let us assume that

$$f(x,\lambda,\mu) = x^5 - \mu x^3 - \lambda x \tag{3.44}$$

and that f_h is a sequence of C^∞ function satisfying (3.4) and

$$f_h(x,.,.) = -f(-x,.,.) \; . \tag{3.45}$$

We may assume that

$$f_h(x,\lambda,\mu) = xg_h(x,\lambda,\mu) \tag{3.46}$$

with g_h *even* in x. Similarly we set $g(x,\lambda,\mu) = x^4 - \mu x^2 - \lambda$ so that $f = xg$. For the sake of simplicity we shall then compare the problems $g(x,\lambda,\mu) = 0$ and $g_h(x,\lambda,\mu) = 0$.

Theorem 3.5 *Let* $g(x,\lambda,\mu) = x^4 - \mu x^2 - \lambda$ *and let* $g_h(x,\lambda,\mu)$ *be a sequence of* C^∞ *functions, even in* x, *converging uniformly to* g *in a neighbourhood of the origin with all the derivatives. Then there exists a neighbourhood* U *of the origin, an* $h_0 > 0$ *and a constant* c *such that for each* $h \leq h_0$ *there exists a unique point* $(0,\lambda_0^h,\mu_0^h)$ *in* U *such that*

$$g_h(x,\lambda + \lambda_0^h,\mu_0^h) \overset{c.e.}{\simeq} g(x,\lambda,0) \ , \tag{3.47}$$

and moreover

$$|\lambda_0^h| + |\mu_0^h| \leq c(|g_h(0)| + |\partial_{xx}g_h(0)|) \ . \tag{3.48}$$

Proof Consider the mapping

$$\Phi:(x,\lambda,\mu) \to (g,g_{xx},g_{xxx}) \tag{3.49}$$

and its approximation

$$\Phi_h:(x,\lambda,\mu) \to (g_h,\partial_{xx}g_h,\partial_{xxx}g_h) \ . \tag{3.50}$$

Clearly $\Phi(0) = 0$ and $\det(D\Phi(0)) = 48 \neq 0$. Hence we may apply Lemma 3.1 and get a *unique* point $(x_0^h,\lambda_0^h,\mu_0^h)$ such that $\Phi_h = 0$. However, since g_h is even in x, the uniqueness will imply that $x_0^h = 0$. Hence $\partial_x g_h(0,\lambda_0^h,\mu_0^h)$ is also zero, since g_h is even. In conclusion we have that $g_h(x,\lambda_0^h,\mu_0^h) \simeq x^4 + \text{h.o.t.}$ and (3.47) follows from [9] prop. 4.1. On the other hand (3.48) follows from (3.22).

Let us analyse in more detail the behaviour of the unfolding $g_h(x,\lambda + \lambda_0^h,\mu + \mu_0^h)$ when compared to $g(x,\lambda,\mu)$. It is easy to see, for instance, that for $\overline{\mu} > 0$ fixed problem $g(x,\lambda,\overline{\mu}) = 0$ has three limit points while for $\overline{\mu} < 0$ we have just one limit point. From another point of view, we could consider for any given $(\overline{\lambda},\overline{\mu})$ the problem $g(x,\overline{\lambda},\overline{\mu}) = 0$ and count the number of different solutions in x. It is easy to see that, in our case, the number of solutions changes at those $(\overline{\lambda},\overline{\mu})$ such that

$$g(x,\overline{\lambda},\overline{\mu}) = g_x(x,\overline{\lambda},\overline{\mu}) = 0 \text{ for some } x \ . \tag{3.51}$$

An easy computation shows that the pairs $(\overline{\lambda},\overline{\mu})$ which satisfy (3.51) lie on the two curves

$$C_1 = \{(0,\mu) | \ \mu \in R\} \ , \tag{3.52}$$

$$C_2 = \{(\lambda,\mu) \mid \mu \geq 0 \ , \ \ \lambda = -\frac{\mu_2}{4}\} \ . \tag{3.53}$$

The two curves C_1 and C_2 split the (λ,μ) plane into three regions R_0, R_2, R_4 as in Fig. 1, where

$$R_k = \{(\overline{\lambda},\overline{\mu}) \mid \text{the equation } g(x,\overline{\lambda},\overline{\mu}) = 0 \text{ has k distinct solutions for } x\} \tag{3.54}$$

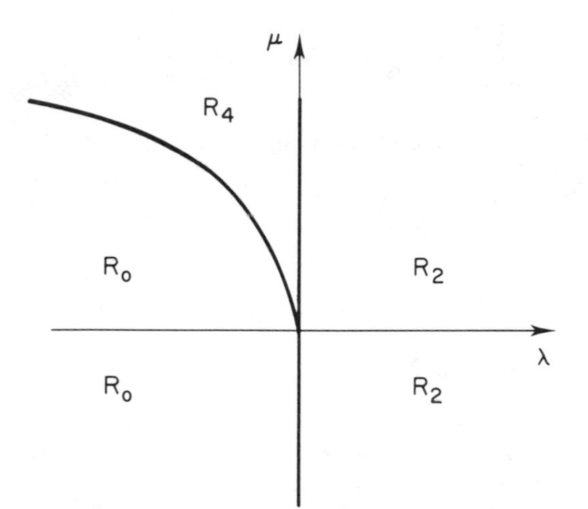

FIG. 1

We would like to show that $g_h(x,\lambda + \lambda_0^h, \mu + \mu_0^h)$ has similar behaviour. For this we remark first that since $g_\lambda(0) \neq 0$ we have the existence of two mappings

$$\lambda = \lambda^1(x,\mu) \ \ (\equiv x^4 - \mu x^2) \ , \tag{3.55}$$

$$\lambda = \lambda_h^1(x,\mu) \ , \tag{3.56}$$

such that

$$g(x,\lambda^1(x,\mu),\mu) \equiv 0 \ , \tag{3.57}$$

$$g_h(x,\lambda_h^1(x,\mu),\mu) \equiv 0 \ . \tag{3.58}$$

For obvious reasons $\lambda_h^1(x,\mu)$ is an even function in x. Consider now

$$\lambda_h^2(\mu) = \lambda_h^1(0,\mu) \ . \tag{3.59}$$

We easily check that on the curve

$$C_1^h = \{(\lambda,\mu) \mid \lambda = \lambda_h^2(\mu)\} \tag{3.60}$$

we have $g_h = \partial_x g_h = 0$. Moreover $\lambda_h^2(\mu)$ converges uniformly to zero with all the derivatives. In order to find a suitable curve C_2^h we consider the mapping

$$\Phi:(x,\lambda,\mu) \to (g(x,\lambda,\mu),\ x^{-1}g_x(x,\lambda,\mu)) \tag{3.61}$$

and its approximation

$$\Phi_h:(x,\lambda,\mu) \to (g_h, x^{-1}\partial_x g_h) \tag{3.62}$$

(allowed since $\partial_x g_h$ is odd in x). From Lemma 3.2 we get

$$\lambda = \lambda_h^3(x)\ ,\ \mu = \mu_h(x) \tag{3.63}$$

such that

$$g_h(x,\lambda_h^3(x),\mu_h(x)) \equiv 0\ , \tag{3.64}$$

$$\partial_x g_h(x,\lambda_h^3(x),\mu_h(x)) \equiv 0 \tag{3.65}$$

and $\mu_h(x) \to 2x^2$, $\lambda_h^3(x) \to -x^4$, uniformly with all the derivatives and with the usual error estimates. Clearly (3.63) can be interpreted as the parametric equations of a curve C_2^h converging to C_2. Our aim is now to analyse the intersection (if any) of the two curves C_1^h and C_2^h. For that let us consider first the mapping

$$\Phi:\mu \to \lambda_{xx}^1(0,\mu)$$

and its approximation

$$\Phi_h:\mu \to \partial_{xx}\lambda_h^1(0,\mu)\ .$$

Since $\Phi(0) = 0$ and $\Phi'(0) \neq 0$, Lemma 3.1 ensures the existence of a unique $\overline{\mu}_h$ such that

$$\partial_{xx}\lambda_h^1(0,\overline{\mu}_h) = 0\ .$$

Differentiating (3.58) with respect to x twice we get

$$\partial_{xx}g_h(0,\lambda_h^1(0,\overline{\mu}_h),\overline{\mu}_h) = 0$$

and hence $\overline{\mu}_h = \mu_0^h$ and $\lambda_h^1(0,\overline{\mu}_h) = \lambda_0^h$ (we used again the symmetry of g_h). Let us look now at C_2^h. We remark first that for x = 0 we get at $\overline{\lambda} = \lambda_h^3(0)$ $\overline{\mu} = \mu_h(0)$

$$g_h(0,\overline{\lambda},\overline{\mu}) = 0\ ,$$

$$\partial_x g_h(0,\overline{\lambda},\overline{\mu}) = 0 \ ,$$

$$\partial_{xxx} g_h(0,\overline{\lambda},\overline{\mu}) = 0 \ .$$

On the other hand, differentiating (3.65) with respect to x and using the fact that both $\lambda_h^3(x)$ and $\mu_h(x)$ are even functions we get

$$\partial_{xx} g_h(0,\overline{\lambda},\overline{\mu})+\partial_{x\lambda} g_h(0,\overline{\lambda},\overline{\mu}).\partial_x\lambda_h^3(0)+\partial_{x\mu} g_h(0,\overline{\lambda},\overline{\mu}).\partial_x\mu_h(0) = 0$$

and therefore

$$\partial_{xx} g_h(0,\overline{\lambda},\overline{\mu}) = 0 \ ,$$

which shows that $\overline{\lambda} = \lambda_0^h$, $\overline{\mu} = \mu_0^h$. Hence C_1^h and C_2^h cross at (λ_0^h,μ_0^h). Assume now that there exists another value $\overline{x} \neq 0$ such that $\lambda_h^3(\overline{x}) = \lambda_h^2(\mu_h(\overline{x}))$. We may assume that $\mu_h(\overline{x}) > \mu_0^h$ since $\mu_h(x)$ is an even function which converges to $2x^2$ uniformly with all the derivatives. We have

$$\lambda_h^1(\overline{x},\mu_h(\overline{x})) = \lambda_h^3(\overline{x}) = \lambda_h^1(0,\mu_h(\overline{x}))$$

and obviously $\lambda_h^1(-\overline{x},\mu_h(\overline{x})) = \lambda_h^1(\overline{x},\mu_h(\overline{x}))$. However, differentiating (3.58) we have

$$\partial_x g_h(x,\lambda_h^1(x,\mu),\mu)+\partial_\lambda g_h(x,\lambda_h^1(x,\mu),\mu).\partial_x\lambda_h(x,\mu) \equiv 0$$

and using (3.65) we get

$$\partial_x\lambda_h^1(\overline{x},\mu_h(\overline{x})) = \partial_x\lambda_h^1(-\overline{x},\mu_h(\overline{x})) = \partial_x\lambda_h(0,\mu_h(\overline{x})) = 0 \ .$$

Hence for $\overline{\mu} = \mu_h(\overline{x})$ the function $\lambda_h^1(x,\overline{\mu})$ has three extrema with the same value, which contradicts the fact that $\lambda_h^1(x,\overline{\mu})$ converges uniformly with all the derivatives to a polynomial of degree 4. Hence (λ_0^h,μ_0^h) is the unique intersection of C_1^h and C_2^h. Let us consider now the tangents to the two curves at (λ_0^h,μ_0^h). Since both $\lambda_h^3(x)$ and $\mu_h(x)$ are even functions, $\partial_x\lambda_h^3(0)=\partial_x\mu_h(0)=0$ and we must consider the second derivatives. Since

$$\lambda_h^3(x) = \lambda_h^1(x,\mu_h(x))$$

we have, differentiating twice at x = 0,

$$\begin{aligned}\partial_{xx}\lambda_h^3(0)&=\partial_{xx}\lambda_h^1(0,\mu_0^h)+2\partial_{x\mu}\lambda_h^1(0,\mu_0^h)\partial_x\mu_h(0)+\\ &+\partial_{\mu\mu}\lambda_h^1(0,\mu_0^h)(\partial_x\mu_h(0))^2+\partial_\mu\lambda_h^1(0,\mu_0^h)\partial_{xx}\mu_h(0) =\\ &=\partial_\mu\lambda_h^1(0,\mu_0^h)\partial_{xx}\mu_h(0) = \partial_\mu\lambda_h^2(\mu_0^h)\partial_{xx}\mu_h(0) \ ,\end{aligned}$$

which shows that C_1^h and C_2^h have the same tangent since $\partial_{xx}\mu_h(0)\neq 0$.

We have proved the following theorem.

Theorem 3.6 *With the same hypotheses as Theorem 3.5 there exists a neighbourhood* U *of the origin and an* $h_0 > 0$ *such that for each* $h \leq h_0$ *there exists in* $U \cap \{x=0\}$ *two curves* C_1^h *and* C_2^h *converging uniformly to* C_1 *(see (3.52)) and to* C_2 *(see (3.53)) respectively.* C_1^h *and* C_2^h *have a unique intersection at* (λ_0^h, μ_0^h) *(defined by Theorem 3.5) with a common tangent. The two curves split* $U \cap \{x=0\}$ *into three regions* R_0^h, R_2^h, R_4^h *as in* Fig. 2. *If* $(\overline{\lambda}, \overline{\mu}) \in R_k^h (k = 0,2,4)$ *then the equation*

$$g_h(x, \overline{\lambda}, \overline{\mu}) = 0$$

has k *distinct solutions in* $U \cap \{\lambda = \overline{\lambda}, \mu = \overline{\mu}\}$.

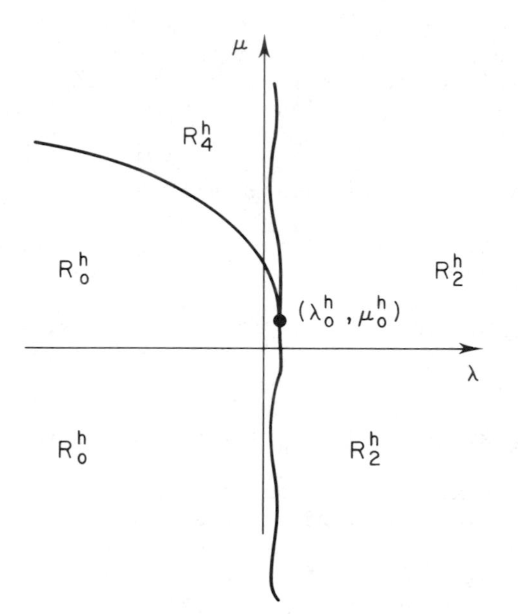

FIG. 2

Remark As in the previous theorems we can also estimate the rate of convergence of C_1^h and C_2^h to C_1 and C_2 respectively.

Remark The results of Theorem 3.6 show that for $\mu = \overline{\mu} > \mu_0^h$ the problem

$$g_h(x, \lambda, \overline{\mu}) = 0$$

has three limit points, while for $\overline{\mu} < \mu_0^h$ it has one (nondegenerated) limit point. Hence the behaviour of the continuous problem is well reproduced by the discrete problem, although for different values of μ. The siutation would not be the same if the roles of λ and μ were interchanged ($x^5 - \mu x^3$ has infinite codimension!).

4. A SPECIAL CASE OF CODIMENSION TWO

We shall consider in this section another *genuine two parameter* problem, namely

$$f(x,\lambda,\mu) \equiv x^3 - \lambda x + \mu = 0 \ . \tag{4.1}$$

A reduced problem of the form (4.1) is met for instance in the study of the nonlinear bending of an elastic thin plate (see [4] for more details). Here μ represents, in some sense, the transversal load parameter while λ represents the boundary force parameter. As before we shall assume that we are given a sequence $f_h(x,\lambda,\mu)$ of C^∞ functions which converges uniformly to $f(x,\lambda,\mu)$ in a neighbourhood of the origin with all the derivatives. Clearly (4.1) may be seen either as a perturbation of the nondegenerated hysteresis $x^3 + \mu = 0$ or as a perturbation of the pitchfork $x^3 - \lambda x = 0$. In the first case, Theorem 3.3 ensures the existence of a unique point $(x_0^h, \lambda_0^h, \mu_0^h)$ such that $f_h(x + x_0^h, \lambda_0^h, \mu + \mu_0^h)$ is contact equivalent to $x^3 + \mu$. However, in general, $f_h(x + x_0^h, \lambda + \lambda_0^h, \mu_0^h)$ will not be contact equivalent to $x^3 - \lambda x$. Hence we *need* as in the previous section, an additional *perturbation* parameter. We assume that we are given the problem

$$f(x,\lambda,\mu,\alpha) \equiv x^3 - \lambda x + \mu + \alpha x^2 = 0 \tag{4.2}$$

and a sequence $f_h(x,\lambda,\mu,\alpha)$ of C^∞ functions such that

$$f_h(x,\lambda,\mu,\alpha) \to f(x,\lambda,\mu,\alpha) \tag{4.3}$$

uniformly in a neighbourhood of the origin with all the derivatives.

Theorem 4.1 Assume that f *and* f_h *are given as in* (4.2),(4.3). *Then there exists a neighbourhood* U *of the origin in* R^4 *and an* $h_0 > 0$ *such that for all* $h \le h_0$ *there exists a unique point* $(x_0^h, \lambda_0^h, \mu_0^h, \alpha_0^h)$ *in* U *such that*

$$f_h(x + x_0^h, \lambda_0^h, \mu + \mu_0^h, \alpha_0^h) \overset{c.e.}{\simeq} x^3 + \mu \ , \tag{4.4}$$

and

$$f_h(x + x_0^h, \lambda + \lambda_0^h, \mu_0^h, \alpha_0^h) \overset{c.e.}{\simeq} x^3 - \lambda x \ . \tag{4.5}$$

Proof Consider the mapping

$$\Phi : (x,\lambda,\mu,\alpha) \to (f, f_x, f_\lambda, f_{xx}) \tag{4.6}$$

and its approximation

$$\Phi_h : (x,\lambda,\mu,\alpha) \to (f_h, \partial_x f_h, \partial_\lambda f_h, \partial_{xx} f_h) \ . \tag{4.7}$$

We have $\Phi(0) = 0$ and

$$|D\phi(0)| = \begin{vmatrix} 0 & 0 & 1 & 0 \\ 0 & -1 & 0 & 0 \\ -1 & 0 & 0 & 0 \\ 6 & 0 & 0 & 2 \end{vmatrix} \neq 0 . \tag{4.8}$$

We can then apply Lemma 3.1 which gives a unique point $(x_0^h, \lambda_0^h, \mu_0^h, \alpha_0^h)$, where

$$f_h = \partial_x f_h = \partial_\lambda f_h = \partial_{xx} f_h = 0 . \tag{4.9}$$

Since $\partial_{xxx} f_h$, $\partial_{x\lambda} f_h$ and $\partial_\mu f_h$ are all different from zero at $(x_0^h, \lambda_0^h, \mu_0^h, \alpha_0^h)$ for h small enough, we can apply [9] prop. 4.1 and get the result.

Remark There is to our knowledge no definition of contact equivalence for mappings depending on two parameters. The result of Theorem 4.1 suggests that a requirement of contact equivalence separately in each parameter would not be unreasonable.

We want now to look at (4.2) as a two-parameter perturbation of the pitchfork $x^3 - \lambda x$; this is consistent with the previous section since $x^3 - \lambda x$ has codimension 2. In the framework of [9] it is easy to see that one can define, in the plane of perturbations (μ,α) a *curve of bifurcation points*

$$B : \mu = 0 , \tag{4.10}$$

and a *curve of hysteresis points*

$$H : \mu = \alpha^3/27 , \tag{4.11}$$

which split the plane into four regions of *stable* bifurcation diagrams as in Fig. 3.

We want to show that the same is true for the approximate problem (4.3).

Theorem 4.2 Assume that f *and* f_h *are given as in* (4.2),(4.3). *Then there exists a neighbourhood* U *of the origin in* R^2 *and an* $h_0 > 0$ *such that for each* $h \leq h_0$ *there exist two curves uniquely defined in* U

$$B_h : \mu = \mu_h^2(\alpha) , \tag{4.12}$$

$$H_h : \mu = \mu_h^1(\alpha) , \tag{4.13}$$

such that for all $\bar{\alpha}$ we have

$$f_h(x,\lambda,\mu_h^2(\bar{\alpha}),\bar{\alpha}) \text{ has a bifurcation point,} \tag{4.14}$$

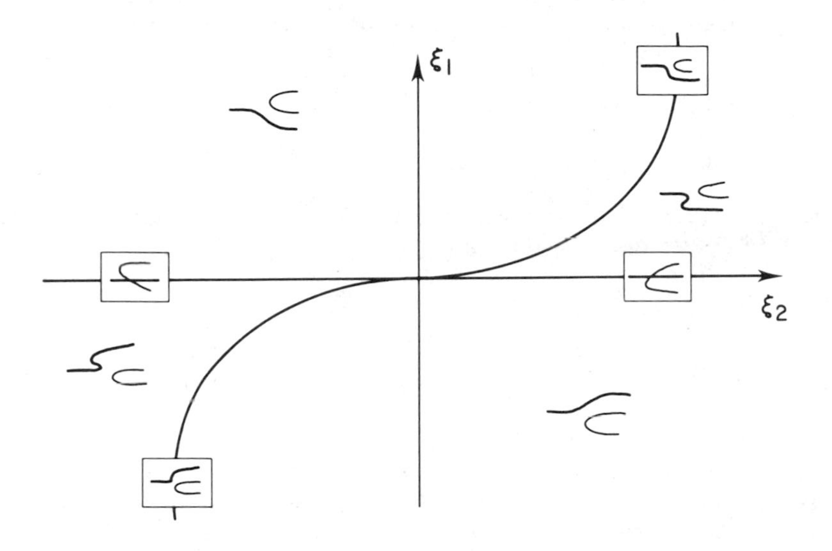

FIG. 3

$$f_h(x,\lambda,\mu_h^1(\bar{\alpha}),\bar{\alpha}) \text{ has a hysteresis point,} \tag{4.15}$$

$$\mu_h^2(\alpha) \to 0; \ \mu_h^1(\alpha) \to \alpha^3/27 \tag{4.16}$$

uniformly with all the derivatives. Moreover the two curves cross only once; the crossing point is at (μ_0^h,α_0^h) *(defined by Theorem 4.1) and the two curves have a third order contact.*

Proof It suffices to consider the mappings

$$\Phi:(x,\lambda,\mu,\alpha) \to (f,f_x,f_{xx}) \ ,$$

$$\Phi_h:(x,\lambda,\mu,\alpha) \to (f_h,\partial_x f_h,\partial_{xx} f_h) \ ,$$

$$\Psi:(x,\lambda,\mu,\alpha) \to (f,f_x,f_\lambda) \ ,$$

$$\Psi_h:(x,\lambda,\mu,\alpha) \to (f_h,\partial_x f_h,\partial_\lambda f_h) \ .$$

Since $\Phi(0) = 0$ and $D_{(x,\lambda,\mu)}\Phi(0)$ is nonsingular we may apply Lemma 3.2 and get $x = x_h^1(\alpha)$, $\lambda = \lambda_h^1(\alpha)$ and, more importantly, $\mu = \mu_h^1(\alpha)$ which satisfies (4.15). Since $\Psi(0) = 0$ and $D_{(x,\lambda,\mu)}$ $\Psi(0)$ is nonsingular we get, always from Lemma 3.2 $x = x_h^2(\alpha)$, $\lambda = \lambda_h^2(\alpha)$ and $\mu = \mu_h^2(\alpha)$ which satisfies (4.14). The uniform convergence of (4.16) is also a consequence of Lemma 3.2. The uniqueness of the intersection follows from simple geometrical considerations and from Theorem 4.1. It remains to show that $\mathcal{B}_h$ and $\mathcal{H}_h$ have a third order contact. Thanks to (4.16) and to the uniqueness of the intersection it is enough to show that $d/d\alpha\, \mu^1{}_h = d/d\alpha\, \mu^2{}_h$ for $\alpha = \alpha_0^h$. For this we note that

$$f_h(x_h^1(\alpha),\ \lambda_h^1(\alpha),\ \mu_h^1(\alpha),\alpha) \equiv 0 \tag{4.17}$$

$$f_h(x_h^2(\alpha),\ \lambda_h^2(\alpha),\ \mu_h^2(\alpha),\alpha) \equiv 0\ . \tag{4.18}$$

Differentiating both expressions with respect to α at $\alpha = \alpha_0^h$ and using (4.9) we get

$$\partial_\mu f_h \cdot \frac{d}{d\alpha}\mu_h^1 + \partial_\alpha f_h = 0\ , \tag{4.19}$$

$$\partial_\mu f_h \cdot \frac{d}{d\alpha}\mu_h^2 + \partial_\alpha f_h = 0 \tag{4.20}$$

and the result follows since $\partial_\mu f_h \neq 0$ for h small enough.

Remark It is clear that $\mathcal{B}_h$ and $\mathcal{H}_h$ split the (μ,α) plane into four regions of stable bifurcation diagrams similar to those of Fig. 3. Hence we may see that, at least for that particular case, the introduction of *two* parameters into a problem of co-dimension 2 allows us to recover the whole bifurcation diagram in the approximate problem. We conjecture that this is true in general, for any finite codimension,

In a similar way to the one we used in the last example of Section 3 we want finally to check the behaviour, in the (λ,μ) plane, of the regions $(k \in N)$

$$\mathcal{R}_k = \{(\bar\lambda,\bar\mu)\,|\,f(x,\bar\lambda,\bar\mu) = 0 \text{ has k distinct solutions in x}\} \tag{4.21}$$

and of their analogous regions

$$\mathcal{R}_k^h = \{(\bar\lambda,\bar\mu)\,|\,f_h(x,\bar\lambda,\bar\mu) = 0 \text{ has k distinct solutions in x}\}\ . \tag{4.22}$$

Surprisingly enough, we shall see that the behaviour of $\mathcal{R}_k^h$ matches reasonably the limit case already for the *unperturbed* problem $\alpha = 0$, although the introduction of the parameter α as in (4.2) and (4.3) still gives some improvement.

It is an easy matter to check that for the problem (4.1) we have two regions, $\mathcal{R}_1$ and $\mathcal{R}_3$, divided by the curve

$$C : \begin{cases} \lambda = 3t^2 \\ \mu = 2t^3 \end{cases}\ , \quad t \in R \tag{4.23}$$

which has a cusp at the origin with horizontal tangent (Fig.4).

Let us now consider the mappings

$$\Phi:(x,\lambda,\mu) \to (f,f_x)\ , \qquad \Phi_h:(x,\lambda,\mu) \to (f_h,\partial_x f_h)\ .$$

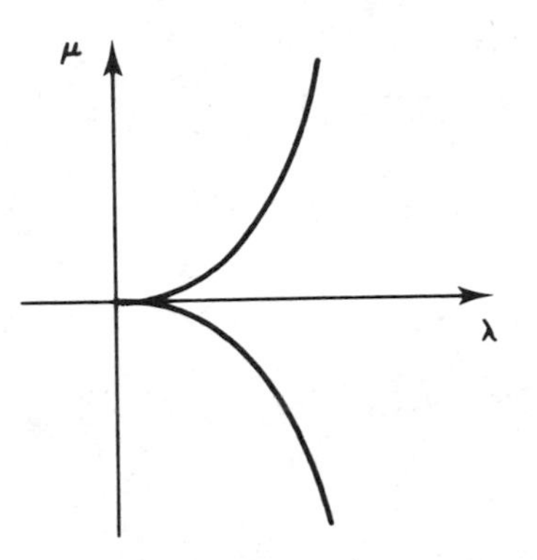

FIG. 4

Applying Lemma 3.2 we get a curve

$$C_h : \begin{cases} \lambda = \lambda_h(x) \\ \mu = \mu_h(x) \end{cases} \tag{4.24}$$

defined in a neighbourhood of the origin for h small enough with

$$\lambda_h(x) \to 3x^2 , \qquad x \in U(0) , \tag{4.25}$$

$$\mu_h(x) \to 2x^3 , \tag{4.26}$$

uniformly with all the derivatives. It is easy to see that C_h has a unique cusp: in fact (4.25) imples that $d_x\lambda_h$ vanishes at a unique point, say $\bar{x}_h$. Since

$$f_h(x,\lambda_h(x),\mu_h(x)) \equiv 0 , \tag{4.27}$$

$$(\partial_x f_h)(x,\lambda_h(x),\mu_h(x)) \equiv 0 , \tag{4.28}$$

differentiating (4.27) at $x = \bar{x}_h$ and using (4.28) we have that

$$\partial_\lambda f_h d_x\lambda_h + \partial_\mu f_h d_x\mu_h = 0 \tag{4.29}$$

and therefore $d_x\mu_h = 0$ at $x = \bar{x}_h$. In general, however $d_{xx}\mu_h \neq 0$ at $\bar{x}_h$ so that the cusp has an oblique tangent; its behaviour will be, qualitatively, as in the two cases of Fig. 5. The introduction of the parameter α eliminates this *imperfection*.

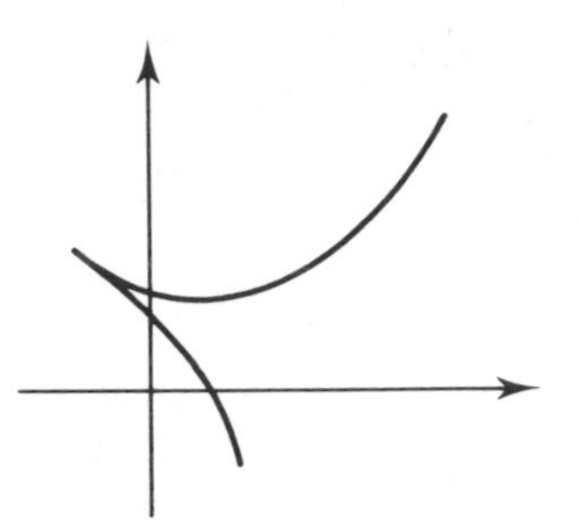

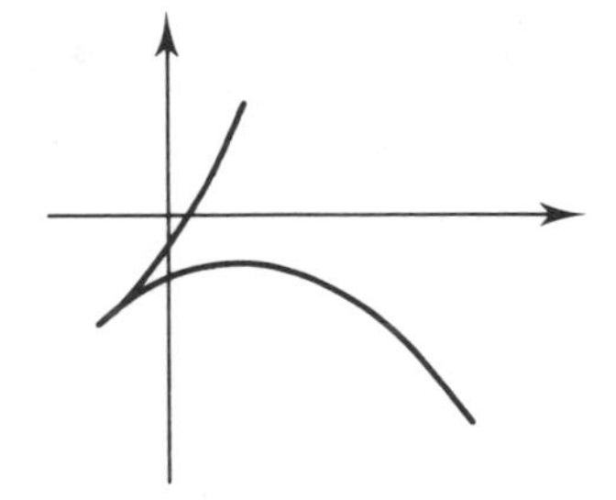

FIG. 5

Theorem 4.3 *Let* f *and* f_h *be given as in* (4.2),(4.3) *and consider for each* $\alpha \in R$ *the two curves* $C(\alpha)$ *and* $C_h(\alpha)$ *defined as before. There exists a neighbourhood* U *of the origin in* R *and an* $h_0 > 0$ *such that, for each* $\alpha \in U$ *and for each* $h \leq h_0$ *we have: if* $\alpha > 0$ *(resp.* <0*)* $C(\alpha)$ *has a cusp-tangent with negative (positive) slope and if* $\alpha > \alpha_0^h$ *(resp.* $<\alpha_0^h$*) then* $C_h(\alpha)$ *has a cusp-tangent with negative (positive) slope. For* $\alpha = \alpha_0^h$ *(given by Theorem 4.1)* C_h *has an horizontal cusp-tangent.*

Proof The result for $C(\alpha)$ is obvious since

$$C(\alpha) \equiv \begin{cases} \lambda = 3t^2 + 2\alpha t \ , \\ \mu = 2t^3 + \alpha t^2 \ . \end{cases} \tag{4.30}$$

Let now, as before, $\lambda_h(x,\alpha)$ and $\mu_h(x,\alpha)$ be defined by

$$f_h(x,\lambda_h(x,\alpha),\mu_h(x,\alpha),\alpha) \equiv 0 \ , \tag{4.31}$$

$$(\partial_x f_h)(x,\lambda_h(x,\alpha),\mu_h(x,\alpha),\alpha) \equiv 0 \ , \tag{4.32}$$

Again $\partial_x\lambda_h$ vanishes at a unique point $\overline{x}_h(\alpha)$ and $\partial_x\mu_h$ also vanishes at the same point. Hence we have just to check that $\partial_{xx}\mu_h(\overline{x}_h(\alpha_0^h),\alpha_0^h) = 0$ and that $\partial_{xx\alpha}\mu_h \neq 0$. The last inequality is obvious since $\mu_h \to 2x^3 + \alpha x^2$. To prove that $\partial_{xx}\mu_h(\overline{x}_h(\alpha_0^h),\alpha_0^h) = 0$ we remark that, on setting $\overline{\mu}_h = \mu_h(\overline{x}_h(\alpha_0^h),\alpha_0^h)$ and $\overline{\lambda}_h = \lambda_h(\overline{x}_h(\alpha_0^h),\alpha_0^h)$, we have at $(\overline{x}_h,\overline{\lambda}_h,\overline{\mu}_h,\alpha_0^h)$

$$f_h = \partial_x f_h = \partial_x\lambda_h = \partial_x\mu_h = 0 \ . \tag{4.33}$$

Differentiating (4.32) with respect to x we also get

$$\partial_{xx}f_h = 0 \ . \tag{4.34}$$

As in the proof of Theorem 4.2 equation (4.34) together with $f_h = \partial_x f_h = 0$ implies that

$$\bar{x}_h = x_0^h \; ; \quad \bar{\lambda}_h = \lambda_0^h \; ; \quad \bar{\mu}_h = \mu_0^h \tag{4.35}$$

and therefore that

$$\partial_\lambda f_h = 0 \; . \tag{4.36}$$

Differentiating (4.31) twice with respect to x and using (4.33), (4.34), (4.36) we get $\partial_{xx}\mu_h = 0$.

Remark As in the previous section, we could estimate the *rate of convergence* of $(x_0^h, \lambda_0^h, \mu_0^h, \alpha_0^h)$ to (0,0,0,0) in terms of the values of f_h and of some of its derivatives at the origin. We find, easily enough, that

$$|\mu_0^h| \leq c(|f_h(0)| + |\partial_x f_h(0)|^2 + |\partial_\lambda f_h(0)|^2 + |\partial_{xx} f_h(0)|^2) \; ,$$

$$|\lambda_0^h| + |x_0^h| \leq c(|f_h(0)| + |\partial_x f_h(0)| + |\partial_\lambda f_h(0)| + |\partial_{xx} f_h(0)|^2) \; ,$$

$$|\alpha_0^h| \leq c(|f_h(0)| + |\partial_x f_h(0)| + |\partial_\lambda f_h(0)| + |\partial_{xx} f_h(0)|) \; .$$

Similar estimates can also be found for other related quantities such as $x_h^i(\alpha)$, $\lambda_h^i(\alpha)$, $\mu_h^i(\alpha)$, $C_h(\alpha)$.

REFERENCES

1. BREZZI, F., RAPPAZ, J. and RAVIART, P.-A.,Finite element approximations of nonlinear problems; I. Branches of non-singular solutions. *Numer. Math.* 36, 1-25 (1980).
2. BREZZI, F., RAPPAZ, J. and RAVIART, P.-A.,Finite element approximations of nonlinear problems; II. Limit Points. *Numer. Math.* 37, 1-28 (1981).
3. BREZZI, F., RAPPAZ, J. and RAVIART, P.-A.,Finite element approximations of nonlinear problems; III. *Numer. Math.* 38, 1-30 (1981).
4. CIARLET, P.G. and RABIER, P., *Les equations de von Kármán*, Lecture Notes in Mathematics 826. Springer (Berlin) (1980).
5. FUJII, H., Numerical analysis of global bifurcations in reaction-diffusion equations (to appear).
6. FUJII, H., MIMURA, M. and NISHIURA, Y., A picture of global bifurcation diagram in ecological interacting and diffusion systems. *Res. Rep. KSU-ICS* 79-11 Kyoto Sangyo University (1976).
7. FUJII, H. and YAMAGUTI, M., Structure of singularities and its numerical realization in nonlinear elasticity. *J. Math. Kyoto University* 20, 489-590 (1980).
8. KIKUCHI, F., Finite element approximations to bifurcation problems of turning point type.*INRIA Meeting "Méthodes de Calcul Scientifique et Technique III" 1977*(to appear-Springer).

9. GOLUBITSKY, M. and SCHAEFFER, D., A theory for imperfect bifurcation via singularity theory. *Comm. Pure Appl. Math.* 32, 21-98 (1979).
10. GOLUBITSKY, M. and SCHAEFFER, D., Imperfect bifurcation in the presence of symmetry. *Comm. Math. Phys.* 67, 205-232 (1979).

APPROXIMATION OF DOUBLE BIFURCATION POINTS FOR NONLINEAR EIGENVALUE PROBLEMS

*J. Rappaz and †G. Raugel

*EPF-Lausanne, †Université de Rennes

1. INTRODUCTION

The typical equation with which we shall be concerned in this paper is

$$F(\lambda,u) \equiv u + TG(\lambda,u) = o \tag{1.1}$$

where T is a compact linear operator from W into V for some real Banach spaces V and W, λ is a real parameter and G is a nonlinear C^p mapping from $\mathbb{R}\times V$ into W such that $G(\lambda,o) \equiv o$.

To study the numerical approximation of (1.1), we introduce a family of finite dimensional subspaces $\{V_h\}_{o<h\leq 1}$ of V and a family of linear operators $\{T_h\}_{o<h\leq 1}$ from W into V_h which converge uniformly to T when h tends to zero; we solve the approximate problem

$$F_h(\lambda,u_h) \equiv u_h + T_hG(\lambda,u_h) = o\,. \tag{1.2}$$

Remark here that, in section 4, we give a model example which is the *numerical approximation by a finite element method of the nonlinear eigenvalue problem* $-\Delta u - \lambda u + u^3 = o$ in Ω, $u = o$ on $\partial\Omega$, where Ω is a bounded domain of $\mathbb{R}^2$; we show that this problem and its numerical approximation are of type (1.1) and (1.2) respectively.

In [2], the approximation of solutions of (1.1) by those of (1.2) in a neighborhood of a nonsingular solution, of a simple limit point and a simple bifurcation point of (1.1) has been considered. In this paper, we shall study the numerical approximation of solution branches of (1.1) which *bifurcate at a double eigenvalue;* we shall give some results obtained in [6]. Note that our analysis includes a variety of classical approximation schemes such as *conforming finite element methods, mixed finite element methods, spectral methods...* . Our results appear as a nontrivial generalization of [2, part III] and can be extended to bifurcation at multiple eigenvalue.

In the following, if X,Y and Z are three Banach spaces and if $f: X\times Y \to Z$ is a C^p nonlinear mapping, we denote by $D^m_{x^s y^{m-s}} f(x_o,y_o)$, $o\leq s\leq m\leq p$ the m-th Frechet derivative of f at the point $(x_o,y_o) \in X\times Y$, s times with respect to x and (m-s) times with respect to y. We denote by $\|\cdot\|$ the various norms in X,Y,Z, $L(X\times Y,Z)$,... etc.

2. THE CONTINUOUS PROBLEM

Throughout this paper we consider an equation of the form

$$F(\lambda,u) \equiv u + TG(\lambda,u) = o \tag{2.1}$$

with the following assumptions on T and G:

(i) $T \in L(W,V)$ is a linear compact operator from W to V where V and W are two (real) Banach spaces;

(ii) $G: \mathbb{R}\times V \to W$ is a C^p nonlinear mapping of the form

$$G(\lambda,u) = -\lambda Lu + R(u) \tag{2.2}$$

where $L \in L(V,W)$ is a continuous linear operator and $R:V \to W$ is a C^p mapping ($p\geq 4$) such that, for an integer k with $2\leq k<p-1$, $D^\ell_{u^\ell} R(o) = o$ for $o\leq \ell<k$.

We first remark that the trivial solution $u=o$ is a solution of (2.1) for all values of λ and we are interested by nontrivial solutions of (2.1) in a neighborhood of a solution (λ_o,o) which is a double critical point in the following sense:

$$\begin{cases} D_uF(\lambda_o,o) \equiv I-\lambda_oTL \text{ is singular and } \lambda_o^{-1} \text{ is an eigenvalue} \\ \text{of the compact operator } TL \in L(V,V) \text{ with geometric} \\ \text{and algebraic multiplicity two.} \end{cases} \tag{2.3}$$

We denote by $V_1 = \ker(I-\lambda_oTL)$, $V_2 = \text{Range } (I-\lambda_oTL)$ and we have $V = V_1 \oplus V_2$ and $\dim V_1 = 2$. Moreover $(I-\lambda_oTL)$ is an isomorphism of V_2 and we denote by H its inverse. In the sequel we set Q the projection from V to V_2 such that $Q(V_1) = \{o\}$.Then the equation (2.1) is equivalent to the system

$$QF(\lambda,u) = o \; ; \qquad (I-Q)F(\lambda,u) = o \; . \tag{2.4}$$

It is classical to check, by using the implicit function theorem, there exist a positive constant ξ_o, a neighborhood B of zero in V_1 and a unique C^p mapping $v: (-\xi_o,\xi_o)\times B \to V_2$ such that for all $\xi \in (-\xi_o,\xi_o)$ and $\sigma \in B$:

$$QF(\lambda_o+\xi, \sigma+v(\xi,\sigma)) = o \; ; \quad v(o,o) = o \; . \tag{2.5}$$

Hence, solving equation (2.1) in a neighborhood of (λ_o, o) amounts to solve *the bifurcation equation*

$$f(\xi,\sigma) \equiv (I-Q)F(\lambda_o+\xi,\ \sigma+v(\xi,\sigma)) = o \tag{2.6}$$

in $(-\xi_o,\xi_o)\times B$ (see the Lyapounov-Schmidt procedure). By using the form (2.2) and by differentiating the first equation of (2.5) we obtain by elementary calculations: $D^\ell v(o,o) = o$ for $o\leq\ell\leq k-1$, $D^k_{\xi^\ell\sigma^{k-\ell}}\, v(o,o) = o$ for $1\leq\ell\leq k$, $D^k_{\sigma^k}\, v(o,o) = -HQTD^k_{u^k}\, R(o)$. Hence, by developping the function f around zero we obtain:

$$f(\xi,\sigma) = -\ \lambda_o^{-1}\xi\sigma + 1/k!\ (I-Q)TD^k_{u^k}\ R(o)\sigma^k + r(\sigma) \tag{2.7}$$

where $r(\sigma) = O(||\sigma||^{k+1})$ when $\sigma\to o$.

Let now $\sigma_o \in V_1$ be such that

$$-\lambda_o^{-1}\varepsilon\sigma_o + 1/k!\ (I-Q)TD^k_{u^k}\ R(o)\sigma_o^k = o \tag{2.8}$$

where $\varepsilon = 1$ if k is even and $\varepsilon = \pm 1$ if k is odd.

In the following, we assume that:

$$\sigma\in V_1 \text{ and } -\lambda_o^{-1}\varepsilon\sigma + 1/(k-1)!(I-Q)TD^k_{u^k}\ R(o)\sigma_o^{(k-1)}\sigma = o \text{ imply } \sigma = o. \tag{2.9}$$

Remark that if we choose a basis φ_1,φ_2 of V_1 and if $\varphi_1^*\ \varphi_2^*$ belong to $V_2^\perp = \{x^*\in V' : \langle x,x^*\rangle = o,\ \forall x\in V_2\}$ with $\langle\varphi_i,\varphi_j^*\rangle = \delta_{ij}$, $1\leq i,j\leq 2$, where V' is the dual space of V and $\langle\cdot,\cdot\rangle$ is the duality pairing between V and V', then we can define for i = 1,2:

$$g_i(\xi,\alpha_1,\alpha_2) = -\ \lambda_o^{-1}\xi\alpha_i + 1/k!\ \langle TD^k_{u^k}\ R(o)(\alpha_1\varphi_1+\alpha_2\varphi_2)^k, \varphi_i^*\rangle \tag{2.10}$$

If we set $\sigma_o = \alpha_o^1\varphi_1 + \alpha_o^2\varphi_2$, (2.8) means that (α_o^1,α_o^2) is a point of intersection of the two curves in $\mathbb{R}^2$ given by $g_i(\varepsilon,\alpha_1,\alpha_2) = o$, $i = 1,2$ and hypothesis (2.9) means that these two curves do not have a point of tangency at (α_o^1,α_o^2).

In a same way as [4] we have:

Proposition 1: We assume hypothesis (2.9). Then σ_o gives rise to a unique branch of solutions of (2.6) in $(-\xi_o,\xi_o)\times B$ of the form:

$$\xi(t) = \varepsilon t^{k-1}\ ; \qquad \sigma(t) = t\ \tilde\sigma(t) \tag{2.11}$$

where $\tilde\sigma$ is a C^{p-k} function of t defined in a neighborhood $(-t_o,t_o)$ of zero and $\tilde\sigma(o) = \sigma_o$.

Proof: By considering the expansion (2.7) of f we can define the C^{p-k} function $\ell: \mathbb{R}\times V_1 \to V_1$ by

$$\ell(t,\tilde{\sigma}) = t^{-k} f(\varepsilon t^{k-1}, t\tilde{\sigma}) \tag{2.12}$$

We have $\ell(t,\tilde{\sigma}) = -\lambda_o^{-1}\varepsilon\tilde{\sigma} + 1/k!\ (I-Q)TD^k_{u^k}\ R(o)\tilde{\sigma}^k + O(t)$ (2.13)

when $t \to o$ and consequently $\ell(o,\sigma_o) = o$.

We verify that $D_{\tilde{\sigma}}\ell(o,\sigma_o) = -\lambda_o^{-1}\varepsilon. + 1/(k-1)!(I-Q)TD^k_{u^k}\ R(o)\sigma_o^{k-1}$. and the hypothesis (2.9) implies that $D_{\tilde{\sigma}}\ell(o,\sigma_o)$ is an isomorphism of V_1.

Proposition 1 is a consequence of the implicit function theorem.

□

Hence, under hypothesis (2.9) σ_o gives rise to a unique branch of solutions of (2.1) in a neighborhood of (λ_o,o) in $\mathbb{R}\times V$ of the form

$$\lambda(t) = \lambda_o + \xi(t) \ ; \qquad u(t) = \sigma(t) + v(\xi(t),\sigma(t)) \tag{2.14}$$

where $\xi(t)$ and $\sigma(t)$ are given by Proposition 1. We can easily see that $\| v(\xi(t),\sigma(t))\| = O(t^k)$ when $t \to o$.

Remark 2.1: If $\sigma_o = o$, we obtain the trivial branch for (2.14). The only interesting case is $\sigma_o \neq o$ which yields a nontrivial branch; in particular, a necessary condition to get $\sigma_o \neq o$ is $D^k_{u^k}\ R(o) \neq o$.

Remark 2.2: With an additional hypothesis, Mc Leod and Sattinger [4] prove that if, for any σ_o solution of (2.8), we have assumption (2.9), then this procedure gives all the branches of solutions of (2.1) in a neighborhood of (λ_o,o).

3. THE APPROXIMATE PROBLEM

For each value of a parameter $h > o$ which tends to zero, we introduce a finite-dimensional subspace V_h of V and an operator $T_h \in L(W,V_h)$. We are interested by the solutions in $\mathbb{R}\times V_h$ of

$$F_h(\lambda,u) \equiv u + T_h\ G(\lambda,u) = o \tag{3.1}$$

in a neighborhood of (λ_o,o) when we assume

$$\lim_{h\to o} \| T-T_h\|_{L(W,V)} = o \ . \tag{3.2}$$

Remark that if $(\lambda,u) \in \mathbb{R}\times V$ is a solution of (3.1), then $u \in V_h$ and consequently we can solve (3.1) in $\mathbb{R}\times V$ instead of $\mathbb{R}\times V_h$.

As in section 2, the equation (3.1) is equivalent to the system

$$QF_h(\lambda,u) = o \ , \qquad (I-Q)F_h(\lambda,u) = o \ . \tag{3.3}$$

By assuming hypothesis (3.2) and that

$$D^pR \text{ is bounded on all bounded subsets of } V, \tag{3.4}$$

we can use the results obtained in [2, part II, thm.2] to prove there exist positive constants ξ_o, a and h_o, a neighborhood B of zero in V_1 and, for $h \leqslant h_o$, a unique C^p mapping $v_h:(-\xi_o,\xi_o)\times B\to V_2$ such that for all $\xi \in (-\xi_o,\xi_o)$ and $\sigma \in B$:

$$QF_h(\lambda_o+\xi,\ \sigma+v_h(\xi,\sigma)) = o\ , \qquad \| v_h(\xi,\sigma)\| \leqslant a\ . \tag{3.5}$$

Hence, solving problem (3.1) in the neighborhood of (λ_o,o) amounts to solve the approximate bifurcation equation

$$f_h(\xi,\sigma) \equiv (I-Q)F_h(\lambda_o+\xi,\ \sigma+v_h(\xi,\sigma)) = o \tag{3.6}$$

in a neighborhood of the origin of $\mathbb{R}\times V_1$.

By using the form (2.2) and by differentiating the first equation of (3.5) we get $D^\ell_{\xi^\ell}\, v_h(o,o)=o$, $o\leqslant\ell\leqslant k$, $D^\ell_{\sigma^\ell}\, v_h(o,o) = o$ for $2\leqslant\ell\leqslant k-1$ if $k\geqslant 3$ and by elementary calculations we obtain $D^\ell_{\xi^\ell}\, f_h(o,o) = o$ for $o\leqslant\ell\leqslant k$ and $D^\ell_{\sigma^\ell}\, f_h(o,o) = o$ for $2\leqslant\ell\leqslant k-1$ if $k\geqslant 3$. Generally $D_\sigma f_h(o,o)$ is not zero but tends to zero when $h\to o$. We set

$$\delta_h \equiv \| D_\sigma f_h(o,o)\| \tag{3.7}$$

and we obtain:

Proposition 2: Assume hypotheses (3.2) and (3.4). Then there exists a positive constant K (independent of h) such that for $h\leqslant h_o$ small enough we have:

$$\begin{cases} \delta_h \leqslant K\{\sup\limits_{\sigma\in S}\ \sup\limits_{\sigma^*\in S^*} |\langle (T-T_h)L\sigma,\sigma^*\rangle| \\ \qquad + \sup\limits_{\sigma\in S} \|(T-T_h)L\sigma\| \cdot \sup\limits_{\sigma^*\in S^*}\|((T-T_h)L)^*\sigma^*\|\}\ , \end{cases} \tag{3.8}$$

where $((T-T_h)L)^* \in L(V',V')$ is the adjoint operator of $(T-T_h)L \in$ $\in L(V,V)$, $S = \{\sigma\in V_1 : \|\sigma\| = 1\}$, $S^* = \{x^*\in V_2^\perp : \|x^*\| = 1\}$ with $V_2^\perp = \{x^* \in V' : \langle x,x^*\rangle = o,\ \forall x \in V_2\}$.

Proof: By using (3.6) we check for $\sigma \in V_1$: $D_\sigma f_h(o,o)\sigma = (I-Q)D_uF_h(\lambda_o,o)(\sigma+D_\sigma v_h(o,o)\sigma)$ and consequently

$$\delta_h \leqslant K \sup_{\sigma\in S}\ \sup_{\sigma^*\in S^*} |\langle D_uF_h(\lambda_o,o)(\sigma+D_\sigma v_h(o,o)\sigma),\sigma^*\rangle|. \tag{3.9}$$

For $\sigma \in V_1$, $\sigma^* \in V_2^\perp$ we have

$$|\langle D_uF_h(\lambda_o,o)(\sigma+D_\sigma v_h(o,o)\sigma),\sigma^*\rangle| = |\lambda_o\langle (T-T_h)L(\sigma+D_\sigma v_h(o,o),\sigma^*\rangle|\ ; \tag{3.10}$$

furthermore, by differentiating the first equation of (3.5) and by using the fact that $QD_uF_h(\lambda_o,o)$ is an isomorphism of V_2 with its inverse which is uniformly bounded for $h \leq h_o$ small enough, we can prove that

$$\| D_\sigma v_h(o,o) \| \leq K \sup_{\sigma \in S} \| (T-T_h)L\sigma \| . \tag{3.11}$$

Inequality (3.8) is a direct consequence of (3.9)-(3.11). □

Proposition 3: Assume hypotheses (2.9), (3.2) and (3.4). Then there exist positive constants χ,α,t_o,h_o and, for $h \leq h_o$, a unique branch of solutions of (3.6) of the form

$$\xi_h(t) = \xi(t) = \varepsilon t^{k-1}, \quad \sigma_h(t) = t\tilde{\sigma}_h(t); \quad \chi^{-1}\delta_h^{1/k-1} \leq |t| \leq t_o \tag{3.12}$$

where $\tilde{\sigma}_h$ is a C^{p-k} function of t defined for $\chi^{-1}\delta_h^{1/k-1} \leq |t| \leq t_o$ with $\|\tilde{\sigma}(t) - \tilde{\sigma}_h(t)\| \leq \alpha$, and $\tilde{\sigma}(t)$ is given by Proposition 1.

Proof: We can not give all the details of the proof which are in [6]; however we show here the idea. Let us consider

$$\hat{f}_h(\xi,\sigma) \equiv f_h(\xi,\sigma) - D_\sigma f_h(o,o)\sigma ; \tag{3.13}$$

we can verify that $\| \hat{f}_h(\varepsilon t^{k-1}, t\tilde{\sigma})\| = O(t^k)$ when $t \to o$. We set for $t \neq o$

$$\ell_h(t,\tilde{\sigma}) \equiv t^{-k} f_h(\varepsilon t^{k-1}, t\tilde{\sigma}) = \hat{\ell}_h(t,\tilde{\sigma}) + t^{1-k} D_\sigma f_h(o,o)\tilde{\sigma} \tag{3.14}$$

$$\text{where } \hat{\ell}_h(t,\tilde{\sigma}) \equiv t^{-k}\hat{f}_h(\varepsilon t^{k-1}, t\tilde{\sigma}), \tag{3.15}$$

and by using results of [2, part II] we prove that, for $o \leq s < p-k$, $D^s\hat{\ell}_h$ converge to $D^s\ell$ uniformly on all bounded subset of $\mathbb{R} \times V_1$ and $D^{p-k}\hat{\ell}_h$ is bounded on all bounded subset of $\mathbb{R} \times V_1$. By choosing t_o sufficiently small in Proposition 1 we obtain that $D_{\tilde{\sigma}}\ell(t,\tilde{\sigma}(t))$ is an isomorphism of V_1 with $\| D_{\tilde{\sigma}}\ell(t,\tilde{\sigma}(t))^{-1} \|$ bounded for $|t| \leq t_o$. Consequently, for $h \leq h_o$ small enough, $D_{\tilde{\sigma}}\hat{\ell}_h(t,\tilde{\sigma}(t))$ is an isomorphism of V_1 with a uniformly bounded inverse. If $|t| \geq \chi^{-1}\delta_h^{1/k-1}$ we have $\| t^{1-k}D_\sigma f_h(o,o) \| \leq \chi^{k-1}$; by choosing χ small enough and looking (3.14) we check that $D_{\tilde{\sigma}}\ell_h(t,\tilde{\sigma}(t))$ is an isomorphism of V_1 with $\sup_{t_o \geq |t| \geq \chi^{-1}\delta_h^{1/k-1}} \| D_{\tilde{\sigma}}\ell_h(t,\tilde{\sigma}(t))^{-1} \| \leq c$ for $h \leq h_o$, where c is independent of h.

We use a fixed point theorem to show that for $t_o \geq |t| \geq \chi^{-1}\delta_h^{1/k-1}$, there exists a unique point $\tilde{\sigma}_h(t) \in V_1$ such that $\ell_h(t,\tilde{\sigma}_h(t)) = o$ and $\| \tilde{\sigma}(t) - \tilde{\sigma}_h(t) \| \leq \alpha$ where α is small enough. The implicit function theorem implies that $\tilde{\sigma}_h$ is of class C^{p-k}. □

Hence, under hypotheses of Proposition 3, we obtain a unique branch of solutions of (3.1) of the form

$$\lambda(t) = \lambda_o + \xi(t); \qquad u_h(t) = \sigma_h(t) + v_h(\xi(t),\sigma_h(t)), \tag{3.16}$$

with $\chi^{-1}\delta_h^{1/k-1} \leqslant |t| \leqslant t_o$ and $\|u(t)-u_h(t)\| \leqslant |t|\alpha$ where α is small enough; $\xi(t)$ and $\sigma_h(t)$ are given by Proposition 3. Hence we get our

Main theorem: Assume hypotheses (2.9),(3.2) and (3.4). Then there exist positive constants $\alpha,\chi,\xi_o,t_o,h_o$ such that, for all $h \leqslant h_o$, the set of solutions (λ,u_h) of (3.1) having the properties $\xi_o \geqslant |\lambda-\lambda_o| \geqslant \chi^{1-k}\delta_h$, $u_h \in \{u \in V: \|u(t)-u\| \leqslant |t|\alpha, |t| \leqslant t_o\}$, consists of a C^{p-k} unique branch $\{(\lambda(t),u_h(t)): \chi^{-1}\delta_h^{1/k-1} \leqslant |t| \leqslant t_o\}$ of the form (3.16). Moreover we obtain the error estimate:

$$\begin{cases} \displaystyle\sup_{\chi^{-1}\delta_h^{1/k-1}\leqslant|t|\leqslant t_o} \|u(t)-u_h(t)\| \leqslant \\ \displaystyle\leqslant K\{\delta_h^{1/k-1} + \sum_{\ell=o}^{k-1} \sup_{|t|\leqslant t_o} \|(T-T_h)\frac{d^\ell}{dt^\ell} G(\lambda(t),u(t))\|\}. \end{cases} \tag{3.17}$$

Proof: The first part of this theorem is a direct consequence of Propositions 2 and 3. To obtain the error estimate (3.17) we begin to establish the estimate:

$$\|\tilde{\sigma}_h(t)-\tilde{\sigma}(t)\| \leqslant K\|\ell_h(t,\tilde{\sigma}(t))\|, \ \forall t \text{ with } \chi^{-1}\delta_h^{1/k-1} \leqslant |t| \leqslant t_o \tag{3.18}$$

which is a consequence of the relation

$$o = \ell_h(t,\tilde{\sigma}_h(t)) = \ell_h(t,\tilde{\sigma}(t)) + \int_o^1 D_{\tilde{\sigma}}\ell_h(t,s\tilde{\sigma}_h(t)+(1-s)\tilde{\sigma}(t)).\,.(\tilde{\sigma}_h(t)-\tilde{\sigma}(t))ds. \tag{3.19}$$

From (3.18) we have $\|\sigma(t)-\sigma_h(t)\| \leqslant K\|t\ell_h(t,\tilde{\sigma}(t))\|$ and by straightforward calculations we prove that this last quantity is bounded by the right member of (3.17). In a same way, by using the first equation of (3.5), we establish the error estimate for $\|v(\xi(t),\sigma(t))-v_h(\xi(t),\sigma_h(t))\|$. (see [6] for the details).

□

Remark 3.1: If $T_h = \Pi_h T$ where Π_h is a projection from V into V_h (Galerkin approximation), the second term of the right member of (3.17) is equal to $\sum_{\ell=o}^{k-1} \sup_{|t|\leqslant t_o} \|(I-\Pi_h)u^{(\ell)}(t)\|$ where $u^{(\ell)}$ is the ℓ-th. derivative of u with respect to t.

Remark 3.2: In [6] we analyze, in particular cases (k=2,3), the solutions (λ,u_h) of (3.1) for $|\lambda-\lambda_o| \leqslant \chi^{1-k}\delta_h$, $\|u_h\|$ small enough, by using a scaling technique; we obtain numerical imperfect bifurcations in the same way as [1,3,5,7,8].

Remark 3.3: Under some conditions of symmetry (see [6]) we can prove there exists ξ_h^o such that $D_\sigma f_h(\xi_h^o,o) = o$ and $\xi_h^o \to o$ when $h \to o$. In this case we obtain for $|\xi_h^o|$ the same error estimate as for δ_h (in Proposition 2) and the point $(\lambda_o+\xi_h^o,o)$ is a double bifurcation point of the approximate problem (3.1). By setting $\ell_h(t,\tilde{\sigma}) = t^{-k} f_h(\xi_h^o + \varepsilon t^{k-1}, t\tilde{\sigma})$ we obtain a full branch of solutions $\{(\lambda_h(t), u_h(t)) : |t| \leq t_o\}$ of problem (3.1) which converges to the branch $\{(\lambda(t), u(t)) : |t| \leq t_o\}$.

4. AN EXAMPLE

Let $\Omega = (o,1)\times(o,1)$ be the unit square in $\mathbb{R}^2$; we consider the problem: find $\lambda \in \mathbb{R}$, $u \in H_o^1(\Omega)$ such that

$$-\Delta u - \lambda u + u^3 = o \quad \text{in } \Omega . \tag{4.1}$$

We set $V = H_o^1(\Omega)$, $W = L^2(\Omega)$; $T \in L(W,V)$ is the linear operator such that $-\Delta(Tf) = f$ in Ω, $Tf = o$ on the boundary $\partial\Omega$ of Ω; L is the injection from V into W; $R(u) = u^3$; $G(\lambda,u) = -\lambda u + u^3$. The problem (4.1) is quite equivalent to

$$F(\lambda,u) \equiv u + TG(\lambda,u) = o , \qquad (\lambda,u) \in \mathbb{R}\times V . \tag{4.2}$$

Let now λ_o be the second eigenvalue of the problem $-\Delta\varphi = \lambda\varphi$ in Ω, $\varphi = o$ on $\partial\Omega$. In this case λ_o^{-1} is a double eigenvalue of T and the hypotheses (2.8),(2.9) are satisfied for $k = 3$, $\varepsilon = 1$. Let $\{(\lambda(t), u(t)) : |t| \leq t_o\}$ be a branch of solutions of (4.1) or (4.2) which bifurcates from the point (λ_o, o); this branch is of the form (2.14).

In order to give an approximation of problem (4.1) we take a finite element method with piecewise polynomials of degree 1. Let τ_h be a regular family of triangulations of $\bar{\Omega}$ and $V_h \subset H_o^1(\Omega)$ be the subspace of piecewise polynomials of degree 1 on each triangle of the triangulation. We approximate problem (4.1) by $(\lambda, u_h) \in \mathbb{R}\times V_h$ such that: for all $v_h \in V_h$,

$$\int_\Omega \text{grad } u_h \text{ grad } v_h dx - \lambda\int_\Omega u_h v_h dx + \int_\Omega u_h^3 v_h dx = o, \tag{4.3}$$

or equivalently by

$$F_h(\lambda,u_h) \equiv u_h + T_h G(\lambda,u_h) = o , \tag{4.4}$$

where $T_h \in L(W,V_h)$ is the operator defined by $\int_\Omega \text{grad}(T_h f)\text{grad } v_h dx = \int_\Omega f v_h dx$, $\forall v_h \in V_h$.

In this case we obtain, by using Proposition 2, $\delta_h = O(h^2)$. By considering our Main Theorem and Remark 3.1 we get a branch of solutions $\{(\lambda_h(t), u_h(t)) : \chi^{-1}h \leq |t| \leq t_o\}$ of (4.3) or (4.4) with $\sup_{\chi^{-1}h \leq |t| \leq t_o} \| u(t) - u_h(t) \|_{H_o^1(\Omega)} = O(h)$. Naturally, this branch of solutions is obtained for values λ_h such that $|\lambda_h - \lambda_o| \geq \chi^{-\frac{1}{2}} h^2$. Remark that $O(h^2)$ is the order of convergence of eigenvalues for the linearized problem.

REFERENCES

1. BAUER L., KELLER H.B., REISS E.L., Multiple eigenvalues lead to secondary bifurcation. *SIAM J. Appl.Math.* 17,101-122 (1975)
2. BREZZI F., RAPPAZ J., RAVIART P.A., Finite dimensional approximation of nonlinear problems.
   Part I: Branches of nonsingular solutions. *Numer.Math.*36,1-25
   Part II: Limit Points. *Numer. Math.*(to appear) (1980)
   Part III: Simple Bifurcation Points. *Numer.Math.* (to appear).
3. GOLUBITSKI M., SCHAEFFER D., A theory of imperfect bifurcation via singularity theory. *Commun.Pure Appl.Math.* 32,21-98. (1979).
4. Mc LEOD J.B., SATTINGER D.H., Loss of stability and bifurcation at a double eigenvalue. *J.Funct.Anal.* 14, 62-84,(1973).
5. MATKOWSKY B.J., REISS E.L., Singular perturbations of bifurcations. *SIAM J. Appl.Math.* 33, 230-255, (1977).
6. RAPPAZ J., RAUGEL G., Finite dimensional approximation of double bifurcation points for nonlinear problems. *Rapport du Centre de Math.Appl.,Ecole Polytechnique Palaiseau* (1981).
7. REISS E.L., Imperfect bifurcation. *Applications of Bifurcation Theory* (P.H. Rabinowitz ed.), Academic Press, New-York (1977).
8. SHEARER M., Secondary bifurcation near a double eigenvalue. *SIAM J. Math.Anal.* 11, 365-389, (1980).

NEW RESULTS IN THE FINITE ELEMENT SOLUTION OF STEADY VISCOUS FLOWS

*M.D. Gunzburger and †R.A. Nicolaides

**University of Tennessee, Knoxville,* †*Carnegie-Mellon University.*

1. INTRODUCTION

We consider conforming finite element methods for the approximate solution of the stationary Stokes equations whose weak form is given by: seek the velocity-pressure pair $(\underline{u}, p) \in \vec{H}_0^1(\Omega) \times \bar{L}^2(\Omega)$ such that

$$\int_\Omega \nabla\underline{u} : \nabla\,\underline{v} + \int_\Omega p \operatorname{div} \underline{v} = \langle \underline{f}, \underline{v} \rangle \quad \forall \underline{v} \in \vec{H}_0^1(\Omega) \tag{1.1}$$

$$\int_\Omega q \operatorname{div} \underline{u} = [g, q] \quad \forall q \in \bar{L}_2(\Omega) . \tag{1.2}$$

Here $\vec{H}_0^1(\Omega)$ denotes the Hilbert space of real valued two dimensional velocity fields whose components have a distributional derivative in $L^2(\Omega)$ in each variable and which have zero trace on $\partial\Omega$, the boundary of the domain $\Omega \in \mathbb{R}^2$ where $(\underline{u}, p)$ is sought. $\vec{H}_0^1(\Omega)$ is normed by

$$\|\underline{v}\|_1^2 = \int_\Omega \nabla\underline{v} : \nabla\underline{v}$$

where the colon denotes the scalar product of the two tensors standing on each side of it. In (1.1) - (1.2), $\bar{L}^2(\Omega)$ denotes the linear subspace of $L^2(\Omega)$ of elements orthogonal to constants and $\underline{f}$ and g denote bounded linear functionals on $\vec{H}_0^1(\Omega)$ and $\bar{L}^2(\Omega)$, respectively. The viscosity parameter has been absorbed into the pressure field and also affects the scaling of $\underline{f}$.

Approximations to $(\underline{u}, p)$ are defined in the usual manner. We have subspaces $\mathcal{V}^h \subset \vec{H}_0^1(\Omega)$ and $\mathcal{S}^h \subset \bar{L}^2(\Omega)$ and seek a pair $(\underline{u}^h, p^h) \in \mathcal{V}^h \times \mathcal{S}^h$ satisfying (1.1) and (1.2) for all

$\underline{v}^h \in \mathcal{V}^h$ and $q^h \in \mathcal{S}^h$, respectively. The standard theory for the error in this approximation [1, 2] yields that $(\underline{u}^h, p^h)$ converges to $(\underline{u}, p)$ at an optimal rate if the error is measured in the graph norm

$$|||\underline{v}, q||| = \|\underline{v}\|_1 + \|q\|_0 ,$$

i.e.

$$|||\underline{u} - \underline{u}_h, p - p_h||| \leq C \inf |||\underline{u} - \underline{u}^h, p - p^h||| \tag{1.3}$$

where the infimum is over all $(\hat{\underline{u}}^h, \hat{p}^h) \in \mathcal{V}^h \times \mathcal{S}^h$.

In many applications one is interested in measuring the error in other norms, in particular the $\vec{L}^2$ and $\vec{H}^1$-norms of the velocity and the L^2-norm of the pressure. From (1.3) we immediately obtain

$$\|\underline{u} - \underline{u}^h\|_1 \leq C(\inf_{\hat{u}^h \in \mathcal{V}^h} \|\underline{u} - \hat{\underline{u}}^h\|_1 + \inf_{\hat{p}^h \in \mathcal{S}^h} \|p - \hat{p}^h\|_0) \tag{1.4}$$

with a similar estimate for the error in p. More precise estimates exist [6] which contain more information and which, for certain subspaces, decouple the error in the velocity from that in the pressure. Finally, if $\underline{u}$ is sufficiently smooth, then a standard duality argument yields that

$$\|\underline{u} - \underline{u}^h\|_0 \leq Ch\|\underline{u} - \underline{u}^h\|_1$$

where h is a measure of the grid size.

2. FINITE ELEMENT PAIRS

In this section we consider some specific choices for the subspaces $\mathcal{V}^h$ and $\mathcal{S}^h$. Some care must be exercised in making this choice since discrete approximations to the non-positive definite problem (1.1) - (1.2) are, in general, unstable. There are several conditions which must hold for stability to be guaranteed; a discussion of these may be found in [1, 5, 6]. The particular choices for $\mathcal{V}^h$ and $\mathcal{S}^h$ presented here are known to result in stable approximations [3].

In choosing $\mathcal{V}^h$ and $\mathcal{S}^h$, some dicision must be made about the degrees of the element polynomials. In fact, one must first establish triangulations for $\mathcal{V}^h$ and $\mathcal{S}^h$. These triangulations need not be coincident and for some of the elements described below, different but related triangulations are used for the velocity and pressure fields. Estimates such as (1.4) show that if the mesh sizes for $\mathcal{V}^h$ and $\mathcal{S}^h$ are comparable, then the degree of $\mathcal{S}^h$ should be one less than that of $\mathcal{V}^h$. In this case, the two error terms on the right hand side of (1.4) are of the same order in h. This "comparability" condition is not necessary for $\vec{H}_0^1(\Omega)$ convergence of the approximate velocity field, but is probably necessary for optimal $\vec{H}_0^1(\Omega)$ convergence.

Our first scheme, apparently first suggested in [2], is based on subdividing Ω, which we take to be a polygonal region, into triangles in the usual manner. We then choose $\mathcal{S}^h$ to be all piecewise constant functions on the triangles with mean zero over Ω and choose for $\mathcal{V}^h$ all continuous piecewise quadratic vector fields vanishing on $\partial\Omega$. Clearly, the degree of the polynomials in $\mathcal{S}^h$ are two less than those in $\mathcal{V}^h$, thus violating the requirement for optimal $\vec{H}_0^1(\Omega)$ accuracy. This is borne out by the numerical results reported below.

The second scheme [7] is based on subdividing Ω into quadrilaterals and choosing for $\mathcal{S}^h$ all piecewise constant functions on these quadrilaterals with mean zero on Ω. Subsequently, we subdivide each quadrilateral into two triangles by drawing a diagonal and choose for $\mathcal{V}^h$ all continuous piecewise linear vector fields (over the triangles) vahishing on $\partial\Omega$. This is an example of distinct, although closely related, triangulations being used for $\mathcal{V}^h$ and $\mathcal{S}^h$. Evidently, the comparability condition on the degrees of the polynomials in $\mathcal{S}^h$ and $\mathcal{V}^h$ is satisfied. To render this element pair stable, a further constraint must be imposed on $\mathcal{S}^h$ due to possible uncouplings in the discrete equations [3]. We discuss this in greater detail below in the context of a specific example. We note that if $\mathcal{S}^h$ is chosen to be all piecewise constant functions over the triangles, with mean zero on Ω, then the resulting scheme is hopelessly unstable [5].

The third scheme [4] is based on first subdividing Ω into quadrilaterals and subsequently subdividing each quadrilateral into four triangles by drawing both diagonals. We then choose $\mathcal{V}^h$ to be all continuous piecewise linear vectors fields over the triangles which vanish on $\partial\Omega$ and choose $\mathcal{S}^h \equiv \text{div } \mathcal{V}^h$. $\mathcal{S}^h$ is a certain subspace of the class of all piecewise constants on the triangles. See [4] for details. This particular choice for $\mathcal{S}^h$ and $\mathcal{V}^h$ is one for which the $\vec{H}_0^1(\Omega)$ error in the velocity approximation is uncoupled from the pressure approximation [5]. Again, this scheme satisfies the comparability condition and optimal $\vec{H}_0^1(\Omega)$ accuracy can evidently be achieved.

3. COMPUTATIONAL RESULTS

We wish to compare the asymptotic error estimates as a function of h for the schemes of the previous section. To do so we solve (1.1) - (1.2) in the region $\Omega = \{x, y | 0 < x, y < 1\}$ where $\underline{f}$ and g are chosen so that the exact solution is given by $\underline{u} = [\sin \pi x \sin \pi y , \ x^2(1 - x) \sin \pi y]$ and $p = (1 - 4y + y^3) \cos \pi x$. For this choice of $\underline{u}$ and p, $\underline{u}$ on $\partial\Omega$ vanishes but both $\underline{f}$ and g are nonzero. Generally, in incompressible flows $g = 0$, but the fact that here it is nonzero does not affect the validity of the results for the case of $g = 0$. This is because for smooth solutions the rates of convergence are independent of $\underline{f}$ and g [5].

The results presented in the tables below are for norms of differences between the computed solutions $\underline{u}^h$, p^h and certain interpolates $\hat{\underline{u}}^h$, $\hat{p}^h$ since these errors can be computed exactly (except, of course, for roundoff errors). Estimates of the actual error of the computations can then be obtained via the triangle inequality, e.g.,

$$\|\underline{u} - \underline{u}^h\| \leq \|\underline{u} - \hat{\underline{u}}^h\| + \|\underline{u}^h - \hat{\underline{u}}^h\| . \tag{3.1}$$

The first term on the right is purely approximation theoretical and the second term is estimated in the tables below. For all element pairs, $\hat{\underline{u}}^h$ is the pointwise interpolant of $\underline{u}$ in $\mathcal{V}^h$. For the first linear-constant (where there are two triangles per quadrilateral) element pair, $\hat{p}^h$ is the interpolant of in the space of piecewise constant functions on the quadrilaterals, with the interpolation points being the centroids of quadrilaterals. For the quadratic-constant and the second linear-constant (where there are four triangles per quadrilateral) element pairs, $\hat{p}^h$ is the interpolant of p in the space of piecewise constant functions over the triangles, with the interpolation points being the centroids of triangles.

Tables 1-3 contain the computed convergence rates for the $L^2(\Omega)$ error in the pressure and the $\vec{L}^2(\Omega)$ and $\vec{H}_0^1(\Omega)$ errors for the velocity for each of the element pairs. The rates are computed by comparing the errors from successive grids.

TABLE 1

Rates of Convergence for the Quadratic-Constant Element Pair

h	L^2-pressure rate	$\vec{L}^2$-velocity rate	$\vec{H}_0^1$-velocity rate
1/11			
	1.703	1.913	.973
1/12			
	1.715	1.917	.970
1/13			
	1.725	1.921	.969
1/14			

TABLE 2

Rates of Convergence for the Linear-Constant Element Pair with Two Triangles per Quadrilateral

h	L^2-pressure rate	$\vec{L}^2$-velocity rate	$\vec{H}_0^1$-velocity rate
1/11			
	2.045	1.990	2.021
1/12			
	2.124	2.017	2.041
1/13			
	2.007	2.001	2.024
1/14			

For all the element pairs except the quadratic-constant pair, the rates given in the tables are at least as great as the corresponding rates for the approximation error and therefore by (3.1) the rate of convergence of the error will be no worse than that given by the approximation theoretic part, i.e., convergence is at optimal rates. On the other hand, the rates given in Table 1 show that the errors in (3.1) are dominated by the second term on the right hand side. For example, the

$\vec{H}_0^1$ approximation error is $O(h^2)$ while by Table 1, the second term in (3.1) is at best $O(h)$. Strictly speaking, this does not imply that $\|\underline{u} - \underline{u}^h\|_1$ is not $O(h^2)$; however direct computation of this error based on the use of high order quadrature formulas show that the error is at best $O(h)$. In fact, for all three element pairs we are led to the estimates

$$\|p-p^h\|_0 = O(h), \ \|\underline{u}-\underline{u}^h\|_1 = O(h) \text{ and } \|\underline{u}-\underline{u}^h\|_0 = O(h^2) \qquad (3.2)$$

in spite of the fact that the quadratic-constant pair is considerably more complex to compute with than the other element pairs (see section 4).

TABLE 3

Rates of Convergence for the Linear-Constant Element Pair with Four Triangles per Quadrilateral

h	L^2-pressure rate	$\vec{L}^2$-velocity rate	$\vec{H}_0^1$-velocity rate
1/11			
	1.099	2.062	1.008
1/12			
	1.096	2.043	1.008
1/13			
	1.069	2.042	1.006
1/14			

We note that in some instances the rates measured relative to the interpolant are one order higher than the corresponding rates for the approximation error. This occurs for the $\vec{H}_0^1$ error in the velocity in Table 2 and for the pressure error in Tables 1 and 2. (In fact, with appropriate averaging, this phenomena occurs for the pressure error for the element of Table 3.) This form of "superconvergence" is potentially useful for example when linear functionals of the true solution are to be approximated.

4. SOLUTION OF LINEAR SYSTEM

The discrete linear systems were solved in each case by a banded elimination solver with partial pivoting; this is necessary since the assembled coefficient matrix, although symmetric, is indefinite. For all element pairs the computed pressure should be normalized so that it has zero mean. Except for the quadratic-constant element pair, an additional orthogonality condition must be specified (see section 2 above). For the specific example considered in section 3, this condition is simply that not only should the pressure have mean zero over the whole unit square, but should also separately have zero mean over red and black rectangles of the checkerboard pattern of the grid. See [3] for details. It was found convenient not to impose these normalization and orthogonality conditions on the approximating spaces. Instead, during the elimination process, zero pivot elements are encountered. In the actual computations, these are detected whenever a pivot is encountered whose magnitude is smaller than a prescribed tolerance which depends on themachine precision. That particular step of the elimination may then be skipped since the column to be eliminated is already in reduced form. Then, during the backsolve, the components of the solution vector corresponding to the zero pivots are arbitrarily set to a constant, say unity. Finally, the orthogonality and normalization conditions are imposed on the solution of the discrete linear system by an obvious post-processing. We note that regardless of what orthogonalities are required of the pressure field, once the velocity field $\underline{u}^h$ is found, the pressure can always be computed, at some cost, as the least squares solution to the algebraic equivalent of the first differential equation (1.1).

The three finite element schemes described in section 2 yield approximate solutions whose asymptotic error rates are the same; indeed, they are given by (3.2). However, the storage and computing time required by the various schemes differ sharply. In Table 4 we tabulate the number of unknowns and half-band width of the coefficient matrices resulting from each scheme applied to the example of section 3 where the unit square is subdivided into an $n \times n$ grid, i.e., $h = 1/n$. The tabulated expressions are valid for large n, i.e. we only give the leading term in n. We also give the storage and computing time requirements for each element pair relative to those for the quadratic-constant element pair. Clearly the linear-constant (on boxes) element pair achieves the same rates as the other pairs with considerably less storage and computing time requirements.

TABLE 4

Storage and Computing Time Requirements

Element Pair	(a)	(b)	(c)	(d)
Quadratic-constant	$10n^2$	10n	1	1
Linear-constant (on boxes)	$3n^2$	3n	.09	.027
Linear-constant (on 4 triangles)	$7n^2$	7n	.49	.343

(a) Number of unknowns; (b) Half-bandwidth; (c) Relative storage; (d) Relative computing time.

ACKNOWLEDGEMENTS

This work was supported by the Air Force Office of Scientific Research under Grant Nos. AF-AFOSR-80-0083 and AF-AFOSR-80-0091.

REFERENCES

1. BREZZI, F., On the Existence, Uniqueness and Approximation of Saddle Point Problems Arising From Lagrange Multipliers. *RAIRO* 8 - R2, 129 - 151 (1974).
2. CROUZEIX, M. and P.A. RAVIART, Conforming and Non-conforming Finite Element Methods for Solving the Stationary Stokes Equations. *RAIRO* 7-R2, 33 - 76 (1973).
3. FIX, G., A General Theory of Mixed Finite Element Methods. To appear.
4. FIX, G., M.D. GUNZBURGER and R.A. NICOLAIDES, On Mixed Finite Element Methods for First Order Elliptic Systems. To appear.
5. GUNZBURGER, M.D. and R.A. NICOLAIDES, The Computational Accuracy of Some Finite Element Methods for Incompressible Viscous Flow Problems. To appear.
6. NICOLAIDES, R.A., Existence, Uniqueness and Approximation for Generalized Saddle Point Problems. To appear.
7. NICOLAIDES, R.A., Co-energy Methods for Elliptic Flow and Related Problems. To appear.

CHARACTERISTICS AND FINITE ELEMENTS METHODS APPLIED TO THE EQUATION OF FLUIDS

M. Bercovier*, O. Pironneau**
Y. Hasbani*, E. Livne*

(*) *The Hebrew University of Jerusalem*
(**) *University of Paris 13 and INRIA.*

1. INTRODUCTION

Let Ω be a bounded domain of R^n, $(n=1,2,3)$, Γ its boundary. Given a velocity field $\underset{\sim}{u}$, div $\underset{\sim}{u} = 0$, $\underset{\sim}{u}.\underset{\sim}{n}|_\Gamma = 0$, we are interested in the numerical solution of the advection-diffusion equation :

$$\frac{\partial\rho}{\partial t} + \underset{\sim}{u}.\,\underset{\sim}{\nabla}\rho - \nu\,\Delta\rho = f \quad \text{in} \quad \Omega \times]\,0,T\,[\tag{1}$$

$$\rho(\underset{\sim}{x},0) = \rho_o(x) \quad \text{in} \quad \Omega \tag{2}$$

$$\rho(\underset{\sim}{x},t) = g(\underset{\sim}{x},t), \quad \text{on} \quad \Gamma \times]\,0,T\,[\tag{3}$$

Standard Galerkin approximations in the space variables yield nonsymmetric stiffness matrices. Moreover it is well known that numerical instabilities (oscillations) can be observed, unless one takes a very fine grid (in the Finite Element Method) and/or a small time step.

These instabilities come from the presence of boundary layers (or shocks) when $\nu \to 0$. The latest Finite Element developments for the solution of (1), (2), (3) were dealing with upwinding techniques [1] , [2] , [3]. Recently an alternative approach has been proposed, [4],[5],[6],[7],[8]. It consists of rewriting

$$\frac{\partial\rho}{\partial t} + \underset{\sim}{u}.\,\underset{\sim}{\nabla}\,\rho$$

as the <u>total derivative</u> $D\rho/Dt$, i.e. :

$$\frac{D\rho}{Dt} - \nu\,\Delta\,\rho = f \tag{4}$$

and applying the usual finite difference techniques to the (total) time derivative and a Galerkin method in space variables.

Our aim is to present the general algorithm one derives from (4). This gives rise to different numerical implementation. We recall some error estimates for those methods and give in the one dimensional case a corresponding finite-difference truncation order. We conclude by a survey of some one and two dimensional examples.

2. THE GENERAL ALGORITHM

To solve (1) - (3) we rewrite (4) in variational form

$$\left(\frac{D\rho}{Dt},w\right)+\nu(\nabla\rho,\nabla w)=(f,w)\quad \forall w\in H_0^1(\Omega) \tag{5}$$

The (,) denotes the L^2-scalar product.
Let $X(x,t;\tau)$ be the trajectory which goes through x at time t, i.e. :

$$\frac{dX}{d\tau}(\tau)=u(X,\tau)\quad ;\quad X(t)=x\quad ; \tag{6}$$

then we have

$$\frac{D\rho}{Dt}=\frac{\partial}{\partial\tau}\rho(X(x,t;\tau),\tau)\Big|_{\tau=t} \tag{7}$$

Therefore this time derivative can be approximated by any of the usual finite difference techniques (e.g. Crank-Nicholson, Gear) ; for clarity let us use the fully implicit scheme, then (5) is approximated by

$$\frac{1}{k}(\rho_h^{n+1},w_h)+\nu(\nabla\rho_h^{n+1},\nabla w_h)=(f,w_h)+\frac{1}{k}(\rho_h^n(X_k(.,(n+1)k;nk)),w_h) \tag{8}$$

$$\forall w_h\in V_{oh}\ ;\ \rho_h^{n+1}\in\{g_h\}+V_{oh}$$

where V_{oh} is any finite element space (or even Galerkin) approximating $H_0^1(\Omega)$;

X_k is a numerical approximation of X, (6).

Except in the one dimensional case the last integral in (8) is usually too expensive to compute exactly. Hence the last integral in (8) is approximated by a numerical quadrature

$$\int_\Omega \rho_h^n(X_k(x,(n+1)k;nk))w_h(x)dx \simeq \sum_i \rho_h^n(X_k(\xi^i,(n+1)k;nk))w_h(\xi^i)\pi_i \tag{9}$$

Each time step requires the following calculation
1. Solution of (6) for each ξ^i
2. Solution of the <u>symetric</u> linear system (8).

3. NUMERICAL IMPLEMENTATIONS

It is clear that one must be consistent in choosing V_h, X_k and $\{\xi^i,\pi_i\}$.

Up to now the following choices have been studied

Choice 1 [4][8]

$V_h = \{w_h$ continuous piecewise linear on triangles$\}$ (10)

$X_k(x,(n+1)k;nk) = x - u^{n+1}(x)k$ (Euler) (11)

ξ^i = random point in triangle T_i ; $\pi_i = \text{area}(T_i)$ (12)

In principle it is better to use the same quadrature formula on the first integral of (8) also (mass lumping).

Choice 2 [5][6][7]

V_h as in (10)

$X_k(x,(n+1)k,nk) = x-k\, u^{n+\frac{1}{2}}(x-\frac{k}{2}u^{n+1}(x))$ (Runge-Kutta degree 2) (13)

ξ^i = vertices of the triangulation ; π_i = support of $(w^i)/3$ (14)

where w^i is the basis function associated with vertex i.

Variants can naturally be constructed by switching from piecewise linear to piecewise bilinear (Q^1) isoparametric ; Gaussian integration can be used for (9). Non conforming piecewise linear elements have also been proposed [6] .

Choice 3 [5] [8] [9]

$V_h = \{ w_h$ continuous, piecewise quadratic on triangles , or biquadratic on quadrangles $\}$

X_k as in (13).

ξ^i at the nodes of V_h, or at the Gaussian integration points for Q^2.

4. ERROR ESTIMATES

The following properties can be established :

. Choice 1 gives an L^∞ stable scheme of order $k + \frac{h}{k}$ thus the optimal order is $\sqrt{h}$. The scheme is conservative statistically.

. Choice 2 gives an L^∞ stable scheme of order $h + k + h^2/k$ thus of optimal order h. The scheme is not conservative because of the numerical integration. As an alternative to (9) one can use

$$\int_\Omega \rho_h^n(X_k(x,(n+1)k;nk))w_h(x)\,dx \tag{15}$$

$$= \int_\Omega \rho_h^n(y)w_h(X_k(y,nk;(n+1)k))\,dy$$

$$\simeq \sum_i \rho_h^n(\xi^i)w_h(X_k(\xi^i,nk;(n+1)k)\ \pi_i$$

With these formulas all schemes are conservative but convergence is an open problem.

. Choice 3 is numerically good but convergence and stability have not been shown.
Another interpretation of (9) is

$$\int_\Omega \rho_h^n(X_k(x,(n+1)k;nk))w_h(x)dx \simeq (\tilde{\pi}_h\rho_h^n(X_k(.,(n+1)k;nk)),\tilde{\pi}_h w_h(.)) \tag{16}$$

where $\tilde{\pi}_h$ is polynomial interpolation operator.

Then (8) is replaced by

$$\frac{1}{k}(\tilde{\pi}_h\ \rho_h^{n+1},\ \tilde{\pi}_h w_h) + \nu(\nabla\rho_h^{n+1},\nabla w_h) \tag{17}$$

$$= (f,w_h) + \frac{1}{k}(\tilde{\pi}_h\rho_h(X_k(.,(n+1)k;nk)),\ \tilde{\pi}_h w_h) \qquad \forall\ w_h \in V_{oh}$$

If $\tilde{\pi}_h$ is the quadratic interpolation based on the values of the 3 second derivatives at the center of each triangle and on the 3 values at the vertices, then the scheme becomes $O(h^2+k^2+h^3/k)$ (see [8]) and L^∞-stable.

It is of interest to look at the finite difference schemes that one gets when Ω is one dimensional and u is constant.

For choice 1 with exact integration with $\nu = 0$, $f = 0$ and $uk \leq h$ one gets

$$\frac{1}{6}[\rho_{i-1}^{n+1} + 4\rho_i^{n+1} + \rho_{i+1}^{n+1}] = \frac{1}{6}[\rho_{i-1}^n + 4\rho_i^n + \rho_{i+1}^n] - uk\,\frac{\rho_{i+1}^n - \rho_{i-1}^n}{2h} \tag{18}$$

$$- u^2k^2\,\frac{\rho_i^n - 2\rho_i^n + \rho_{i+1}^n}{2h^2} - u^3k^3\,\frac{\rho_{i+1}^n - 3\rho_i^n + 3\rho_{i-1}^n - \rho_{i-2}^n}{6h^3}$$

It can be shown ([9]) that the scheme is in fact of order $k^2h^2 + k^3h$ (superconvergence) when ρ is 4 times continuously differentiable.

5. NUMERICAL TESTS

For one dimensional model problems we took $\Omega =]0,40[$, and $h = .625$. We first took the pure transport problem ($\nu = 0.$) with

$$\rho_o = \exp\ [.(x - 3.75)^2/2]$$

and ρ_{oh} its linear interpolate. Table 1. gives for different values of uk/h^2 (Courant-Friedrichs-Lewy number) the computed peak for t = 9.375, using a 3 node Gauss Legendre quadrature rule for (15).

CFL number	0.1	0.3	0.7	0.9	1.875
(max,t=9.375)	1.308	.9480	.9290	1.035	1.0122

- Table 1 -

As could be expected result are best when uk/h^2 is large, since we do less time steps and thus introduce less numerical diffusivity. (Note that the scheme is optimal in that we have an "exact solution for $uk/h^2 = 1$").

Next we compared the analytical solution

$$\rho(\underset{\sim}{x},t) = \exp\left[-(x - 3.75 - t)^2/2(1+x\nu t)\right]/\sqrt{1+2\nu t}$$

for $\nu = .1$ and $k = 20/79$, with the solution of (5), (6), (7). Actually instead of (8) a Crank-Nicholson scheme was used and exact integration of all scalar products. In figure 1 one sees that after 80 time steps there is no visible difference between the numerical and analytical solution.

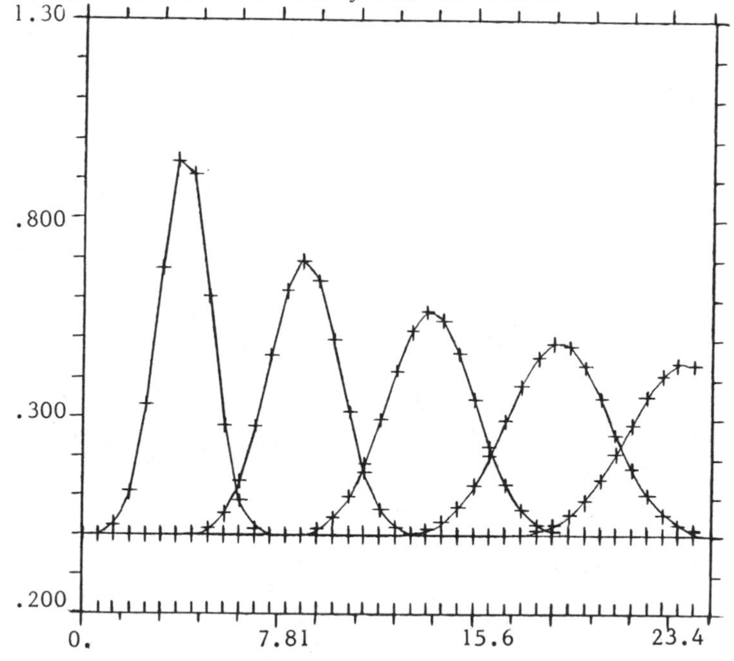

FIG.1.

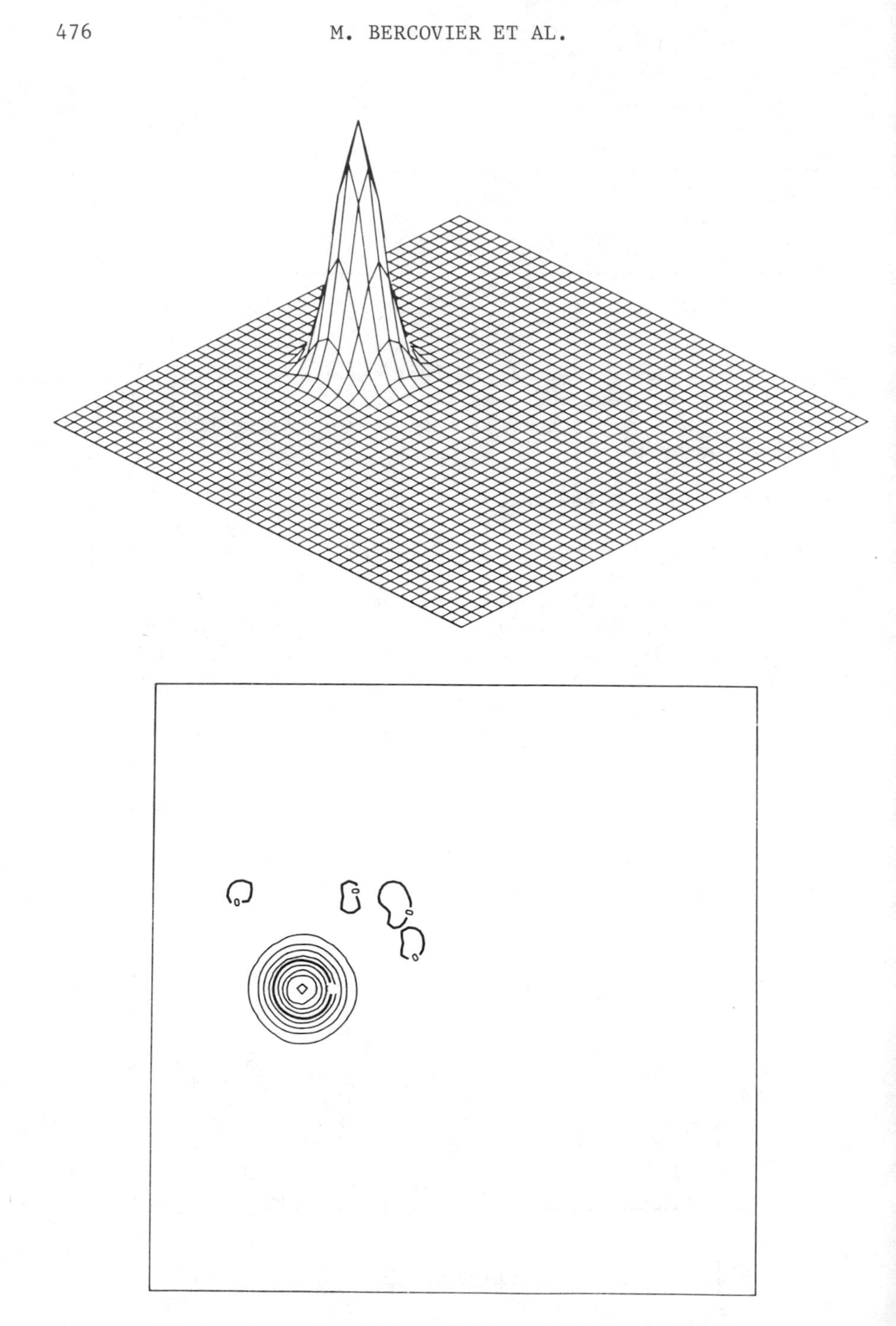

FIG.2. Gaussian-hill - Full revolution. 32 Time steps.

Next we consider a uniformly rotating flow in a square with a 40 X 40 mesh, (Q_1 elements), and $\nu = 0$.
We used a 4 X 4 Gauss Legendre numerical integration. Figure 2 shows the level curves and the gaussian peak after one revolution in 32 uniform steps.
We cannot discern in the 3D view any numerical instability. Very small disturbances appear behind the peak as can be seen on the level curves, (negative values up to 1 %). Nevertheless the scheme is diffusive as shown by
Table 2 where the computed peak is given after 1 to 4 revolutions.

nb. of revolutions	1	2	3	4
peak value	95.07	91.06	87.62	84.53

- Table 2 - (One revolution in 32 steps)

Finally one must emphasize that there is no phase error in all our computations.

CONCLUSION

To summarize these transport-projection algorithms have the following features :

. They are L^∞-stable and convergent

. They require at each time step the solution of a symetric positive definite system only

. They have a very small phase error and no numerical instabilities or wiggler.

. Their main draw-back however is that they are not conservative if a numerical quadrature is performed.

. Of course they are diffusive within their spacial order of convergence.

ACKNOWLEDGEMENT

The numerical tests shown in the paper have been obtained by Y. HASBANI and E. LIVNE.

REFERENCES

1. M. TABATA. A finite element approximation corresponding to the upwinding differencing. Memoirs of Numerical Mathematics, vol. 1, (1977) p. 47-63.

2. O. ZIENCKIEWICZ, J. HEINRICH. The finite element method and convection problem in fluid mechanics. Finite Elements in

Fluids (vol. 3), Gallagher ed. Wiley (1978).

3. A. BROOKS, T.J. HUGHES. Steeam line upwind Petrov Galerkin upwinding. Proceedings third Conf. on Finite Elements in flow problems. Banff Alberta, June 1980.

4. C. BARDOS, M. BERCOVIER, O. PIRONNEAU. The Vortex Method with Finite Elements, Rapport de Recherche INRIA n° 15, (1980).

5. J.P. BENQUE, B. IBLER, A. KERAMSI, G. LABADIE. A Finite Element Method for Navier-Stokes Equations. Proceedings of the third International Conference on Finite Elements in flow problems, Banff Alberta, Canada, 10-13 June, 1980.

6. O. PIRONNEAU. On the Transport-diffusion algorithm and its applications to the Navier-Stokes equations (to appear).

7. J. DOUGLAS, T.F. RUSSEL. Numerical Methods for Convection Dominated Diffusion Problems Based on Combining the Method of Characteristics with Finite Element or Finite Difference Procedures, to appear.

8. O.PIRONNEAU, P.L. RAVIART, V.M. SASTRY. Characteristics and finite elements for the convection diffusion equation.

9. M. BERCOVIER, Y. HASBANI, E. LIVNE : (to appear).

10. M.J. FRITTS, J.P. BORIS. The Lagrangian solution of transient problems in hydronamics using a triangular mesh. J. Comp. Ph., 31, 172-215 (1979)

11. M. CROUZEIX, P.A. RAVIART. Conforming and non-conforming finite element methods for solving the stationary Stokes equations, RAIRO, R-3, 33-76 (1973)

12. F. THOMASSET. Finite Element Methods for Navier-Stokes Equations, VKI Lecture Series, March 25-29, 1980

13. M. BERCOVIER, O. PIRONNEAU. Error Estimates for Finite Element Method Solution of the Stokes Problem in the Primitive variables, Numerische Mathematik, 33, 211-224 (1979)

14. P. LESAINT, P.A. RAVIART. Résolution Numérique de l'Equation de Continuité par une Méthode du Type Elements Finis. Proc. Conference on Finite Elements in Rennes (1976).

15. J.P. BORIS, D.L. BOOK. Flux corrected transport. SHASTA J. Comput. Phys. II, 36-69 (1973).

QUASI-NEWTON METHODS IN FLUID DYNAMICS

M. S. Engelman

CIRES, University of Colorado

1. INTRODUCTION

Over the last few years, the finite element method has found increased use and wider acceptance for the solution of the equations governing the flow of viscous incompressible fluids, including the effects of heat transfer. Inevitably, the analysis requires the solution of a discrete system of non-linear equations, and, particularly for large systems, this represents the most time-consuming stage of the analysis. In steady state simulations a system of nonlinear equations need only be solved once, while for transient flow simulations which use an implicit time integration scheme, such a system must be solved at each time step. Since the reliability and cost-effectiveness of the algorithm used for the solution phase can ultimately be governing factors for the size of the finite element system that can be handled, there is interest in the development of improved procedures for the solution of the non-linear equations.

To date, the methods of "successive substitution" and Newton-Raphson seem to have been the most widely used in finite element programs for the analysis of fluid response. Over the past few years a new group of algorithms has emerged for the solution of nonlinear equations. These are known as quasi-Newton, variable metric, matrix update or modification methods. They have been used extensively in the field of optimisation, and were first applied in finite element analysis by Matthies and Strang [10] in collaboration with Bathe and Cimento [1]. In a previous paper [8], we have presented a comparative study of the successive substitution, Newton-Raphson and quasi-Newton methods for stationary Newtonian incompressible flows; here we extend these results to include non-Newtonian, temperature dependent and transient fluid flow simulations. We begin with a description of the solution strategies that we studied followed by a discussion of the implementation of quasi-Newton methods in a finite element framework. We then present our numerical experiments and discuss their implications for the finite element analysis of fluids.

2. SOLUTION TECHNIQUES

The solution techniques and their implementation are quite general. We consider these techniques as applied to the time dependent flow of a Newtonian (or non-Newtonian) viscous incompressible fluid including the effects of heat transfer (Boussinesq approximation). The discretised form of these equations is obtained by employing a Galerkin finite element procedure together with a penalty function formulation. Space limitations prevent us from going into more detail regarding the above formulation; complete details can be found in [2], [7].

However, no matter what type of flow is considered, at some stage of the analysis a set of global discretised equations of motion for the fluid of the form

$$K(\underline{u})\underline{u} = \underline{F} \ . \tag{2.1}$$

must be solved. The vector $\underline{u}$ represents the unknown nodal degrees of freedom and $\underline{F}$ includes prescribed body forces, surface tractions, heat fluxes, etc. The coefficient matrix $K(\underline{u})$ is a sparse banded matrix whose composition depends to some extent on the problem being considered. Since the nonlinear character of (2.1) is usually dominant, the choice of algorithm to solve (2.1), and its convergence is a key issue. Two of the most common choices are:

(i) Successive substitution (Picard iteration), which is a fixed point iteration of the form

$$K(\underline{u}_i)\underline{u}_{i+1} = \underline{F} \tag{2.2}$$

The nonlinearity is evaluated at the known iterate $\underline{u}_i$, and a linear system is solved at each step

(ii) Newton-Raphson, which may be described as follows:

$$\underline{u}_{i+1} = \underline{u}_i - J(\underline{u}_i)^{-1} R(\underline{u}_i) \tag{2.3}$$

where $R(\underline{u}) = K(\underline{u})\underline{u} - \underline{F}$ is the residual force vector and $J(\underline{u})$ is the Jacobian of the system of equations. In practice, this iteration is organised as

$$J(\underline{u}_i)\,\Delta\underline{u}_i = -\,R(\underline{u}_i)\ ,\quad \underline{u}_{i+1} = \underline{u}_i + \Delta\underline{u}_i$$

Both methods converge for a fair range of problems. Sucessive substitution is a linearly convergent algorithm, whereas Newton-Raphson is quadratic, but the latter may have a much smaller radius of convergence. A major disadvantage of both methods is that a complete factorisation and solution of an unsymmetric system of equations is required at each iteration.

The Newton-Raphson method belongs to a more general class of iterative procedures for solving the n equations $f(\underline{u}) = 0$:

$$\underline{u}_{i+1} = \underline{u}_i - s_i H_i \underline{f}_i\ , \tag{2.4}$$

where $\underline{f}_i = f(\underline{u}_i)$, H_i is an nxn matrix determined by the particular method employed and s_i is a scaling (or acceleration) factor which may be introduced to reduce $\underline{f}_{i+1}$. In this case an additional iterative procedure (line search) is necessary to determine s_i.

In our context, $f(\underline{u}) = R(\underline{u}) = K(\underline{u})\underline{u} - \underline{F}$ so that (2.4) becomes

$$\underline{u}_{i+1} = \underline{u}_i - s_i H_i R(\underline{u}_i)\ . \tag{2.5}$$

Setting $H_i = J(\underline{u}_i)^{-1}$ and $s_i = 1$ yields the Newton-Raphson method. If we set $H_i = J(\underline{u}_0)^{-1}$ and $s_i = 1$ we obtain the modified Newton method. In this algorithm the need to refactor at each iteration is obviated. However, the savings in operations is at the expense of slower convergence, or, in many cases, divergence.

Finally we come to a sub-class of the algorithms described by (2.4), known as quasi-Newton methods. They derive from the suggestion of Davidon [5] to update H_i in a simple manner after each iteration rather than recompute it entirely (Newton-Raphson) or leave it unchanged (modified Newton); see the survey by Dennis and More [6].

We write these algorithms in the form

$$\underline{u}_{i+1} = \underline{u}_i - s_i K_i^{-1} \underline{f}_i\ ;\quad K_{i+1} = K_i + \Delta K_i \tag{2.6}$$

and require that the K_i satisfy the "quasi-Newton equation" or "secant condition":

$$K_i(\underline{u}_i - \underline{u}_{i-1}) = \underline{f}_i - \underline{f}_{i-1} \ . \qquad (2.7)$$

Introducing the notation $\underline{\delta}_i = \underline{u}_i - \underline{u}_{i-1}$, $\underline{y}_i = \underline{f}_i - \underline{f}_{i-1}$ and $d_i = K_i^{-1} f_i$ we can write (2.6) and (2.7) as

$$\begin{aligned} \underline{u}_{i+1} &= \underline{u}_i - s_i \underline{d}_i \\ K_{i+1} &= K_i + \Delta K_i \\ K_i \, \underline{\delta}_i &= \underline{y}_i \ . \end{aligned} \qquad (2.8)$$

$\underline{d}_i$ is the search direction; for the Newton-Raphson method the search direction is $\underline{d}_i = J(\underline{u}_i)^{-1} R(\underline{u}_i)$ and for modified Newton it is $\underline{d}_i = J(\underline{u}_0)^{-1} R(\underline{u}_i)$. The name quasi-Newton arises from the fact that the K_i can be thought of as approximations to the Jacobian at $\underline{u} = \underline{u}_i$. If the rank of ΔK is r, then the scheme in (2.8) is called a direct update of rank r; in practice the update is either in rank one ($K_i = \underline{w}_i \underline{z}_i^T$ for some vectors $\underline{w}_i$ and $\underline{z}_i$) or rank two. Although (2.8) still requires the solution of n equations at each iteration, there are a number of ways to avoid refactorisation and a new solution. In fact, the update can be expressed directly as a correction to the inverse:

$$K_{i+1}^{-1} = K_i^{-1} + \Delta K^{-1}; \quad \text{rank} \quad \Delta K_i^{-1} = \text{rank} \quad \Delta K_i \ . \qquad (2.9)$$

For a complete discussion of the choice and the calculation of direct and inverse updates we refer the reader to Dennis and More [6]. For convenience, we shall use only inverse update forms. A natural update for nonsymmetric coefficient matrices is that of Broyden [4], which was one of the first quasi-Newton formulas to be proposed. Broyden's update in inverse form is

$$K_i^{-1} = K_{i-1}^{-1} + \frac{(\underline{\delta}_i - K_{i-1}^{-1} \underline{y}_i) \, \underline{\delta}_i^T}{\underline{\delta}_i^T K_{i-1}^{-1} \underline{y}_i} K_{i-1}^{-1} \ . \qquad (2.10)$$

3. IMPLEMENTATION

A number of difficulties would be encountered with the standard quasi-Newton implementation in a finite element framework. If problems of realistic size are to be analysed, the Jacobian cannot be retained entirely in core and must be stored in blocks on low-speed storage in a reduced storage mode. The update in (2.10) becomes difficult, and even if performed out of core it could destroy the sparseness pattern and the storage mode of the original matrix.

The basic idea which enables us to overcome these difficulties is to calculate at each iterative cycle the updated K_i^{-1} from the original K_0^{-1}. This is less cumbersome than it appears; in fact, it leads to a comparatively simple and concise implementation.

The algorithm begins with the choice of K_0. This may be the initial Jacobian $J(\underline{u}_0)$, where u_0 is an initial approximation -- generally chosen to be the solution of the associated Stokes problem. The LU factorisation of K_0 is computed and stored in the process of solving $K_0\underline{d}_0 = R(\underline{u}_0)$, which also gives the initial search direction $\underline{d}_0$. We then estimate, with a line search if necessary, the acceleration factor s_0. The new iterate is $\underline{u}_1 = \underline{u}_0 - s_0\underline{d}_0$. In general, let us assume that we have just computed $\underline{u}_i$ and $R(\underline{u}_i)$ using $\underline{u}_{i-1}$, s_{i-1}, $\underline{d}_{i-1}$. The algorithm for the new search direction d_i is as follows:

Given $\underline{u}_i$, $\underline{u}_{i-1}$, s_{i-1}, $\underline{d}_{i-1}$ and $R(\underline{u}_i)$,

a) Back substitute to form $\underline{q}_1 = K_0^{-1}R(\underline{u}_i)$

b) For $j = 1,\ldots, i-1$

Retrieve p_j, $\underline{\delta}_j$ and $\underline{r}_j$

Compute $\underline{q}_{j+1} = \underline{q}_j + p_j(\underline{\delta}_j - \underline{r}_j)\underline{\delta}_j^T\underline{q}_j$

c) Form and store

$$\underline{r}_i = \underline{q}_i - \underline{d}_{i-1}$$

$$\underline{\delta}_i = \underline{u}_i - \underline{u}_{i-1} = - s_{i-1}\,\underline{d}_{i-1}$$

$$p_i = \frac{1}{\underline{\delta}_i^T\underline{r}_i}$$

d) Form $\underline{d}_i = \underline{q}_i + p_i(\underline{\delta}_i - \underline{r}_i)\underline{\delta}_i^T\underline{q}_i$.

This algorithm requires the calculation and storage of two vectors of dimension n for each iteration. Most quasi-Newton programs would actually carry out the updating at each step -- not on the K_i which are never explicitly calculated, but on the L_i and U_i of the Gauss factorisation $K_i = L_i U_i$. In this case there is no need to save the vectors $\underline{u}_i$ and r_i after the update, but a very serious drawback is that the new factors L_i and U_i have to be computed and their original sparsity can be lost. Therefore, we prefer to keep and reintroduce the updating vectors, up to a limit imposed by the user. We can estimate the number of operations when m is the mean bandwidth of K_0 (in practice we use a skyline storage scheme with variable bandwidth). The operation count to compute Broyden's new search direction d_N, at iteration N, is $(2m + 3N + 2)n$. In practice m greatly dominates N, since we set an upper limit on the number of factors stored, say 5 or 10. When this upper limit is reached we have one of three options:

a) to update the Jacobian matrix and restart the algorithm with $\underline{u}_0 = \underline{u}_i$; or

b) to shift the updating vectors one position downwards (thus losing the first pair) and continue; or

c) to discard the updating vectors and continue the updating algorithm, retaining the original Jacobian K_0.

Experience has indicated (cf. [8]) that the first option results in far more rapid convergence than option b) or c), provided the Jacobian is updated every 5 to 10 iterations.

4. NUMERICAL EXPERIMENTS

We present here the results of three numerical experiments. Each simulation was performed using successive substitution, Newton-Raphson and Broyden's update (reforming the Jacobian each 5 iterations) as the solution algorithm. An important aspect of the iterative solution is the choice of an appropriate convergence criterion. This choice is more difficult when line searches are part of the algorithm. The standard test

$$\frac{||\underline{u}_i - \underline{u}_{i-1}||}{||\underline{u}_i||} \leq \varepsilon_u , \qquad ||\underline{u}_i|| = \text{Euclidean norm}$$

is no longer sufficient since $||\underline{u}_i - \underline{u}_{i-1}||$ may be small only because s_i is small. This criterion is supplemented by another on the residual forces

$$\frac{||R(\underline{u}_i)||}{||R_0||} \leq \varepsilon_r$$

where R_0 is some reference residual vector, e.g. $R(\underline{u}_0)$. In the following examples we used $\varepsilon_u = \varepsilon_r = 10^{-5}$. This small tolerance was chosen so as to differentiate more clearly between the different solution algorithms.

The first simulation was of the transient flow of a lid driven cavity with Reynolds number 250, on an 8x8 mesh of nine node quadrilaterals. A variable step backward Euler time integration scheme with error control was employed. Table 1 shows the relative execution times and number of iterations for each solution algorithm.

TABLE 1

Transient Lid Driven Cavity (21 time steps)

Method	No. of iterations	CP time (Cyber 172)
Succ. Substitution	88	3343 sec.s
Newton-Raphson	49	1839 sec.s
Quasi-Newton	43	1334 sec.s

The second set of simulations was of stationary 'Stokes' flow of a non-Newtonian fluid within a channel driven by a pressure gradient. Both Power Law and Bingham constitutive relations were used; for a complete description of this flow and the constitutive relations see [9],[3]. Table 2 presents the results of 3 simulations using different initial solution vectors. A single column of 5 nine node elements was used to model the domain. The sensitivity of the Newton-based methods to the initial solution vector is well illustrated by these results.

TABLE 2a

Power Law Flow $(\partial p/\partial x = 7.5)$

Method	Initial solution vector $\mathbf{u}_0$						
	Power Law n=.5			Power Law n=.25			
	Stokes	SS_1	SS_2	Stokes	SS_4	SS_7	SS_8
Succ. Subst.	11	10	9	26	22	19	18
Newton- -Raphson	18	4	3	rapid div.	slow div.	no conv.	4
Quasi-Newton	7(41)	6(31)	5(0)	18(22)	14(13)	10(3)	4(2)

() = number of line searches; SS_i = i iterations of succ. substitution

TABLE 2b
Bingham Fluid Flow ($\partial p/\partial x = 7.5$)

Method	Initial solution vector $\underline{u}_0$			
	Stokes	SS_3	SS_5	SS_7
Succ. Subst. Newton-Raphson Quasi-Newton	22 oscillation divergence	19 about 24 (102)	17 exact 18 (41)	15 solution 8 (26)

Finally in order to demonstrate the cost-effectiveness of quasi-Newton methods for large problems a three dimensional temperature driven cavity problem was run. A mesh of 6x6x6 27 node bricks was used to model a unit box with u=0 on the entire boundary, T=0 on the x=0 face T=1 on the x=1 face with the remaining faces being thermally insulated. The Rayleigh number of the flow was 2500. and the Prandtl number 1000. The results are presented in Table 3.

TABLE 3
3-D Temperature Driven Cavity

Method	Iterations	CP time (CRAY-1)
Succ. Subst.	8	1472 sec.s
Newton-Raphson	4	832 sec.s
Quasi-Newton	5	375 sec.s

1493 mean bandwidth
2197 nodes
216 elements
5852 equations

5. CONCLUSION

We believe that we have demonstrated that the use of a quasi-Newton algorithm for the solution of the discrete non-linear matrix equations arising in finite element analyses can result in substantial savings in computer time and resources. Our implementation of Broyden's update can be introduced with relative ease into existing computer programs and can constitute a very effective solution procedure by itself or in combination with other basic techniques. An area of further research is the development of techniques which would automatically determine a solution strategy.

ACKNOWLEDGEMENTS

The author wishes to thank Professor G. Strang for many fruitful discussions on quasi-Newton methods. This research was supported in part by the Army Research Office (DAAG2P-80-K0033).

REFERENCES

1. BATHE, K.-J. and CIMENTO, A., Some practical procedures for the solution of nonlinear finite element equations, *Comp. Meth. in Applied Mech. Eng.*, 22,59,(1979).

2. BERCOVIER ,M. and ENGELMAN, M.S., A finite element for the numerical solution of viscous incompressible flows, *J. of Comp. Physics*. 30, 181 (1979).

3. BERCOVIER, M.and ENGELMAN, M.S., A finite element method for non-Newtonian flows, to appear *J. Comp. Phys.*, (1981).

4. BROYDEN, C.G., A class of methods for solving nonlinear simultaneous equations, *Math. Comp.*, 19, 577 (1965).

5. DAVIDON, W.C., Variable metric method for minimisation, Rep. ANL-5990, Rev. 1959, Argonne National Laboratory, Argonne, IL.

6. DENNIS, J.E., and MORE, J., Quasi-Newton methods, motivation and theory, *SIAM Rev.* , 19, 46 (1977).

7. ENGELMAN, M.S., *FIDAP User and Theoretical Manuals*, 1981.

8. ENGELMAN, M.S., STRANG, G. and BATHE, K.-J., An application of Quasi-Newton methods in fluid mechanics, to appear *Int. J. Num. Meth. Eng.*, (1981).

9. LYNESS, J.F., OWEN, D.R. and ZIENKIEWICZ, O.C., Finite Element analysis of the steady flow of non-Newtonian fluids through parallel sided conditions, *Finite Element Methods in Flow Problems*, Ed. J.T. Oden et al., UAH Press, Alabama (1974).

10. MATTHIES, H., and STRANG, G., The solution of nonlinear finite element equations, *Int. J. Num. Meth. Eng.*, 14, 1613 (1979).

THE APPLICATION OF THE FINITE ELEMENT METHOD TO PROBLEMS RELEVANT TO THE GLASS MAKING PROCESS

*+C.A. Gare, *J. Ashworth, *J.C. Carling,
+D.M. Burley, *H. Rawson and +G.T. Vickers.

**Department of Ceramics, Glasses and Polymers*
+Department of Applied and Computational Mathematics
University of Sheffield

1. INTRODUCTION

In the industrial production of glass articles there are many areas which are amenable to study by numerical methods. One example is the melting furnace, a rectangular cavity in which molten glass is driven by the combined effect of buoyancy and throughput. In addition, the viscosity and thermal conductivity vary considerably with temperature, and convection can be an important mode of heat transfer. Thus the equations governing the flow and the temperature are strongly coupled and furthermore little work of practical value can be done without three-dimensional models. The application of finite difference methods to these coupled, three-dimensional problems is well established (for example, Carling et al [3], Suzuki [7] and Carling [2]) and perhaps, without wishing to be provocative, finite difference methods are better suited to furnace modelling. Partly for this reason, partly because the other topics considered here are more interesting and pertinent to the finite element approach, but mainly because of space limitations, it has been decided to omit consideration of buoyancy-driven flow.

Another problem of relevance to the glass making process concerns the heat transfer in a glass container mould in which the temperature on the inside surface of the mould changes with time, as successive containers are shaped and passed through the moulding machine. Capurso et al [1] and Petropoulos et al [6] considered the transient problem in one-dimension using finite difference techniques, but their finite element work was restricted to steady state behaviour only. In section 3 the two-dimensional and time-dependent problem is solved in which the inside surface temperature is a specified cyclic function of time.

The third area of interest, and perhaps the one most suited to solution by finite elements, concerns glass fabrication processes in which the glass forms a free surface. These situations occur frequently in the glass industry. For example, at the start of the float glass process, glass flows over a weir into a bath of molten tin; in a fibre, tube or sheet manufacturing process, glass is drawn either upwards from the surface or downwards through an orifice; in making containers a small mass of glass (graphically described as a gob) is formed and falls under gravity into the container mould for pressing. Previous papers on the application of finite element methods to viscous free surface flows include those of Nickell et al [4] and Orr and Scriven [5]. Section 4 deals with two viscous flows in which there are free surfaces. The first considers gravity-driven flow down a slope consisting of two planar sections inclined at different angles. This problem is an idealised version of the flow at the start of the float glass process. The other is the axisymmetric equivalent of this plane flow: the flow down a cone in which the angle of the conical surface changes. The industrial significance in this case is that, if the conical body is inverted and the direction of flow reversed, the problem becomes an idealised tube drawing process. In addition, free surface flow down a cone has a non-trivial analytic solution, which relies on a lubrication approximation, and this can be compared to corresponding finite element predictions.

2. FINITE ELEMENT METHOD

There are certain features of the application of the finite element method which are common to all the problems considered here and these features are described in this section. Other particular features are covered in the appropriate sections.

The primitive variable version of the two-dimensional Navier-Stokes equations, the incompressible continuity equation and the energy equation form the basis of the numerical models. In general, viscosity and thermal conductivity are temperature dependent and buoyancy is modelled by the Boussinesq approximation.

Straight-sided triangular elements have been used throughout the work and mesh generation has been performed manually. Quadratic trial functions (N_i) have been chosen for velocities and temperature, in conjunction with nodal values at each vertex and mid-side. Linear trial functions (L_i) have been used for pressure, viscosity and thermal conductivity.

Weighted residual techniques have been applied to the non-dimensional form of the governing equations. In the case of the velocity and temperature equations, a weighting function N_i has been chosen, whilst for the continuity equation the function L_i has been used. After substituting the approximate solutions for the field variables into the governing equations, the familiar set of integral equations connecting trial functions and nodal values can be derived. These equations contain second order terms which have been reduced to first order by application of Green's theorem. By integrating analytically over each element, the equations are converted to a final system matrix. It is necessary to iterate on these equations, because of the presence of either a buoyancy term, convection terms, variable properties or a free surface whose shape has to be determined as part of the solution. At each stage, the matrix equation in the nodal values of the field variables is inverted using a block substitution direct method applicable to sparse matrices. The NAG library routine FO4AXF has been used.

3. TIME-DEPENDENT HEAT TRANSFER

A two-dimensional cross-section of a glass container mould, external radius a, is shown in figure 1a. The mould consists of

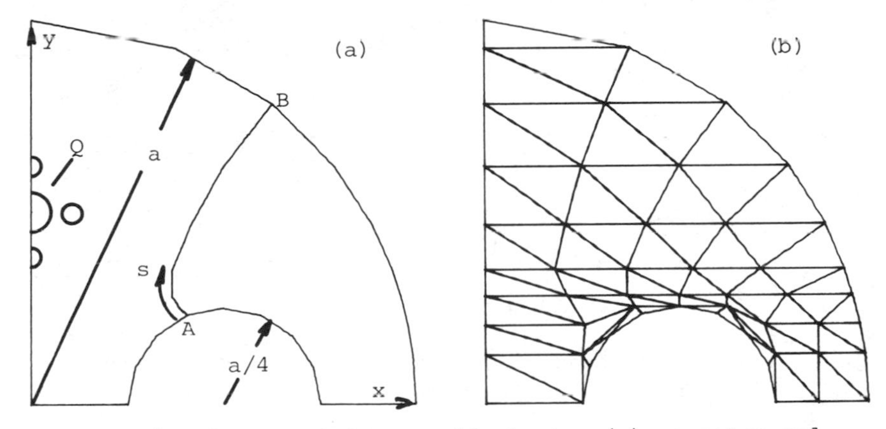

FIG. 1. The glass container mould showing (a) geometry and (b) mesh

two cylindrical cavities, radius a/4, placed symmetrically about the y axis. There are, therefore, two axes of symmetry and the computation is performed in the first quadrant only. There are internal passages which allow cooling fluid to circulate and the resulting heat loss from the mould is represented by a heat sink. The equation governing the dimensionless temperature, θ, is

$$\frac{\partial\theta}{\partial t} = \frac{\partial}{\partial x}\left(\bar{k}\,\frac{\partial\theta}{\partial x}\right) + \frac{\partial}{\partial y}\left(\bar{k}\,\frac{\partial\theta}{\partial y}\right) - Q, \tag{3.1}$$

where the dimensionless thermal conductivity of the metal mould, $\bar{k}$, varies with temperature and material composition, and the sink term, Q, varies with position,

$$Q = Q_o \exp(-\,8r/a),$$

where r is the distance measured radially from point $y = a/2$. Q is effectively zero for $r \gtrsim 0.35a$.

On the external boundary of the mould a heat loss boundary condition is imposed:

$$-\bar{k}\ \partial\theta/\partial n = h(\theta - \theta_o),$$

where n is measured in the direction of the outward normal, h is a heat transfer coefficient and θ_o is the external ambient temperature. On the internal boundary the temperature, θ_A, is specified as a cyclic function of time. A typical variation of T_A is shown in figure 2b.

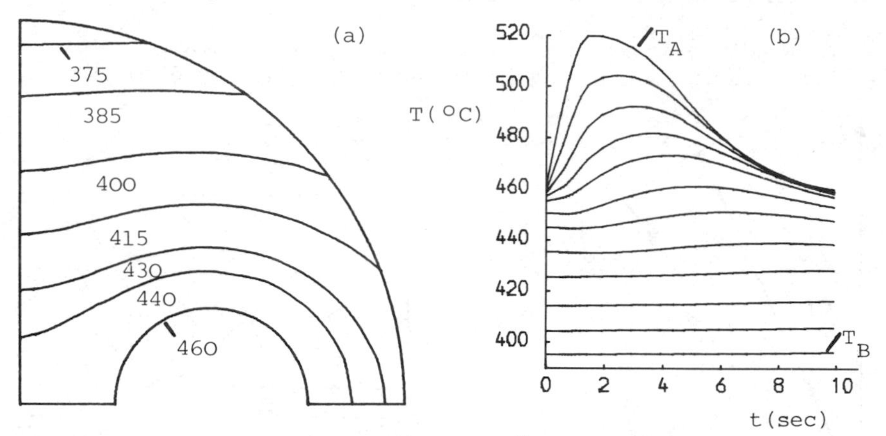

FIG. 2. (a) Steady state isotherms (°C). (b) Temperature against time at various points on line s.

The right hand side of equation 3.1 is represented by finite elements, whereas the time derivative on the left hand side is treated as a finite difference. The computation is advanced by incorporating the difference terms in the finite element formulation as follows:

$$f(\theta^*) - \theta^*/\Delta t + \theta/\Delta t = 0,$$

where f is the right hand side of equation 3.1, θ^* is the unknown temperature field at time t + Δt, and θ is the known temperature field at time t. Thus the method is fully implicit. The value of Δt varies according to the cyclic variation of T_A. Δt is set equal to the value which corresponds to a change in the internal boundary temperature of 5°C. The steady-state temperature distribution is used for starting the time-dependent calculation. Steady-state isotherms are shown in figure 2a. After about 3 cycles a repeated cyclic pattern emerges and figure 2b shows temperature against time at various positions along the line s. Figure 3 shows temperature against distance along s at various times in the cycle. It can be seen that, with this mould, the depth of influence of the cyclic boundary variation is approx-

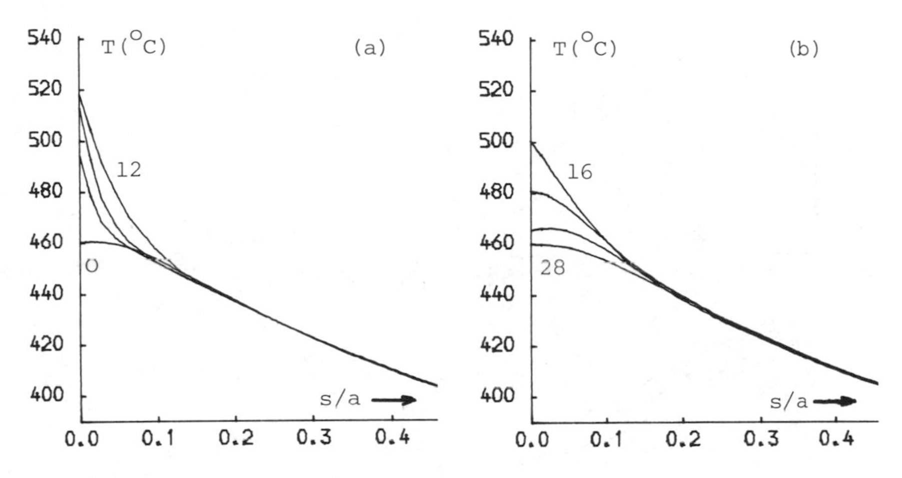

FIG. 3. Temperature against distance along s at time steps in the cycle (a) 0,4,8,12 (b) 16,20,24,28.

imately a/5. Mould design can now be investigated by studying the effect on the temperature distribution of variations in mould parameters.

4. FREE SURFACE FLOW

4.1 *Flow over two intersecting planes*

Figure 4a shows the geometry of the problem being considered. The non-slip base is formed by two planes inclined at angles α and β as shown. Lengths are made dimensionless with h_o and velocities with u_o. At large distances upstream and downstream it is possible to apply the well known analytic solution for gravity induced flow down a plane. This solution gives a parabolic velocity profile and the relationship between u_α and h_α of

$$h_\alpha^2/u_\alpha = 3 \operatorname{cosec} \alpha \, Fr/Re \tag{4.1}$$

where $Re = u_o h_o/\nu$ is the Reynolds number and $Fr = u_o^2/(gh_o)$ is the Froude number. Having freely chosen either h_α or u_α, equation 4.1 enables the other to be determined. This height, h_α, and the parabolic profile corresponding to u_α are specified as inlet conditions. The exit height, h_β, of the free surface is given asymptotically by

$$h_\beta = (3h_\alpha u_\alpha \operatorname{cosec} \beta \, Fr/Re)^{1/3}, \tag{4.2}$$

and u_β follows from continuity. These have to be determined as part of the calculation and are *not* imposed as boundary conditions.

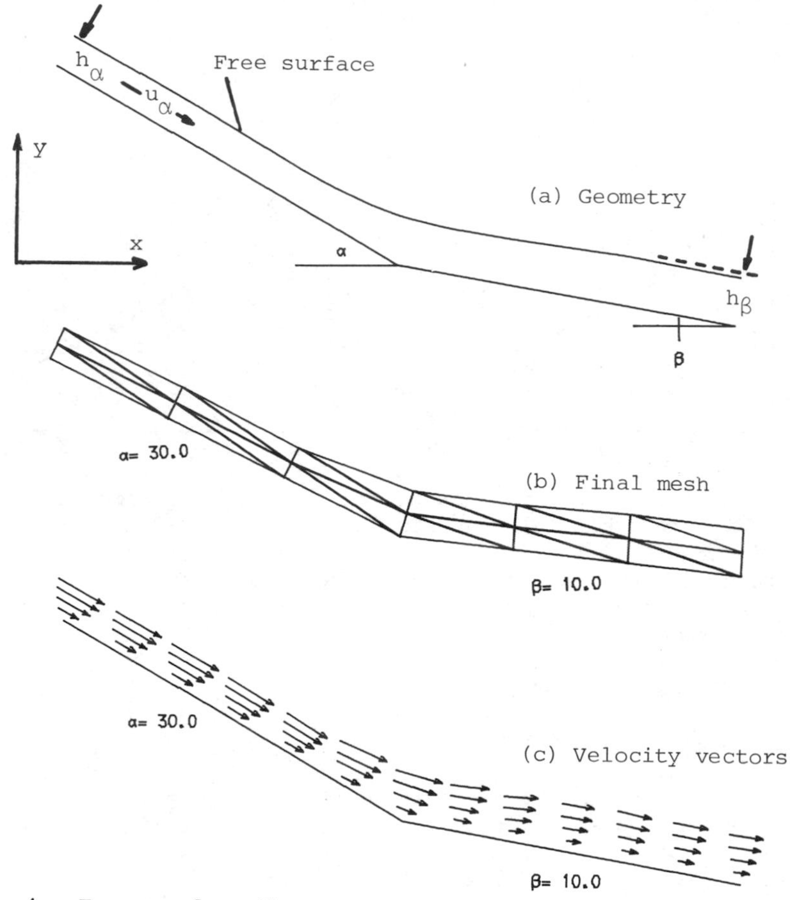

FIG. 4. Free surface flow over two planes (Re = 0.22; Fr = 0.036)

It is assumed that the flow is isothermal, incompressible and slow, and accordingly the governing equations are

$$\frac{\partial p}{\partial x} - \frac{\partial^2 u}{\partial x^2} - \frac{\partial^2 u}{\partial y^2} = 0 \,,$$

$$\frac{\partial p}{\partial y} - \frac{\partial^2 v}{\partial x^2} - \frac{\partial^2 v}{\partial y^2} + \frac{Re}{Fr} = 0 \,, \qquad (4.3)$$

$$\frac{\partial u}{\partial x} + \frac{\partial v}{\partial y} = 0.$$

These equations are cast in finite element form and solved according to the method outlined in section 2.

The free surface is determined by applying the three conditions of zero tangential stress, constant normal stress and the condition of zero velocity normal to the free boundary. The initial shape of the free surface is assumed to lie parallel to the base planes. The velocities and pressures are then computed using the two stress conditions, and, at this stage, the predicted velocity normal to the free surface may not be zero. The position of the free surface is then adjusted by moving each boundary node, in the y direction, until each segment of the boundary becomes parallel to the local mean surface velocity vector. The position of the intermediate nodes between free surface and base are scaled appropriately. This procedure is then repeated until the position of the free surface converges. Convergence occurs rapidly in, typically, six iterations.

At the downstream boundary it is assumed that the velocity gradients in the streamwise direction are zero. These conditions allow the exit height and velocities to be calculated rather than artificially imposed.

As an example of these computations, figures 4b, c show the final form of the finite element mesh and the computed velocity vectors. At the downstream end, the height of the free surface is within 1 % of its asymptotic value.

4.2 *Flow over the surface of a cone*

This problem is the axisymmetric equivalent of the planar flow of section 4.1. The geometry is similar to that in figure 4a provided that x is replaced by r and y by z. One reason for using this axisymmetric flow to test the finite element program is that there exists an analytic solution, based on a lubrication approximation, which gives velocities and predicts a free boundary shape of a non-trivial form. At large distances upstream or downstream of the point at which the cone surface changes angle, there is an asymptotic free surface performing a function similar to the heights h_α or h_β in the planar flow. The difference in the case of the cone flow is that the height of the asymptotic surface varies with distance along the surface, thus providing a more stringent check on the finite element predictions.

The analytic solution for the asymptotic surface can be derived by considering a cone of base angle α and setting up a co-ordinate system s along the cone surface and n perpendicular. After performing a co-ordinate transformation from the r,z system to the s,n system it can be shown that the flow equations become

$$\frac{\partial p}{\partial s} = \frac{Re}{Fr}\sin\alpha + \frac{1}{r}\frac{\partial\omega}{\partial n} ,$$

$$\frac{\partial p}{\partial n} = -\frac{Re}{Fr}\cos\alpha - \frac{1}{r}\frac{\partial\omega}{\partial s} , \qquad (4.4)$$

continued overleaf

$$\frac{\partial}{\partial s}(r\, v_s) + \frac{\partial}{\partial n}(r\, v_n) = 0\,, \qquad \text{(4.4 cont.)}$$

where $r = s\cos\alpha + n\sin\alpha$, $\omega = r(\partial v_s/\partial n - \partial v_n/\partial s)$ and all quantities are dimensionless as before. If it is assumed that $\partial/\partial s << \partial/\partial n$, then equations 4.4 can be integrated directly and after applying the conditions $v_s = 0$ at $n = 0$ and $\partial v_s/\partial n = 0$ at $n = h$ the velocity becomes

$$v_s = \frac{Re\cos^2\alpha}{Fr\sin\alpha}\,\frac{s^2}{4}\,(2f^2\log(\bar{n}) - (\bar{n})^2 + 1)\,,$$

where $f = 1 + (h/s)\tan\alpha$ and $\bar{n} = 1 + (n/s)\tan\alpha$. Integrating $2\pi r v_s$ with respect to n from the cone surface ($n = 0$) to the free boundary ($n = h$) gives the volume flow, Q. The resulting expression connecting h and s is as follows,

$$\frac{8\,Fr\sin^2\alpha}{\pi\,Re\cos^4\alpha}\,Q = (4\,f^4\log f - 3\,f^4 + 4\,f^2 - 1)s^4. \qquad (4.5)$$

For small values of the parameter $(h/s)\tan\alpha$, the variation of h with s simplifies to

$$h = \left(\frac{3\,Fr\,Q}{2\,\pi\,Re\sin\alpha\cos\alpha}\right)^{1/3} s^{-1/3}\,, \qquad (4.6)$$

and also under these circumstances

$$v_s = \frac{Re}{2Fr}\sin\alpha\, h^2(2(n/h) - (n/h)^2).$$

The finite element solution of the cone flow is performed in the r,z polar co-ordinate system in much the same way as the plane flow of section 4.1. Clearly the equations differ in order to account for the axial symmetry, but otherwise the treatment, for example, of boundary conditions and the method of adjusting the free surface, remains similar. The difference in the finite element formulation is that a weighted residual function rN_i is used in place of N_i, similarly rL_i for L_i. In addition the equations must be integrated over a volume $2\pi r\,dr\,dz$. The use of the modified weighting functions enables solutions to be obtained for cases which include the axis of symmetry in the solution domain.

Figure 5a shows computed velocity vectors for $\alpha = 45^\circ$ and figure 5b compares computed boundary shapes with the analytical equivalent (equation 4.5). Also shown in the figure is the variation of h given by equation 4.6, which is applicable only at small values of $(h/s)\tan\alpha$. The two numerical predictions are for two values of overall length. In general, the agreement between analytic solution and numerical prediction is very good.

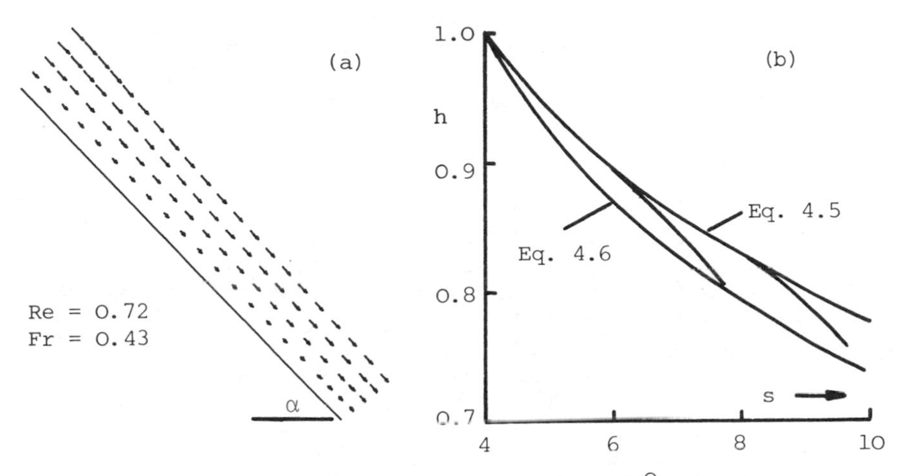

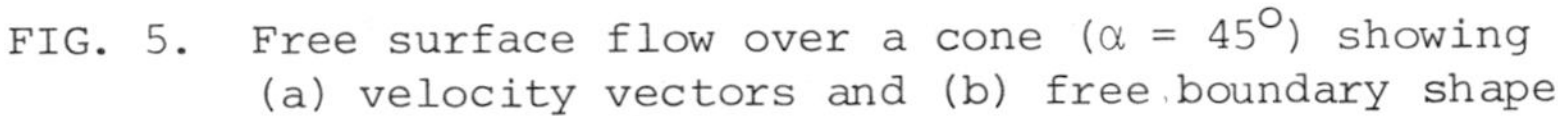

FIG. 5. Free surface flow over a cone ($\alpha = 45^{\circ}$) showing (a) velocity vectors and (b) free boundary shape

The discrepancy which occurs near the downstream boundary is caused by the application of conditions of zero streamwise derivatives on the exit plane. In this example such a condition is not strictly justified, but the error appears to be small and to move downstream as the computational region is extended.

Finally an example is given of flow down the outside of an inverted funnel. Figure 6 shows the finite element mesh, velocity vectors and the position of the free surface. The reader

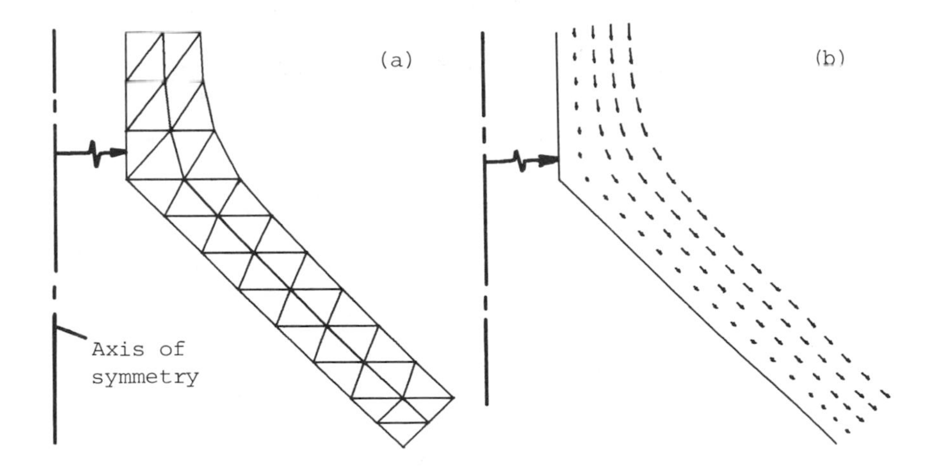

FIG. 6. Free surface flown down the outside of an inverted funnel showing (a) finite element mesh and (b) velocity vectors.

is reminded that this somewhat eccentric use of a funnel bears some relation to the manufacture of glass tubing. If the direction of flow is reversed and the annular region drawn upwards in a continuous manner, an idealised tube drawing process is obtained.

REFERENCES

1. CAPURSO, T., PETROPOULOS, J., MEUNIER, H. and HENRIETTE, J., Heat transfer through glass and mould during the glass forming process. Part 1 - Heat transfer at the glass-mould interface during periodic forming processes. *Glass International* 56, 48-59 (1979).
2. CARLING, J.C., Two and three-dimensional mathematical models of the flow and heat transfer in forehearths. *Glastech. Ber.* 49, 269-277 (1976).
3. CARLING, J.C., BURLEY, D.M., HENDERSON, W.D., RAWSON, H., McSHEEHY, C.J.M. and SUZUKI, J., Results of mathematical model studies of tank forehearths and three-dimensional convection-driven flow in rectangular cells. *Proc. XIth Int. Congress on Glass, Prague IV*, 1-10 (1977).
4. NICKELL, R.E., TANNER, R.I. and CASWELL, B. The solution of viscous incompressible jet and free surface flows using finite element methods. *J. Fluid Mech.* 65, 189-206 (1974).
5. ORR, F.M. and SCRIVEN, L.E., Rimming flow: numerical simulation of steady viscous free surface flow with surface tension. *J. Fluid Mech.* 84, 145-165 (1978).
6. PETROPOULOS, J., MEUNIER, H., HENRIETTE, J. and WELVAERT, M., Heat transfer through glass and mould during the glass forming process. Part 2 - Computation of temperature fields in glass forming moulds by finite element method. *Glass International* 57, 50-58 (1980).
7. SUZUKI, J., Three-dimensional flow and temperature distribution in rectangular cavities. *M.Sc. Thesis, Department of Ceramics, Glasses and Polymers, University of Sheffield.* (1976).

SURFACE REPRESENTATION IN COMPUTER-AIDED GEOMETRIC DESIGN

John A. Gregory

Brunel University

1. INTRODUCTION

A precursor to a finite element analysis of a curved surface such as an aircraft or car body, is that of the mathematical description of the surface itself. This topic is a subject of the theory of *computer-aided geometric design* or *computational geometry* and is related to finite element methods through its use of piecewise defined interpolation and approximation theory. In particular, a curved surface in *c.a.g.d.* is usually represented as a piecewise defined vector valued function $[x(u,v), y(u,v), z(u,v)]$, where the parametric variables (u,v) are defined over rectangular elements.

The use of vector valued, rather than scalar valued, functions gives the designer much greater flexibility in the representation of the curved surface. Also, the almost exclusive use of rectangular elements enables continuity conditions between those elements to be handled relatively easily, whilst simplifying the data structure required for the complete surface and simplifying routines such as those required for plots and cross-sections. Indeed, the special problems associated with vector valued surface representation seems to have precluded the use of triangular elements favoured by the finite element analyst and more research is needed in this area. There are, however, situations peculiar to vector valued representations, where a surface patch requiring a non-rectangular domain of definition, such as a triangle or pentagon, can occur within a rectangular patch framework. This paper, after reviewing the subject of rectangular patch representations, will consider a recent development in the representation of such surfaces.

A starting point for the development of curved surface representations is to consider a wire frame model as illustrated in the simple example of Fig. 1. Each portion of the surface bounded by four curved sides is to be defined by a rectangular patch, that is a vector valued function with a rectangular domain of definition. The example also shows how a triangular patch can occur within a rectangular patch framework. The next section gives a brief review of local rectangular patch representations.

For more details the interested reader should consult references [2],[6], and [7].

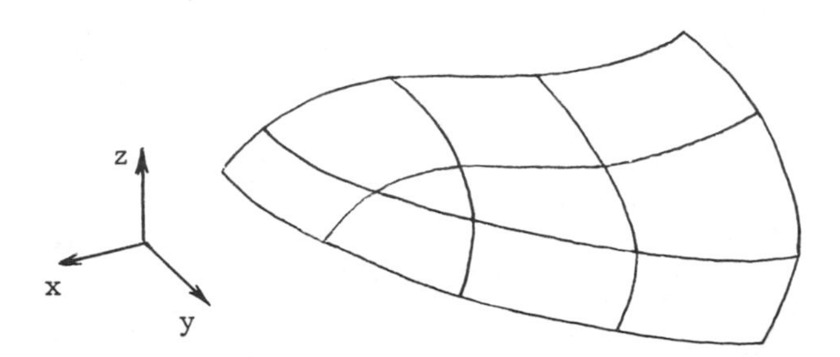

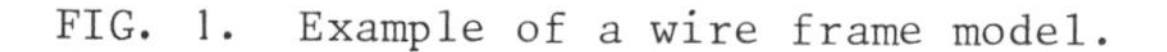
FIG. 1. Example of a wire frame model.

2. SURFACE REPRESENTATION OVER RECTANGLES

In briefly reviewing the subject of rectangular surface patch representations, the following structure can be discerned: Let univariate curves be defined by

$$\underline{r}(u) = \sum_{i=0}^{m} \alpha_i(u)\, \underline{a}_i \,, \quad u \in [a,b] \,, \tag{2.1}$$

$$\underline{r}(v) = \sum_{j=0}^{n} \beta_j(v)\, \underline{b}_j \,, \quad v \in [c,d] \,, \tag{2.2}$$

where $\underline{r} \equiv [x,y,z]$, $\underline{a}_i, \underline{b}_j \in R^3$, and the $\{\alpha_i\}$ and $\{\beta_j\}$ are polynomial basis functions of degree m and n respectively. Then, for $(u,v) \in [a,b]\times[c,d]$, three types of bivariate surfaces can be defined:

2.1. *The Lofted Surfaces*

$$\underline{r}_1(u,v) = \sum_{i=0}^{m} \alpha_i(u)\, \underline{a}_i(v) \,, \quad \underline{r}_2(u,v) = \sum_{j=0}^{n} \beta_j(v)\, \underline{b}_j(u) \,. \tag{2.3}$$

2.2. *The Tensor Product Surface*

$$\underline{t}(u,v) = \sum_{i=0}^{m}\sum_{j=0}^{n} \alpha_i(u)\beta_j(v)\, \underline{c}_{i,j} \,. \tag{2.4}$$

2.3. *The Boolean Sum Blended Surface*

$$\underline{p}(u,v) = \underline{r}_1(u,v) + \underline{r}_2(u,v) - \underline{t}(u,v) \tag{2.5}$$

The tensor product surface will be familiar to the finite element analyst, where the $\{\alpha_i\}$ and $\{\beta_j\}$ are defined by Hermite interpolation bases. The Boolean sum blended surface is a generalization introduced by Coons [4], its application being restricted to interpolatory bases. In this case the lofted surfaces can be written as

$$\underline{r}_1 = P_1[\underline{f}] \quad \text{and} \quad \underline{r}_2 = P_2[\underline{f}] , \tag{2.6}$$

P_1 denoting a linear operator which acts on a bivariate vector valued function $\underline{f}(u,v)$ considered as a function of u, and P_2 denoting an operator which acts on $\underline{f}(u,v)$ considered as a function of v. The tensor product and Boolean sum blended surfaces are then defined by

$$\underline{t} = P_1P_2[\underline{f}] , \quad \underline{p} = (P_1 \oplus P_2)[\underline{f}] \equiv (P_1 + P_2 - P_1P_2)[\underline{f}] . \tag{2.7}$$

Such operators have interesting approximation theoretic properties which have been studied by Gordon [8]. The Boolean sum blended form has been used in finite element theory, for example in the derivation of the mapping techniques of Gordon and Hall [9] and in the systematic derivation of serendipty type elements [5]. For the purposes of *c.a.g.d.*, the Boolean sum blended surface has the useful property of combining the interpolation properties of the two lofted surfaces, where it is assumed that $\underline{f}$ is such that P_1 and P_2 commute.

2.4. *Rational Forms*

Extra degrees of freedom can be introduced by replacing (2.1) and (2.2) with the rational forms

$$\underline{r}(u) = \frac{1}{\upsilon(u)} \sum_{i=0}^{m} \alpha_i(u)\, \upsilon_i \underline{a}_i , \quad \upsilon(u) = \sum_{i=0}^{m} \alpha_i(u)\, \upsilon_i , \tag{2.8}$$

$$\underline{r}(v) = \frac{1}{\omega(v)} \sum_{j=0}^{n} \beta_j(v)\, \omega_j \underline{b}_j , \quad \omega(v) = \sum_{j=0}^{n} \beta_j(v)\, \omega_j . \tag{2.9}$$

Appropriate modifications can then be made to the bivariate surface descriptions. Quadratic rational curves and surfaces are useful in that they can be used to represent conics and quadrics.

2.5. *Continuity Between Surface Patches*

Suppose $\underline{p}(u,v)$ and $\underline{q}(u,v)$ define two regular parametric representations on $[0,1]\times[0,1]$. Then the patches join with C^0 continuity of position if, for example,

$$\underline{p}(1,v) = \underline{q}(0,v) \ . \tag{2.10}$$

Conditions sufficient for C^1 (tangent plane) and C^2 (curvature) continuity across this common boundary are then given by

$$c_1\, \underline{p}_{1,0}(1,v) = c_2\, \underline{q}_{1,0}(0,v) \ , \quad c_1, c_2 > 0 \ , \tag{2.11}$$

$$c_1^{\,2}\, \underline{p}_{2,0}(1,v) + k_1\, \underline{p}_{1,0}(1,v) = c_2^{\,2}\, \underline{q}_{2,0}(0,v) + k_2\, q_{1,0}(0,v) \ . \tag{2.12}$$

The above conditions match tangent and curvature along the v = constant direction and clearly more general continuity conditions could be formulated. However, the above conditions, and their duals across other boundaries, suffice for most practical cases.

2.6. Examples

2.6.1. Linear interpolation

The lofted surfaces are defined in operator form on the square $[0,1]\times[0,1]$ by

$$\begin{aligned} P_1[\underline{f}](u,v) &= (1-u)\, \underline{f}(0,v) + u\, \underline{f}(1,v) \ , \\ P_2[\underline{f}](u,v) &= (1-v)\, \underline{f}(u,0) + v\, \underline{f}(u,1) \ , \end{aligned} \tag{2.13}$$

The tensor product and Boolean sum blended forms are then defined by (2.7). The tensor product surface interpolates $\underline{f}$ at the corners of the square, whereas the Coons' Boolean sum blended surface matches $\underline{f}$ along the entire boundary.

2.6.2. Cubic Hermite interpolation

Let the cubic Hermite basis functions be defined on $[0,1]$ by

$$\begin{aligned} \phi_0(t) &= (1-t)^2(2t+1) \ , \quad \phi_1(t) = (1-t)^2 t \\ \psi_0(t) &= t^2(-2t+3) \ , \quad \psi_1(t) = t^2(t-1) \ . \end{aligned} \tag{2.14}$$

Then the lofted surfaces are defined on $[0,1]\times[0,1]$ by

$$P_1[\underline{f}](u,v) = \sum_{i=0}^{1} [\phi_i(u)\, \underline{f}_{i,0}(0,v) + \psi_i(u)\, \underline{f}_{i,0}(1,v)] \ ,$$

$$P_2[\underline{f}](u,v) = \sum_{j=0}^{1} [\phi_j(v)\, \underline{f}_{0,j}(u,0) + \psi_j(v)\, \underline{f}_{0,j}(u,1)]. \quad (2.15)$$

The tensor product and Boolean sum blended forms are then defined by (2.7), where, for commutivity of the operators, we assume that $\partial^2\underline{f}/\partial u\partial v = \partial^2\underline{f}/\partial v\partial u$ at the corners of the square.

2.6.3. The cubic as a convex combination

For a given constant k, $0 < k \leq 3$, let cubic basis functions be defined on $[0,1]$ by

$$\begin{aligned} \gamma_0(t) &= (1-t)^2(2t-kt+1), \quad \gamma_1(t) = k(1-t)^2 t, \\ \gamma_2(t) &= kt^2(1-t), \quad \gamma_3(t) = t^2(-2t+kt+3-k). \end{aligned} \quad (2.16)$$

Then with $\alpha_i = \gamma_i$, $\beta_j = \gamma_j$, and $m = n = 3$, univariate curves (2.1) and (2.2) can be defined. The lofted surfaces and tensor product surface are then defined by (2.3) and (2.4) respectively. The restriction that $0 < k \leq 3$ ensures that $\gamma_i(t) \geq 0$ on $[0,1]$ and since $\Sigma\gamma_i = 1$ it follows that the curves and surfaces are defined by convex combinations. The basis (2.16) is related to the cubic Hermite basis (2.14) in that, in (2.1),

$$\begin{aligned} \underline{r}(0) &= \underline{a}_0, \quad \underline{r}'(0) = k(\underline{a}_1-\underline{a}_0), \\ \underline{r}(1) &= \underline{a}_3, \quad \underline{r}'(1) = k(\underline{a}_3-\underline{a}_2). \end{aligned} \quad (2.17)$$

The examples *2.6.2* and *2.6.3* can be used to construct piecewise defined C^1 surfaces. The Coons' Boolean sum blended cubic Hermite surface has the property that it matches a function $\underline{f}$, and cross boundary tangent vector $\underline{f}_n$ say, along the entire boundary of the square $[0,1]\times[0,1]$. The tensor product cubic Hermite surface is a special case of the Boolean sum blended form which is used in many *c.a.g.d.* systems. The tensor product form of the cubic convex combination with $k = 3$ is an example of a surface used by Bezier [3], where the basis functions are the Bernstein polynomials. The representation as a convex combination gives the designer some control over the behaviour of the curve or surface, for example, the representation will lie in the convex hull of the coefficients. The case $k = 2$ corresponds to that used by Ball [1] in a rational lofted form, where the variable coefficients of the rational lofted form are also defined by rational forms.

3. SURFACE REPRESENTATION OVER TRIANGLES AND PENTAGONS

We now consider the problem of constructing surface patches with non-rectangular domains of definition but which occur within a rectangular patch framework. More specifically, we consider the need for triangular and pentagonal patches as illustrated by the wire frame model problems of Fig. 2. Such

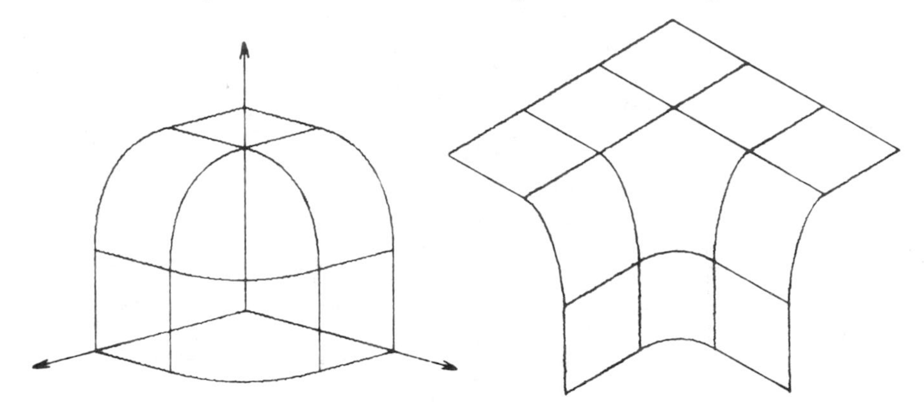

FIG. 2. Triangular and pentagonal patch model problems.

problems are intrinsically vector valued, there being no equivalent scalar valued problems, since the domains of definition of the surface patches cannot lie in a common domain in R^2. One approach to this problem is to divide the non-rectangular patch into a subsystem of rectangular patches and this has been considered by Bezier [3] and Handscomb [12]. Here, however, we briefly describe a method due to the author and P. Charrot which specifically uses non-rectangular domains. Further details can be found in references [10] and [11].

We restrict the discussion to the C^1 case, where continuity of function and tangent plane is required across the boundary curves of the wire frame model. The boundary curve and cross boundary tangent vector are assumed to be defined on each side by an appropriate rectangular patch representation. The method is then to construct Boolean sum blended interpolants which match function and tangent plane conditions on two adjacent sides of the non-rectangular domain. An appropriate convex combination of these interpolants is then formed which is designed to give interpolation to the function and tangent plane along the entire boundary.

3.1. Boolean Sum Blended Taylor Interpolant

Let lofted Taylor interpolants be defined by

$$T_1[\underline{f}](u,v) = \underline{f}(0,v) + u\,\underline{f}_{1,0}(0,v)\ ,$$
$$T_2[\underline{f}](u,v) = \underline{f}(u,0) + v\,\underline{f}_{0,1}(u,0)\ . \tag{3.1}$$

Then the Boolean sum blended surface

$$\underline{p}(u,v) = (T_1 + T_2 - T_1T_2)[\underline{f}](u,v) \tag{3.2}$$

has the property that it interpolates $\underline{f}$ and the tangent plane of $\underline{f}$ on $u=0$ and $v=0$.

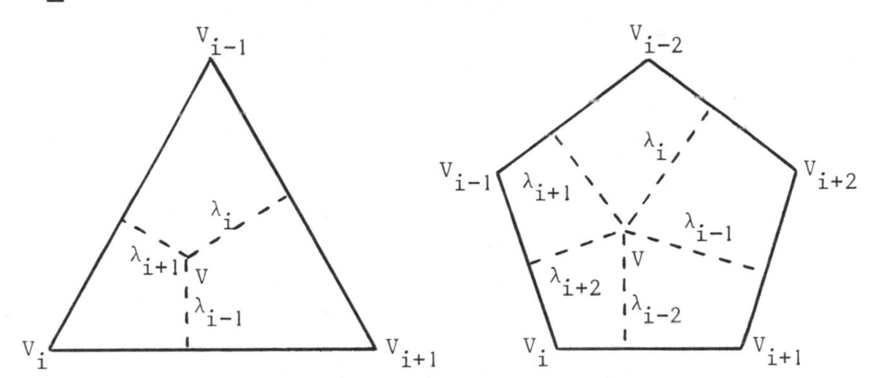

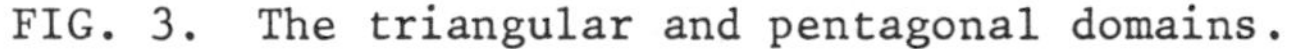

FIG. 3. The triangular and pentagonal domains.

3.2. *The Triangular Surface Patch*

It is convenient to choose the domain as an equilateral triangle of height unity and to define λ_i as the perpendicular distance of a general point V to the side opposite the vertex V_i , $i=1,2,3$. The boundary curve $\underline{f}$ and the cross boundary tangent vector $\underline{f}_{n_i}$ are assumed to be given along each side $\lambda_i = 0$. Choosing $u = \lambda_{i+1}$ and $v = \lambda_{i-1}$ as the independent variables, and interpreting $\underline{f}_{n_{i+1}}$ along λ_{i-1} = constant, $\underline{f}_{n_{i-1}}$ along λ_{i+1} = constant , allows us to define the Boolean sum blended interpolant, $\underline{p}_i(V)$ say, which interpolates the function and tangent plane on $\lambda_{i+1} = 0$ and $\lambda_{i-1} = 0$. The triangular surface patch is then defined by

$$\underline{p}(V) = \sum_{i=1}^{3} \alpha_i(V)\,\underline{p}_i(V)\ , \quad \alpha_i(V) = \lambda_i^{\,2}(3-2\lambda_i+6\lambda_{i+1}\lambda_{i-1})\ . \tag{3.3}$$

The leading term $\lambda_i^{\,2}$ in the definition of $\alpha_i(V)$, together with the property that $\Sigma\alpha_i(V) = 1$, ensures that the function and tangent plane of $\underline{p}(V)$ on $\lambda_i = 0$ is an average of those of $\underline{p}_{i+1}(V)$ and $\underline{p}_{i-1}(V)$. Thus $\underline{p}(V)$ interpolates the function and tangent plane along the entire boundary of the triangle.

3.3. The Pentagonal Surface Patch

The domain is now chosen to be a regular pentagon of height unity and, as before, λ_i is defined as the perpendicular distance of a general point V to the side opposite the vertex V_i, $i=1,..,5$. A Boolean sum blended Taylor interpolant $\underline{p}_i(V)$ could now be defined with respect to the variables λ_{i+2} and λ_{i-2}. We prefer, however, to choose the variables

$$u = \lambda_{i+2}/(\lambda_{i-1}+\lambda_{i+2}) \ , \ v = \lambda_{i-2}/(\lambda_{i+1}+\lambda_{i-2}) \ , \tag{3.4}$$

in the definition of the Boolean sum blended Taylor interpolant. The tangent vector $\underline{f}_{n_{i+2}}$ is then interpreted along v=constant which is along the radial line joining the point of intersection of $\lambda_{i+1} = 0$ and $\lambda_{i-2} = 0$ to the point V. Similarly, the tangent vector $\underline{f}_{n_{i-2}}$ is interpreted along the radial direction u=constant. The resulting surface $\underline{p}_i(V)$ interpolates the function and tangent plane on $\lambda_{i+2} = 0$ and $\lambda_{i-2} = 0$. The pentagonal surface patch is now defined by

$$\underline{p}(V) = \sum_{i=1}^{5} \alpha_i(V)\ \underline{p}_i(V) \ , \tag{3.5}$$

$$\alpha_i(V) = \lambda_{i+1}^2 \lambda_i^2 \lambda_{i-1}^2 \Big/ \sum_{k=1}^{5} \lambda_{k+1}^2 \lambda_k^2 \lambda_{k-1}^2 \ . \tag{3.6}$$

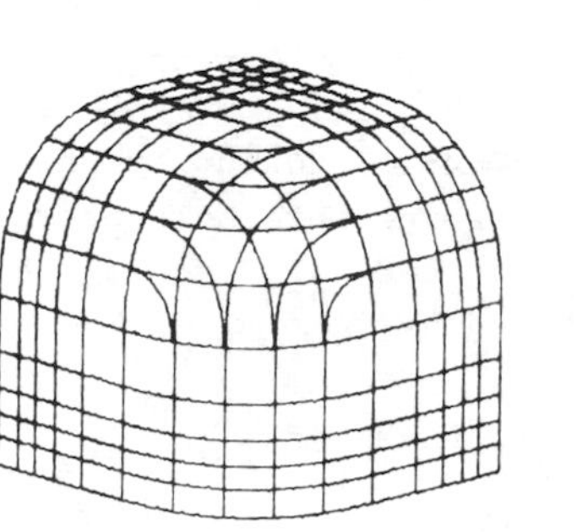

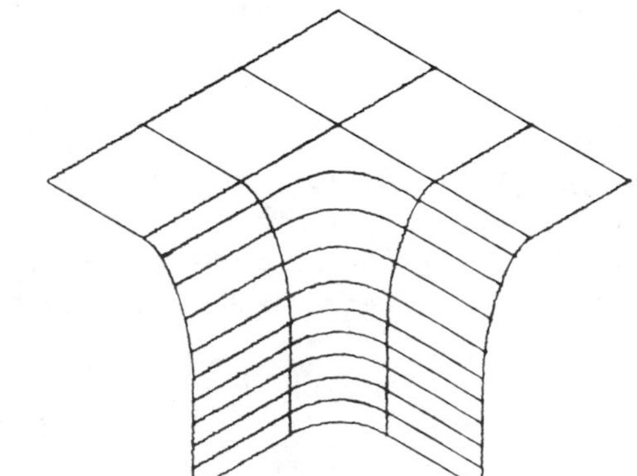

FIG. 4. Cross-sections through surface patches of model problems.

By applying arguments similar to those used for the triangular surface patch, the properties of the rational function (3.6) ensure that $\underline{p}(V)$ interpolates the function and tangent plane along the entire boundary of the pentagon. (An alternative rational weight function for the triangle, corresponding to that used for the pentagon, is $\alpha_i(V) = \lambda_i^2/\Sigma\lambda_k^2$. A polynomial weight function for the pentagon cannot be found.)

The implementation of the triangular and pentagonal surface patches for the model problems of Fig. 2 is shown in Fig. 4, where the plotting lines are cross sections. In these examples, the rectangular patches are tensor product cubic Hermite surfaces. Further examples can be found in references [10] and [11].

ACKNOWLEDGEMENTS

The work of this paper was partially supported by the Science Research Council grant GR/A92238. The author would like to thank P. Charrot for providing the computer generated figures.

REFERENCES

1. BALL, A.A., CONSURF part one: introduction of the conic lofting tile. *Computer Aided Design* 6, 243-249 (1974).

2. BARNHILL, R.E., Representation and approximation of surfaces, pp. 68-119 of J.R. Rice (Ed.), *Mathematical Software III*. Academic Press, New York (1977).

3. BEZIER, P., Mathematical and practical possibilties of UNISURF, pp. 127-152 of R.E. Barnhill and R.F. Riesenfeld (Eds.), *Computer-Aided Geometric Design*. Academic Press, New York (1974)

4. COONS, S.A., Surfaces for the computer-aided design of space forms, M.I.T. MAC-TR-41 (1967), available from NTIS, U.S. Department of Commerce, Springfield, VA 22151.

5. DELVOS, F.J., POSDORF, H., and SCHEMPP, W., Serendipity type bivariate interpolation, pp. 47-56 of D.C. Handscomb (Ed.), *Multivariate Approximation*. Academic Press, London (1978).

6. FAUX, I.D., and PRATT, M.J., *Computational Geometry for Design and Manufacture*. Ellis Horwood, Chishester (1979).

7. FORREST, A.R., On Coons and other methods for the representation of curved surfaces. *Computer Graphics and Image Processing* 1, 341-359 (1972).

8. GORDON, W.J., Blending-function methods of bivariate and multivariate interpolation and approximation. *SIAM J. Numer. Anal.* 8, 158-177 (1971).

9. GORDON, W.J., and HALL, C.A., Transfinite element methods: blending-function interpolation over arbitrary curved element domains. *Numer. Math.* 21, 109-129 (1973).

10. GREGORY, J.A., and CHARROT, P., A C^1 triangular interpolation patch for computer-aided geometric design. *Computer Graphics and Image Processing* 13, 80-87 (1980).

11. GREGORY, J.A., and CHARROT, P., A pentagonal surface patch for computer-aided geometric design. To appear.

12. HANDSCOMB, D., Conditions for a smooth junction between three quasi-rectangular patches. *Numerische Methoden der Approximationstheorie*, Oberwolfach Conference (1981).

INTERACTIVE GENERATION OF THREE-DIMENSIONAL MESHES

Y. Depeursinge*, Ph. Caussignac**, S. Jeandrevin**
and W. Voirol**

* *Laboratoire Suisse de Recherches Horlogères, Neuchâtel, Switzerland*
** *Ecole Polytechnique Fédérale de Lausanne, Lausanne, Switzerland*

1. INTRODUCTION

The first step of a finite element calculation consists of the generation of a mesh over the domain of integration. The quality of the results obtained is strongly dependent on that of the mesh created. In the two-dimensional case, the mesh generation phase no longer presents major difficulties and can be achieved with a high degree of reliability. This is, however, not the case for three-dimensional problems. In many such problems the surface of the domain of integration can be very complicated and thus tends to make any automatic mesh generation algorithm (AMGA) fail. Manual generation cannot apparently help at all. The method consists of splitting up the domain in subdomains in order to treat each one separately and finally of assembling the various parts. To date attention has rested mainly on the realization of very efficient AMGAs (see e.g. [3] and [7]), but not specifically on the structure of the programming system which remains generally relatively simple. Recent work, [1] and [2], has however shown the importance of this latter point. In a general way the domain is split up preferably into as larger parts as possible in order to minimize and simplify the manipulations for these subdomains. In the present work, we propose the opposite procedure; i.e. the global domain is split up in subdomains which are sufficiently small to be meshed easily. The difficulties encountered in the production of sophisticated AMGAs are reported in the conception of an efficient system of assemblage of the various subdomains. A solution to this problem consists of an editor-type programming system [4]. The global mesh is generated interactively piece by piece, each part being easily moved, modified or even replaced by one which is better suited. This method enables us to treat very complicated problems. The tree-like structure of the editor improves the ability to produce graphic representations of three-dimensional meshes, which is in practice essential. Moreover this structure facilitates automatic checks of the mesh created, since the origin of possible trouble is well defined: trouble can only occur during the assemblage of the various parts of the domain. This property enables us to improve the reliability of the mesh. Finally it should be noticed that the system produced can be used with any finite element program or any graphic software.

2. GENERAL STRUCTURE

An editor-type system is based on the concept of "object", [4]. These are individual entities on which certain operations can act. In the present case the objects can correspond to any subdomains previously meshed; i.e. to any set of finite elements. Specifically an object is described by two data structures, the former being related to the topology and the latter to the metric of the mesh, [5], [6]. The operations correspond to particular commands which can be activated interactively, such as e.g. the assemblage of two parts of the domain. The objects can be stored in a "library of objects" (LØ) in order to be used later. They are characterized by a particular name, which allows us to seek them again later on. Several LØs can exist simultaneously and their management constitutes the *first level* of the system. In order to simplify this management all these LØs are assembled in a single large data base. Once the access to the LØs has been achieved, the *second level* can be reached. This

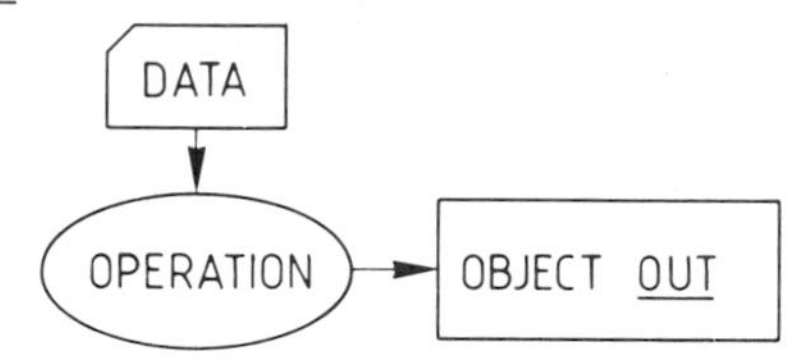

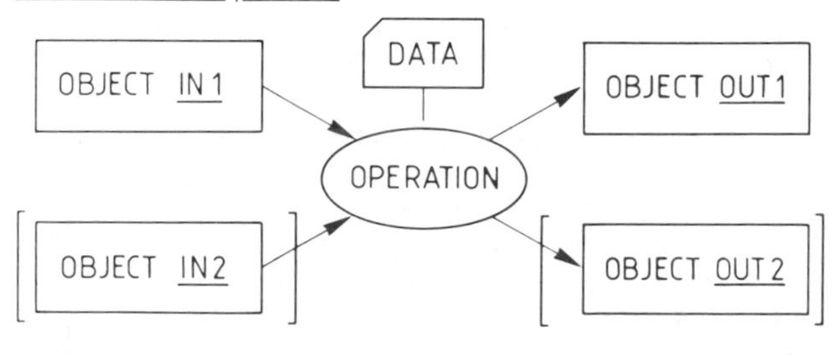

Fig. 1. Basis operations

level corresponds to the management of the objects. It enables us to extract/store the objects from/in the LØs, and to "prepare" them for the *third level* which corresponds to the work on the objects. This preparation consists in informatic type manipulations (held entirely by the system) in order to put the objects at the disposal of the operations to be executed at the lower level.

In a general way two kinds of operations can be distinguished (see Fig. 1). The first kind consists of the creation of new objects, and the second in the transformation of one or two objects into one or two others. If the name of one of the output objects is the same as that of one of the input objects, the former is replaced by the latter. This can be done when the objects on which the operations act do not have to be kept for the continuation of the process and when these operations are expected to be infallible. Conversely, if these names are different, the former is preserved. The work on the objects consists thus of a succession of steps, each one corresponding either to the creation of new objects, or to the application of an operation to objects originating from any accessible LØ or from any preceding step. After each step the choice of the following one is made interactively on the basis of various checks on the objects built, such as for example graphic representations. If the result is satisfactory, the object so produced can either be stored in one of the LØs, or be available for the following steps. Conversely, if need be, this object can be eliminated and any of the preceding steps undertaken again, and this without destroying the work done in the former steps. As a consequence, it is relatively easy at each step to correct the errors or to modify the objects produced, thus enabling us to treat very complicated problems.

The structure of the system is presented schematically in Fig. 2 for a typical case. The three characteristic levels are clearly distinguishable. The first level, which corresponds to the management of the LØs, acts here only on two LØs, named "LØ1" and "LØ2". This is clearly not a limitation and other LØs could be taken into account. The second level, which corresponds to the management of the objects, allows us to extract successively the objects named A, C, D and H from LØ1 and B from LØ2. The work on these objects can then be undertaken. As said before, this consists of a succession of steps, each one corresponding to the application of a particular operation. In Fig. 2 the name of the objects is denoted inside the squares and the order of the successive steps inside the circles. After each step, the objects obtained can be checked by a test operation (e.g. the drawing of the object). This figure allows us to illustrate the main features of the system.

First let us show how to correct the errors committed at any step of the process (steps 5, 6 and 7). The assemblage of the objects named E and D obtained at the 5th step has not furnished

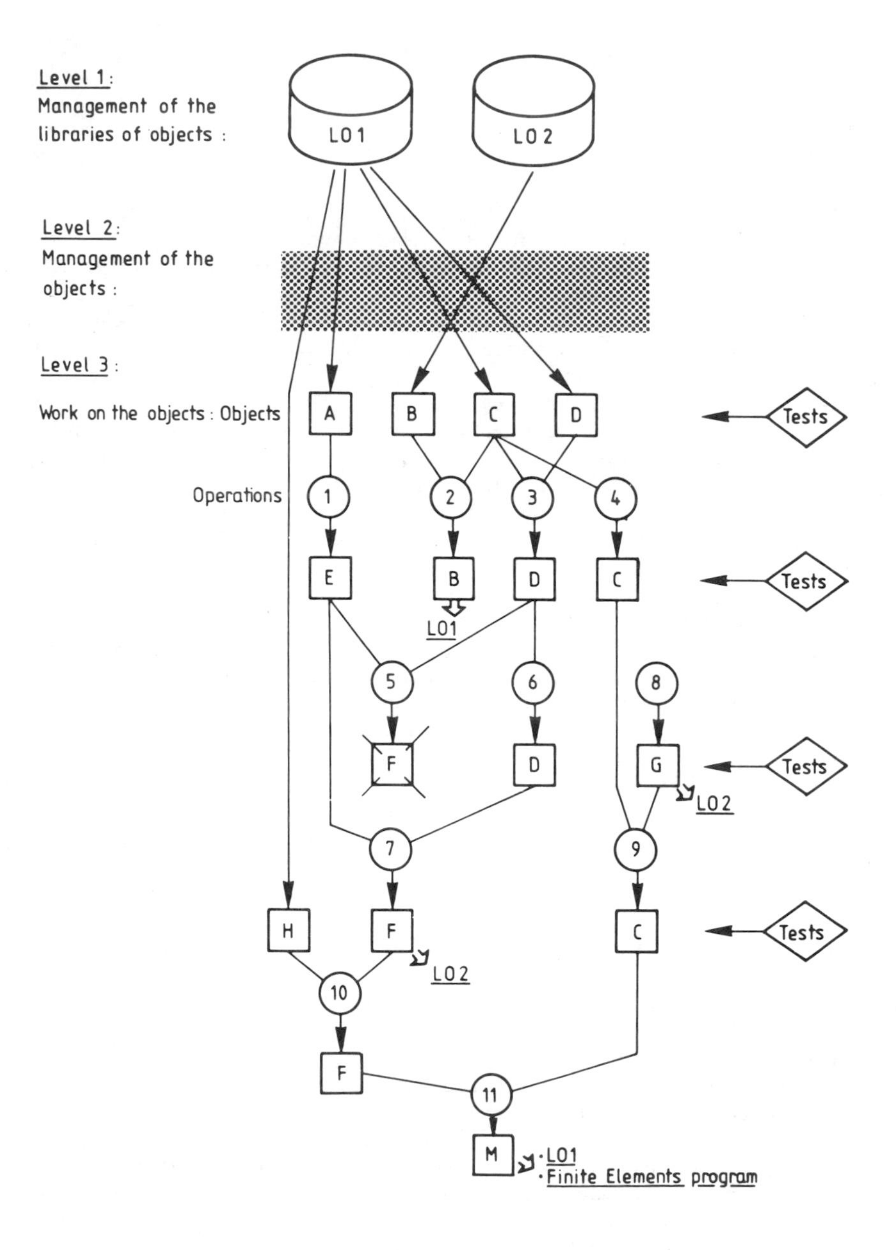

FIG. 2. Functional diagram of the system

the desired results, because the object D was misplaced relative to the object E. As D has been preserved during the 5th step, this object can be positioned again (6th step) and thereafter assembled with E (7th step). The object obtained is then named F, thus overriding the object built at the 5th step. Furthermore at any time a new object can be created, e.g. by an automatic mesh generator (step 8), or looked for in an LØ (e.g. object named H for the step 10); it is also possible during the sequence to save in a specified LØ the objects built at any step (steps 2, 7, 8 and 11). When an object has been saved in an LØ, further modifications on it cannot alter the version stored in the library (e.g. at step 10). Finally, when the global mesh has been generated (step 11), it can be saved in an LØ, and prepared by a suitable interface for the use in any finite element program.

To conclude this section it should be noticed that the structure described is typically *tree-like*. The work consists of a succession of steps, the results of which can be checked interactively. After every step it is thus possible to delete or to keep the objects produced for further operations and also to save them in a specified LØ. In the same way the objects available for any further step can originate either from any preceding one or directly from any accessible LØ. The parallelism of this structure makes this system particularly *flexible*. It enables us to mesh step by step many diverse and very complicated domains. Moreover, the meshed parts which have been saved in an LØ during the generation can be used for other applications than those previously planned. Another advantage of the system is its *rapidity*. On the one hand all the symmetries of the problem can be used extensively, whilst on the other hand errors can be corrected after any step, and this without altering other objects. Moreover, as the basis objects, i.e. those which are created directly, are generally small and simple, errors can only occur at the assemblage and are therefore well localized. Hence there exists the possibility of making automatic checks and, consequently, of improving *reliability*. Finally, let us note that the interactivity of the system makes its *utilization relatively easy*, which is not its least advantage.

3. DESCRIPTION OF THE SYSTEM

Each level contains a main "menu" composed of several items. Each menu contains one command which allows us to go down to the lower level and another one, to go up to the upper level (always named [END]). Such transitions can be done at any time point, without altering the work done up to that time. The other commands activate specific operations depending on data furnished by the user in answer to:

- subordinate menus - when several options are available
- particular questions - when precise information must be

provided (e.g. the name of an object or the number of a point). In order to point out the main possibilities of the system, we give in Fig. 3 the characteristic menus of the three levels described in the preceding section.

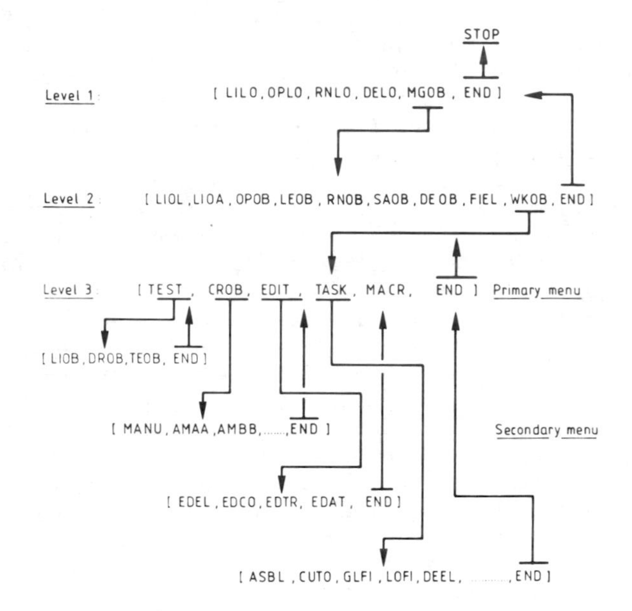

Fig. 3. Main menus

The main menu of the first level, which corresponds with the "management of the libraries of objects" (LØ), enables us first to list the names of all the available LØs ([LILØ]). The next operation consists of the aperture of one or several LØs ([ØPLØ]). If the specified LØs already contain some objects, this item makes them accessible to the operations of the lower levels. Conversely, if the LØ requested does not yet exist, [ØPLØ] creates it and prepares it to receive the objects to be saved. [RNLØ] and [DELØ] respectively rename and delete some LØs and [MGØB] controls the access to the lower level, which corresponds with the "management of objects". [END] enables us to leave the system.

The two first commands of the main menu of the second level ([LIØL] and [LIØA]) list respectively the objects of a particular LØ and those which are available at a given moment from either any accessible LØ or the third level, The command [ØPØB] opens some objects specified by their name and by the name of the LØ wherein they are stored. This means that these objects are prepared in order to be directly manipulated by the operations of the third level. The opposite command ([LEØB]) leaves certain previously opened objects; i.e. these objects become inaccessible to the operations of the third level. If need be these objects can be taken again from some LØs, except for those which have not been previously saved. The command [RNØB] allows us to rename an object specified by its name and if need be the name of the LØ from which it originates. [SAØB] and [DEØB] respectively save and delete some specified objects in a given LØ. The command [FIEL] prepares the chosen objects for a finite element calculation. This operation "cleans" all the information which were specifically necessary to the mesh generation system. [WKØB] controls the access to the third level, which corresponds with the "work on the objects", and [END] to the upper level.

The main menu of the third level is split into primary and secondary menus. The former corresponds only to a switch in direction of the latter. The item [TEST] allows us access to the various available check means of the objects. [LIØB] lists and [DRØB] draws a specified object. The graphic representation modes have been particularly emphasized: in order to improve the clearness of the drawing of three-dimensional meshes, it is possible to represent only parts of them, chosen depending on physical (e.g. the elements which correspond with copper in an electrical engine) or geometrical (e.g. the elements situated on an intersection plane of the domain) conditions. Finally, several modules of automatic checks (e.g. verification of the assemblage of two objects) can also be activated ([TEØB]). The item [CRØB] in the primary menu allows us to create new objects which can be generated either manually ([MANU]), i.e. by definition of the finite elements one by one, or automatically ([AM..] where the two points correspond to the identifier of any particular automatic mesh algorithm). The operations which can act on the objects belong to two groups. On the one hand, the "editors" which are highly interactive, i.e. requiring only few calculations but with sophisticated dialogues such as graphic dialogues (e.g. interactive modification of the position of a point), and on the other hand, the "tasks" which need relatively long calculations (e.g. the assemblage of two objects) and consequently with a low degree of interactivity. The access to the editors is obtained by the item [EDIT]. The two first editors, the "editor of elements" ([EDEL]) and the "editor of coordinates" ([EDCØ]) allow us to modify respectively the topology (e.g. the transformation of hexahedrons into tetrahedrons) and the metric (e.g. the position of a point) of a given mesh. The "editor of transformations" ([EDTR]) applies geometric transformations on a specified object. The last editor is the "editor of attributes" ([EDAT]) which can characterize or modify the characterization of any surfaces or volumes inside the objects. The characterization seeks to concentrate the physical data to be furnished later on to the finite element program (e.g. all the elements with the same magnetic permeability will have the same characterization). The command [TASK] controls the access to the various tasks available in the system. The first one corresponds with the assemblage of two objects ([ASBL]). The opposite task ([CUTØ]) cuts an object into two parts by a plane defined by the user. [GLFI] and [LØFI] enable us to define an object globally and locally around a point. The deletion of some finite elements (in order to make a hole inside the domain) can be done by the command [DEEL]. If need be other items can clearly be added to this menu. Finally, it is also possible to link macro-operations ([MACR]), which means that several operations (from among the editors and the tasks) can be linked together in a specified order to be executed sequentially with a given repetition factor.

4. CONCLUSION

The presented system allows us to generate interactively three-dimensional meshes for the finite element method. Its structure is of editor-type and is typically tree-like, inducing thus a high degree of flexibility and reliability. Displacements in the tree can be done either towards the top or towards the bottom. One of the most important features is the possibility for checking the mesh produced along all the successive steps of the generation process, and to correct locally at any moment the errors committed without altering the rest of the mesh. To conclude, let us recall that, due to the structure of the editor, the graphic representation of the mesh becomes really efficient: the flexibility of the system allows us to draw only chosen parts (surfaces or volumes) of the global mesh and this depending on either geometric or physical conditions.

REFERENCES

1. GRIEGER, I., Geometry elements in computer-aided design. *Comp. and Struct.* 8, 371- (1978).
2. KANARACHOS, A. and ROPER, O., Computer aided net generation by means of Coon's image. *VDI. Z.* 121, 297- (1979).
3. KRAUSE, F.L., The uses of CAD software in engineering design and production planning. *ZWF Z. Wirtsch. Fertigung* 75, 72- (1980).
4. NEWMAN, W.M. and SPROULL, R.F., *Principles of Interactive Computer Graphics, 2nd Edition*. McGraw-Hill, New York (1979).
5. PERRONNET, A. and DEPEURSINGE, Y., La structure de donnée TØPØ. *Publications Modulef*, INRIA, Le Chesnay (1980).
6. PERRONNET, A., Descriptions des structures de données du Club Modulef. *Publications Modulef*, INRIA, Le Chesnay (1979).
7. WILSON, P.R., Data generation for finite element analysis. *Proc. 3rd Int. Conf. and Exhib. on Comp. in Eng. and Building Design*. IPC Scientific and Technology Press, London (1978).

ABSTRACTS OF POSTER SESSIONS

1. MINIMUM PRINCIPLES FOR SOME COUPLED SYSTEMS

R.T. Ackroyd
UKAEA, Risley, Warrington, Cheshire, WA3 6AT, England.

The fission and scattering processes couple the directions and energies of neutrons in transport. Spatial boundary conditions are difficult to satisfy. For self-adjoint systems boundary-free minimum principles, generalised least squares and Galerkin methods are derived from a functional identity. By extending the concept of an approximate solution a minimum principle is given for some non self-adjoint systems.

2. THE RELATIONSHIP BETWEEN THE PATCH TEST FOR NON-CONFORMING FINITE ELEMENTS AND THE CONSISTENCY OF THE CORRESPONDING FINITE DIFFERENCE EQUATIONS

J.M. Aitchison
Oxford University Computing Laboratory, 19 Parks Road, Oxford, OX1 3PL, England.

It has often been observed that the application of the patch test to non-conforming finite elements corresponds to checking the consistency of the corresponding finite difference equations. The idea has not been pursued in the past because of the complexity of the manipulative algebra involved in the consistency calculation for irregular grids or for nodes with several degrees of freedom. Here we present a technique for performing this algebraic manipulation on a computer, which removes previous objections to the use of this test.

We consider the solution of a particular differential equation using a specific element. The mass and stiffness matrices are formed in the usual way to give a set of algebraic equations for the nodal values of the variables. We now consider this set of algebraic equations to be a finite difference representation of the original differential equation, and test for its consistency and truncation error. In most cases of interest there will be several variables at each node and a corresponding number of equations centred at the node. In order to check the consistency of the equations it is necessary to perform an elimination on them, usually involving differentiation, which is performed automatically. Results for a variety of conforming and non-conforming elements are presented.

3. A VARIABLE ORDER GLOBAL ELEMENT METHOD FOR NONLINEAR PARABOLIC PARTIAL DIFFERENTIAL EQUATIONS

M. Berzins and P.M. Dew
Department of Computer Science, University of Leeds, Leeds, LS2 9JT, England.

An algorithm based on a global element Chebyshev method is described for the solution of nonlinear parabolic partial differential equations of the form

$$\sigma(u,x,t)\,\frac{\partial u}{\partial t} = \frac{1}{x^m}\frac{\partial}{\partial x}\left\{x^m\phi(u,x,t)\,\frac{\partial u}{\partial x}\right\} + f\left(u,\frac{\partial u}{\partial x},x,t\right)$$

$$(x,t) \in [a,b] \times (0,T)\ .$$

The algorithm is designed to handle equations with quite general boundary conditions and can readily be extended to systems of parabolic equations.

The interval [a,b] is partitioned into a number of elements and the solution on each element is approximated by a Chebyshev Series expansion, the degree of which may be varied from element to element. Continuity of u and $\phi\,\partial u/\partial x$ is preserved across the element interface. The system of ordinary differential equations that results from this discretisation is of the form (in standard notation)

$$A\,\frac{d\underline{u}}{dt} = \underline{F}(\underline{u},t)\ ,$$

where A is a banded matrix, and the numerical solution is computed using a modified version of Gear's method. It can be shown that this system of ordinary differential equations can be derived from a perturbed form of the original partial differential equation. Error estimates are obtained from the perturbed equation and strategies for automatically selecting the *optimum* degree Chebyshev series expansion for each element are discussed.

4. A NEW APPROACH TO THE ANALYSIS OF AN EXTENDED CLASS OF LINEAR MULTISTEP METHODS

F. Brancaleoni and V. Ciampi
Istituto di Scienza delle Costruzioni, Universita di Roma, Via Eudossiana 18, 00184 Roma, Italy.

Linear multistep methods (ℓ.m.m) are a well known general class of algorithms for solving, by direct time integration, stiff systems of ordinary differential equations which arise from the discretization of continuum problems via finite element techniques.

Forms of ℓ.m.m specialised to 1st and 2nd order ordinary differential equations have been proposed in the literature. The second class comprises a wide choice of methods with favourable and tunable properties (e.g. Newmark, Houbolt, Wilson, Hilber methods, Geradin and Zienkiewicz general formulae), while the former comprises methods which are fewer in number and have non-tunable properties (Gear, Jensen, Park methods), but which are characterized by a better error propagation behaviour. The advantages of the two classes have both been achieved here through the development of a new general procedure for the stability and accuracy analysis of ℓ.m.m which, providing an excellent insight of their algorithmic properties, has also produced a methodology for determining 1st order methods with the same favourable characteristic of existing 2nd order ones, plus improved error propagation behaviour. Applications to Newmark and Hilber methods are presented, together with numerical evidence of the reduced error amplification.

5. THE CAD-FEM INTERFACE

G.A. Butlin
FEGS Limited. 5 Jesus Lane, Cambridge, CB5 8BA, England.

The two key aspects of a CAD system that concern the finite element analyst are (a) the means available for simplifying geometrical models in the CAD system to suit finite element analysis purposes; (b) the geometrical and topological content available for transfer into the f.e.m. model. These two aspects are discussed in the light of experience with interfacing both drafting orientated and geometric-model orientated CAD systems with major finite element analysis systems through the FEMGEN finite element mesh generator.

6. UPPER BOUND PLANE STRESS LIMIT ANALYSIS USING AN UNCONSTRAINED MINIMISATION METHOD

J.S. Campbell, M.J. Creed and J.B. Delaney
Department of Civil Engineering, University College, Cork, Ireland.

In recent years design philosophy has changed from the permissable stress approach to the limit state concept. This has led to the adoption of the CP110 concrete design code and to the drafting of the new proposed steel design code B/20. Whilst these codes require the limit load for the structure, current methods of determining this load are restricted to simplified shapes and load distributions. Clearly, techniques for determining the limit load for more general cases are required and this provided the motivation for the current research reported here.

In the paper, upper bound limit loads for plane stress type

structures are determined. The method utilizes both finite element modelling and an unconstrained minimisation technique. Two minimisation algorithms are studied, namely the direct Powell method and the gradient method of Davidson-Fletcher-Powell. Finite elements used include the simple 3 noded triangle and the 8 noded isoparametric quadrilateral. For the latter case, the convergence of the numerical integrated power dissipation was studied. Applications which were treated include an end-loaded deep cantilever and weakened slabs having square, oblique square and slot cut-outs.

7. A FINITE ELEMENT APPROACH TO SOLVE A NONLINEAR HYPERBOLIC PROBLEM

A. Cardillo, M. Lucchesi, A, Pagni and G. Pasquinelli
Istituto del Consiglio Nazionale delle Ricerche, C.N.U.C.E., 36 Via S. Maria, 56100 Pisa, Italy.

In their pioneer numerical study of nonlinear vibrations of strings, Fermi, Pasta and Ulam used an inappropriate finite difference method of approximate integration that led them to some surprising results. These results have, however, now been completely explained; actually the work on the F.P.U. problem has opened, almost by accident, an entire new field of analysis (the study of soliton type solutions in partial differential equations).

We have considered the original problem which, with an obvious notation, can be formulated as follows:

$$u_{tt} - (1 + \varepsilon u_x)u_{xx} = 0 \,, \qquad \forall \ (x,t) \in [0,1] \times [0,T] \,,$$

$$u(x,0) = u_0(x) \,, \qquad u_t(x,0) = \dot{u}_0(x) \,, \ u(0,t) = u(1-t) = 0,$$

and we have tested on it a new and appropriate numerical method; the method must be such as not to fail (as the simple procedure of F.P.U. does) even though the solution becomes at a certain point in time less regular, i.e. a shock is formed. In order to apply the method the problem must first be given a weak formulation, i.e.

$$- \int_0^1 u_0(x)v(x,0)dx - \int_0^1\int_0^T u_t v_t dxdt + \int_0^1\int_0^T F(u_x)v_x dxdt = 0 \,,$$

$$u(0,t) = u(1,t) = 0 \,, \quad u(0,x) = u_0(x) \,, \quad u_t(0,x) = \dot{u}_0(x) \,,$$

where $F(y)$ is a primitive function of $(1 + \varepsilon y)$. The above formulation must be valid for any function $v(x,t)$ such that

$$v(x,t) \in C^0([0,1] \times [0,T], \ v(0,t) = v(1,t) = 0, \ v(x,T) = 0 \,.$$

The functions v_x and v_t are continuous in $[0,1] \times [0,T]$ except, at most, along a curve γ where they are bounded. In this formulation the use of an appropriate space-time finite element appears natural and leads to a numerical recursive formula.

The method has already been used with success in a study of linear hyperbolic problems, but appears particularly suited for dealing with shock phenomena arising due to nonlinearities. We find in particular that the approximate solutions satisfy automatically the so called entropy condition. We have used the method also on a problem which is not *genuinely nonlinear*.

8. AN OPTIMAL MESH PROCEDURE IN FINITE ELEMENT METHOD

J. Carnet*, P. Ladeveze** and D. Leguillon***
**Département de Mathématiques, Université de Haute Normandie, place E. Blondel, 76130 Mont Saint Aignan, France.*
***Laboratoire de Mécanique et Technologie, ENSET, Université P. et M. Curie, 61 bd Wilson, 94230 Cachan, France.*
****Laboratoire de Mécanique Théorique, CNRS, Université P. et M. Curie, 4 place Jussieu, 75230 Paris Cedex 05, France.*

The task of minimising computer requirements is of great interest in the use of finite element (f.e.) techniques. The main task on which this optimisation can be performed is the choice of the mesh. We propose herein a method for generating two-dimensional optimal grids, while providing accurate information on the reliability of the computed solution.

The displacement finite element method is currently that most used in structural computations. Unfortunately, the stress or heat flux fields thus obtained do not generally satisfy the equilibrium equation. However, it is possible to improve the quality of the finite element solution in order to have admissible fields by completely explicit calculations. Therefore, once optimal extensions to the finite element solution are defined, the classical comparison theorem provides a close upper bound of the finite element error. This error and the corresponding error density are obtained by computing elementary error matrices which depend only on the data and on the geometrical properties of the triangles. As a first approximation, it is possible to define the optimal surfaces of the triangles of a new mesh leading to an *a priori* given error. This is performed using reduction (or eventually expansion) rates of the triangles of the current mesh. The automatic mesh procedure used is founded on heuristic rules. It is completely automatic and does not need any interactive operation. The localisation of subregions, where a finer mesh is required, is performed in a very flexible way by the data of expansion coefficients which are in our case the reduction rates defined above. The grid is generated by successive layers of triangles starting from the boundary. Smoothness is obtained by such local rules as, overlapping two vertices, creating or moving one vertex, and by a global relaxation algorithm.

9. NUMERICAL SOLUTION OF NONLINEAR ELLIPTIC PARTIAL DIFFERENTIAL EQUATIONS USING THE GLOBAL ELEMENT METHOD

R. Cook
Department of Computational and Statistical Science, University of Liverpool, P.O. Box 147, Liverpool L69 3BX, England

A quasi-Newton iterative scheme is presented for the numerical solution of nonlinear elliptic partial differential equations, using the *global element method* to solve the linearized equations. The global element method leads to a highly structured system of algebraic equations. In particular the Jacobian matrix in Newton's method is block sparse with diagonal blocks of full rank and non-zero off-diagonal blocks of low rank. An updating procedure for the approximate Jacobian matrix is given, which preserves the special structure of the true Jacobian matrix. Finally, some comparisons are made with other iterative schemes, namely, a simple functional iterative method and a quasi-Newton method where no account is taken of the structure present in the system.

10. FRACTURE CRITERION FOR BRITTLE MATERIALS BASED ON FINITE STATISTICAL CELLS

H. Cords, G. Kleist, G. Kraus and R. Zimmermann
Institute für Reaktorwerkstoffe, Kernforschungsanlage Jülich GmbH, Postfach 1913, D-5170 Jülich 1, West Germany.

The Weibull statistical analysis of brittle materials has been reconsidered with the inclusion of one additional material constant so that a more comprehensive description of the failure behaviour may be obtained. In order to be able to apply the differential calculus, Weibull was restricted to the use of infinitesimal elements of volume. It was found that these infinitesimally small elements are in conflict with the basic statistical assumption. However, it is now not necessary to employ differential calculus, since most stress analyses are based on finite element calculations and these are most suitable for a subsequent statistical analysis of strength. In the present work the size of a finite element for the statistical calculations has been introduced as a third material parameter. This represents the minimum volume containing all statistical features of the material, such as the distribution of flaws and the grain sizes. Sixteen different tensile specimens have been investigated experimentally and analysed with the probabilistic fracture criterion. As a result it can be stated that the failure rates of all types of specimens are predictable within the range of one standard deviation.

11. PARABOLIC VARIATIONAL INEQUALITIES APPLIED TO TIME DEPENDENT GROUNDWATER FLOW WITH PARTICULAR REFERENCE TO THE EFFECT OF TIDES ON A COASTAL AQUIFER

A.W. Craig and W.L. Wood
Department of Mathematics, University of Reading, Whiteknights, Reading, RG6 2AX, England.

The study of the time dependent fluid flow through porous media leads to free boundary problems for linear parabolic equations. In 1972 Baiocchi introduced a new method for the solution of the steady state elliptic problem in which he transformed the dependent variable and then showed that the transformed variable was the solution of a variational inequality. Using this technique he managed to prove existence and uniqueness for the weak form of the original problem and also numerical algorithms were suggested for the solution which were rigorous and efficient in practice.

The problem considered here has time dependent boundary conditions reflecting the effects of tidal variation and this introduces several new features into the problem.

12. GENERAL SOLUTION OF PARTIAL DIFFERENTIAL EQUATIONS - A FINITE ELEMENT APPROACH

Z.J. Csendes and N. Nassif
Department of Electrical Engineering, McGill University, 3480 University Street, Montreal, P.Q., Canada H3A 2A7

In this paper we establish the following theorems:
Theorem 1 Let S^N denote an N-dimensional simplex and suppose that M^N is a mesh of s arbitrary finite elements S^N. If an nth order Silvester polynomial

$$\alpha^{(n)}_{k_1 k_2 \ldots k_{N+1}} = \prod_{i=1}^{N+1} P_{k_i}(S_i) \ , \quad \sum_{i=1}^{N+1} k_i = n \ ,$$

where $P_m(Z) = \sum_{i=1}^{m} (\frac{nz-i+1}{i})$ is defined on S^N, then it is possible to approximate the general solution of a PDE defined on M^N provided that the range space approximation is of order m=n-N-1.
Theorem 2 Let $a\phi = f$ be an arbitrary partial differential equation on M^N, and let A and $\underline{f}$ be the inner products

$$A = (\tilde{\alpha}^{(m)^T}, a\tilde{\alpha}^{(n)}) \ , \qquad \underline{f} = (\tilde{\alpha}^{(m)^T}, f) \ .$$

An approximation to the general solution of the partial differential equation is given by $\alpha = \tilde{\alpha}^{(n)}\underline{\Phi}$, where $\underline{\Phi} = P^+(AP^+)^{-1}\underline{f} + (N-P^+(AP^+)^{-1}AN)\psi$. In the last equation P is the projection matrix $P = (\tilde{\alpha}^{(m)^T}, \tilde{\alpha}^{(m)})^{-1}(\tilde{\alpha}^{(m)^T}, \tilde{\alpha}^{(n)})$, P^+ is the generalised

inverse of P, N is the nullmatrix of P, and the vector $\underline{\psi}$ is arbitrary.
Theorem 3 If N = 2, the number of normal derivative conditions required to make the s polynomials $\alpha^{(n)}$ in M^2 have C^1 continuity is C = (3ns-B)/2, where B is the number of nodes in M^2 restricted by boundary conditions.

13. A NEW APPROACH TO THE NUMERICAL ANALYSIS OF LINEAR ELASTIC SHELL PROBLEMS

Ph. Destuynder and A. Lutoborski
Centre de Mathematiques Appliquees, Ecole Polytechnique, 91128 Palaiseau Cedex, France.

We consider a mixed formulation of a linear thin shell problem known as the Budianski-Sanders (B-S) model. This classical model involves first order derivatives of tangential displacement u_t, but also second order derivatives of the normal displacement u_3. Thus the finite element methods based on this model require expensive elements. In order to avoid this, we introduce the rotation Θ of the normal N to the medium surface of the shell ω.

Next using penalty and duality arguments, we obtain the perturbed B-S model depending on a small parameter η. This formulation requires only linear C^0 elements in the approximation. We establish an error estimate between the solution of the classical and perturbed B-S models. In order to construct the finite element approximation of the perturbed B-S model, we define two triangulations of the reference domain - the first being a subtriangulation of the second. The coarser one is used in the approximation of u_3 and u_t and the refined one is used for the rotation Θ. We establish the existence of the finite element solution assuming additionally that the mesh size is small in comparison with the local radius of curvature of ω.

We prove an error estimate of O(h) for the displacement components and the stresses, and of $O(h^2)$ for the rotations. Finally using an asymptotic method we analyse the solution of the resulting linear system depending on the perturbation parameter. We prove that it is sufficient to solve a small size linear system with a positive definite matrix, so that direct methods can be used. More precisely the dimension of the matrix is three times the number of vertices in the coarser triangulation.

14. THE SINGLE ELEMENT INTEGRATION APPROACH TO THE FINITE ELEMENT METHOD

L.C.W. Dixon and D.J. Morgan
Numerical Optimisation Centre, Hatfield Polytechnic, Hatfield, Herts., England.

In this paper it is demonstrated that, by adopting the *single element integration* process introduced in Dixon, Harrison and Morgan [1] and introducting into it optimally chosen parameters, it is possible (a) to symmetrise non-selfadjoint systems, (b) to demonstrate that the solution obtained is optimal with respect to a selected norm and (c) to satisfy automatically side constraints of the form div $\underline{u} = 0$. The paper can be considered as a development of ideas in recent publications by Temam, Raviart, Griffiths, Morton and Barrett; as well as those given in [1].

1. Dixon, L.C.W., Harrison, D. and Morgan, D.J., On singular cases arising from Galerkin's method. pp.217-225 of J.R. Whiteman (ed.), The Mathematics of Finite Elements and Applications II, MAFELAP 1978. Academic Press, London, 1979.

15. OPTIMAL CONVERGENCE PETROV-GALERKIN METHODS BASED ON QUADRATIC ELEMENTS

D. Duncan
Department of Mathematics, University of Dundee, Dundee, DD1 4HN, Scotland.

The problem $u_t + u_x = 0$ is solved by a Petrov-Galerkin method based on quadratic elements. By Fourier transposition of the system of difference equations obtained we produce a 2×2 eigensystem from which we can derive much information about the properties of the method. The method is found to produce a solution composed of two waves with speeds 1 and -s. In the semi-discrete case s = 4. To achieve optimal convergence this second or *parasitic* wave must be filtered out by means of a correction applied to the initial data [1]. By a suitable choice of test functions in the semi-discrete and fully-discrete cases we can achieve $O(h^6)$ convergence after the parasitic wave has been filtered out.

1. Hedstrom, G.W., The Galerkin method based on Hermite cubes. SIAM J. Numer. Anal. 16, No.3, June 1979.

16. BLENDING-SPLINES AS A MEANS OF SOLVING INTEGRAL EQUATIONS APPROXIMATELY

G. Hämmerlin
Mathematisches Institut, Ludwig-Maximilians Universität, Theresienstrasse 39, D 8 München 2, West Germany.

Fredholm integral equations of the second kind with degenerate kernels can always be solved explicitly. In order to solve a linear integral equation with a general kernel, we construct a substitution kernel using blending-splines which can be conceived as a degenerate kernal. Since blending-splines interpolate a given kernel not only at the mesh points of a given grid but also along the mesh-lines, we have to expect improved convergence properties compared with ordinary tensor product splines. The method, utilizing in this way a special kind of finite element over a quadratic grid, is applied particularly to eigenvalue problems. Several error estimates for the precision of approximate eigenvalues and eigenfunctions are given.

17. THE PENALTY METHOD APPLIED TO THE NON-STATIONARY STOKES EQUATIONS

F.-K. Hebeker
Fachbereich 17 Mathematik - Informatik, Universität Gesamthochschule Paderborn, Postfach 1621, 4790 Paderborn, West Germany.

A mathematical analysis of the penalty method applied to the initial boundary value problem for the Stokes equations is given. This extends the study of Reddy [2] to the case of non-stationary flow.

We first prove a regularity result for the penalty equations by means of a special Galerkin approximation developed by Heywood [1]. The remarkable convergence result (when the penalty parameter $\varepsilon \to 0$)

$$\| \operatorname{div} u_\varepsilon \| \leq O(\sqrt{\varepsilon}) \text{ in } L_\infty(0,T,L_2(\Omega))$$

is implied which together with

$$\| u_\varepsilon - u \| \leq O(\sqrt{\varepsilon}) \text{ in } L_\infty(0,T,L_2(\Omega)) \cap L_2(0,T,H_0^1(\Omega))$$

justifies the penalty method for this problem. Based upon the regularity result the penalty equations's discretization in time (with an A-stable multistep integration formula of the second order due to Zlamal [4]) and in space, (with a finite element method with "S_m^h-family") is shown to give an optimal L_2-error estimate.

1. Heywood, J.G., The Navier-Stokes equations: On the existence, regularity and decay of solutions. Indiana U. Math.J.29, 639-681, 1980.
2. Reddy, J.N., On the F.E.M. with penalty for incompressible fluid flow problems. pp.227-235 of J.R. Whiteman (ed.), The Mathematics of Finite Elements and Applications III, MAFELAP 1978. Academic Press, London, 1979.
3. Zlamal, M., Finite element methods in heat conduction problems. pp.85-104 of J.R. Whiteman (ed.), The Mathematics of Finite Elements and Applications II, MAFELAP 1975. Academic Press, London, 1976.
4. Zlamal, M., Finite element methods for nonlinear parabolic equations. RAIRO Anal. Numer. 11, 93-107, 1977.

18. ERRORS INTRODUCED BY THE MAPPING IN ISOPARAMETRIC FINITE ELEMENTS

V. Hoppe
B & W Engineering, 2 Torvegade, DK-1400 Copenhagen K, Denmark.

Four cases of error in finite element solutions of field- and elasticity-problems may be identified: round-off, interior discretization (element-size), boundary discretization (accuracy of modelling the boundary) and mapping (element shape). A method is described which separates the last error type from the others by means of the so-called mapping eigenvalues of a matrix which is the product of two projection matrices, one from the space of global monomials $x^i.y^j.z^k$ to the space of interpolation polynomials in the local curvilinear co-ordinates, (ξ,η,ζ). The other represents the reverse projection. For a linear mapping the mapping error is zero and the total projection is an identity and the matrix is the unit matrix. For nonlinear mappings the deviations of the eigenvalues $0 \leq \lambda_i \leq 1$ from unity are a measure of the mapping error. Examples can be found in which any one of the four errors is dominating.

19. ON EVALUATION OF TWO THICK PLATE THEORIES - FINITE ELEMENT SOLUTIONS

T. Kant
Department of Civil Engineering, University College of Swansea, Singleton Park, Swansea, SA2 8PP, Wales.

Finite element formulations of two-shear-deformation plate bending theories based on the displacement models,

I II

$$U(x,y,z) = z\theta_x(x,y) \text{ and } z\theta_x(x,y) + z^3\theta_x^*(x,y) \;;$$

$$V(x,y,z) = z\theta_y(x,y) \text{ and } z\theta_y(x,y) + z^3\theta_y^*(x,y) \;;$$

$$W(x,y,z) = z(x,y) \text{ and } w(x,y) + z^2w^*(x,y) \;;$$

are presented. While model I is well-known in classical mechanics, model II represents the lowest order correction for the out-of-plane deformation effects to the theories relying on model I. Isoparametric quadrilateral finite elements using C^0 continuous shape functions are developed. Numerical evaluations based on the use of 9-node Lagrangian and heterosis elements are made for a number of critical problems which figure prominently in the literature.

20. A FINITE ELEMENT METHOD FOR SOLVING THE KLEIN-GORDON EQUATION

V.S. Manoranjan
Department of Mathematics, University of Dundee, Dundee, DD1 4HN, Scotland.

A discrete-time Galerkin method is considered for approximating solutions of the nonlinear Klein-Gordon equation $\partial^2u/\partial x^2 - \partial^2u/\partial t^2 = F(u)$. The collisions of non-periodic solitary waves for various choices of F are studied. It is found that this method performs better than the 'stabilized' leapfrog scheme proposed by Ablowitz et al.

21. UNIVERSITY CONSORTIUM FOR INDUSTRIAL NUMERICAL ANALYSIS

S. McKee
Oxford University Computing Laboratories, 19 Parks Road, Oxford, Ox1 3PL, England.

General information on the University Consortium for Industrial Numerical Analysis (UCINA) is presented. Typical problems considered by the Consortium and their resolution or partial resolution are presented.

22. THE FINITE ELEMENT METHOD FOR DYNAMIC VISCOELASTIC ANALYSIS

A, Némethy
Department of Mathematics, Kuwait University, P.O. Box 5969, Kuwait.

The constribution deals with the solution of the vibration equation of a viscoelastic anisotropic plate by the finite element method. Making use of the Laplace transform a weak solution is defined and its existence and uniqueness is proved. We show that this weak solution will be a limit of approximate solutions defined by the finite element method. The rate of convergence is established. A problem of the inverse Laplace transform leads to the generalized eigenvalue problem.

23. FINITE ELEMENT ERROR ESTIMATES FOR SOME VARIATIONAL INEQUALITIES

M.A. Noor
Mathematics Department, Islamia University, Bahawalpur, Pakistan.

Variational inequalities are now fundamental in continuum mechanics and related areas. Recently much attention has been given to the numerical approximation of variational inequalities by finite element methods. In all these approximations, the choice of the finite dimensional convex set plays a very important role in deriving error estimates. In this paper we show how the choice of the finite dimensional convex sets suggest the methods to be adopted. We also derive L_2-estimates for the finite element approximation of some variational inequalities. Various open problems are also discussed.

24. MIXED FINITE ELEMENT METHODS FOR MILDLY NONLINEAR ELLIPTIC PROBLEMS

M.A. Noor
Mathematics Department, Islamia University, Bahawalpur, Pakistan.

The main aim of this paper is to consider the numerical approximation of mildly nonlinear elliptic problems by means of finite element methods of the mixed type. The technique is based on an extended variational principle, in which the constraint of interelement continuity has been removed at the expense of introducing a Lagrange multiplier. We have shown that the saddle point, which minimizes the energy functional over the product space, is characterised by the variational equations. This equivalence is used in deriving the error estimate for the finite element approximations of the mildly nonlinear elliptic problems. We give an example of a mildly nonlinear boundary value elliptic problem to show how the error estimates can be obtained from the general results.

25. THE SINGULARITY OF CRITICAL IDEAL FLOW IN OPEN CHANNELS

M.J. O'Carroll and E.F. Toro
Department of Mathematics and Statistics, Teesside Polytechnic, Middlesbrough, Cleveland, TS1 3BA, England.

Several finite element calculations have been published recently for critical open channel flows. The critical flow is always approached indirectly. Calculations are made for near-critical flow rates and the critical flow is supposed to correspond to non-convergence or waviness of solutions. An algebraic version of the problem, for uniform flows, has a cusp catastrophe at the critical flow where non-uniqueness changes to nonexistence of solutions. Here we examine a phase portrait of a one-dimensional version of the problem and show a correspondence between finite element calculations and the presence and type of singular points in the phase plane.

26. A NESTING TECHNIQUE FOR EFFICIENT DETERMINATION OF LOCALISED STRESS CONCENTRATIONS

C. Patterson and N.I. Bagdatlioglu
Department of Mechanical Engineering, University of Sheffield, Mappin Street, Sheffield, S1 3JD, England.

In stress design analyses, while it is very important to determine peak stress concentrations reliably, it is usually prohibitive to use an adequately refined global model for such local features, However, the stress concentration may, more feasibly, be determined using a nesting technique whereby a relatively coarse global model is used to give adequately accurate data near the concentration. This then supplies boundary data for a refined local model of the concentration region. In a test problem, consisting of a rectangular plate in tension and having a circular hole with a notch at the stress concentration, it is shown that a global model neglecting the notch, together with a local model of the notch region, can be used to determine the peak stress advantageously.

27. A REGULAR BOUNDARY METHOD USING NON-CONFORMING ELEMENTS FOR STRESS AND HARMONIC PROBLEMS

C. Patterson and M.A. Sheikh
Department of Mechanical Engineering, University of Sheffield, Mappin Street, Sheffield, S1 3JD, Sngland.

In its usual form the boundary element method is based on singular integral equations resulting from use of a fundamental solution with singularity on the boundary. However, regular integral equations may also be used, which are obtained simply, by moving the singularity outside the domain of the problem. Also, on discretizing the equations after the manner of finite elements, interelement continuity is usually imposed. While necessary in the finite element method, to maintain positive definiteness, interelement continuity is not necessary with boundary elements. Satisfactory numerical results have been obtained, using the regular method with non-conforming elements, for a variety of two-dimensional harmonic and stress problems, having regular and singular solutions.

28. ON THE USE OF FUNDAMENTAL SOLUTIONS IN THE TREFFZ METHOD FOR HARMONIC PROBLEMS

C. Patterson and M.A. Sheikh
Department of Mechanical Engineering, University of Sheffield, Mappin Street, Sheffield, S1 3JD, England.

The manifest success of the finite element method in design

applications is creating an increasing demand for detailed three dimensional modelling. Because of the high cost implied this has stimulated interest in boundary domain methods. The boundary integral method is under intensive investigation but a number of other techniques are available. In the method of Treffz the approximate solution has usually been sought as a power series expansion. Here the Treffz method is used for harmonic problems and the solution is sought as a linear combination of fundamental solutions with a singularity outside the domain of the problem. A number of regular and singular two dimensional problems have been analysed with favourable results.

29. ON A MIXED FINITE ELEMENT METHOD FOR TWO-DIMENSIONAL ELASTICITY EQUATIONS

J. Pitkäranta and R. Stenberg
Institute of Mathematics, Helsinki University of Technology, 02150 Espoo 15, Finland.

We consider a penalty method with reduced/selective integration, proposed in [1] for solving the two-dimensional elasticity equations. The method is equivalent to a mixed method with bilinear approximations for the stresses and piecewise constant approximations for the displacements on rectangular elements. We prove that the method gives an $O(h^{3/2})$ convergence in L_2-norm for the stresses. The same convergence rate is obtained for the displacements at the midpoints of the rectangles. These convergence rates are verified by numerical calculations.

1. Taylor, R. and Zienkiewicz, O.C., Complementary energy with penalty functions in finite element alaysis. pp.153-174 of R. Glowinski, E.Y. Rodin and O.C. Zienkiewicz (eds.), Energy Methods in Finite Element Analysis. Wiley, London, 1979.

30. ON ADAPTIVE DISCRETIZATION TECHNIQUES IN CURVILINEAR COORDINATES FOR CONVECTION-DOMINATED FLOWS

R. Piva, A. Di Carlo and B. Favini
Istituto di Meccanica Applicata alle Macchine, Universita degli Studi di Roma, Via Eudossiana 18, 00184 Roma, Italy.

The numerical techniques for the generation of boundary fitted curvilinear coordinates may be used, not only to map a given domain onto a simpler one, but also to obtain a set of difference equations on the transformed domain *not* dominated by odd-order derivatives. In fact the loss of accuracy suffered with increasing Reynolds number in the Navier-Stokes equations, on a uniform rectangular grid, is ultimately due to the increasing relative weight of the convective terms. In general curvilinear coordinates, the second *covariant* derivative of velocity contains also terms proportional to its first partial derivative. Hence, for a given boundary problem at a given Reynolds number, a suitable coordinate transformation may be devised, which keeps control of terms containing first derivatives. It should be observed, however, that the numerical solution of the required mapping equations *over the entire domain*, becomes more and more critical as the Reynolds number increases and that local adaptive convection driven mappings have to be used.

31. A THREE DIMENSIONAL HEAT CONDUCTION PROBLEM, WITH INTERNAL DISCONTINUITY OF THE TEMPERATURE, LEADING TO THE USE OF A SUBSTRUCTURING TECHNIQUE

C. Rose
Electricite de France, 1 Ave du Général de Gaulle, BP 408, 92141 Clamart, France.

It is required to compute the temperature in an electricty plant alternator shaft. This shaft consists of two jointed cylinders, one of which is subjected to sliding friction, bringing some thermic energy to the system. The shaft is divided into three parts, two made of copper and one of steel, which are separated in certain places by a thin sheet of insulator. From a mathematical point of view, an elliptic problem must be solved, whose coefficients are step-functions. The problem is:

Find $u \in H^1(\Omega)$ satisfying

$$- \operatorname{div} (K \operatorname{grad} u) = 0 \text{ on } \Omega,$$

$$K \frac{\partial u}{\partial \nu} + h(u - uo) = g \text{ on } \partial\Omega ,$$

where h is the transfer coefficient and uo is the outside temperature. From a technical point of view, the modelling of this problem with a finite element method involves difficulties due to the thinness of the insulator sheet and to the complex geometry of the domain. This has caused us to solve the problem in the following way:

1. the domain Ω is divided in three parts Ω_1, Ω_2, Ω_3, separated by discontinuity surfaces instead of insulator sheets.
2. On each Ω_i, $i = 1,3$, the temperature has to satisfy the conduction equations.
3. Moreover discontinuity conditions on the interfaces are expressed in terms of insulator thickness and conductivity.

A substructuring technique fits this approach well, and so

1. On each Ω_i : the conduction equations are partially solved, with conservation of the unknowns on the interfaces. Three condensed conductivity matrices are obtained this way.
2. These three matrices are assembled with the discontinuity matrices computed separately.
3. The system obtained is solved to give the temperature on the interfaces.

4. These values are carried into the three systems giving the temperature inside the three domains Ω_i.

32. CURVED FINITE ELEMENTS FOR CONICAL SHELLS

A.B. Sabir and J.M. Hafid
Department of Civil and Structural Engineering, University College, Newport Road, Cardiff, CF2 1TA, Wales.

While considering the analysis of silos, the authors became aware of an apparent lack of information about stresses and displacements in the conical part of the silos. Research on the problem of bending of conical shells has been mainly concerned with the derivation of equations and it appears that relatively little progress has been made in finding solutions from which numerical results can be obtained. We therefore undertook the development of conical shell finite elements. Two such elements are presented. The first of these has four corner nodes, each with only five external degrees of freedom resulting in a 20×20 stiffness matrix. The second element is of high order and contains additional internal degrees of freedom.

33. THE SOLUTION OF VISCOUS FLOW FREE BOUNDARY PROBLEMS USING THE FINITE ELEMENT METHOD

C. Saxelby
Oxford University Computing Laboratory, 19 Parks Road, Oxford, OX1 3PL, England.

We consider the problem of finding a numerical approximation to the solution of the Navier-Stokes equations for steady isotropic incompressible viscous flow. The equations are used in a velocity/pressure formulation rather than that of stream-function/vorticity. This leads to easily manageable boundary conditions in *free boundary* problems. The method employed is to use the finite element method with a Galerkin (weak) form of the equations. An iterative technique is used to locate the free boundary, whose position is initially unknown.

We present a problem concerning the drawing of a glass fibre out of a bath of molten glass under gravity. Various solutions are obtained, corresponding to different configurations of the inlet and outlet conditions.

34. DESIGN PHILOSOPHY OF A GENERAL PURPOSE FINITE ELEMENT PROGRAM

G. Sewell
International Mathematical & Statistical Libraries Inc., Sixth Floor, NBC Building, 7500 Bellaire Boulevard, Houston. Texas 77036, U.S.A.

International Mathematical & Statistical Libraries Inc.'s

TWODEEP is a small, inexpensive, easy to use, general purpose finite element program with the following design priorities:
1. Generality TWODEEP solves nonlinear, steady state, time-dependent and eigenvalue systems of PDEs on general two dimensional regions, with fairly general boundary conditions.
2. Ease of Use A preprocessor allows the PDEs to be input in a higher level language and an automatic mesh refinement and grading capability makes mesh generation quite easy.
3. Storage Efficiency A special bandwidth reduction scheme generally reduces the Cuthill-McKee bandwidth by 20% or more. The algorithm used to handle out-of-core storage reduces the amount of core storage required to a multiple of the number of unknowns, while increasing total costs by typically less than 50%.
4. Speed TWODEEP does not generally take advantage of special features of simpler problems but uses standard, proven techniques which efficiently handle more general problems.

35. FINITE ELEMENT COMPUTATION OF TISSUE TEMPERATURE IN THE HUMAN TORSO

P. Smith* and E.H. Twizell**
**Department of mathematics and Computer Studies, Sunderland Polytechnic, Sunderland, SR1 3SD, England.*
*** Department of Mathematics, Brunel University, Uxbridge, Middlesex, UB8 3PH, England.*

The finite element method is used to compute steady state temperature distribution throughout the human torso. The torso is represented by a cylinder and longitudinal heat conduction is neglected thus reducing the problem to that of calculating tissue temperatures within a circular cross-section of the torso. This cross-section is discretized in a symmetric manner, the resultant elements having one curved and two straight sides. Tissue temperature within each element is approximated using the rational basis functions, the functions being generated from the geometric properties of the elements.

The model is used to simulate the reaction of the nude human torso to a split boundary condition, where one half of the torso is subjected to one environment and the other half to a different environment.

36. FEATURES FACILITATING THE DEVELOPMENT OF THE PROGRAM PACKAGE SITU FOR COMPUTER AIDED DESIGN PURPOSES

H. Tängfors, K. Rusesson and N.-E. Wiberg
Department of Structural Mechanics, Chalmers University of Technology, Sven Hultins Gata 8, Fack, 412 96 Göteborg, Sweden.

A finite element program package SITU (SImple To Use) has been developed, mainly for nonlinear and time dependent problems.

Each calculation involves one or more separate steps for problem definition, for generation of necessary arrays, for FE-calculation and for display of results. Any step can be executed in interactive or batch mode. No separate pre- and post-processing programs are needed, as they are integrated in SITU.

Data are administrated by the hierarchic data base management system PRISEC, which has been developed for SITU but can be used in any other application program.

Special attention is paid to the efficient generation of the layout of a page for display, which is important in a CAD-process. The text to be displayed on a page is given as input data to the special program PRADIS (PRogram for Automatic DISplay) by which a massive amount of problem dependent FORTRAN-code containing text strings can be avoided.

37. CURVED FINITE ELEMENT MESH VERIFICATION

*R. Wait and **N.G. Brown

* *Department of Computational & Statistical Science, Liverpool University, P.O. Box 147, Liverpool L69 3BX, England.*

** *Department of Building & Civil Engineering, Liverpool Polytechnic, 2 Rodney Street, Liverpool, L1 2UA, England.*

In a computer program a finite element mesh is invariably represented as a set of lists of numbers. It is the aim here to provide techniques whereby it is possible to verify that given sets of data correctly define a well posed finite element grid. Thus we wish to check that the data corresponding to an individual element represents a single region with a simply connected interior. In addition it is necessary to check that when the elements are pieced together there are no overlaps and no gaps left in the representation of the domain of the problem. In addition it is useful to check that nodes are only duplicated or pathologically close together when this is necessary, as for example when discretizing one region with a crack.

It is our intention to develop a procedure for verifying the suitability of meshes in both two and three dimensions when it is assumed that the meshes are defined in terms of curved isoparametric elements or curved blended elements.

38. A COMPARISON BETWEEN METHODS OF SUBSTRUCTURATION

C. Wielgosz

Laboratoire Mécanique des Structures, Ecole Nationale Superieure de Mécanique, Universite de Nantes, 1 Rue de la Noë, 44072 Nantes Cedex, France.

Our purpose is to compare the left quotient operation (LQO) introduced by Nayroles, to a classical method of substructuration. First we show that for linear problems the LQO leads to the construction of special shape functions which allow exact results for the displacements and the forces to be obtained. In a classical method one has to choose shape functions, then to write a variational principle and generally the results are approximate because they depend on the choice of these shape functions.

We then formulate the necessary conditions which the shape functions must satisfy in order that the classical method gives the same exact results as the LQO. Examples in which Galerkin and Petrov-Galerkin methods are used to obtain exact results are finally given.

39. THE TREATMENT OF THE SINGULARITY AT THE CREST IN THE INTEGRAL EQUATION SOLUTION OF LIMITING GRAVITY WAVES IN WATER

J.M. Williams
Hydraulics Research Station, Wallingford, Oxon, OX10 8BA, England.

Progressive irrotational gravity waves in water have a limiting form in which, in their steady flow representation, the crest reaches the total energy line. The crest is then angled rather than rounded, with an included angle of 120°. It has been shown, Grant [1], that the expansion of the flow near the crest in the complex potential plane has leading terms with exponents 2/3 and 1.469 respectively. These leading terms have been incorporated by the author, Williams [2], into an integral equation technique which has yielded definitive solutions for limiting waves over the full range of depth: wavelength ratios from zero (solitary wave) to infinity (deep-water-wave).

The display outlines the method and presents specimen results, with indications of the accuracy achieved and details of the computing strategy.

1. Grant, M.A., The singularity at the crest of a finite emplitude progressive Stokes wave. J. Fluid Mech. 59, 257-262, 1973.
2. Williams, J.M., Limiting gravity waves in water of finite depth. Phil. Trans. R. Soc. London. A, in press, 1981.

SUBJECT INDEX

F = figure, T = table, *passim* = scattered references